Springer Biographies

The books published in the Springer Biographies tell of the life and work of scholars, innovators, and pioneers in all fields of learning and throughout the ages. Prominent scientists and philosophers will feature, but so too will lesser known personalities whose significant contributions deserve greater recognition and whose remarkable life stories will stir and motivate readers. Authored by historians and other academic writers, the volumes describe and analyse the main achievements of their subjects in manner accessible to nonspecialists, interweaving these with salient aspects of the protagonists' personal lives. Autobiographies and memoirs also fall into the scope of the series.

More information about this series at https://link.springer.com/bookseries/13617

Jost Lemmerich

Max von Laue

Intrepid and True: A Biography of the Physics
Nobel Laureate

 Springer

Jost Lemmerich (1929-2018)
Berlin, Germany

Translated by
Ann M. Hentschel
Stuttgart, Baden-Württemberg, Germany

ISSN 2365-0613   ISSN 2365-0621 (electronic)
Springer Biographies
ISBN 978-3-030-94701-9   ISBN 978-3-030-94699-9 (eBook)
https://doi.org/10.1007/978-3-030-94699-9

Translation by Ann M. Hentschel from the German language edition: Jost Lemmerich: *Max von Laue - Furchtlos und treu*, © Basilisken-Presse, Rangsdorf 2020. Published by Basilisken-Presse, Rangsdorf. All Rights Reserved.

This Springer imprint is published by the registered company Springer Nature Switzerland AG
The registered company address is: Gewerbestrasse 11, 6330 Cham, Switzerland

The von Laue coat of arms bears the inscription "Intrepid and True." Max's father was knighted in 1913, also granting his heirs the privilege to use the noble prefix *von* (GSta PK, Herolds Amt, I HA, rep. 176, no. 5833)

In memory of
Hans-Joachim Queisser's
"The Physics of the Twenties"
1977

Preface

One would expect that biographies of world-famous lifelong friends and colleagues would duly appreciate the bond of friendship between them. Albert Einstein (1879–1955) and Max von Laue (1879–1960) were good friends for over forty-five years. In the vast majority of Einstein biographies, this fact is not even mentioned. Any allusion to their mutual trust and solicitude tends to underrate it.[1]

No comprehensive biography of Max von Laue exists apart from the present work. Katharina Zeitz's dissertation, published in German in 2006 under the title: 'Max von Laue (1879–1960)—His Importance in the Restoration of German Science after World War II,' purposefully offers little of a biographical nature: "This treatise," she specified, "is explicitly not intended to be a biography of Max von Laue." The topics she addressed: Max von Laue's contribution toward reconstructing science after World War II; Max von Laue in Berlin—the establishment and expansion of the Fritz Haber Institute; and Max von Laue—the science organizer, are covered in the present volume only in connection with supplemental material. There are many obituaries,[2] some of them quite detailed, and one exhibition in his honor (Boeters and Lemmerich 1979). However, Laue's scientific oeuvre, the contemporary reception of it, and its significance today call for a more comprehensive biographical account.[3]

Max von Laue himself, whose father was ennobled in 1913, wrote an essay in 1944 on 'My career in physics—a self-portrayal.' The war prevented its immediate

---

[1] The Einstein biography by Armin Hermann (1994) refers to Laue several times, but he is not even mentioned in the one by Abraham Pais; likewise for the one by Philipp Frank. See also Cassidy 1995: 132: "As curator of the state instrument collection, Sommerfeld was also obliged to make experimental research possible, at least to a small extent—a job for assistants whom he banished to the dark cellar. Despite this subordinate status, it was there that Max von Laue and the outsiders working with him made a significant discovery. They produced evidence that X-rays behave like electromagnetic waves."

[2] Obituaries and laudatory articles on Max von Laue are by Friedrich Beck (1989: 24), Armin Hermann (1980), Peter Paul Ewald (1960), Gerhard Hildebrand (1987), Marion Kazemi (2006: 170), Walther Meissner (1960: 101), and Lise Meitner (1960: 196). An overview of the sources can be found in Henning 2004: 249.

[3] A biography of Max von Laue filling just under a hundred pages was published in Leipzig by Friedrich Herneck in 1979 but is now out of print.

appearance but he made additions to it in 1951 and the expanded version was reprinted at the head of the third volume of his collected works.[4] Such personal retrospectives generally leave many details and personal friendships unmentioned. The writings in his collected works are Max von Laue's own selection. Some of the omissions are nevertheless important to understand his development and impact as a theoretical physicist. They are therefore included in this biography.

In the present biography Max von Laue's scientific and historical publications have been taken as fully into account as possible, and their content, the messages he sought to convey, are duly assessed to show the importance he placed on clarifying the science. The correspondence and some of the publications are extensively quoted in order to give an idea of his style, his choice of words, and expression. Max von Laue's scientific publications are often very mathematical. Formal derivations could not be appropriately included in a biography such as this. Anyone interested in these important details must study the original papers.

The Max von Laue papers are deposited at various locations. The personnel files are kept at his workplaces: the former Friedrich-Wilhelms-Universität (now named Humboldt-Universität) in Berlin, Ludwig-Maximilians-Universität Munich, University of Zurich, Johann Wolfgang Goethe-Universität Frankfurt am Main, Archive of the Max Planck Society for the Advancement of Science in Berlin-Dahlem. Major portions of the extensive correspondence are held at the archives of the Goethe University in Frankfurt am Main, the Max Planck Society Berlin-Dahlem, and the Deutsches Museum. (See the key to archives and societies.)

Max von Laue was a passionate letter writer. Most are typewritten, even ones composed away from home. His handwriting is often difficult to decipher. Many letters in response have survived. Several addressees kept his correspondence and incorporated it into their estates, for example, his son Theodore, Lise Meitner, Walther Meissner, and Wilhelm Wien, who kept almost all of his letters, whereas Albert Einstein's collection is incomplete. The letters to Max Planck were apparently destroyed during the war. The incoming letters to Max von Laue including ones by Max Planck were added to the estate by the Laue family and entrusted to the archives of the Goethe University in Frankfurt am Main for safekeeping, likewise some of the family correspondence.

The scientific correspondence documents the way he thought and acted in trying to solve a given problem. Most of his textbooks, teaching material, and publications on physics and other branches of science trace the straight path to the "final" result. Laue's letters, on the other hand, relive the frustrations of dead ends, detours, and doubts that a scientist experiences.

The sheer volume of letters from people of all descriptions dictates that a selection be reached. Two multi-layered motives served as a guide: to present Max von Laue as an actor in his field and as a humanist. Important historical events in the twentieth century are briefly reviewed for the modern-day reader.

---

[4] For the autobiographical manuscript, see also DMA, NL 045, "Geschrieben für Dokumente zur Morphologie, Symbolik und Geschichte" (Deutsche Denker/Geistiges Antlitz); the expanded version appeared as Laue 1944/52 and 1961, vol. 3: V–XXXIV; see also Hartmann 1952: 191.

The political environment changed five times throughout his lifetime. This always had repercussions on the further course that von Laue and his family and friends chose to follow. The quotations take this aspect into account as well. All these authentic, often spontaneously written documents form the framework of this biography. What is lacking is important supplementary indirect utterances from conversations, from dialogue. Thus, this portrait certainly cannot claim to fulfill the criteria defining that "biographical truth" which Sigmund Freud and Arnold Zweig discussed when the latter told him of his intention to write his biography. Freud's response on 31 May 1936 was: "No, I love you far too much to permit such a thing. He who becomes a biographer commits himself to lying, to secrecy, hypocrisy, glossing over, and even hiding his misinterpretations—as biographical truth is not to be had, and if it were to be had, it would be useless" (Freud 1968: 137; cf. Bruder 2003). How the following biography fares is left to the reader's own judgment.

Berlin, Germany   Jost Lemmerich

References

Beck, Friedrich. 1989. Max von Laue. In *Physiker und Astronomen in Frankfurt. Geschichte der Johann Wolfgang Goethe-Universitat Frankfurt/Main*, eds. Klaus Bethge and Horst Klein. Frankfurt am Main: Metzner

Boeters, Karl E., and Jost Lemmerich, eds. 1979. *Gedächtnisausstellung zum 100. Geburtstag von Albert Einstein, Otto Hahn, Max von Laue, Lise Meitner in der Staatsbibliothek Berlin Preußischer Kulturbesitz*. Exhibition catalogue. Berlin: Staatsbibliothek

Bruder, Klaus-Jürgen, ed. 2003. *Die biographische Wahrheit ist nicht zu haben*. Giessen: Psychosozial-Verlag

Cassidy, David C. 1992. *The Life and Science of Werner Heisenberg*. New York: Freeman. German trans. Cassidy, D.C. 1995. *Werner Heisenberg*. Heidelberg, Berlin, Oxford: Spektrum Akad.-Verlag

Ewald, Paul Peter. 1960. Max von Laue. *Biographical Memoirs of the Royal Soc*. London, 6

Freud, Ernst L., ed. 1968. *Sigmund Freud—Arnold Zweig. Briefwechsel*. Frankfurt am Main: S. Fischer

Hartmann, Hans.1952. *Schöpfer des neuen Weltbildes. Große Physiker unserer Zeit*. Bonn: Athenäum

Hermann, Armin. 1980. Max von Laue. In *Dictionary of Scientific Biography*, ed. Charles Coulston Gillispie. New York: Scribner

Hermann, Armin. 1994. *Einstein. Der Weltweise und sein Jahrhundert. Eine Biographie*. Munich, Zurich: Piper

Herneck, Friedrich. 1979. *Max von Laue*. Leipzig: Teubner

Hildebrandt, Gerhardt. 1987. Max von Laue, der Ritter ohne Furcht und Tadel. In *Berlinische Lebensbilder*, Vol. 1: *Naturwissenschaftler*, eds. G. Hildebrandt and Wilhelm Treue Einzelveröffentlichungen der Historischen Kommission zu Berlin, no. 60. Berlin: Colloquium

Kazemi, Marion. 2006. *Nobelpreisträger in der Kaiser-Wilhelm-Gesellschaft zur Förderung der Wissenschaften: Max von Laue*. Veröffentlichungen aus dem Archiv der Max-Planck-Gesellschaft 15. Berlin: Max Planck Society

Laue, Max von. 1944/52. Mein physikalischer Werdegang. Eine Selbstdarstellung. (November 1944) Reprinted in Hartmann 1952: 37–70

Laue, Max von. 1961. In *Max von Laue Gesammelte Schriften und Vorträge*. 3 vols, Braunschweig: Vieweg.

Meißner, Walther. 1948. Gedenkrede auf Max Planck. *Sitzungsberichte der Bayerischen Akademie der Wissenschaften*, Math.-Nat. Kl. 1948: 2

Meitner, Lise. 1960. Nachruf auf Max von Laue. *Mitteilungen der Max-Planck-Gesellschaft* 4: 196.

Zeitz, Katharina. 2006. *Max von Laue (1879–1960).Seine Bedeutung für den Wiederaufbau der deutschen Wissenschaft nach dem Zweiten Weltkrieg*. Stuttgart: Steiner

Acknowledgments

It is thanks to the extensive assistance of many archivists in my search for material that this biography is so rich in numerous quotations from letters and other primary documents. It is a pleasure for me to express my gratitude to them first and foremost as well as to their associated institutions:

Archive of the Berlin-Brandenburg Academy of Sciences and Humanities; Archive of the Max Planck Society in Berlin-Dahlem; Archive of the Johann Wolfgang Goethe University in Frankfurt am Main; Churchill Archives Centre in Cambridge, England; Cornell University Library in Ithaca, New York; Archive of the Deutsches Museum in Munich; Archive of the Deutsche Physikalische Gesellschaft; Archive of the German Academy of Sciences Leopoldina; Albert Einstein Archives in Jerusalem; Geheimes Staatsarchiv Preußischer Kulturbesitz in Berlin, Archive of the Humboldt University in Berlin; Joseph Regenstein Library (Special Collections), University of Chicago; Archive of the Ludwig Maximilians University in Munich; The National Archives in Kew, England; Niels Bohr Archives in Copenhagen; The Royal Institution of Great Britain Archives in London; The Royal Society Archives in London; Staatsbibliothek zu Berlin Preußischer Kulturbesitz, Manuscripts Department; Archive of the Royal Swedish Academy of Sciences in Stockholm; City Archive of the Bavarian State, Munich; Archive of the University of Göttingen; Archive of the Center for the History of Science and Technology, University of Hamburg; Archive of the University of Zurich; Company Archive of Axel Springer SE; and the Central and State Library of Berlin.

I am grateful to Christian Matthaei for granting me access to Max von Laue's estate and for providing information on the family history. Dr. habil. Michael Maaser, the director of the University Archives of the Goethe University in Frankfurt, and his staff generously supported my research. Very extensive material can be found in the archives of the Max Planck Society. Only through the help of Dr. Marion Kazemi, Dipl.-Archivar Dirk Ullmann, and Mr. Bernd Hoffmann, was it possible for me to follow up on the multifarious leads. My cordial thanks for this. I owe the selection of photographs to Susanne Uebele. In the archives of the Deutsches Museum I encountered valuable support from Dr. Wilhelm Füssl, Dr. Matthias Röschner, and Mrs. Irene Püttner. I am indebted to Dr. Vera Enke of the Archive of the Berlin-Brandenburg

Academy of Sciences and Humanities and her staff for essential preliminary research. Professor Dr. Siegfried Hess and Professor Dr. Heinz Lübbig kindly took the trouble to critically review the scientific part of the German manuscript, for which I am very grateful. I owe much advice and important improvements and corrections to the manuscript to Mr. Ralf Hahn M.A., Berlin, who also examined the Peter Paul Ewald papers at the Cornell University Library. I would like to thank Professor Dr. Klaus Hentschel, Stuttgart, and Professor Dr. Dieter Hoffmann, Berlin, for references to secondary literature. My thanks go out to all the archivists and library aids for their valuable assistance among the stacks.

Dr. rer. nat. h.c. Jost Lemmerich
2018

The translator would like to thank Mr. Herbert Pusch for his generosity in permitting the present work to appear in English; Basilisken-Presse for the transferral of the rights to Springer Nature; Dr. Angela Lahee as Executive Editor for including this work in the Springer Biographies series, Ashok Arumairaj, and Kamalambal Palani for their assistance in the production. Prof. Frank N. von Hippel provided invaluable support in initiating the translation project and Prof. Klaus Hentschel kindly reviewed the manuscript.

Contents

About the Author

Dr. rer. nat. h.c. Jost Lemmerich (1929–2018) was born and died in Berlin. After earning his degree in physics at the Technical University of Berlin in 1962 as a Siemens employee, he worked on high-voltage insulation development before serving as expert in electrotechnics and control engineering at the German and European Patent Offices. His ten books include the biographies of James Franck and Michael Faraday. Annotated collections of correspondence, notably by Lise Meitner and Max von Laue, were compiled during two fellowships at Churchill College in Cambridge. He was also a gifted organizer of exhibitions in the history of science, among others on physics in the twenties, on important figures such as Einstein, Otto Hahn, Helmholtz, and Röntgen as well as on the Kaiser Wilhelm Society, the laser, microscopy, automation, thermodynamics, metrology, the ozone, quantum theory, and nuclear fission.

He was awarded the Karl Scheel Medal of the Berlin Physical Society, Germany's Distinguished Service Cross, the Justus Liebig Medal of the University of Giessen, and the Prince Bishop Johann von Egloffstein Gold Medal of the University of Würzburg.

Key to Archival Sources and Abbreviations

Key to Archival Sources

AEA	Albert Einstein Archives, Jerusalem
AMPG	Archive of the Max Planck Society, Berlin-Dahlem
APTB	Archive of the Physikalisch-Technische Bundesanstalt
BAB	Federal Archives, Berlin, Germany
BAK	Federal Archives, Koblenz, Germany
BBAW	Berlin-Brandenburg Academy of Sciences, Archive
BDC	Former Berlin Document Center, now transferred to BAB
CAC	Churchill Archives Centre, Cambridge, England
CUL	Cornell University Library, Ithaca, New York
DMA	Deutsches Museum Munich, Archive
DPGA	Deutsche Physikalische Gesellschaft, Archive
DTANL	Deutsche Akademie der Naturforscher Leopoldina, Archive
ETHZ	Swiss Federal Institute of Technology, University Library, Zurich
GNT	Center for History of Science & Technology, Univ. of Hamburg, Archive
GSta PK	Geheimes Staatsarchiv Preußischer Kulturbesitz, Berlin, now in BAB
HUA	Humboldt University Berlin, Archive
JRL	Joseph Regenstein Library, Special Collections, University of Chicago
KVA	Royal Swedish Academy of Sciences, Archives, Stockholm
LMUA	Ludwig Maximilians University Munich, Archive
NARA	US National Archives and Record Administration, Washington D.C.
NBA	Niels Bohr Archive, Copenhagen, Denmark
NL	*Nachlaß* = estate holdings, papers
PA	*Personal-Akte* = staff file
RIGBA	Royal Institution of Great Britain, Archives, London
RSA	Royal Society Archives, London
SB PK	State Library of Berlin Prussian Cultural Heritage
STAM	City archive of the Bavarian State, Munich
SUB	State and University Library of Lower Saxony, Göttingen

TNA   The National Archives, Kew, England
UAF   University Archive Frankfurt/Main
UAS   Corporate archive Axel Springer SE
UGA   University of Göttingen, Archive
UZA   University of Zurich, Archive
ZLB   Zentral- und Landesbibliothek Berlin

Societies and Professional Institutions

AVA   Aerodynamische Versuchsanstalt, Aerodynamics Design Testing
               Station
DPG   Deutsche Physikalische Gesellschaft, German Physical Society
FHI   Fritz Haber Institute
GDCh   Gesellschaft Deutscher Chemiker
KWG   Kaiser Wilhelm Gesellschaft
KWI   Kaiser Wilhelm Institute
MPG   Max Planck Society
MPI   Max Planck Institute
PTA/PTR   Physikalisch-Technische Anstalt/Physikalisch-Technische
               Reichsanstalt (Prussian/imperial bureau of standards in Berlin)
PTB   Physikalisch-Technische Bundesanstalt (German federal bureau of
               standards)

Journals

Ann. Phys.   Annalen der Physik
Jb. Radioakt. Electr.   Jahrbuch der Radioaktivität und der Elec-
                                        tronik
Nachr. Akad. Wiss. Göttingen   *Nachrichten*, communications of the
                                        Academy of Sciences Göttingen
Naturw.   Die Naturwissenschaften
Phil. Mag.   Philosophical Magazine
Phys. Bl.   Physikalische Blätter
Phys. Z.   Physikalische Zeitschrift
Sb. Math.-Phys. Kl. Bayer. Akad. Wiss. *Sitzungsberichte*, proceedings of the
                                        mathematical and physical class of the
                                        Bavarian Academy of Sciences
Sb. Preuss. Akad. Wiss.   *Sitzungsberichte*, proceedings of the
                                        Prussian Academy of Sciences

Verh. DPG	*Verhandlungen der Deutschen Physikalischen Gesellschaft zu Berlin*, proceedings
Z. Crystall./Kristallogr.	Zeitschrift für Kristallographie (formerly Z. für Crystallographie)
Z. Physik	Zeitschrift für Physik

Chapter 1
Introduction

At the turn of the century, when Max Laue, Albert Einstein, Otto Hahn, and Lise Meitner were students, there were only speculations about how matter is structured. How might one conceive crystalline structure? A negatively charged particle had been discovered. It was called an electron and its mathematical description followed soon afterwards.[1] It appeared to be a component of atoms. But what exactly were atoms? Chemists were working successfully with a vague notion of an atom, as if they knew what it was. Robert Bunsen and Gustav Kirchhoff had jointly developed spectral analysis in 1859: each chemical element has a characteristic spectrum—but why?

Mathematics was already highly developed by then. The British physicist James Clerk Maxwell applied it to formulate the electromagnetic phenomena discovered by Michael Faraday in 1855 and to light in 1864. But was radiation actually being emitted by oscillating electrons, as several physicists were assuming?

Another open question was: What were those rays that Wilhelm Conrad Röntgen discovered in Würzburg in 1895 which penetrate through matter? Were they pulses?[2] How do the electromagnetic waves discovered by Heinrich Hertz propagate? The dispersion of light was another issue of debate. The medium that must exist for this was called the "aether". It was neither visible nor could it be analysed. Does it envelope Earth, does it move along with it?

As a rule, basic research in the German Empire was being conducted at its twenty-one universities. In 1886/87 nine of them were employing over a hundred lecturers each. In Berlin, the Friedrich-Wilhelms-Universität, founded in 1810, engaged 288 faculty members to teach 5766 students, followed by the Ludwig-Maximilians-Universität in Munich with 158 lecturers and 3060 students.

The Göttingen mathematician Felix Klein and the expert at the Prussian Ministry of Culture, Friedrich Althoff, introduced a system of priority research for some universities (Sachse 1928: 234, 277; Brocke 1987: 195). Göttingen was assigned

---

[1] Abraham (1903). On the importance of the *Annalen der Physik*, see Pyenson (1985): 195.

[2] On the question of what exactly radiation might be, see Hentschel (2007): 172, 283, 443, 525.

© Basilisken-Presse, Natur+Text GmbH 2022
J. Lemmerich, *Max von Laue*, Springer Biographies,
https://doi.org/10.1007/978-3-030-94699-9_1

mathematics as its main focus. Students were largely given free rein in determining which courses to take. They only had to pass final oral examinations in defense of a submitted written thesis. The natural sciences, including physics, generally belonged to the Faculty of Philosophy. Therefore, an examination in philosophy was compulsory. This external framework defined the path that Max Laue was to follow in academia.

References

Abraham, Max. 1903. Prinzipien der Dynamik des Elektrons. *Annalen der Physik* 315: 105–179

Brocke, Bernhard vom. 1987. Friedrich Althoff. In *Berlinische Lebensbilder. Wissenschaftspolitik in Berlin: Minister, Beamte, Ratgeber*, p. 195. Ed. by Wolfgang Treue and Karlfried Gründer. Berlin: Colloquium

Hentschel, Klaus. 2007. *Unsichtbares Licht? Dunkle Wärme? Chemische Strahlen? Eine wissenschaftshistorische und -theoretische Analyse von Argumenten für das Klassifizieren von Strahlungssorten 1650–1925 mit Schwerpunkt auf den Jahren 1770–1850*. Diepholz, Stuttgart, Berlin: Verlag für Geschichte der Naturwissenschaft und Technik

Pyenson, Lewis. 1985. *The Young Einstein. The Advent of Relativity*. Bristol, Boston: Hilger

Sachse, Arnold. 1928. *Friedrich Althoff und sein Werk*. Berlin: Mittler

Chapter 2
Childhood and Youth

Max Theodor Felix Laue was born on 9 October 1879 in Pfaffendorf (district of Coblenz) in the German Empire. Prussia was one of the four kingdoms in the German Kaiserreich, founded in 1871 after the victorious war against France. The military played an important and respected role in Prussia and the new state. Max Laue's father, Carl Julius Laue (1848–1927), was a military intendant assessor and ranking lieutenant when his son was born. He had fought in the campaign of 1870/71, initially as a noncommissioned officer, then as vice-sergeant, before being promoted to lieutenant of the Landwehr and awarded the Iron Cross 2nd Class.[1]

The large fortress Ehrenbreitenstein looms above Coblenz across the river (Fig. 2.1). Carl Julius Laue was probably stationed there and performed his service at the fortress. He and his brother and two sisters originally came from the Magdeburg area. In 1878 he married Minna, née Zerrenner, the daughter of a local merchant there. His father Theodor owned three distilleries. Carl Julius had met Minna in Magdeburg while doing service there in 1868/69.[2]

---

[1] On the father's military career, see GSta PK, Königl. Heroldsamt, I. HA, rep. 176, no. 5833, fol. 6; Lemmerich (2011): 155.

[2] Lemmerich (2011): 282. Max von Laue wrote in a letter about his mother: "Now she began to speak about her engagement to a rich young merchant who had showered her with attentions, but unfortunately died before the wedding could take place. She didn't tell me his family name. His given name is like mine. Since, she named her son after him".

© Basilisken-Presse, Natur+Text GmbH 2022
J. Lemmerich, *Max von Laue*, Springer Biographies,
https://doi.org/10.1007/978-3-030-94699-9_2

Fig. 2.1 Old Coblenz can be seen on the left, Pfaffendorf, Max von Laue's birthplace, is in the foreground, Ehrenbreitenstein Fortress is on the right. *Source* The author's estate

His son Max, born a year later, probably attended elementary school in Pfaffendorf.[3] But soon the family had to leave because in the military it was customary to change location upon promotion. Meanwhile Minna had given birth to their daughter Elizabeth. Traveling through Brandenburg along the river Havel and by Altona, the young family arrived in Posen (currently known as Poznan in Poland) in 1888, the fortified capital of the Prussian province of the same name. Carl Julius Laue was promoted to captain. Ten-year-old Max attended grammar school there at Friedrich-Wilhelms-Gymnasium. Occasional excursions were made back to the parental home turf and once to the Devil's Wall in the Harz Mountains.[4] As a *Tertianer* pupil in the fourth or fifth grade there, Max had to change schools again when his family relocated to Berlin. His most memorable classroom experience was at the Wilhelms-Gymnasium when he learned about the separation of copper metal from a copper sulphate solution. He recollected in his autobiography:

> The impression that this first contact with physics made was tremendous. For a few days afterwards I wandered about idly lost in thought, causing my mother to ask me anxiously what was wrong with me. When she got to the bottom of it, she saw to it that I often visit the 'Urania' on Taubenstraße, a popular science society where physical apparatus stood ready

---

[3] Max Laue had to fill out forms about himself several times during his academic career, among others for membership into an academy. They differ somewhat, for example, his application for admission to the Akademie Deutscher Naturforscher Leopoldina in Halle (Archive, Max Laue file) and the form in his personnel file at the Humboldt University Archive (HUA, UKP, L51). The present biography uses the most likely figures.

[4] Excursion by car 1–6 Mar. 1932 with his son Theodore. AMPG, III rep. 50, suppl. 1, car touring book [unpaginated, unchronological].

Strassburg.

Fig. 2.2 View of Strasbourg. To secure the city militarily, a total of 14 forts were built in 1872. This made it one of the most heavily armed locations of the German Empire. *Source* The author's estate

in large numbers to demonstrate simple experiments; reading the given instructions, one just had to press a button and then watch the informative process take place.[5]

But soon the family had to move away again to Strasbourg, which Louis XIV had annexed in 1682 and transformed into an anti-German stronghold. The victorious campaign of 1870/71 had reunited Strasbourg with the German Empire.

Strasbourg was a larger city counting just over 100,000 residents, 9485 of whom were attached to the military. The Protestant population outnumbered the Catholics. The city had suffered much destruction during the war of 1870–71, and the Reich had invested heavily in restoring it. The construction of a university building was part of that policy (Fig. 2.2).

The family lived in spacious official accommodations near All Saints' Abbey. Laue's father wanted his son to continue his education in a cadet school with suitable training in subordination, but his mother succeeded in dissuading her husband. In 1893 Max joined the Tertianer class at the famous Protestant Gymnasium, a grammar school for boys only, as was normally the case at the time. In his memoirs

---

[5] Laue (1961), vol. 3: V; also DMA, NL 045 and Hartmann (1952): VI. Laue attended lectures at the Urania as well.

Max Laue devoted a lengthier section to this period and his teachers. He praised the "extremely kindhearted director, H. Veil […], who knew that each individuality has a right to develop itself according to its own law, who, himself an enthusiastic classical philologist with a strong theological bent, acknowledged the legitimacy of mathematics and the natural sciences and defended me on more than one occasion against other teachers when they took offence at my mathematical and physical inclinations, as sometimes happened."[6] He also warmly remembered his teacher of classical languages and religion, Erdmann, quoting his motto: "Anyone in enthusiastic pursuit of a great cause can't go completely astray." Laue himself agreed: "My life goes to show for it (ibid.)." The mathematics and physics teacher, Professor Göhring, was also deemed worthy of mention, not only for his manner of stimulating scientific thinking but also for his advice to read the 'Popular Lectures on Scientific Subjects' by Hermann von Helmholtz (1875), which kindled Laue's lifelong admiration for the author.[7] In the two final years of school he and two of his school friends[8] sharing similar scientific inclinations performed all kinds of experiments with greater or lesser success, including generating X-rays. They studied Adolf Wüllner's textbook (1862) on experimental physics[9] and educated themselves independently in mathematics to the point of trying their hands at differential and integral calculus. Max became the best pupil in the subject. Hiking and bicycle tours[10] with his friends in the vicinity of Strasbourg deepened Laue's interest in the countryside and in the animal kingdom awakened early by a gift from his maternal grandfather, a copy of Brehm's *Animal Life* (1890).

In March 1898 Laue passed his school leaving exams, but only earned top marks ("sehr gut") in mathematics and physics. By then he had reached the age for completing his compulsory military service, which bothered him because he did not like the rifle drills, "*Griffeklopfen*"[11] as he called them (Fig. 2.3). He was also very critical of the way that recruits were treated. He abhorred militaristic comportment and unquestioning obedience throughout his life, as many letters prove.

---

[6] Laue to Theodore, Berlin, 5 Aug. 1954. AMPG, III rep. 50 suppl. 7/14 (car touring book), fol. 31. Laue provides two spellings for this name: Veil or Weil.

[7] Unfinished address by Laue at the Academy, in Göber and Herneck (1960).

[8] Fecht to Laue, 7 Apr. 1954. UAF, Laue papers, 3.3. Hermann Fecht, one of Laue's school friends, nicknamed him "Lehe."

[9] Adolf Wüllner was a professor at the Technical University of Aachen.

[10] Laue to Meitner, no loc., 29 Sep. 1943. Lemmerich (1998): 310; see also: AMPG, III rep. 50 suppl. 1, car touring book: "In August of this year I could have celebrated my 50th bicycle-riding anniversary. […] I got a bicycle. It still had solid tires." When the Laues took a road trip to the Black Forest in June 1933, Laue recalled cycling in the area as a schoolboy. He had relatives on his father's side in Offenburg.

[11] Laue is mentioned frequently in the Einstein correspondence. This allusion appears in Einstein to Zangger, 11 Jan. 1915, Schulmann (2012).

Fig. 2.3 Max Laue in uniform, Strasbourg 1898/99. *Source* Bildagentur bpk, No. 10024545, G. Michel

References

Brehm, Alfred et al. 1890/93. *Brehms Tierleben. Allgemeine Kunde des Tierreichs.* 3rd ed., 10 vols. (1st ed. 1864/69, 2nd ed. 1876/79) Leipzig: Bibliographisches Institut
Göber, Willi, and Friedrich Herneck. 1960. *Forschen und Wirken. Festschrift zur 150-Jahr-Feier der Humboldt-Universität zu Berlin 1810–1960.* Vol. 1, Berlin: VEB Deutscher Verlag der Wissenschaften
Hartmann, Hans. 1952. *Schöpfer des neuen Weltbildes. Große Physiker unserer Zeit.* Bonn: Athenäum
Helmholtz, Hermann von. 1865/76. *Populäre wissenschaftliche Vorträge.* Braunschweig: Vieweg
Laue, Max von. 1961. Mein physikalischer Werdegang. Eine Selbstdarstellung. In *Max von Laue Gesammelte Schriften und Vorträge.* Vol. 3: V–XXXIV, Braunschweig: Vieweg

Lemmerich, Jost, ed. 1998 *Lise Meitner – Max von Laue. Briefwechsel 1938–1948*. Berliner Beiträge zur Geschichte der Naturwissenschaften und der Technik, no. 22. Berlin: ERS-Verlag

Lemmerich, Jost, ed. 2011. *Mein lieber Sohn! Die Briefe Max von Laues an seinen Sohn Theodor in den Vereinigten Staaten von Amerika 1937–1946*. With two contributions by Christian Matthaei. Berliner Beiträge zur Geschichte der Naturwissenschaften und der Technik, no. 33. Berlin, Liebenwalde: ERS-Verlag

Schulmann, Robert, ed. 2012. *Seelenverwandte. Der Briefwechsel zwischen Albert Einstein und Heinrich Zangger 1910–1947*. Ruth Jörg collaborator. Zürich: Verlag Neue Zürcher Zeitung

Wüllner, Adolf. 1862/99. *Lehrbuch der Experimentalphysik mit theilweiser Benutzung von Jamin's Cours de physique de l'école polytechnique*. 4 vols. Leipzig: Teubner

Chapter 3
Studies in Physics

Still on active military duty, Laue was granted permission to study at the Imperial University in Strasbourg. The French university had been dissolved in 1871 to make room for the new Kaiser Wilhelm University. Laue was matriculated on 19 October 1898. According to the student register, from the winter semester of 1898 until the end of the summer semester of 1899 he attended the courses Experimental Physics II taught by Ferdinand Braun, Potential Theory by Reye, Defined Integrals by Krazer, Analytical Geometry of Space, Descriptive Geometry, Exercises in Geometry, and Physical Exercises by Braun, General Experimental Chemistry by Fittig, and Ethics by Ziegler (UAF, Laue papers, 1.10).

At 45 years of age, Ferdinand Braun had been lecturing at Strasbourg as full professor since 1895 (Kurylo and Süsskind 1981). His career had led him there from a teaching position at a high school. It was during this period that he discovered the rectifier effect while analysing crystals. Working with discharge tubes inspired by the discovery of X-rays in 1895, he had succeeded in inventing a display method for electrical processes by means of his device which came to be known as the 'Braun tube.' After the young Italian Guglielmo Marconi had demonstrated the applicability of Hertzian waves in communications technology, Braun began very successful high-frequency investigations in 1898, supported by his assistant Jonathan Zenneck,[1] who also held the practical course for beginners. Laue attended these stimulating lectures in uniform and watched the brilliant experimental demonstrations by Ferdinand Braun. Unlike many theoretical physicists of his day, Max Laue always appreciated experimental research and Braun's lectures may have been one of the reasons for it.

---

[1] Jonathan Zenneck (1871–1954) liked to tell the story that his parents, in wise premonition, had given him the names—Jonathan Adolf Wilhelm—so that he would always be well equipped with the right first name for each of the three political eras in Germany. He was an active promoter of the Deutsches Museum in Munich and was often in contact with Max von Laue, see Lemmerich 2011: 204.

© Basilisken-Presse, Natur+Text GmbH 2022
J. Lemmerich, *Max von Laue*, Springer Biographies,
https://doi.org/10.1007/978-3-030-94699-9_3

Around Christmastime in 1898, his mother was definitively diagnosed as suffering from cancer. An attempt was made to save her life by surgery, but it came too late. Decades later, Laue wrote about this to his son Theodore:

At the beginning of March she asked Father Steinwedel, who had confirmed Elly and me, to give her and all of us Holy Communion. Then her strength rapidly drained away. On the 19th I was sitting alone at her bedside and at her request was giving her some water in a teaspoon. And as I was about to do so again, a slight shudder ran through her body and her sufferings were over. I closed her eyes soon afterwards. My father entered the room only later. She was then laid out in the large livingroom of our apartment. The military had granted me 3 days' leave (but I still had to take part in a shooting exercise), and then I sat by her bier until she was brought away to the freight depot accompanied by a long, solemn train 2 or 3 days later. She was buried in Magdeburg at a later time. My father had forbidden me from also taking part in that ceremony (Lemmerich 2011: 282).

Max became estranged from his father, who remarried a few years later and was living in Berlin. It is not known why no enduring bond developed between Max and his younger sister.

In the autumn of 1899 Laue continued his academic studies at the University of Göttingen. He was registered as "stud. Math.," matriculation no. 387. A physics student by the name of Paul Kirchberger became a good friend of his.[2] At Göttingen Laue was able to attend lectures by eminent scientists. He took part in practical sessions in chemistry and physics as well.

The focus on mathematics at Göttingen attracted extraordinarily brilliant faculty, foremost Felix Klein, David Hilbert (Fig. 3.1) and, from 1902 on, Hermann Minkowski. In his self-portrayal 'My career in physics' Laue wrote about Hilbert:

This man figures in my memory as perhaps the greatest genius I have ever set eyes upon. [...]
    Mathematics conveys the experience of Truth most purely and directly; its essentiality in basic human education rests on this. And a beautiful, intrinsically consistent mathematical proof was one of my greatest joys already at school (Hartmann 1952: XV, Laue 1961).

Carl Runge, a notable representative of applied mathematics born in 1866, joined its ranks in 1904. His family home became a favorite meeting place among the younger generation of scholars. Woldemar Voigt held the chair in physics. He had been studying the fundamental physical properties of crystals since the turn of the century. How much this subject featured in his lectures is not known, but Laue was impressed enough by the way that Voigt presented the material to decide to dedicate himself definitively to theoretical physics. Other academic teachers at included Max Abraham, Ernst Zermelo and Walther Nernst. But for Laue as a student, course attendance was much less significant than his readings. The scientific literature took up the majority of his time. His textbooks probably included Gustav Robert Kirchhoff's 'Lectures on Mathematical Physics', 'Emission and Absorption' as well as the 'Treatises on Mechanical Heat Theory.' In his memoirs he mentions studying

---

[2] Laue (1961) vol. 3: V; UAF, Laue papers, 1.10. Paul Kirchberger also authored two books (1920, 1922). As a target of the Nazi racial laws he committed suicide in 1938, see Lemmerich (1998): 204.

Fig. 3.1 The mathematician David Hilbert (1862–1943) at a lecture in 1932. *Source* Photograph in the author's estate

the equations found by the Scot James Clerk Maxwell on the correlation between electricity and magnetism, which Maxwell had formulated on the basis of Michael Faraday's experiments on induction. About these Maxwell equations, Laue quoted the Austrian physicist Ludwig Boltzmann's exclamation:

> Was it a god who wrote these signs, appeasing my inner turmoil, revealing Nature's forces all around me? (UAF, Laue papers, 1.10).

Max Laue refused to join any student fraternity that was still practising the dueling tradition. Many students chose to study in Munich during the winter term in order to be near the ski slopes in the mountains. In this rite Laue was willing to partake, although as a novice he had to settle for hiking. It was only in 1906 that he truly began to enjoy skiing as a sport. Table 3.1 lists the courses he attended in Munich.

Table 3.1 An overview of the lectures attended by Laue at the Ludwig-Maximilians-Universität in Munich

- General Psychology, Prof. Lipps
- Kant's Critique of Pure Reason,' Dr. Cornelius
- Goethe's 'Faust,' Dr. Bormski
- Theory of Analytic Functions, Prof. Alfred Pringsheim
- Kinetic Theory of Gases, Dr. Korn
- Calculus of Variations, Dr. Korn
- Practicals in Physics, Privy Councillor Röntgen
- Modern Geometrical Optics, Prof. Graetz
- On Electrical Oscillations, Dr. Zehnder

Source UAF, Laue papers, folder 1.10

The tuition fees that Laue had to pay in Munich totalled 105 marks.[3] Conrad Röntgen made a notable appearance during one of the practical sessions and conversed with Laue about physics.

A change of place was again in order for Max, though, when he decided to move to Berlin in 1902 for the summer semester, where he could stay with his two closest school friends. But he arrived well after the semester had begun because he was first obliged to perform compulsory military exercises as reserve officer for eight weeks.

Nevertheless Laue had not yet come to stay in Berlin as a permanent resident. Initially, he was probably unaware of the many scientific institutions that the city had to offer, not only at the university, but also at the polytechnical *Technische Hochschule* and the imperial metrological and research establishment *Physikalisch-Technische Reichsanstalt* (PTR).

Laue knew that Max Planck was teaching theoretical physics at the university. He had read his textbook on thermodynamics from 1897, but Planck's great discovery of 1900, the quantization of radiation, was unknown to him. Max Laue was registered at the Royal Friedrich-Wilhelms University in Berlin on 1 July 1902 as enrollee no. 4671. The lectures he attended are duly recorded in the Anmeldebuch (UAF, Laue papers, 1.10) (Table 3.2).

Theoretical optics broadly conceived became Laue's main field of research for a long time. On the basis of the knowledge he had acquired at Göttingen, Laue successfully participated in the exercises in theoretical physics and "immediately attracted Planck's attention right from the start by offering a particularly elegant solution to the posed problem [...]" (Hartmann 1952: XIII f.). Planck was a disciplined lecturer who prepared himself very carefully but left immediately after each lecture, making it impossible for his auditors to ask any questions (Meißner 1948: 2).

Laue's few notes on Planck's lectures are undated and the sheets are unpaginated. One is headed "Crystal optics," followed by the designations for the quantities employed: electrical force, its orientation, electric potential, dielectric constants, and magnetic force. Poynting's theorem, which describes the strength of the flow of

---

[3] It is not clear how Max Laue was able to afford to pay for his upkeep and tuition. It is likely that his mother had given him some money.

Table 3.2 Lectures attended by Laue at the Königliche Friedrich-Wilhelms-Universität zu Berlin

Summer semester from Easter to Michaelmas (29 September 1902)
• Prof. Planck, Theoretical Optics, enrolled 4 July 1902, terminated 1 Aug. 1902
• Prof. Planck, Exercises in Mathematical Physics, 4 July 1902 to 1 Aug. 1902
• Dr. Starke, On Electrical Discharges and Their Associated Effects, 3 July 1902 to 1 Aug. 1902
Winter semester from October 1902 to April 1903
• Prof. Planck, Theory of Heat, 6 Nov. 1902 to 2 Mar. 1903
• Prof. Planck, Exercises in Mathematical Physics, 6 Nov. to 2 Mar. 1903
• Prof. Schwarz, Applications of the Theory of Ellipt. Functions
• Prof. Schwarz, Fundamentals of Spectral Analysis and the Limits of Its Applicability, 11 Nov. 1902 to 14 Nov. 1902

Source UAF, Laue papers, folder 1.10

energy in the mathematical formulation of vector calculus, was the point of departure of the subsequent mathematical treatment. Planck expounded on polarization, reflection, and refraction; those calculations were some of the things Laue jotted down.

In the summer semester of 1903, Planck offered a course on the 'System of Physics as a Whole.' Laue's notes from this lecture reveal an overarching view of the field of physics. Again Planck began with the mathematical apparatus, the Lagrangian equations of the second kind, then derived the equations of elastic motion using Hamilton's principle. He next presented the derivation of the Maxwell equations from Hamilton's principle, and the kinetic energy of the aether.

Radiation processes then followed and reflection, absorptive media, scattering, adiabatic processes, black-body radiation, Wien's displacement law, and the entropy of radiation. Planck apparently did not mention the regularity he had discovered in 1900 of radiation emitted by a black body and the consequential quantization of radiation (UAF, Laue papers, 1.10).

The entire series of Planck's lectures on theoretical physics supplied six semesters' worth of material; Laue thus attended only a small fraction of them. As was normally the case at universities at that time, Planck had no secretarial services at his disposal. So his teaching assistants had to review the students' assigned problems, occasionally discussing the more interesting solutions with Planck.[4]

It was rather unusual for any aspiring theoretician following Laue's course of study to participate in courses with a more experimental focus. But Max Laue always chose to combine both orientations of physics in his research.

---

[4] Neither Max von Laue nor Lise Meitner reported on the content of such consultations with Max Planck.

Fig. 3.2 Otto Lummer
(1860–1925). *Source*
Archive of the Max Planck
Society, Berlin-Dahlem

Otto Lummer (Fig. 3.2) conducted research at the *Physikalisch-Technische Reich-sanstalt* (PTR). In the field of optics Lummer had invented a very simple interferom-eter together with Ernst Gehrcke. Lummer would bring along to his lectures several optical devices built at the PTR for demonstration purposes. Those experiments made a lasting impression on Laue, who attended his lecture on the Fundamentals of Spec-tral Analysis and the Limits of Its Applicability in the winter semester of 1902/03. Ten years later, at the award of a scientific prize in Breslau, he said:

> Now, I remember having seen an experiment by Lummer to demonstrate Kirchhoff's law, which at that time at least, one did not have much opportunity to see. At the base of a long tube heated evenly to a bright red glow stands a porcelain crucible, on the bottom of which some figure is drawn in ink. This figure is not discernible in the least, as long as the walls of the tube are heated to a uniform temperature; but it immediately appears as soon as this temperature equilibrium is disturbed. According to Kirchhoff's law, at thermal equilibrium the intensity of the radiation is independent of the nature of the walls in all directions, so one only sees an indiscriminate uniform brightness (Laue 1914).

Apparently Laue did not attend any of the lectures on experimental physics in Berlin being offered by Emil Warburg since 1894. Laue probably did not befriend the young experimental physicists James Franck and Richard Pohl there at that time either.

On 19 March 1903, Laue was unceremoniously promoted to the rank of lieutenant as a routine element of his military career in the reserves, as by then he was no longer on active duty.

References

Hartmann, Hans. 1952. *Schöpfer des neuen Weltbildes. Große Physiker unserer Zeit.* Bonn: Athenäum

Kurylo, Friedrich, and Charles Süsskind. 1981. *Ferdinand Braun. A Life of the Nobel Prizewinner and Inventor of the Cathode-Ray Oscilloscope.* Cambridge: MIT Press

Laue, Max von. 1914. Über optische Abbildung. *Naturwissenschaften.* 31: 757–760

Laue, Max von. 1961. Mein physikalischer Werdegang. Eine Selbstdarstellung. In *Max von Laue Gesammelte Schriften und Vorträge.* Vol. 3, Braunschweig: Vieweg, V–XXXIV

Lemmerich, Jost, ed. 1998. *Lise Meitner – Max von Laue. Briefwechsel 1938–1948.* Berliner Beiträge zur Geschichte der Naturwissenschaften und der Technik, no. 22. Berlin: ERS-Verlag

Lemmerich, Jost, ed. 2011. *Mein lieber Sohn! Die Briefe Max von Laues an seinen Sohn Theodor in den Vereinigten Staaten von Amerika 1937–1946.* With two contributions by Christian Matthaei Berliner Beiträge zur Geschichte der Naturwissenschaften und der Technik, no. 33. Berlin, Liebenwalde: ERS-Verlag

Meißner, Walther. 1948. Gedenkrede auf Max Planck. *Sitzungsberichte der Bayerischen Akademie der Wissenschaften,* Math.-Nat. Kl. 1948: 2

Chapter 4
Dissertation and Early Scientific Research

In the summer semester of 1903, Laue asked Professor Planck to propose a topic for his doctoral thesis. Since he had attended his lectures on optics, Planck suggested that he analyse the theory behind interference phenomena produced by a plane-parallel plate (Laue 1961b, 3: V, Hartmann 1952: XVI). A few sheets with notes on this topic can be found among Laue's papers.[1]

Interference studies were not new to Berlin, as Albert A. Michelson had come to visit Hermann von Helmholtz from the USA with his family in the autumn of 1880 to carry out his famous experiment on aether drift (Livingston 1973: 74).

Busy at work on his dissertation, Laue also attended Planck's lecture on thermodynamics, in which the derivation of the radiation law that Planck had found in (1900) was discussed along with his hypothesis of energy quanta.

On 8 June 1903 Laue submitted his completed dissertation with his curriculum vitae to register for the final examination.[2] By then he had completed two semesters in Strasbourg, four in Göttingen, one semester in Munich, and two in Berlin. Max

---

[1] UAF, Laue papers, 1.10: Vorbereitung zur Doktorarbeit; on the influence of slit width on a diffraction spectrum: Einfluß der Spaltbreite auf das Beugungsspektrum [undated, 5 pages: 2 drawings, text, calculations]; and on the light emitted by rotating molecules: Über das Licht welches von rotierenden Molekülen ausgesandt wird [undated, 5 pages, calculations].

[2] HUA, Phil. Fak., no. 4, vol. 175, 384, 65 b, 69v; in his evaluation Max Planck cited his own papers (1901a) and (1902). Laue referred, among others, to Lummer and Gehrcke (1903).

© Basilisken-Presse, Natur+Text GmbH 2022
J. Lemmerich, *Max von Laue*, Springer Biographies,
https://doi.org/10.1007/978-3-030-94699-9_4

Planck promptly wrote a one-and-a-half-page evaluation of the dissertation five days later:

| On Interference Phenomena on Plane-parallel Plates | The fruit of this investigation is the refutation of a conclusion drawn on one occasion by Lummer and Gehrcke; namely, if the addition of a ninth beam with the same path divergence as 8 other interfering beams generates an interference image of even greater definition, one is not therefore justified to conclude that the ninth beam is capable of interference (inherent) with the first beam. The auth[or] shows this not only by his formula, but also by a discrete observation of an elementary nature. |

Excerpt from Max Planck's evaluation of Laue's dissertation, dated 13 June 1903. *Source* HUA, Phil. Fak., no. 4, vol. 17538465b.

Planck also discussed how Laue's dissertation treated the energy principle, as it applies to a plane wave impinging upon a transparent plane-parallel plate. His verdict on the thesis was "Diligentiae et a aminis documentun laudabile." Walther Nernst shared Planck's judgment as the second reviewer. The oral defense took place on July 9th. Planck was the first interrogator. There were questions about the electricity of rigid and fluid bodies, on the equations of electric and magnetic fields, on units of electric charge, on Helmholtz's vortex theorems, and on thermodynamics. His final assessment was: "quite satisfactory knowledge." Emil Warburg also asked about electricity in solids and fluids, about the propagation of sound, Fresnel's wavelength, thermocouple, resistance measurement, and induction. His verdict was: "generally quite satisfactory knowledge." The mathematician H. A. Schwarz tested him on integral equations, partial differential equations, Poisson integrals, and galvanic currents. His report was that Laue had given very well-informed responses that stood out for their "clarity, certitude, and correctness." In the compulsory subject of philosophy, Friedrich Paulsen tested him on Immanuel Kant and gave the grade "magna cum laude."

It was conventional to list and thank the relevant academic teachers at the end of a written thesis. Laue listed Ferdinand Braun, Adolf Krazer, and Theodor Reye for Strasbourg; Max Abraham, Theodor des Coudres, David Hilbert, Felix Klein, Walther Nernst, Eduard Riecke, Woldemar Voigt, and Otto Wallach for Göttingen; Arthur Korn, Alfred Pringsheim, and Conrad Röntgen for Munich; and Otto Lummer, Friedrich Paulsen, and Max Planck for Berlin. The dissertation was 58 pages long (Fig. 4.1). In the introduction Laue discussed the foregoing research on interference by other scientists. He mentioned the work of the French physicist Armand Hippolyte

Fig. 4.1 Title page of Max
Laue's inaugural
dissertation. It is dedicated to
his parents. *Source* SB PK,
Ah 7856, Phil. Diss., 1903

Über die Interferenzerscheinungen an
planparallelen Platten.

INAUGURAL-DISSERTATION
ZUR
ERLANGUNG DER DOCTORWÜRDE

VON DER PHILOSOPHISCHEN FACULTÄT
DER
FRIEDRICH-WILHELMS-UNIVERSITÄT ZU BERLIN
GENEHMIGT
UND
NEBST DEN BEIGEFÜGTEN THESEN
ÖFFENTLICH ZU VERTEIDIGEN
am 25. Juli 1903
VON

Max Laue
aus Ehrenbreitstein bei Koblenz.

OPPONENTEN:
Herr stud. phil. Paul Meth.
- stud. phil. Otto Benecke.
- Dr. phil. Herrmann Fecht.

BERLIN.
BUCHDRUCKEREI VON GUSTAV SCHADE (OTTO FRANCKE).
LINIENSTRASSE 158.

Louis Fizeau around 1850 and asserted: "Michelson brought this method to the
greatest perfection." He went on to refer to the results obtained from a plane-parallel
plate by Otto Lummer and others, distinguishing his own work from that of his
predecessors as follows: "The only question treated by these authors so far is how
great the distance between two infinitely fine spectral lines may be in order to be
considered still separate. One judges the resolving power from this. But a more
detailed investigation must take the finite width of the lines into consideration and
their internal distribution of intensity (Laue 1904a)."

Laue assumed that light consists of an unbounded series of sinusoidal oscillations
not limited in time and not, as other physicists assumed, a series of damped oscillations that are continually regenerated. Thus he was following his teacher Planck
(Lemmerich 2011: 282).

His criticism of a publication by Lummer and Gehrcke (1903) was based on calculation. Laue proved logically and mathematically unequivocally that the conclusion
they had drawn was false. They had claimed that the interference phenomena would
change if the number of interfering waves was increased from eight to nine. According
to Laue's calculations, this was wrong.

This first part of his paper was followed by a second in which Laue investigated
how the conditions change when the beam falling on the plate is constrained by a
slit on the plate. He realized that diffraction occurs at the slit. This results in rays
being diverted from the original path. Laue pointed out: "The rays emerging from the
front surface rather produce the same interference phenomenon as those emerging

Fig. 4.2 Diagram from
Laue's dissertation on
interference effects off a
plane-parallel plate,
illustrating the experimental
arrangement. There aren't
anymore reflections off the
surface. *Source* SB PK, Ah
7856, no. 14

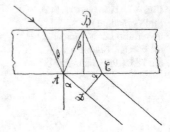

from the rear surface, hence the energy principle demands that in this case it not be just a single point of the visual field that is illuminated but a certain area (Fig. 4.2)." The interference conditions of this arrangement were determined. He closed with the words: "Finally, may I be permitted to express my most respectful thanks to my revered teacher, Professor Planck, for the inspiration to pursue this research and the support he has afforded me."

Laue passed the examination on 25 July 1903 with the grade *magna cum laude*. When he informed his father of this by telegram, it remained unanswered at first. It had terrified him into thinking that something had happened to his daughter, who was traveling (Lemmerich 2011: 68 f.).

The graduation ceremony took place on 8 August 1903 with the oath solemnly read out—in Latin.[3] A joyful celebration with friend and foe alike then followed as a matter of course!

Barely three months after the dissertation was published, an abridged version appeared in the *Annalen der Physik* (Laue 1904a). It reveals the underlying reasons for the work: "The aim of all spectroscopic investigations must be to determine the function that the intensity distribution in a spectral line has as it relates to the oscillation frequency." The fallacy by Lummer and Gehrcke was corrected in public once again. This was not in and of itself uncommon at the time, but for "such a young colleague" to dare to do so was nevertheless somewhat unusual.

Max Laue decided not stay in Berlin after earning his doctorate because, as he wrote in his memoirs, he preferred to live in a smaller town (Laue 1961a: V; Hartmann 1952: XVII). Göttingen attracted him, where he planned to take the state examination just in case he would have to become a secondary school teacher in the end. He also attended lectures on specific problems in physics, such as Max Abraham's lecture on Electron Theory, i.e., the extension of Maxwell's theory to incorporate the electron.[4] He also attended the lecture by the astronomer Karl Schwarzschild on Geometrical

---

[3] The oath formula was, "Te sollemniter interrogo, an fido data polliceri et confirmare religosissime constitueris, te artes honestas pro virili parte tueri, provehere atque ornare velle, non lucri causa neque ad vanam captandam gloriolam, des quo divinae veritatis lumen latius propagatum effulgeat." Freely translated: "I solemnly ask thee whether, so help thee God, thou art resolved to loyally promise and affirm the following, that thou art willing to protect the sciences as far as it be in a man's powers, not for the sake of merit, nor for the sake of vain glory; but rather that thou shalt give so that the light of divine truth, in which it is passed on, shall shine forth in bright splendor."

[4] Laue (1947): 57 f.: "It was realized around 1900, however, that a general reduction of elec-trodynamics to mechanics is impossible. [...] The fact that a charge carrier in motion drags its

Optics. Laue chose to be tested on the subject chemistry for the state examination, which also included mineralogy. Since he had only done some general reading on this, the grilling by the geologist Professor Adolf von Koenen made him almost fail. Laue's reliable knowledge of chemistry barely saved him in the final evaluation for a teaching certificate.

At Göttingen, he joined the meetings among mathematicians together with Professors David Hilbert, Felix Klein, Hermann Minkowski, and Carl Runge. That is where he got to know Max Born, who was three years his junior. But no closer affinity arose between them.[5]

Laue agreed to write a lengthy article 'On the theory of continuous spectra' (Laue 1904b) for Johannes Stark (1874–1957), the successful experimental physicist who had just founded the *Jahrbuch der Radioaktivität und Elektronik* in 1904. The article first summarized the results obtained by eleven physicists over the past ten years, including Planck's papers from 1900 and 1902, thus providing a critical overview of the current state of knowledge. It begins with this introduction:

> A series of observational data shows that even in the case of most homogeneous natural light the process of radiation is not as simple and orderly as, for example, when a tuning fork is sounded. The failure of all interference experiments with light from two different sources is among them. The fact that natural light is mostly unpolarized is also indicative. This led Fresnel to the idea that the succession of regular sinusoidal oscillations that every light-energized particle—to speak in modern terms, every vibrating electron—performs, is interrupted at intervals which are small against the time of an optical observation but large against the period of oscillation, by motive forces that serve to recoup the energy expended by the oscillations for the electron (Laue 1904b).

Commentary on the interference experiments by Michelson followed along with the remark that they tell us nothing about the constitution of light. Then he pointed out another way of looking at light. It shows how uncertain physicists still were about making any predictions: "Schuster has shown by the following example that, in fact, any view about the constitution of light is compatible with this attempt. He imagines the light wave as a disordered sequence of very short pulses. This conception has subsequently drawn special interest because, according to the idea probably first expressed by Wiechert and Stokes, X-rays are a sequence of such pulses, which are formed at impact against the anticathode by the moving electrons in the cathode ray."[6]

Laue then treated Planck's hypothesized "natural radiation." The last part of the article hints at Laue's future research. "The hypothesis of natural radiation plays the

---

electromagnetic field along with it and that this field has momentum suggested the idea of an electromagnetic inertial mass. [...] This found its mathematical expression, for example, in 1902 in the theory by Max Abraham (1825–1922) for the momentum of the electron in motion, conceived as a charged sphere; the mass was found to be dependent on the velocity and for a long time Abraham's formula for this competed with the relativistic formula."

[5] Max Born, then an assistant of Hermann Minkowski, had not yet decided to change his focus to physics. He was David Hilbert's "private assistant." Laue drew Born's attention to Woldemar Voigt's lecture on crystal physics, which they attended together.

[6] The mentioned physicists are: Arthur Schuster (1851–1934), Johann Emil Wiechert (1861–1928), professor at Göttingen since 1898, and George Gabriel Stokes (1819–1903).

same role here as the principle of molecular disorder in the kinetic theory of gases. Only thus has it become possible to attribute entropy and temperature to radiation on an electromagnetic basis, and to derive from the mathematical condition of thermal equilibrium the formula which gives the distribution of energy in the spectrum of cavity radiation (ibid.)."

Woldemar Voigt presented a comprehensive paper by Max Laue (1904c) at the meeting of the Göttingen Academy on 23 July 1904 and a second one on 25 February 1905. Based on his previous publication and a paper by Woldemar Voigt: 'On the propagation of radiation in dispersive and absorptive media' (1896), Laue pursued the question of how the oscillation of natural light alters its form: "The question suggested here seems to me to be of great interest to radiation theory now, and yet, to my knowledge such an effect has hitherto not been taken into account there. In the following investigation, therefore, let us draw the conclusion about the change in oscillatory form from the change in the spectrum, by computational means (Laue 1904c)." It became a very mathematical paper. Laue had, after all, matriculated in mathematics. He alluded to a paper by Max Planck, again with the hypothesis of "natural radiation," and theoretically investigated an idealized resonator with an upstream color filter. His conclusion was that the hypothesis of natural radiation was a necessary condition for the validity of the second law of thermodynamics for radiation processes.

The second publication was submitted in February 1905 under the same title with "II" tacked on at the end (Laue 1905a). The reason for this is pointed out in the introduction: "When my first paper on this question was published, Herr Planck had the kindness to point out to me by letter that objection can be raised against the line of inquiry, calling the result into question. For there the discussion is only about a plane wave; in the case of natural radiation, such a wave cannot be isolated […] (ibid.)." The mathematical treatment was carried out starting from the assumption that there be no difference between a spherical wave and a plane wave. Laue addressed the influence of the slit width in spectral apparatus on the spectrum. The second law of thermodynamics was noted several times. Natural radiation, he said, loses its "disorderliness" when passing through a selectively absorptive medium. The results of his first paper remained sound, Planck's concerns did not apply.

But Laue was probably not completely satisfied with these two publications, because on September 17th another manuscript with the same title was sent out, this time to the *Annalen*. The dispersion and/or selective absorption of a light wave was studied. He pointed out why, by way of introduction: "The purpose of the following investigation is to carry out computationally the inference of a change in the mode of vibration to a change in the spectrum without being constrained by any particular differential equation. […] The method we shall use starts by our representing the oscillations at the origin by a Fourier integral as a superposition of pure sinusoidal oscillations (Laue 1905b)." Again, reference was made to Planck's research from 1900 "On the law of energy distribution in the normal spectrum" (Planck 1901b). A remark by Laue about the interaction with the electrons of the permeated material retrospectively reflects his participation in Hilbert's seminar on the electron: "[…] for here the wave emanating from the region of disturbance influences the motion of the electrons, and the excitations caused by these latter possibly interfere with the wave in such a way that the resultant wave propagates more slowly (Laue 1905b: 550)." This important statement appears in the sixth section, notably prior the publication of Einstein's famous paper 'On the electrodynamics of moving bodies' (Einstein 1905a) within the following context: "Now, an electromagnetic disturbance in the latter [vacuum] propagates with the velocity of light v; nevertheless this cannot be applied to propagation within matter without further analysis; [...] But it [the wave] cannot propagate faster, because it is impossible for the resultant wave to be in places that none of the superposing waves has reached yet. Electron theory thus leads to the conclusion that electrodynamic and optical effects propagate in matter at most with velocity v, and one should also expect that the group velocity always be smaller, at most equal to it (Laue 1905b: 550)." Laue then discusses the conflict with the second law of thermodynamics at high resolving power.

An entirely different issue attracted interest in Göttingen during the summer semester of 1905. David Hilbert organized a seminar on electron theory in collaboration with Hermann Minkowski, Emil Wiechert, and Gustav Herglotz. Max Laue and Max Born were among the participants. But the seminar did not inspire Laue to publish anything on the subject; it was not until fifty years later that he published his 'History of the electron' (Laue 1959). One more short paper was completed while he was still at Göttingen. Gehrcke had drawn his attention to the curving of spectral lines generated by an echelon grating. Laue attacked this imaging problem mathematically and succeeded in calculating the proof of the curvature (Laue 1905c).

Soon after completing this paper in September 1905 Laue returned to Berlin. Planck had offered him an assistantship which he assumed in November.[7] This was when he came into closer contact with his academic teacher.[8] The duties Laue had to perform were few: He was responsible for managing the institute library and for correcting the students' assignments, which he found very instructive.[9] Two lifelong friendships developed out of this employment. The first was with a student and doctoral candidate of Planck's, Walther Meissner, who was three years younger than him.[10] The second was Lise Meitner, who came to Berlin from Vienna in the autumn of 1907 to spend a few semesters attending Planck's courses on theoretical physics. Several years elapsed, though, before the reserve between both parties was abandoned and a very close friendship developed.

---

[7] In Berlin, Max Laue lived at 47 Pariser Straße, see *Verh. Phys. Ges.* Berlin. A teaching assistant's salary at that time was usually 100 Marks.

[8] Laue wrote to Lise Meitner on 25 Sep. 1944: "I am now reading Grillparzer's autobiography. Do you know it? When he reports about his visit with Goethe and about the emotions he felt then, I have to think vividly of my feelings towards Planck, namely in my younger years." Grillparzer describes how he felt approaching Goethe, dressed in black and heavily decorated. But the next day Goethe was much friendlier and ate with him. Grillparzer (1971); see also Lemmerich (1998): 525; CAC, MTNR; and Sime (2001): 42, 54.

[9] Laue (1961b), 3: V; Hartmann (1952): XVIII. Lise Meitner, another teaching assistant of Planck's for a while after Laue, did not pass on any details about this job.

[10] Walther Meissner took his doctorate under Max Planck in 1907 while Laue was Planck's assistant.

In the summer of 1906 Laue was fully occupied. He had to take part in a military exercise for eight weeks but had also taken on the task of editing the fourth volume of Hermann von Helmholtz's lectures, as he informed Einstein on June 2nd, with whom he was discussing the theory of radiation, emission and absorption, and "light quanta" (Einstein 1993: doc. 37; cf. Einstein 1905b).

At the beginning of the winter semester, Max Planck presented the paper 'On the electrodynamics of moving bodies' by Albert Einstein (1905a) from Zurich at the Physics Colloquium. It had appeared in volume 17 of the fourth series of the *Annalen der Physik*, of which Planck was one of the editors. Laue was among the audience. He wrote in his memoirs: "The transformation of space and time which the theory of relativity it expounded undertook felt strange to me, and by no means was I spared the scruples that others later expressed aloud. But I continued to mull over these ideas, especially since Planck subsequently had a number of his own analyses on them published (Laue 1961a: V; Hartmann 1952: XIX)." This was the theory later known as special relativity, which would engage Laue more than a year later.

Arnold Sommerfeld as full professor of theoretical physics at the University of Munich offered Max Laue a position, but after discussing it with Max Planck, Laue decided to stay in Berlin, as he wrote him on 16 October 1907 (DMA, NL 089).

At the meeting of the Physikalische Gesellschaft zu Berlin on November 3rd, Dr. Max Laue was admitted to the society at the suggestion of Max Planck and under his chairmanship.[11] Laue had alluded to the importance of thermodynamics, the second law of thermodynamics, several times in his papers on optical problems. Now he could discuss the problem with Planck directly. Decades later, he recalled this in a letter to Lise Meitner (20 Feb. 1944) written during World War II, in which he also reported about the destruction of Planck's villa: "I can't tell you how many things I experienced in that house, when Planck started discussing the reversibility of reflection and refraction with me, and the non-additivity of entropy came out of it! That was in 1906, while Ludwig Boltzmann was still alive, whose 100th birthday it is today (Lemmerich 1998: 352)."

In April 1906 Laue submitted the manuscript 'On the thermodynamics of interference phenomena' to the editors of the *Annalen der Physik* (Laue 1906a). The wording clearly shows that Laue had learned from earlier papers to present the subject clearly in the introduction:

Until now, the thermodynamics of radiation has not included interference phenomena within the sphere of its considerations; rather, it has always presupposed that all energy quantities, such as the intensity of a beam of rays, the brightness of an illuminated surface, etc., are composed additively of the corresponding energy quantities of all the coinciding beams. The concept of coherence is quite foreign to it; it characterizes a beam solely by its intensity and by the geometrical determinants [...]. On the other hand, it is easy to see that even for thermodynamics it is not equivocal whether coinciding rays are coherent or not. Here lies a gap in the theory, which the following investigation intends to contribute toward filling (p. 365).

For these considerations, Laue proceeded from the acknowledged addition theorem of entropy. He used as an ideal example a Michelson interferometer slightly

---

[11] I would like to thank Mr. Ralf Hahn, M. A., for this communication.

Fig. 4.3 Diagram of the Michelson interferometer. The experimental arrangement "consists [...] of a plane-parallel plate P (not silvered on one side as Michelson has it) of only imperceptibly absorptive material and two absolutely reflective plane mirrors S_1 and S_2 laid out symmetrically to plate P" (Laue 1906a, reprinted 1961, 1: 56)

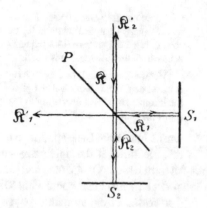

modified without one-sided silvering on the separating plate (Fig. 4.3). He remarked: "The goal of our investigation is to show by means of the mentioned interference phenomena that it [the proposition that the entropy of a system is additively composed of the entropy of its parts] is incompatible with the entropy principle (p. 366)."

After calculating the reflection off the mirrors and considering the transmission time, he arrived at the conclusion: "Interference phenomena exist to which we must ascribe reduced entropy if we are to regard the addition theorem as valid (p. 372)." Then he discussed the impossibility of a *perpetuum mobile* of the second kind. The impossibility of making incoherent rays interfere dictates that such a perpetual motion machine of the second kind be impossible. The last section was devoted to entropy and probability according to Boltzmann.

> If, on the other hand, a system of beams is given, then, before proceeding to calculate its entropy one has to answer the question of the coherence of its members—which in principle is always possible—and then to decide how far one may use the addition theorem of entropy [...].
>
> These considerations also shed light on the question of how to apply probability considerations in the field of radiation theory. Provided, for instance, one starts from the heuristic point of view proposed by Mr. Einstein that monochromatic radiation (at least at low density) behaves thermodynamically as if it consisted of independent energy quanta of a definite size [...] (pp. 376–377).

Laue closed his publication with a bow to Boltzmann: "It speaks for the extraordinary significance of that Boltzmannian notion of a connection between entropy and probability, which in the peculiar field of interference phenomena, where thermodynamics proper leaves us completely in the lurch, we are able to choose so simply between the principle and the addition theorem of entropy (p. 379)."

At the meeting of the Physikalische Gesellschaft zu Berlin on May 4th, he gave a talk on: 'The spectral decomposition of light by dispersion' (Laue 1906b). This paper is also particularly interesting with respect to his discovery and interpretation of the nature of X rays six years later. Laue first reviewed the various studies by other authors on the topic:

[…] they merely want to place a more descriptive, albeit less rigorous method alongside this distinctly analytical method [of decomposition by means of the Fourier integral], by conceiving the light wave as a sequence of very short pulses, and initially investigating the behavior of an individual pulse […].

We want, as I said, to conceive light waves as a succession of pulses of a width that is small compared to the wavelength of light; we may think of electromagnetic impulses, for instance, of which according to the hypothesis by Wiechert and Stokes X-rays consist (pp. 170, 173).

On 12 June 1906 Laue applied to the faculty for the *venia legendi*—the permission to offer lectures at the university—and submitted the required second thesis: his April publication 'On the thermodynamics of interference phenomena' (1906a). The dean asked Max Planck and Paul Drude to pass their verdicts on it, the latter as full professor of experimental physics.[12] Over fifty full professors of the Faculty of Philosophy had to take note of and approve the application. Max Planck gave a very thorough review of the habilitation thesis. He acknowledged Laue's research on interference phenomena on a plane-parallel plate as having completely resolved the observational questions. Then he discussed the publication on the propagation of radiation in dispersive and absorptive media and praised the independent route that Laue had chosen. The thermodynamic publications received Planck's particular praise and attention. "Since Dr. L. has thus shown in more than one way that he is capable of attacking great scientific questions independently and resolving them, I have no reservations about recommending his continued candidacy for habilitation. I am convinced that his lectures will constitute a valuable addition to instruction in theoretical physics (HUA, Phil. Fak., 1228, letter dated 14 Jul. 1906)."

Nernst concurred with Planck's judgment. In order to successfully complete the procedure, Max Laue had to submit three topics each for a colloquium and a lecture. For the colloquium they were: (1) The flow of energy in the theory of electricity and electrodynamics, (2) On electromagnetic motion, and (3) On the dynamics of the electron according to Cohn's theory. For the public inaugural lecture they were: (1) On the propagation of light in dispersive media, (2) On the resolving power of spectroscopic apparatus, and (3) The development of the theories of electricity according to Maxwell and Hertz. The first title was chosen for the colloquium, and the third for the public inaugural lecture, which he delivered in November.

Later letters show that Planck felt responsible for Laue. As, Laue's way of thinking and his mathematical treatment of physics corresponded in many areas to Planck's own, which Laue perceived as encouragement.[13] Their professional and personal affinity was important to them both. Max Planck generally treated people around him with respect. After Paul Drude, the full professor of experimental physics in

---

[12] HUA, Phil. Fak., 1228 with the expert opinions. Paul Drude committed suicide on 15 Jul. 1906. That is why Max Planck's vote is the only one on file, see HUA, UKP, L 51, Laue, vol. 1.2. The public announcement appeared in *Phys. Z.* (1906) 23: 872. Laue (1906a) was accepted as a habilitation thesis. See also *Verh. Dt. Phys. Ges.* (1907) 21: 606. The topic of the inaugural lecture was "Die Energieströmung in der Elektrizitätslehre und Elektrodynamik."

[13] Laue was several years older than Planck's sons Karl (born 1888) and Erwin (born 1893), neither of whom were interested in the natural sciences.

Fig. 4.4 Paul Drude
(1863–1906), director of the
Institute of Physics at the
University of Berlin in
1905/06 and editor of the
Annalen der Physik from
1900 until his death. *Source*
The author's estate

Berlin, took his own life on 15 July 1906, Planck held his funeral oration (Fig. 4.4). In dismay he asked: "How could this happen? Did it have to come to this? It is impossible to pass over this question in silence; again and again it forces itself upon those thinking back, with elemental force, and it weighs unspeakably heavily upon all those who were close to him in life to think of the possibility that this disaster might have been avoidable had it been recognized in time (Planck 1906)."

In the summer holidays of 1906, Laue traveled to Switzerland and also visited Einstein in Bern.[14] Decades later he described their first meeting to Carl Seelig (1952: 20): "As agreed by letter, I went to see him at the Office of Intellectual Property. In the general reception room an official told me to go back into the corridor, Einstein would find me there. I did so, but the impression that the young man who approached me was so unexpected that I didn't believe this could be the father of relativity theory. So I let him pass by, and it was not until he returned from the reception room that we made each other's acquaintance." It was around this time that the discussion between Laue and Einstein began about what radiation might actually be, an important question for Max Laue owing to his intense research on diffraction and interference phenomena. Einstein had published an article in the *Annalen der Physik* in 1905 entitled 'On a heuristic point of view concerning the generation and transformation of light' (1905b) and another one the following year 'On the generation of light and light absorption' (Einstein 1906a). Laue wrote Einstein a letter on 2 June 1906 about this but begged him not to weigh his words too precisely on the "gold scales" yet because he was "somewhat cut off from theoretical physics for the rest of this year and [I] am doing all of this a bit dilettantishly. But you should at least see my good will." He continued:

---

[14] It is no longer establishable whether Laue met his future wife during this vacation in Switzerland.

Fig. 4.5 Sketch to analyse
the reversibility of reflection
and refraction at a
plane-parallel plate. *P* is an
absorption-free
plane-parallel plate, S_1 and
S_2 are plane mirrors. *Source*
Laue (1907a), reprinted
1961, 1: 70

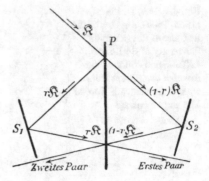

If at the beginning of your last paper you formulate your heuristic point of view to the effect that radiant energy can be absorbed and emitted only in specific quanta, then I have nothing to say against it; all your applications agree with this rendition, too. Now, this is a characteristic not of electromagnetic processes in a vacuum but of absorbing or emitting matter. Radiation, therefore, does not consist of light quanta, as is stated in § 6 of the first paper, but behaves as if it did just while it is exchanging energy with matter. The proof that the former leads to impossible conclusions seems to me to be the essence of my thermodynamics of interference phenomena; but then the derivation of Wien's distribution law given in § 6 falls away too and yields Planck's instead. Your restriction to low radiation densities then becomes superfluous, and in my opinion it is of great importance that the same idea can be used in fields far removed from the theory of radiation proper, such as the Volta effect, the generation of cathode rays, etc. [...] (Einstein 1993: 41).

As an afterthought he added, "I never discussed your heuristic point of view with my boss, by the way (ibid.)".

On 10 November 1906 Max Laue was officially habilitated, certifying him as an academic teacher, and he delivered his inaugural lecture. In the winter term 1907/08 he offered a course on Selected Topics in Theoretical Optics, and for the summer term 1908 and the following winter term his topic was Electron Theory. It was probably Planck who suggested Laue as editor of Hermann von Helmholtz's *Lectures on Electrodynamics and the Theory of Magnetism*, which appeared in Helmholtz (1907). In the same year Laue submitted another paper to the editors of the *Annalen* on the problem of entropy, this time of partially coherent bundles of rays. A kind of "thought experiment" about the splitting of a ray into two rays was used as an illustration (Fig. 4.5).

The reversibility of reflection and refraction was discussed again under the condition of homogeneity. In § 2 he stated: "By this consideration, however, the reversibility of any regular, absorption-free reflection and refraction is proved. [...] Nevertheless, in the following section we shall separately prove the reversibility of this process for the boundary between two isotropic media. Considering the novelty of the subject, a double safeguarding of the bases may seem useful (Laue 1907a: 5)." In the fourth section, the measure of interference capacity was discussed. The calculation was then performed for two applicable waves. It demonstrated that coherence decreases with each reflection. In § 9 entropy is similarly analysed mathematically, but Laue had to confess: "In this field, there still seems to me to be some problems, though (p. 39)." In closing he discussed cavity radiation:

> Furthermore, in a cavity enclosed by perfectly reflecting walls and containing absorbing and diathermanous bodies, temperature equilibrium of all bodies and ray bundles is not yet a sufficient condition for the state of maximum entropy. Rather, it is conceivable that partial coherence still exists between some ray bundles despite the completed temperature equalization. With each absorption and dispersion, however, this coherence is reduced, while new coherences are not generated [...]. Therefore, with time, the state surely approaches the entropy maximum characterized by absolute incoherence. So the theory is perfectly right when it applies the addition theorem to states of equilibrium. Only if the cavity contains no absorbing substances at all can coherences persist in it permanently—just as other unstable states of radiation (pp. 39 f.).

In the same year Laue addressed mathematically the problem of a light wave diffracted off an electrically conducting body. Henry du Bois and Heinrich Rubens had found that very short-wave polarized light is absorbed more strongly on a wire grating if the orientation of polarization of the radiation coincides with the orientation of the grating rods.[15]

In 1906 Kurd von Mosengeil wrote his doctoral thesis under Max Planck on the theory of stationary radiation in a uniformly moving cavity. It strongly criticized a paper by the Viennese theoretical physicist Fritz Hasenöhrl on radiation in moving bodies, asserting that the energy of the radiation corresponded to a mass. This had to be set right. When Mosengeil was killed in a skiing accident, his dissertation was published in the *Annalen* post mortem, "abridged and with a correction by M. Planck" (Mosengeil 1907). Planck corresponded about this with Wilhelm Wien, the co-editor, at the end of January 1907: "At your request, I have arranged for Mr. Laue, my assistant, to shorten the Mosengeil dissertation somewhat."[16]

The entropy of radiation was not the only topic that preoccupied Laue. Einstein's theory of relativity also interested him greatly. In July 1907 he submitted a short paper to the *Annalen der Physik* entitled 'The entrainment of light by moving bodies according to the principle of relativity' as his contribution to the relativity debate (Laue 1907b). He cited Einstein's *Annalen* paper (1905a) and mentioned the relation between the velocities of bodies and light (see also Einstein 1906b; in reply Einstein 1907: 414 and Einstein and Laub 1909). With reference to Fresnel's experiment he stated: "According to the principle of relativity, then, light is perfectly carried along with the body, but for this very reason its velocity relative to an observer not participating in the motion is not equal to the vector sum of its velocity relative to the body and the body's velocity relative to the observer. We are thus relieved of the necessity to introduce an 'aether' into optics that permeates the bodies without taking part in their motion (Laue 1907b: 990)."

In the same month Laue finished his critique 'On Cohn's electrodynamics' (1907c) concerning a paper by Richard Gans about the longitudinal and transversal mass of

---

[15] Laue (1906b): 173. Laue mentions that Wiechert and Stokes had hypothesized that X-rays were electromagnetic pulses.

[16] Planck, Max, to Wien, Berlin, 26 Jan. 1907. SB PK, W. Wien papers. Laue's assistance is not mentioned. Einstein (1916) also addressed this problem.

an electron. His treatment of a paper by Orso Mario Corbino (1908: 344) on the Zeeman effect was equally critical.[17] There was some correspondence with Einstein (1993: 72) about "natural and not natural radiation," as Laue's letter to Einstein from 4 September 1907 indicates. He reported having discussed this once with Woldemar Voigt: "He said that it ought to be possible to detect light energy that is imperceptible off hand through accumulation over long periods of time, e.g., photographically. I simply replied that no one knows how non-natural radiation acts on a plate, without feeling quite satisfied with this rebuttal myself. For I'm not certain that the disorderliness of radiation is important in photochemical processes. […] It seems to me that it's too early to discuss the question of what happens to these small quanta of energy in the absorptive body anyway." If he ever returned to the vicinity of Bern, he added, he would visit Einstein again. "The inspirations I wreaped from our conversation are so valuable to me that I would like to have more." Two days earlier Laue had written enthusiastically to Jacob Laub about his visit with Einstein: "This is a revolutionary! Within the first two hours of our conversation, he succeeded in toppling the whole of mechanics and electromechanics, namely, on statistical grounds (Einstein 1993: 74)."

In September 1907, Laue and his coauthor, Professor Friedrich Franz Martens, submitted to the *Annalen* an article about experiments conducted at the physics department of the local commercial college.[18] They studied the polarization of radiation emanating from heated sheets such as platinum. A historical introduction was prefaced. Laue and Martens (1907) presented the results at the 79th Assembly of Natural Scientists in Dresden. Heinrich Rubens praised this research in the subsequent discussion. At this conference Arnold Sommerfeld gave a lecture on the topic 'An objection against the relativity theory of electrodynamics and its elimination.' He mentioned Laue's paper on 'The propagation of radiation in dispersive and absorptive media' (Laue 1905b).

In Dresden Laue spoke with Sommerfeld about the option of getting him a position as private lecturer at Munich, as Laue's letter to him from Berlin dated 16 October 1907 reveals:

Esteemed Professor:

Forgive me please if this letter arrives a few days later than you expected. I missed Professor Planck several times before managing to speak to him.

In full appreciation of the reasons you presented, I have decided to stay here in Berlin. The investigations in metal optics by Professor Martens and me weigh particularly heavily in this decision; they are not due to be completed for a long time and I should not like to abort them prematurely.

I feel most deeply indebted to you for the advice you gave me in Dresden, as well as for your willingness to take me on as private lecturer in Munich, even if I persisted (DMA, NL 089).

---

[17] The concluding sentence in Laue (1908a, 1908b) was: "The Zeeman phenomenon therefore does not contradict the entropy principle at all, indeed it remains questionable whether it has anything to do with it at all".

[18] The surviving documentation reveals nothing about how this collaboration with the *Handelshochschule* came about.

The definition of the velocity of rays in strongly absorptive media then followed (ibid.).

Within a matter of four years, Max Laue had succeeded in "making a name for himself" with his own research and critical scrutiny of the work of others. In 1908 Laue published a paper in the proceedings of the German Physical Society as well as in the following year in *Annalen der Physik*, entitled 'The irradiation of a moving point charge by waves according to the principle of relativity' (Laue 1908b, 1909a). Laue cited the publications by Oliver Heaviside and Max Abraham, which agreed in principle. However, he declared the form given to the moving charge to be inappropriate. For, at distances that are large against the dimension of the moving charge, the assertions by Heaviside and Abraham no longer applied. This issue could be solved much more simply in relativity theory using the transformation formulas for velocity, as he was able to show. He referred to a mathematical treatment by Planck (1908a). He was still viewing Einstein's theory of relativity somewhat skeptically: "It must, however, be emphasized that this theory should by no means be regarded as secure currently. Since its discrepancies from the other applicable theories are confined to extremely small terms, one may at any rate say that they can be regarded as correct apart from those discrepancies [...] (Planck 1908b: 829)." At the assembly of German natural scientists and physicians in Cologne at the end of September 1908, Laue presented his results in summary under the title 'The addition theorem of entropy' (1909b). Hermann Minkowski presented his exposition of the theory of relativity to his colleagues, which treated location and time coordinates equally. The utterance of his that was later to become famous was made on this occasion: "From this hour on, space on its own and time on its own shall completely sink down into the shadows, and only a kind of union of the two shall maintain independence (Minkowski 1908: 1)."

Planck knew about Laue's wish to leave for Munich. He wrote to Laue on 6 March 1909 that he was in Baden-Baden with his wife for treatment of her lung condition and was very pleased "that Mr. Sommerfeld agrees to your moving to Munich" (UAF, Laue papers, 8.8). He advised Laue not to report his change of residence until his registration in Munich was quite certain. As a private lecturer, Laue was required to reapply for habilitation. This normally involved the submission of a new thesis. Sommerfeld's vote exempted him from this, however.

A new friendship in addition to Max Planck's developed in Berlin around 1908/09 with Wilhelm Wien, born in 1864, who had studied physics with Hermann von Helmholtz (Fig. 4.6). While employed at the *Physikalisch-Technische Reichsanstalt* Wien had conducted fundamental experimental and theoretical research on radiation before becoming Röntgen's successor at Würzburg when the latter accepted a call to Munich. Wien owned a house—a forest hut—in Mittenwald and invited his colleagues to join him there in wintertime to go skiing. Laue probably took part for the first time during the winter of 1908/09.

On 3 May 1909 Laue sent back from Munich two offprints that he had borrowed from Wilhelm Wien, enclosing a note with his ideas about the interference capability of spectrally decomposed light, which he intended to include in similar form in an encyclopaedia article. He asked Wien for his opinion and advice, also requesting

Fig. 4.6 Wilhelm Wien
(1864–1928), a long-time
friend of Max von Laue, ca.
1911. *Source* Archive of the
Max Planck Society,
Berlin-Dahlem

the return of his note (DMA, NL 056/409). The invitation to write an encyclopaedia article came at the instigation of Arnold Sommerfeld as full professor of theoretical physics at the Ludwig Maximilian University in Munich. Sommerfeld was editing the fifth volume on physics of the *Encyclopädie der mathematischen Wissenschaften mit Einschluss ihrer Anwendungen*. Laue was supposed to work on the chapter 'Wave optics.' Wilhelm Wien edited 'Electromagnetic theory of light' and 'Theory of radiation' and completed the manuscript in 1909. He also alluded to the "analysis by Laue on the alteration of radiation as it traverses absorptive and dispersive materials." In article no. 11 on 'Interference phenomena' Wien cited Laue's papers in the *Annalen der Physik* Volumes 20 and 23 (1907b).[19]

In December Laue officially resigned as Planck's assistant in Berlin, unexpectedly for health reasons, to work with Sommerfeld in Munich.[20]

References

Corbino, O. M. 1908. Das Zeeman-Phänomen und der zweite Hauptsatz der Thermodynamik. *Phys. Z.* 9: 344
Einstein, Albert. 1905a. Zur Elektrodynamik bewegter Körper. *Ann. Phys.* 17: 891–921
Einstein, Albert. 1905b. Über einen die Erzeugung und Verwandlung des Lichtes betreffenden heuristischen Gesichtspunkt. *Ann. Phys.* 6: 132–148

---

[19] The complete volumes of the encyclopaedia did not appear until 1920; the individual contributions were published earlier as booklets.

[20] Curriculum vitae of Laue, Berlin, 4 Mar. 1909. LMUA, E-II-2224, 1764, 13,275. Reasons for the change were not provided.

Einstein, Albert. 1906a. Zur Theorie der Lichterzeugung und Lichtabsorption. *Ann. Phys.* 20: 199–206

Einstein, Albert. 1906b. Über eine Methode zur Bestimmung des Verhältnisses der transversalen und longitudinalen Masse des Elektrons. *Ann. Phys.* 13: 583–586

Einstein, Albert. 1907. Über das Relativitätsprinzip und die aus demselben gezogenen Folgerungen. *Jahrbuch der Radioaktivität und der Electronik* 4: 411–464

Einstein, Albert. 1916. Zur Quantisierung der Strahlung. *Verh. DPG* Berlin 18: 318

Einstein, Albert. 1993. *The Collected Papers of Albert Einstein.* Vol. 5: *The Swiss Years, Correspondence 1902–1914.* Edited by Martin J. Klein, A.J. Kox, and Robert Schulmann. Princeton: Univ. Press

Einstein, A. and Jakob Laub. 1909. Bemerkungen zu unserer Arbeit: 'Über die elektromagnetischen Grundgleichungen für bewegte Körper.' *Ann. Phys.* 2: 445–447

Grillparzer, Franz. 1971. *Werke.* Munich: Winkler

Hartmann, Hans. 1952. *Schöpfer des neuen Weltbildes. Große Physiker unserer Zeit.* Bonn: Athenäum

Helmholtz, Hermann von. 1907. *Vorlesungen über Elektrodynamik und Theorie des Magnetismus.* Edited by Max Laue. Leipzig: Barth

Laue, Max.1904a. Über die Interferenzerscheinungen an planparallelen Platten. *Ann. Phys.* 1: 163–181

Laue, Max. 1904b. Zur Theorie der kontinuierlichen Spektren. *Jahrbuch der Radioactivität und Electronik* 1: 400–412

Laue, Max. 1904c. Über die Fortpflanzung der Strahlung in dispergierenden und absorbierenden Medien. *Göttinger Nachrichten* 1904: 480

Laue, Max. 1905a. Über die Fortpflanzung der Strahlung in dispergierenden und absorbierenden Medien II. *Göttinger Nachr.* 1905: 117

Laue, Max. 1905b. Über die Fortpflanzung der Strahlung in dispergierenden und absorbierenden Medien. *Ann. Phys.* 18: 523–566

Laue, Max. 1905c. Die Krümmung der Interferenzlinien beim Stufengitter. *Phys. Z.* 9: 283

Laue, Max. 1906a. Zur Thermodynamik der Interferenzerscheinungen. *Ann. Phys.* 7: 365–378

Laue, Max. 1906b. Die spektrale Zerlegung des Lichtes durch Dispersion. *Verh. DPG* 8: 170–180

Laue, Max. 1907a. Die Entropie von partiell kohärenten Strahlenbündeln, *Ann. Phys.* 6: 1–43, with a postscript in *Ann. Phys.* 9: 795 ff.

Laue, Max. 1907b. Die Mitführung des Lichtes durch bewegte Körper nach dem Relativitätsprinzip. *Ann. Phys.* 10 (4) 23: 989 f

Laue, Max. 1907c. Zur Cohnschen Elektrodynamik. *Ann. Phys.* 10: 991 ff

Laue, Max. 1908a. Das Zeeman-Phänomen und der zweite Hauptsatz der Thermodynamik. *Phys. Z.* 9: 617

Laue, Max. 1908b. Die Wellenstrahlung einer bewegten Punktladung nach dem Relativitätsprinzip. *Verh. DPG* 10: 838–844

Laue, Max. 1909a. Die Wellenstrahlung einer bewegten Punktladung nach dem Relativitätsprinzip. *Ann. Phys.* 2: 436–442

Laue, Max. 1909b. Das Additionstheorem der Entropie. *Phys. Z.* 22: 778 ff

Laue, Max von. 1959. Geschichte des Elektrons. *Phys. Bl.* 3: 105–111

Laue, Max von. 1961. *Gesammelte Schriften und Vorträge.* 3 vols., Braunschweig: Vieweg

Laue, Max von. 1961a. Mein physikalischer Werdegang. Eine Selbstdarstellung. In Laue 1961, 3: V–XXXIV

Laue, Max, and F. F. Martens. 1907. Beiträge zur Metalloptik I. Über die Polarisation der von glühenden Metallen seitlich emittierten Strahlung. *Berichte der DPG* 5: 522

Lemmerich, Jost, ed. 1998. *Lise Meitner – Max von Laue. Briefwechsel 1938–1948.* Berliner Beiträge zur Geschichte der Naturwissenschaften und der Technik 22. Berlin: ERS-Verlag

Lemmerich, Jost, ed. 2011. *Mein lieber Sohn! Die Briefe Max von Laues an seinen Sohn Theodor in den Vereinigten Staaten von Amerika 1937–1946.* With two contributions by Christian

Matthaei. Berliner Beiträge zur Geschichte der Naturwissenschaften und der Technik 33. Berlin, Liebenwalde: ERS-Verlag

Livingston, Dorothy Michelson. 1973. *The Master of Light. A Biography of Albert A. Michelsohn.* Chicago: Scribner

Lummer, Otto, and E. Gehrcke. 1903. Über die Anwendung der Interferenz an planparallelen Platten zur Analyse feinster Spektrallinien. *Ann. Phys.* 3: 457–477

Minkowski, Hermann. 1908. Raum und Zeit. *Jb. Dt. Math.-Verein.* 1908: 1 (publ. 1909)

Mosengeil, Kurd von. 1907. Theorie der stationären Strahlung in einem gleichförmig bewegten Hohlraum. *Ann. Phys.* 5: 867–904

Planck, Max. 1900. Ueber irreversible Strahlungsvorgänge. *Ann. Phys.*1: 69–122

Planck, Max. 1901a. Ueber irreversible Strahlungsvorgänge. *Ann. Phys.* 12: 818–831

Planck, Max. 1901b. Über das Gesetz der Energieverteilung im Normalspectrum. *Ann. Phys.* 3: 553–563; cf. *Verh. DPG* 2 (1900) 237–245

Planck, Max. 1902. Über die Natur des weißen Lichtes. *Ann. Phys.* 2: 390–400

Planck, Max. 1906. Paul Drude. Gedächtnisrede, gehalten in der Sitzung der Deutschen Physikalischen Gesellschaft am 30. November 1906. *Verh. DPG* 8: 599

Planck, Max.1908a. Zur Dynamik bewegter Systeme. *Ann. Phys.* 26: 1–34

Planck, Max. 1908b. Bemerkungen zum Prinzip der Aktion und Reaktion in der allgemeinen Dynamik. *Phys. Z.* 9: 828 ff

Sime, Ruth Lewin. 2001. *Lise Meitner. Ein Leben für die Physik.* Frankfurt/M., Leipzig: Insel-Verlag. English version: Sime, R.L. 1996: *Lise Meitner. A Life in Physics.* Berkeley: Univ. of California Press

Voigt, Woldemar. 1896. Ueber die Aenderung der Schwingungsform des Lichtes beim Fortschreiten in einem dispergierenden und absorbierenden Mittel. *Göttinger Nachrichten* 1896: 186

Chapter 5
Private Lecturer at the University of Munich

Up to this point, Max Laue's academic career had followed a direct course. At Munich, he had become eligible to receive an appointment to a professorship. His chances were good, because there were very few younger theoretical physicists who had studied under various academic teachers and were as versed as Laue in tackling unresolved problems, as his publications demonstrated. Moreover, the Ludwig Maximilian University in Munich had a chair in Theoretical Physics. Around 1910 this was not yet the rule at German universities. The leaders in this field were Berlin, Göttingen, and Munich, and Laue's contributions to the field were well known.

The chairholder in theoretical physics at Munich, Arnold Sommerfeld (Fig. 5.1), was an outstanding academic teacher (Eckert 2013: Chap. 6). After completing his studies in mathematics at Göttingen, he had begun his career as an academic teacher at the Mining Academy in Clausthal.[1] He had then followed a call to the polytechnic in Aachen, where he was more of a mathematician—with research on the theory of gyroscopes and contributions to statics—than a physicist (Klein and Sommerfeld 1965). In 1900, at Röntgen's instigation, he had been appointed to the chair in the field of theoretical physics at the Ludwig Maximilian University in Munich, which had been vacated by Ludwig Boltzmann in order to return home to Vienna. Sommerfeld mainly studied the nature of X-rays and relativity theory. Sommerfeld's institute had its own workshop and mechanic, which meant experimental research was possible there.

Sommerfeld developed a theory for the origin and properties of X-rays based on Planck's formula $E = \mathrm{h} \cdot \nu$ by setting $E \cdot \tau = \mathrm{h}$, that is, the quantum of action, where τ represents the duration of the collision. But it was not possible to use this to describe the two types of X-rays—characteristic radiation, which depended on the anode material, and bremsstrahlung or the continuous spectrum, which was determined by the energy of the electrons (Sommerfeld 1909: 969).

---

[1] Sommerfeld was already working on the nature of X-rays at the *Bergakademie* in Clausthal, albeit without being able to come up with any clear results (Sommerfeld 1899). Other research on the same topic by M. Maier from slightly earlier appeared in the same issue on p. 57.

© Basilisken-Presse, Natur+Text GmbH 2022
J. Lemmerich, *Max von Laue*, Springer Biographies,
https://doi.org/10.1007/978-3-030-94699-9_5

Fig. 5.1 Arnold
Sommerfeld (1868–1951),
mathematician and physicist,
founder of the so-called
Sommerfeld school of
theoretical physics. *Source*
DMA, reproduction by the
Archives of the Max Planck
Society, Berlin-Dahlem

In the year of his move to Munich, Laue submitted two papers for publication on the thermodynamics of diffraction (Laue 1909a, 1910a).[2] In the first publication about diffraction at a grating, Laue took Wilhelm Wien's research on the thermodynamics of radiation as a basis, which considered diffraction irreversible. "Things are quite different ever since we know that in order to calculate entropy, the coherence relations between different parts of the radiation process must be taken into account. [...] Diffraction is—according to our assumption—generally irreversible. An irregular arrangement of many, like, diffracting bodies disperses the radiation in the same way as a single one, only much more strongly, therefore increasing the entropy much more (Laue 1909a: 225, 239—Laue often used the first person plural in his texts as a self-reference.)." Five months later his paper 'On the thermodynamics

---

[2] Planck's correspondence with Lorentz reveals how difficult it was for physicists to understand what radiation was. On 16 Jun. 1909 Planck wrote to Lorentz: "But now to radiation. That it doesn't work without the assumption of light quanta hν, on that we do agree (in contrast to Jeans). That furthermore light quanta *can't possibly* retain their individuality while propagating in the free aether, you have explained so convincingly in the second part of your letter that I believe there can be no doubt about that either. The whole of optics, especially the theory of interference and refraction and diffraction, would have to be overturned if one wanted to ascribe specific existence to light quanta in the free aether." AMPG, V rep. 11 no. 568, orig. emphasis.

Paul Peter Ewald described a lecture at the colloquium prior to Laue's appointment to Munich on 'The behavior of light waves at the focal point.' He thought that was a prerequisite for Laue's move from Berlin to Munich. See Ewald (1968).

of diffraction' was published (1910a). It was a theoretical study of diffraction by small transparent bodies and large apertures. His basic conclusion was: "Therefore, diffraction at large apertures, just as at small particles, is an irreversible process (p. 557)." The subject of diffraction and thermodynamics did not let him go; too many fundamental things still had to be resolved. He presented a paper under the title 'Thermodynamic considerations on diffraction' at the conference of German Natural Scientists and Physicians in Salzburg and completed the manuscript in October. Laue expounded the problem for his colleagues using very little mathematics, but much explanatory, defining text: "Let it be said in advance that in the question of reversibility only diffraction from perfectly specular or completely transparent bodies is to be considered; for any absorption of radiation renders the process irreversible, because the fraction of energy absorbed can never be recovered in a form coherent to the original oscillation. Even the processes of geometrical optics are reversible only in the absence of absorption (Laue 1909b)." He first talked about Rayleigh diffraction on small transparent particles. One statement was: "We conclude that there cannot possibly be any process which completely reverses the diffraction of radiation by small transparent particles, that this is rather an irreversible process (ibid.)." This was followed by a treatment of grating diffraction: "Therefore, grating diffraction is a reversible process in the sense that no entropy increase occurs in the absence of absorption [...] (ibid.)."

In April 1910 Laue submitted to the editors of the *Annalen der Physik* a manuscript entitled 'The radiation of heat in absorbing bodies' (Laue 1910b), in which he discussed and expanded on the previous results of Kirchhoff, Clausius, and Planck. His paper was probably connected with Professor Martens's investigations (Laue and Martens 1907). More important was the paper submitted in May, 'Is the Michelson experiment evidential?' (Laue 1910c), a rebuttal to a publication by Emil Kohl. "Since Michelson's interference experiment, considered the foundation of relativity theory, has gained very unique importance for the development of physics, it is necessary to establish its interpretation against all doubts, including the concerns recently expressed by Mr. Kohl (p. 186)." Very scrupulously Laue analysed the counter-arguments, the interferometer design, the mirror tilts, and the imaging of the interference fringes. He quoted Michelson's book from 1903 and the later experiments with Morley and Miller (1905). Kohl's error, he contended, was the mistaken assumption that the interferences were equally inclined. In closing he asserted: "With the permission of Mr. Kohl, I add that upon perusal of the present observations he fully concurs with the representations offered here concerning the formation of the interference system (Laue 1910c: 191)."

The scientific dialogue between Max Planck and Max Laue continued by mail. We see this from a letter by Planck in Grunewald, Berlin, to Laue on 4 July 1910 about an obviously unpublished consideration of Laue's concerning an assertion by Heinrich Hertz about the moving aether as well as about entropy in turbulent moving systems. (AMPG, V rep. 11 no. 593. There is no evidence of any related publication by Laue.)

Max Laue's private life had not yet led to any closer ties with a woman. He had not found his future wife among his friends and acquaintances in Berlin, or had not even

Fig. 5.2 Magdalena von
Milkau, Max Laue's future
wife. *Source* UAF

looked for one; and there were anyway very few female physics students. Lise Meitner
from Vienna was a major exception. There is no record of Laue having made the
acquaintance of the woman he would marry, during his travels to Switzerland, but it is
conceivable, because after visiting Einstein in Bern, Laue continued to travel through
Switzerland. When he was back in Munich, he wrote a letter on 26 April 1909 asking
"Hochverehrtes Fräulein Magdalena" von Degen-Milkau for permission to visit her
in her apartment for a brief meeting (UAF, Laue papers, 3.11). But concurrence was
yet to be reached about wanting to spend their lives together (Fig. 5.2).

Max Laue may have spent time at the hospitable home of Professor Runge during
his student days in Göttingen. Runge's daughter Iris was a student of mathematics
and physics. The Runge couple were friends with the Ewald couple; Paul Peter Ewald
was a student of Sommerfeld in Munich. Iris Runge visited them in 1909 and got to
know Laue on excursions to the Ammersee with Mr. and Mrs. Ewald. His somewhat
awkward manner, perhaps also his speech impairment—especially when emotionally
moved—kept vivacious Iris on her guard. When Max Laue went to Göttingen in July
to see Runge and apply for the hand of his daughter, without first asking Iris herself,
rejection was the reward. Neither the parents nor their daughter welcomed Laue's

proposal (Tobies 2010: 69 ff., 74, 78 f.). Since Iris Runge was studying in Munich, it was probably not out of the question that they saw each other occasionally.

The private address in Munich that Max Laue indicated in a letter to Wilhelm Wien was Heßstraße no. 3, later redesignated Ludwigstraße no. 22. As the summer of 1909 approached, he informed Wilhelm Wien of his engagement to the baroness, Magdalena Caroline Freiin von Milkau-Degen.[3] She was born in 1891 in Merano in South Tyrol, which at that time still belonged to the Austrian Empire. This daughter of a senior military officer had been living in Switzerland for several years and she never forgot the beauty of the Swiss landscape. Her mother Emilie had become widowed.

Max Planck congratulated the couple on their upcoming wedding in a letter dated 5 August 1910 from Grunderhof near Gmund on the shore of Tegernsee: "Esteemed Dr. First of all, once again my most cordial congratulations to you and your Miss Bride. You know that I honestly mean it when I express my delight at your having now found the richest recompense and *lasting happiness in life* for so much that you had to deprive yourself of in the past. My wife would certainly also have gladly shared your joy, as she always was affectionately attached to you. But now thanks for your letter of the 31st. One should not rob a *bridegroom's* precious time, but I would nevertheless like to confirm to you briefly at least that we are in complete agreement about *Hertz's* theory (AMPG, V rep. 0011 no. 597, orig. emphasis)." He also mentioned his talk in Königsberg and necessary critique of Ernst Mach:

> *Mach's turn* comes in October; I have already written the rel[evant] note, but would like to let it mature a little. Naturally, I do want to avoid anything that might hurt the worthy old gentleman personally, but I do have to *make one jab* at his 'anti-metaphysical theory,' I owe that to my own *conviction*. You'll get a correction proof of that too.
>
> Of course I quite agree with your convictions concerning the propagation velocity of waves, I would just like to leave open whether the formula for such waves also reappears according to the theory of relativity (although it does seem likely to me). I want to now have a student work out the complete hydrodynamics of a perfect fluid (without friction) on the basis of *rel[ativity] theory* (ibid.).

The civil ceremony of the Laue couple took place on 6 October 1910 in Munich.[4] The witnesses were a cousin of the bride, the physician Dr. Robert Born, and the lawyer Rudolf Pixto. Invitations were also extended to his younger colleagues for the celebration. The couple lived at Bismarckstraße no. 22.

The theory of relativity generated so much interest in the five years after Einstein's first article that publishers were scrambling to find an author to write a book about it.[5] In his 'My career in physics,' Laue, mentioning his 1907 paper, recalled: "I don't

---

[3] Laue to Wien, Munich, 3 Apr. 1909. DMA, NL 056/409-452 and HUA, UKP, L 51, vol. 4.

[4] STAM, marriage register, Standesamt München I, 1189/1910. Laue's handwritten communication to the university rectorate reads: "Daughter of the late Captain Erwin Freiherr Milkau and his wife Emmy, née Degen." See also HUA, UKP, L 51 and LMUA, PA E-II, 2224.

[5] Hermann (1994: 157). At the end of 1908 Einstein received an inquiry from the Hirzel publishing house in Leipzig. He agreed at first but later retracted again: "Unfortunately it is quite impossible for me to write that book, because it is impossible for me to find the time for it." Comp. Fölsing (2004: 262). There aren't any files at Friedrich Vieweg & Sohn from this period. Very likely, several

Fig. 5.3 Magda, as Max Laue called his wife, in front of their boathouse in Feldafing on Lake Starnberg, 1911. *Source* UAF

know, was it just that, or were there other factors leading to the invitation by the publishers F. Vieweg & Sohn in 1910 to write a monograph about the theory of relativity? I did so and thus became the author of the first summary account of it that was well received by the general public […]. I wrote it—I had transferred my habilitation from Berlin to Munich in 1909—in a small boathouse on the shore of Lake Starnberg in the ducal park in Feldafing; it stood on pilings in the water and afforded a magnificent view of the Herzogstand, Heimgarten, Benediktenwand, and Karwendel mountains. Never again did I hit the mark so well (Laue 1961b: XX f.)." (Fig. 5.3).

Laue put the finishing touches to the manuscript as early as May 1911. It became the 38th volume in a series called 'Science—A collection of scientific and mathematical monographs' published by the Braunschweig publishing house Vieweg & Sohn (Fig. 5.4). In the preface Laue attempted to assuage any future critics.

In the 5 ½ years that have elapsed since Einstein expounded the theory of relativity, the theory has attracted mounting attention. Admittedly, this attention is not always applause. Some scientists, including bearers of very famous names, consider its empirical grounds to be insufficiently firm. Objections of this kind can, of course, only be resolved by more experiments; however, the present booklet does attach some importance, for instance, to the proof that not a single counterargument exists against the theory's empirical basis. The number of those who cannot befriend themselves with its intellectual content is far greater,

---

physicists were invited by this publisher to write a book on relativity. Who suggested Laue is no longer ascertainable. For remarks by Laue about relativity see Staley (2008).

DIE WISSENSCHAFT

SAMMLUNG
NATURWISSENSCHAFTLICHER UND MATHEMATISCHER
MONOGRAPHIEN

ACHTUNDDREISSIGSTES HEFT

DAS RELATIVITÄTSPRINZIP

VON

DR. M. LAUE
PRIVATDOZENT FÜR THEORETISCHE PHYSIK
AN DER UNIVERSITÄT MÜNCHEN

MIT 14 IN DEN TEXT EINGEDRUCKTEN ABBILDUNGEN

BRAUNSCHWEIG
DRUCK UND VERLAG VON FRIEDR. VIEWEG & SOHN
1911

DAS

RELATIVITÄTSPRINZIP

VON

DR. M. LAUE
PRIVATDOZENT FÜR THEORETISCHE PHYSIK
AN DER UNIVERSITÄT MÜNCHEN

MIT 14 IN DEN TEXT EINGEDRUCKTEN ABBILDUNGEN

BRAUNSCHWEIG
DRUCK UND VERLAG VON FRIEDR. VIEWEG & SOHN
1911

Fig. 5.4 Title page of the book *Das Relativitätsprinzip* by Max Laue, here in its first edition from 1911

though, who perceive the relativity of time in particular as unacceptable with its sometimes apparently indeed quite paradoxical consequences (Laue 1911a: V).

Laue mentioned the fundamental publications by Einstein, Planck, and Minkowski as well as Sommerfeld's most recent ones. Then he pointed out the prerequisites for understanding his book: "In addition to some knowledge of Maxwell's theory [...], this exposition only presupposes the mathematical tools commonly used by theoretical physicists, the infinitesimal calculus, and vector analysis (p. VI)." At the end of the preface Laue thanked Dr. A. Rosenthal for his assistance with the correction proofs. The content of this book of 208 pages was strictly divided into the following seven chapters: I. The problem; II. The older theories of the electro-dynamics of moving bodies; III. The theory of relativity, kinematics section; IV. World vectors and tensors; V. The electrodynamics of empty space according to the principle of relativity; VI. Minkowskian electrodynamics of ponderable bodies; and VII. Dynamics. Despite the presumption in the preface that the reader have some familiarity with the "mathematical tools," Laue occasionally wove in simple, easily visualizable examples. Thus, for instance, in the fourth, expanded edition of the book: "Imagine, for instance, that two astronomers on different mutually approaching celes-tial bodies could exchange thoughts with each other. They would establish that they were recording substantially different times for all astronomical events; and this despite (let's assume) both of them applying the same physical principles, solely because each of them presumes to be stationary (Laue 1921: 52)."

The book contained no philosophical considerations on space and time, no mention of Kant's a priori human conceptions of space. Not all of Laue's colleagues approved of his representations. Rigorous theoreticians criticized that such a book did not need experiment descriptions. Some decades later, Max von Laue alluded to this when a revised edition was in preparation, in a letter to Wilhelm Westphal on 21 January 1950: "I am surprised, by the way, that there is such a demand for the book now, after having been cast into such bad light so often before. When I gave Planck the manuscript for his critique in 1911, he returned it to me with the remark that under no circumstances may I let it be known that it had been in his hands (Private holdings)."

While Laue was still working on the book, he intervened in the dispute between relativists and anti-relativists with a publication in the *Physikalische Zeitschrift*: "Since the old kinematic conditions of rigidity are not compatible with the Lorentz transformation, the first thing to do was to create a new definition for it. It goes to the credit of M. Born for being the first to put forward such a definition, and thereby present the problem (Laue 1911b)." The argumentation ended up with ten degrees of freedom. Laue assumed in his work that there is no such thing as superluminal velocity and based his considerations on Minkowski's four-dimensional world. He reached the conclusion that "the number of kinematic degrees of freedom of a body thus has no upper limit" (ibid.). In a footnote he noted: "I should not like to leave unmentioned that I was able to discuss these matters with Prof. Sommerfeld on several occasions (ibid.)."

Max Born wrote a review for the *Physikalische Zeitschrift* soon after Laue's book appeared. "The book addresses a mathematically trained readership; it is no light reading given the extraordinary brevity and precision of the exposition (Born 1912)." Being a collaborator of Minkowski, Born faulted the mathematical treatment for rather following Sommerfeld's method than Minkowski's matrix calculus. Born also noticed deviations from Minkowski's approach elsewhere. At the end of the review he wrote: "The book closes with an application of general dynamics to cavity radiation, in which Mosengeil's results are elegantly retrieved. The wealth of inspirations that this book contains on minor and major questions can *scarcely be broached in a brief review.*"

The second, improved edition of Laue's book on relativity (1911a) was released as early as 1913. Despite being occupied with his book manuscript Laue was still able to submit a paper 'On the dynamics of relativity' at the end of April 1911. In it he examined the question of how the electron is deformed during motion. Laue cited the publications by Einstein and Planck as well as Ehrenfest and Born. The latter attributed spherical symmetry to the electron, which Laue critically assessed. Laue also discussed the experiment performed by Trouton and Noble in 1903: "According to electron theory, a uniformly moving charged condenser experiences a torque from the electromagnetic forces. Trouton and Noble tried to prove this with a bifilar suspended condenser as a result of the Earth's motion, but could find no rotation of it from the state of rest. The theory of relativity can, of course, explain the result very simply by the fact that the Earth, relative to which the condenser is at rest, is a justified frame of reference. But what does the theory look like if a different reference frame is chosen? Electrodynamic torque is present also according to relativity. Why does no rotation occur? (Laue 1911c; Trouton and Noble 1903)". He then asked which kinds of energy should be included, and thought: "All the ones exhibiting no disturbance in a justified system, the system at rest, therefore also [can]not provide impetus. [...] Furthermore, the condenser of the Trouton-Noble experiment along with its field is a complete, static system. The entire system at uniform velocity has no more need for torque than does a mass point (Laue 1911c)."

In the summer semester of 1911, Max Laue lectured on Thermodynamics with Special Consideration of Chemical Applications. There was no lecture on relativity during this term. On 1 July 1911, Sommerfeld, as a member of the mathematical-physical class of the Royal Bavarian Academy of Sciences, presented a paper by Laue: 'On an experiment on the optics of moving bodies' (1911d). Based on the Michelson experiments, Laue investigated with regard to relativity theory whether or not the aether participates in the rotation of the Earth. For this purpose he assumed a circulating beam of light in a mirror system. According to the classical theory, the center point of the circle it describes is not altered by terrestial motion, but according to the theory of relativity, the circle becomes elliptical. Laue encouraged that related experiments be performed.

Debates about the theory of relativity continued among physicists, and Laue felt called upon to help his colleagues understand Einstein's reasoning. In December 1911, he submitted the following paper to the *Physikalische Zeitschrift*: 'Two objections to the theory of relativity and their refutation' (1911e). He took the seemingly paradoxical statement in relativity, the clock paradox, and tested the objections raised against it using, among other things, Minkowski's four-dimensional world with its equivalence of all temporal orientations. He countered Emil Wiechert's assertions about the existence of an aether. "Besides, it seems to me that the whole question of the existence of the aether and of absolute time can be banished from the physical debate without harm, as long as entirely new facts, such as the existence of superluminal velocities or a contradiction between gravitational phenomena and the principle of relativity, are not established by experiment [...] (Laue 1911e)."

During this time Laue was corresponding with Einstein, as he was studying his work on gravitation, because he wanted to give a talk on it at the Munich colloquium. The form of address he employed in his letter of December 27th was still the semi-formal "Lieber Herr Professor." Laue raised some objections about the equivalence of the two systems K and K'. "It also seems to me extraordinarily characteristic that the gravitational potential acquires a physical significance which the electrostatic potential completely lacks (Einstein 1993: 384)." But despite all this he did not believe in the correctness of the theory. Laue discussed the difficulties of measuring the deflection of light by the Sun and mentioned that Michelson had already proposed interference experiments on the problem with long light paths at two different heights above the ground (ibid.).

As a theoretical physicist, Max Laue saw where the dispute was leading. He attended the conference of German natural scientists and physicians in Karlsruhe in

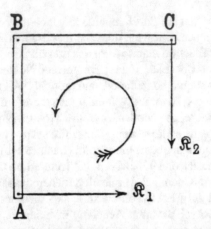

Fig. 5.5 Angle lever observed in two frames of reference: "In the reference system K^0, which is justified according to the principle of relativity, an angle lever ABC is at rest with two equally long arms perpendicular to each other $\left(AB = BC = l^0\right)$. At B it is rotatable about an axis perpendicular to its plane. Two forces K_1^0 and K_2^0 of equal magnitude act at A and C; K_1^0 is parallel to BC, K_2^0 is parallel to BA. The torques of these forces are $\pm|K_1^0|l^0$ and cancel each other out." *Source* Laue (1911f, reprinted 1961a, 1: 163)

1911 and heard his colleague Emil Budde's talk 'On the theory of the Michelson experiment,' which prompted him to make a very critical remark during the discussion. He only addressed three points to correct Budde's conclusions. A detailed article on this under the same title as Budde's talk followed in the *Physikalische Zeitschrift*, in which Laue focussed on the optical processes at the divider plate of the interferometer (Laue 1912a). Laue himself presented his 'Remarks on the lever law in the theory of relativity' (Laue 1911f). At the beginning of the text, one senses Laue's enthusiasm about physics: "One of the finest results of the theory of relativity is probably the theorem of the inertia of energy; its most general formulation was delivered by Mr. Planck 3 years ago in Cologne. It asserts that wherever a flow of energy of density S occurs, a quantity of motion is always attached which, with reference to the unit volume, has the magnitude $g = S/c^2$ (p. 1008)." (Fig. 5.5).

In March 1912 two more papers by Laue on relativity appeared, including another very detailed one on the experiment by Trouton and Noble from 1903 about whether a charged plate capacitor experiences torque as a result of its motion with the rotating Earth (Laue 1912b). "Michelson's interference experiment is often regarded as the sole support that relativity theory has against the 'absolute theory.' But this is by no means the case. The Trouton-Noble experiment also reaches the same verdict; and even though it may have fallen into oblivion in spite of H. A. Lorentz's acknowledgment of it in 1904, this is probably essentially due to its theory lacking the intuitiveness that characterizes Michelson's experiment so much (Laue 1912b: 370). Then Laue expanded on the theory. He first calculated the forces exerted on the capacitor at rest and also discussed the courses of the force lines at its edges, which complicate things greatly: "It will only be possible to convey our result quantitatively that, in addition to the attraction of the plate, forces are acting at the edge of the plates and at the boundary of the dielectric, which at the state of rest exert tangential tension on the plates, and that in the case of motion the total of these forces produces the torque (ibid.)."[6]

Another longer essay was still lying on Laue's desk: He had to write the chapter 'Wave optics' for the fifth volume, third part, of the 'Encyclopaedia of the Mathematical Sciences along with Their Applications,' which was not published until 1913a, in order to be able to include a treatise on interference experiments with X-rays. Sommerfeld, who had probably chosen Laue to author the chapter, was the editor of the volume.

As private lecturer, Laue had to offer lecture courses. During the summer semester of 1912, he lectured on The Principle of Relativity and Its Consequences. As a rhetorician, Laue was rather weak, and since the old rule *"Trius sunt collegium"* applied, the course depended on the attendance of at least three students. Peter Paul Ewald, a doctoral student of Sommerfeld, reported that on sunny days they used to draw lots to determine who had to go to Laue's lecture to meet this quorum (Ewald 1979: 339).

At universities it was customary not to hold lectures at lunch time. The University of Munich was located near the palatial Court Garden. The auditors of Sommerfeld's lectures and the participants of the seminar and colloquium used to meet there at Café Lutz to discuss the material. It became a very important part of life at the institute for the small circle of students, in order to get a grasp of the material and understand Sommerfeld's lectures. An unofficial seminar "without the big-wigs" (*bonzen-freies Seminar*) was also organized. There were bowling evenings and visits to the pub for a thirst-quenching mug of beer (Fig. 5.6).

The nature of X-rays remained unclear. Sommerfeld had corresponded about this with Einstein in early 1909. Einstein's reply in a long and partly witty letter from Zurich on 19 January 1909 was unable to offer any physically unambiguous explanation (Eckert and Pricha 1984; Eckert 2013: 217, 233).

At the end of October 1911, the first Solvay Congress on the subject of 'Radiation' took place in Brussels (Fig. 5.7). A wealthy Belgian industrialist, Ernest Solvay, on the advice of the Berlin physico-chemist Walther Nernst, had donated the funds for an international congress. Sommerfeld was among the invitees including Marie Curie, Einstein, Lorentz, Nernst, Planck, and Rutherford. Sommerfeld presented his ideas about X-rays, but in the lengthy ensuing debate—discussion was the main purpose of this congress—no agreement could be reached as to whether they were pulses or waves. Their wavelength although unknown was considered by all to be very short (Mehra 1975: 75 f.).

Sommerfeld wanted to find a good experimental physicist capable of testing his ideas about X-rays at his own institute. Walter Friedrich had just completed his dissertation under Röntgen on the 'Spatial intensity distribution of X-rays emanating from a platinum anticathode.' He was thus a suitable experimental physicist for the

---

[6] The place of origin of the research is indicated as the Institute for Theoretical Physics in Munich and dated March 1912.

Fig. 5.6 The Munich *Bierkeller* where the physicists went bowling; *far left*: Max Laue, *far right*:
Paul Peter Ewald, 1912. *Source* Archive of the Max Planck Society, Berlin-Dahlem

Fig. 5.7 The first Solvay conference in physics, 1911. *Standing from left to right:* Robert Goldschmidt, Max Planck, Heinrich Rubens, Arnold Sommerfeld, Frederick Lindemann, Maurice de Broglie, Martin Knudsen, Friedrich Hasenöhrl, Georges Hostelet, Édouard Herzen, James Jeans, Ernest Rutherford, Heike Kamerlingh Onnes, Albert Einstein, Paul Langevin. *Seated from left to right:* Walther Nernst, Marcel Brillouin, Ernest Solvay, Hendrik Antoon Lorentz, Emil Warburg, Jean-Baptiste Perrin, Wilhelm Wien, Marie Curie, Henri Poncaré. *Source* The author's estate

task, and Sommerfeld endeavored to recruit him for his institute. But Röntgen was reluctant to release Friedrich, which Sommerfeld refused to accept. Röntgen finally gave in and Friedrich was free to do experiments guided by Sommerfeld's ideas.

Röntgen had discovered X-rays, also known as 'roentgen radiation,' in Würzburg in 1895 and had carried out numerous fundamental experimental investigations into the nature of the rays, leaving little more to be explored (Fig. 5.8). Röntgen believed that he had discovered the longitudinal aether waves that Hermann von Helmholtz had already thought about.[7] An important contribution on the nature of X-rays was made by the Englishman Charles G. Barkla in 1904, who experimentally discovered the polarization of X-rays by scattering off substances of different atomic weights, which gave a clue to the character of the waves. He was awarded the Nobel Prize in 1917. In 1908, he had published a review article on the current research on the subject including his own in *Jahrbuch der Radioaktivität and Elektronik.*

---

[7] Röntgen (1895: 141): "Now it has been known for a long time that, in addition to transverse light oscillations, longitudinal oscillations can also occur in the aether and, in the opinion of various physicists, must occur. Obviously, their existence has not yet been proven and therefore their properties have not yet been experimentally investigated. Shouldn't these new rays be longitudinal swinging motions of the aether? I must confess that during the course of the investigations I have become more and more attached to this thought and thus also permit myself to express this conjecture here, although I am very well aware that the explanation given still requires further substantiation".

Fig. 5.8 Wilhelm Conrad
Röntgen (1845–1923),
undated portrait, with a
dedication from the
Deutsche Röntgen-Museum
on the occasion of the 25th
anniversary of the MPI for
Biophysics. *Source* Archives
of the Max Planck Society,
Berlin-Dahlem

The usual method used to establish the wave nature of a radiation were diffraction experiments. Röntgen had already tried this with his X-rays, but without success. After the discovery of radioactivity, there was another unexplained phenomenon: What do γ-rays observed in radioactive decay signify? Were they identical to X-rays? Do they originate from X-rays, as William HenryBragg (1911) assumed? Two young physicists, Robert Wichard Pohl and Bernhard Walter, experimented with a wedge-shaped slit arrangement in 1908 to demonstrate the diffraction of X-rays. Visual analysis of the images obtained on photographic plates did not yield any clear explanation. (Walter and Pohl 1908; an earlier analysis by Haga and Wind 1903 had yielded nothing.) Sommerfeld sent his critical overview entitled 'On the diffraction of X-rays' to the editors of *Annalen der Physik* on 1 March 1912. The concluding sentence of this paper reads: "In any case, it will be worthwhile to continue to take careful diffraction images with X-rays, since this research has shown that they can be theoretically evaluated, and after our colleague Mr. Koch, in particular, has demonstrated in the subsequent work that they can be measured with all the necessary precision (Sommerfeld 1912: 506)."[8]

---

[8] Peter Paul Koch built a recording microphotometer at the Physical Institute in Munich to conduct objective analysis of the results (Wolfke 1912). However, those measurements of the diffraction

Nobody left any contemporary record of what was really discussed in the two physics institutes at Munich University and in Café Lutz from February to April 1912 (Eckert 2012: 19, see also Authier 2013, Kubbinga 2012). All accounts were written later, some of them decades later. Those many years spent working on optical interference problems appear to be purposeful preparation for Laue's future discovery in his retrospective sketch 'My career in physics—a selfportrait':

> It became specially important to me that the space lattice hypothesis for crystals was still conventional at Munich, while it was hardly mentioned anymore elsewhere. This was partly because lattice models were on display in the university institute collections since Leonhardt Sohncke's day, who had been active in Munich until 1897 and had contributed much toward their mathematical elaboration. The mineralogist Paul von Groth deserves the main credit for always pointing this out in his lectures. And then in February 1912 it happened that P. P. Ewald, a doctoral student of Sommerfeld's who was supposed to analyse the behavior of light waves in a space lattice of polarizable atoms mathematically but was having difficulty with it at first, visited me in my apartment and asked for advice. I did not know how to help him, of course; but during that conversation, an idea popped out of my mouth almost by chance that one ought to send much shorter waves, namely X-rays, through crystals. If atoms really do form a space lattice, this must produce interference effects, similar to interferences of light on optical gratings. Word of this spread among Munich's younger physicists, who were meeting every weekday after lunch at Café Lutz. One among this crowd, Walter Friedrich, who had earned his doctorate shortly before under Röntgen with a thesis on X-ray scattering and had subsequently become Sommerfeld's assistant, volunteered to test this. The only problem was that Sommerfeld initially did not think much of the idea and would have preferred to set Friedrich to work on an experiment about the orientational distribution of radiation emitted by an anode. But this hurdle too was overcome when Paul Knipping, another doctoral student of Roentgen's, offered to help. And so approaching Easter 1912 the experiments began (Laue 1961b; Hartmann 1952: 178–207).

The recollections of Paul Peter Ewald, who had sought Laue's advice at the time, differ from this account (1979).[9] In his introduction he contended that at that time—i.e., prior to Laue's considerations—Sommerfeld was already convinced of the wave nature of X-rays from the experiments that Pohl and Walter had performed and the measurements taken by Peter Paul Koch and assumed the wavelength to be about 0.4 Å.

> This is one foundation of Laue's idea, which is grounded in Sommerfeld's researches; the second is this: through the Göttingen Mathematical Society, Sommerfeld was friends with the mathematician Arthur Schönflies, who had derived the 230 possible periodic arrangements of symmetry elements in applying the then still new geometrical group theory and had submitted it to Felix Klein as a habilitation thesis. The reaction of [the mathematician] Klein was, as Sommerfeld told me, 'What you have there is crystallography.' At a time when eminent investigators among crystallographers were still debating the whole idea of an atomic structure of matter apart from in its most primitive form of Bravais's 14 lattices, the structural theory of crystals was not highly regarded.

---

recordings yielded no clear proof of any wavelike character either. Potential fluorescence radiation due to the metallic slit emitting radiation upon irradiation was not discussed.

[9] Ewald (1968: 541) describes a lecture by Laue before his appointment, 'On the behavior of light waves at the focal point' (Das Verhalten von Lichtwellen im Brennpunkt). See also Boeters and Lemmerich 1979: 56.

Fig. 5.9 Paul Peter Ewald
(1888–1985) developed a
theoretical basis for the
X-ray interference of crystals
and was thus able to explain
the details of Laue's X-ray
scattering experiments.
Source Göttingen University
Library, scanned by the
Archive of the Max Planck
Society, Berlin-Dahlem

One exception was the Munich mineralogist Paul Groth, a friend of L. Sohncke's, the
physics professor at the Munich polytechnic (Ewald 1979: 345 f.).

Ewald then mentioned the crystal model exhibits, which Laue had also mentioned.
"When I asked Sommerfeld in 1910 for a topic for a doctoral thesis, the one he recommended the least among many other suggestions because he doubted its feasibility
inspired me the most, for which he granted his approval nonetheless; it was: to investigate whether resonators, such as have been assumed since Drude, Lorentz, and
Planck inside bodies to explain the refractive power of light, could produce birefringence if instead of being distributed randomly in the refractive medium they were
periodically distributed on a lattice of suitable symmetry (ibid., p. 346)."

Details followed about the then still unknown structure of crystals. The initial
results of his calculations worried Ewald (Fig. 5.9). They deviated substantially from
the known theories, "which frightened me, inexperienced as I was in my expertise,
[so] I wanted to hear Laue's opinion. On the way from the Institute to his apartment
on Bismarckstraße—he and his young wife had invited me to dinner, and we were
walking through the English Garden—I began to present my case to him when I
realized that Laue was unfamiliar with the subject of my thesis. So I had to go back
to the beginning and start with the model of resonators arranged in a space lattice.

Laue asked, 'Why make this assumption?' I replied that crystallographers regard it as the essential difference between crystals and amorphous bodies, such as glass. Laue's response was, 'Oh, I see.' This reply leaves open whether or not Laue had known the lattice notion before; but it is certain that it was not very familiar to him. The word 'space lattice' [*Raumgitter*] most probably contributed toward triggering the association with diffraction gratings [*Beugungsgitter*], because Laue had just recently written the chapter about one- and two-dimensional diffraction gratings as his contribution to the *Enzykopädie der Mathematischen Wissenschaften*. As I was about to proceed with the statement of my problem, Laue interrupted me with the question: 'What are the distances between the resonators?' My answer was evasive: It depended, I said, on whether one assumed atoms at the lattice points, or molecules, or groups of such. The spacing doesn't matter to me, since in any case it's just 10^{-3} or 10^{-4} of the wavelength of light. This certainly was the point at which Laue remembered Sommerfeld's determination of the X-ray wavelength, which is smaller than the wavelength of visible light by the same factor.

As the conversation continued, Laue only half-listened to me. Instead of taking an interest in my scruples, he interrupted me at least twice with the question: 'What happens to very short waves inside the crystal?' I do not recall that he ever dropped the word roentgen rays, which might have made clear to me the significance of his question, although I do not know whether I knew of Sommerfeld's result at the time. My answer was: 'I know nothing about the behavior of very short waves, but one can figure that out with the formulas in my paper, in which the sum of spherical waves is transformed into the sum of plane waves. Those formulas are strictly valid for all wavelengths; I shall be glad to leave them here for you—but I myself must first finish and submit my dissertation and can't spend time on your question right now' (Ewald 1979: 346 f.)."

Ewald submitted his thesis on 16 February 1912 and left Munich after the oral defense. He was therefore not involved in the experiments and discussions leading up to the discovery. Such was the state of affairs when Laue probably started urging that someone test his idea by experiment. Who determined the experimental arrangement, whether it was Laue or Friedrich or Knipping, or whether it was the fruit of joint discussion, has also not come down to posterity. The experiments were conducted in Friedrich's room (Fig. 5.10).

A fine beam of the X-rays emitted by the anticathode was screened out by a kind of collimator made of two lead plates with bore holes. A goniometer was used as the fixture for the crystal in order to be able to adjust the angle of incidence on the crystal surface systematically. Since the direction of the diffracted beam was unknown, five photographic plates in light-proof packing were set up at different locations. X-ray tubes with a cold cathode were used. The high voltage was supplied by a spark coil. As a result of the continuous exposure—the exposure time was many hours—the beam power diminished as the exposure progressed. The degree of wear on the tubes was high, as the vacuum inside the tube deteriorated. No information was given about the constancy of the radiated wavelength. The first crystal used was copper sulfate $CuSO_4 \cdot 5\,H_2O$, which was available at the Institute. From the scientific point of view, it was not such a good idea to choose such a complex compound (Fig. 5.11).

Fig. 5.10 A photograph of the replica of the apparatus, with the plate stand for the first recordings of X-ray interference in crystals at the Deutsches Museum in Munich. *Source* The author's estate

Laue recalled the first results:

Not the first [experiment], but the second certainly did lead to a result. The radiograph of a piece of copper sulphate showed a ring of diffracted lattice spectra next to the primary X-ray beam. I was walking home down Leopoldstraße deep in thought after Friedrich had shown me this exposure. And close to my apartment, Bismarckstraße no. 22, in front of the building at Siegfriedstraße no. 10, the idea struck me about the mathematical theory for this effect. Just shortly before, I had had to reformulate the theory of diffraction at an optical grating going back to Schwerd (1835) for an article in the Encyclopaedia of Mathematical Sciences; so applied twice it also covered the theory of a cross grating. I only had to write it three times, corresponding to the three periods of the space lattice, in order to interpret the new discovery. In particular, the observed halo of rays could be immediately related to the cones that each of the three interference conditions defines on its own. When, a few weeks later, I was able to test this theory quantitatively on another, clearer photograph and found it confirmed, that was the decisive day for me (Laue 1961b: XXIII).

In the 1913 publication Laue still had reservations: "Now this theory is to be subjected to a first quantitative test. It may be that there will be no success at penetrating to the full truth, that especially the values for the wavelengths of the X-rays must later be replaced by others in simple rational relations to them (Laue 1913b: 989)."

On the 10th anniversary of the discovery from the spring of 1912, Walter Friedrich recalled his predecessors at the Institute. He referred in detail to the preliminary researches by Haga and Wind, Walther and Pohl, and the experiments by Barkla and W. H. Bragg. The ideas by physicists as to whether pulses might be involved or waves were discussed, likewise the investigations under Roentgen. About Sommerfeld's contribution Friedrich remarked: "Precisely during the period before Laue's

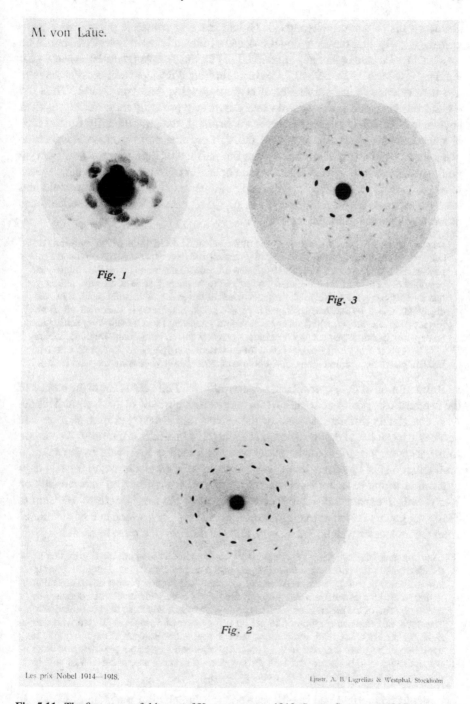

M. von Laue.

Fig. 1

Fig. 3

Fig. 2

Les prix Nobel 1914—1918.   Ljustr. A. B. Lagrelius & Westphal. Stockholm

Fig. 5.11 The first successful image of X-ray structure, 1912. *Source* Santessen (1920)

discovery, his papers on a hypothesis he had made to complete pulse theory were written, namely, that the slow-down process of the cathode ray electron should be regulated by Planck's quantum of action. [...] Optics had its qualified expert in M. v. Laue (Friedrich 1922: 364 f.)." He then mentioned Röntgen's interest in crystals and the presence of the old master of crystallography, Paul von Groth: "Thus the ground was prepared for Laue's discovery in a way probably more fortuitous than anywhere else. The apparent trigger for the brilliant, so extraordinarily fruitful idea of Laue's, that the passage of X-rays through crystals must produce interference phenomena similar in kind to those from the passage of light through a diffraction grating, was research by P. P. Ewald on the behavior of long electromagnetic waves in a space lattice, which the latter had begun at the instigation of Sommerfeld and later published as a doctoral thesis (ibid., p. 365)." Friedrich added the magnitudes of the wavelength and the lattice constants.

> I myself learned of Laue's idea during one of the scientific discussions which followed our colloquium and mediated the so worthwhile contact between experimental and theoretical physics, and which yielded such an abundance of mutual inspiration. With the enthusiasm peculiar to youth, I at once declared myself ready to make a serious attempt, although that initially triggered a lively debate about whether the idea was feasible at all. Even the objections raised by the most competent people could not dissuade Laue and me from planning the experiment. Although as Sommerfeld's assistant I was already busy setting up an experimental arrangement to investigate a problem concerning bremsstrahlung, which took up a great deal of my time, we soon found welcome support in P. Knipping, who had just completed his doctoral thesis, and who therefore had more time at his disposal (ibid.).

A description of the experimental set-up followed. Full details about the results of the five variously positioned plates were described extensively in the first publication.

A completely different account of the events was given by Abram F. Ioffe five decades later. He was doing research at Röntgen's institute at the time. According to his account, the origin of the discovery was a lecture by Ewald in the seminar on Oscillations of Electrons in a Crystal Lattice, which was discussed for two days afterwards in the café. Ioffe mistakenly attributed the diffraction experiments on X-rays with a narrow slit and the determination of the wavelength of 10^{-9} cm to Wolfgang Pauli's father and the lattice constant of 10^{-8} cm, which can be calculated from the molecular weight, the density, and the number of atoms, to Ewald:

> Laue compared the two figures and expressed the thought that the periodic distribution of the electrons in the lattice must be expressed in the propagation of the X-rays, similar to how a diffraction grating produces an optical spectrum. The skeptic Wagner raised objection to this: What kind of spectrum can this be if the grating is periodic in all three dimensions and the spectra of the different orientations cross each other? Von Laue did not back down. The circle of café friends proposed to bet on it for a box of chocolates. To see who was right, Walter Friedrich, who was working with X-rays under Röntgen at the time, took it upon himself to place a crystal in the path of the rays and to set up a photographic plate perpendicular to it, in accordance with the suggestions of those assembled. It was able to detect rays scattered at right angles (Ioffe 1967: 40).

Arnold Sommerfeld had briefly returned to Munich after the skiing holidays in March 1912 and then traveled to Lake Garda with his wife before continuing onwards to Vienna to deliver a talk. He was absent when the experiments were performed, so the results of the diffraction exposures were already available upon his return (Eckert 2013: 246).

The young Privatdozent Max Laue had settled two fundamental questions in a single experiment: first, that X-rays are transverse electromagnetic waves; and second, that atoms in crystals have an ordered structure. Sommerfeld fully acknowledged the importance of the discoveries, but he kept to himself how he managed to resign himself to the fact that Laue's idea had been so successful. No letter of lamentation about it has survived, no regretful "if only I had ..." is known.

Now the results had to be published as quickly as possible. Laue sent a postcard to Wilhelm Westphal from Munich on 1 May 1912, and his scrawl, which was anyway difficult to read, became even messier: "Dear Colleague: Some local coworkers and I are now working on an experimental paper, which we would like to publish imminently, as soon as the result is reasonably certain. Therefore I would like to ask three questions: (1) On which days are the next meetings of the German Phys. Society? (2) Would you perhaps be so kind as to present the work and also say a few words about it? It involves very simple things—Of course, I cannot promise you yet that we are not going to publish the matter elsewhere. It is all a question of speed. Yours, M. Laue" (DMA, NL 080/93/1) Westphal then received another letter from Laue in Munich, written on 23 May 1912: "Dear Colleague: I intend to speak before the German Physical Society in June—namely, at the first meeting following the 8th of June. (The latter is the date that the paper will be presented to the Academy here.) I don't want to spell out the subject today. But I do want to tell you that the enclosed photograph plays a major role in it. I recommend that you and our colleagues get busy guessing what it means. Whoever figures it out gets one thaler. But don't ask Planck. He knows but won't say a thing. Short and sweet: You can see that I am very pleased about it now (and virtually the whole Institute, too)."[10] Later he wrote another postcard showing an X-ray diffraction pattern: Dear Colleague: I shall, of course, be there at the beginning of the meeting; but I am [...] grateful to you for your remark. As I recall, the sessions used to begin at 8:15. How quickly one forgets! The [...] interfering rays are certainly not pulses, but [...] homogeneous oscillations. We don't yet know whether they already exist in the primary radiation or can only form inside the crystal as fluorescence [...]. I am looking forward to the post-session, too. Goodbye, Yours, M. Laue" (Private holding, no location, postmark illegible) Laue spoke at the meeting of the Physical Society in Berlin on 14 June 1912 (*Verh. DPG* 14, no. 12: 303). On the 25th anniversary of Laue's discovery, Max Planck recalled this memorable meeting:

> It was here in this room at this place on the 14th of June 1912; Mr. Rubens was presiding. All of us were in great suspense. I still remember the details of what happened clearly. After the theoretical introduction, when Mr. v. Laue showed the first exposures, which depicted the passage of a beam of rays through a rather arbitrarily oriented piece of triclinic copper vitriol—one could see a few strange spots on the photographic plate next to the central penetration point of the primary rays—the audience looked at the image on the board in

---

[10] Laue to Westphal, Munich, 23 [25?] May 1912. Private holding. Quoted without any source by Gerhardt Hildebrandt in Treue and Hildebrandt (1987: 229). See also Boeters and Lemmerich (1979: 59), there dated 23 May 1912.

suspense and full of expectation, but probably not entirely convinced. But when that Figure 5 was shown, the first typical Laue diagram, registering the radiation traversing a regular zinc-blende crystal oriented exactly in the direction of the primary radiation, with its interference points arranged regularly and neatly at different distances from the center, a general but soft 'ah' ran through the assembly. Each one of us felt that a great deed had been done here [...] (Planck 1937: 77).

During the discussion Planck expressed the opinion that perhaps considerable portions of the space lattice had almost the same oscillations in amplitude and phase and thus did not shift relative to one another or only insignificantly so. Then perhaps it would not be the case for the whole crystal, insofar as it is irradiated, to act as a uniform space lattice, but certainly the case for any such portion. Rubens responded with the remark that the curvature of the incident waves could have an influence on the shape of the interference spots.

At that meeting two other speakers spoke after Laue. First, Kurt Herrmann on 'Temporal change in photoelectric electron emission', and then Robert A. Millikan on the 'Influence of the character of the light source on the photoelectric effect.' Max Laue proposed Friedrich and Knipping for membership in the German Physical Society, which was then approved.

Sommerfeld, as a member of the Mathematical-Physical Class of the Royal Bavarian Academy of Sciences, presented the results of the experiments by Laue, Friedrich, and Knipping on 4 May 1912 (Friedrich et al. 1912). Sommerfeld's submission of their more detailed paper at the meeting of the Bavarian Academy took place on June 8th. Laue filled eight pages developing the theory of diffraction in a space lattice. He still assumed that the atoms of the crystal were where the interfering radiation originated from, even though he had already suggested the possibility of interference by incident rays in a letter to Westphal. "If X-rays really do consist of electromagnetic waves, then it had to be assumed that the space lattice structures give rise to interference effects upon excitation by the atoms into free or forced vibrations; that is, interference effects of the same character as the ones known in optics as grating spectra (ibid.)." Laue did not provide any information about the type of excitation of the atoms. He applied the theory of diffraction to the three periods of the crystal's space lattice. The intersecting lines of the radiation cones produce the concentrically arranged interference points that are clearly visible on the zinc blende photograph. From these images Laue calculated the wavelength of X-rays to be between 1.3×10^{-9} and 5.2×10^{-9} cm. In the section 'General consequences' he discussed the corpuscular and wave theories of X-rays again, mentioning Sommerfeld's considerations: "According to Sommerfeld's conceptions, the primary radiation, insofar as it is 'bremsstrahlung,' will probably have to be assumed to consist of entirely aperiodic pulse waves [...] (ibid.)." Sommerfeld organized a party to mark the discovery. Max Laue was not invited.[11] The reason for this remains unclear; Sommerfeld never provided any information about this incident, even later. This slight grieved Laue for the rest of his life.

---

[11] Laue to Sommerfeld, no location, 3 Aug. 1920. DMA, NL 056 (former shelfmark HS 1977-28/A 197). Related letters from Laue indicate how offended he was.

Fig. 5.12 William Henry Bragg (1862–1942) and his son William Lawrence Bragg (1890–1971). Together they received the Nobel Prize in Physics in 1915 for their achievements in studying crystalline structure by means of X-ray spectroscopy. *Source* Press article, in the author's estate

On 6 June 1912 Laue informed Wilhelm Wien about the experiments in Munich: "Highly esteemed Privy Councillor: The attempt to use the space lattices of crystals for interference phenomena in X-rays has succeeded. On the back is a photogram taken by Friedrich and Knipping. Regular ZnS is irradiated in the direction of a fourfold symmetry axis. The day after tomorrow, Professor Sommerfeld will present the work to the Academy (DMA, NL 056: 409–452)." Six days later, Wien had apparently invited him to his colloquium, Laue replied to him from Munich: "Highly esteemed Privy Councillor: I shall gladly attend the colloquium on the 18th of this month and shall be happy to give a talk (ibid.)."

On 10 June 1912 Einstein sent his congratulations on a post card from Prague where he had accepted a professorship for a while. "Dear Mr. Laue: I heartily congratulate you on your wonderful success. Your experiment is among the very best that physics has ever witnessed (UAF, Laue papers, 9.1; cf. Boeters and Lemmerich 1979: 58)." Einstein also asked about possibly seeing him in Munich (ibid.).

For the meeting of the Bavarian Academy on July 6th, Laue wrote a detailed paper (1912c) for Sommerfeld to present on how to interpret the photograms. Taking the zinc blende photograph as his basis he calculated the lattice constant to be $3.38 \cdot 10^{-8}$ cm. Then he calculated the interference points resulting from the radiation cones. The calculations agreed so well with the diagrams that he wrote: "The reader may check the agreement of the figure for himself by taking measurements with a compass on the photogram (Laue 1912c: 373)." But some questions about the details remained unresolved, as Laue noted (ibid.).

Very soon a correct explanation for the details was found in England by William Henry Bragg and especially his son, William Lawrence Bragg at the Cavendish Laboratory in Cambridge (Fig. 5.12), as well as by Henry G. J. Moseley and Charles

Galton Darwin, based on theoretical considerations and experiments: The interferences are produced by the X-rays emitted from the anticathode of the X-ray tube.[12] Neither Laue nor Sommerfeld's institute could keep pace with the English teams of researchers.

To an uninformed physicist or reporter the recently found details would seem to diminish the importance of the original idea and Laue's accomplishment. The simple path to a discovery is often dismissed with the remark, "someone else would have found that just as easily." On the contrary, after Röntgen's discovery and many tries at determining the wavelength of these rays, Laue was the one able to propose the proper experiment for the dimension of the radiation and not, as one would have expected of a theorist, merely posited a theory with auxiliary assumptions!

From the middle of 1912 on and for the following decade, research at several universities was predominantly limited to inorganic substances. Research on crystals experienced an unimagined upswing. But it was initially 'unthinkable' that one could use this method to elucidate the structure of organic compounds or even macromolecular structures. Laue had the good fortune to be able to witness this successful development of the field. However, fundamental problems were always what interested him, as he often emphasized.

Unbeknownst to Laue, Planck was trying to arrange for an academic position for Laue at the University of Berlin. There was a faculty meeting on this topic on 27 June 1912, and an application was sent to the Ministry of Culture on 1 July 1912 requesting the establishment of an extraordinary professorship in theoretical physics naming Laue as the desirable representative.[13] But other, unrelated activities defined Laue's further scientific career. The Dutchman Peter Debye had been Sommerfeld's assistant in Aachen, later also in Munich, before accepting an appointment by Alfred Kleiner, the full professor of experimental physics at the University of Zurich, to join him there to represent theoretical physics. But Debye was not happy at this post and wanted to return to his home country. He corresponded with Sommerfeld about this from Maastricht on 23 March 1912, informing him that he had accepted a call to Utrecht. As to his reasons, he explained: "I saw then that I would surely have to sacrifice years in Zurich to reorganize a laboratory reasonably, while at the same time being exposed to the great danger of not being able to do anything decent in a field of my own for want of unspent energy (DMA, 1977-28, 8A, 61: 6 f.)." He named three people as his possible successors. First choice was Max Laue. Only six days later Debye answered Sommerfeld from Zurich and thanked him for having suggested they start using the familiar *Du* address between themselves. He conveyed to him his very personal impression of Laue:

---

[12] The archives of the Royal Institution of Great Britain in London have letters from William Lawrence Bragg on the timing of the joint research beginning in the summer of 1912: W.L. Bragg to John Desmond Bernal, 30 Oct. 1942. RI MS WIB/49b/42p.; W.L. Bragg to Lord Rayleigh, 9 Sep. 1942. RI MS, WLB/49b/36: 1.

[13] Betrifft Gesuch des Herrn Prof. Dr. v. Laue. GSta PK, I HA, Rep. 76, Kultusministerium, Va, Sekt. 2, Tit IV, no. 68 D, vol. 1, fol. 208.

The candidacy of Laue has the very best prospects. Kleiner is completely partial to him and just this morning said that among all the candidates we discussed he was evidently the very first to fall under consideration. The words you quote from Laue's letter are ultimately understandable to me. If his wife can't manage with the money and she is not graciously enough disposed not to let domestic discomfort arise from it, he would feel as if someone (external, of course) were robbing him of the very best to which he is entitled, his happiness. And once he has this feeling, he desperately needs someone to be cross with and becomes unsound of mind and then (we must say thank God) eventually finds a demon as well. I am convinced that the position in Zurich will offer him everything that is necessary for his recovery (ibid.).

Alfred Kleiner asked Sommerfeld on 1 April 1912 whether Max Laue could be considered for the position: "We are thinking of Laue, but I would be very grateful if you would give your general a[nd] unreserved opinion on the matter, for ex[ample], also about whether you find it justified that the local private lecturer Dr. Greinacher now be passed over, against whose view of the decision I would like to be legitimated toward the higher authority by objective votes. Concerning Laue, we would of course very much like to know something about his qualities as a lecturer, furthermore about his personal character, a[nd] poss[ibly] about the prospects that he could a[nd] would like to follow a call to Zurich (DMA, NL 089, cited after Sommerfeld 2000: 411)." Sommerfeld replied right away and extensively so on April 3rd. He was surprised that Debye, his former assistant and private lecturer in Munich, wanted to leave Zurich now:

As far as Laue is concerned, he is without question the most capable of the younger researchers in the field of theoretical physics. His book on relativity was a great success and is universally acknowledged. He was one of the very first to recognize the significance of Einstein's discovery; for inst., he immediately derived from it Fresnel's drag coefficient. His judgment is of great certitude, e. g., when he discovered errors in the paper by Kohl on the Michelson experiment. Very interesting are his earlier papers on the thermodynamics of coherent rays, to which Planck repeatedly refers in general essays (speech in Leyden and lecture at Columbia University).

Laue has specifically lectured in Munich on optics (with some original experiments), thermodynamics, relativity, electron theory, and has assisted me with exercises on physical differential equations. In the past two semesters he has organized his own colloquium following his lecture course, in which the participants gave talks, enthusiastically and carefully prepared ones, on modern topics. This is decidedly a great teaching success for a private lecturer, considering that even the full professor finds it difficult to get students to speak. Laue has always addressed a small circle of students who really want to learn something, and has encouraged them extraordinarily. For ex[ample], they were always willing to tolerate additional hours of work.

I am writing this to show that Laue is indeed an impressive lecturer, vividly interested in teaching. He sets high standards for himself and his listeners. He will never recite anything that he has only half thought through or has not reduced to the simplest form. Nor will he conceal the principal difficulties of a subject from his auditors for the sake of so-called popularization.

In personal conversation he speaks somewhat quickly and hastily, but ex cathedra his lecture is always well ordered and receives only praise by diligent listeners. (During his first Munich semesters his lecturing was hard to grasp. This has been cast aside in the last three semesters.) He is also experimentally interested and stimulating, not to such a high degree as Debye but enough for Röntgen's practical students and assistants to repeatedly seek his counsel, and two of them are now conducting practical research at my institute at his arrangement.

Personally he is impeccable; I have never had any difference or problem with him in the past 3 years. I am not as close friends with Laue as I am with Debye, of course, whom I know since his first steps in science. His character is also more reserved. Laue has a nice young wife, no children, and pecuniarily speaking is quite independent.

He certainly will be delighted to come to Zurich and is also immediately available (DMA, NL 089).

Sommerfeld then mentioned that Laue had not been considered for an appointment to Tübingen and how he evaluated Heinrich Greinacher. He drew another comparison with Debye. Kleiner then asked whether he could audit one of Laue's lectures incognito. Sommerfeld's reply on May 13th, i.e., after the discovery of X-ray interference, was:

Esteemed Colleague:

Dr. Laue's lectures are: Tuesday, Thursday, Friday, 3–4 o'clock, Relativity Theory, and Saturday 10–12 related practicals. There are about 10 auditors (quite a decent number for this topic). So it is out of the question that you will go unnoticed. If I saw two strange elder gentlemen sitting among all the familiar faces at a small course, I would also lose the thread a bit, just because I would feel obliged to go a little further out of my way for them. I remember how agitated I became by a strange face in Aachen for a whole hour—there were candidacy issues underway for me then, too. So I would highly recommend that you introduce yourself to Laue beforehand, or directly authorize me to prepare him for your visit in general. This week Thursday (Feast of Ascension) is cancelled, Laue has rescheduled that lecture for today. On Wednesday, 26.V., at 6–8 in the evening, Laue happens to be giving a talk at our colloquium. The practical session on Saturday (assigned problems and presentations by the participants) ought to be particularly characteristic of Laue's instruction. In any case, Laue will be happy to take the opportunity to show you the wonderful interference images of X-radiation in crystals, which have recently been taken here at his instigation and are keeping us all in suspense.

Please do inform me briefly in advance of your arrival and the duration of your stay. I would be very pleased if you and Director Keller would find time to visit me in the evening or at noon, and I would also like to invite Röntgen, whom you know from Zurich. Perhaps you could inform me about this in advance, also about whether you would like me to invite Laue. I suspect that you might not want the latter in the interest of informal discussion and, of course, I shall only say anything at all to Laue if you authorize me (DMA, NL 089).

In retrospect, one wonders why, after such an important discovery, no German university offered Laue a full professorship. Was the appointment process so rigid then that it was out of the question to even discuss the possibility? After his visit to Munich, Alfred Kleiner wrote to Sommerfeld sometime in June: "Highly esteemed Colleague! I would like to thank you very much for your friendly reception in Munich a[nd] for all your efforts regarding our appointment business. I shall keep you informed of the progress in this matter, as I am well aware of the tensions attached to such things. [...] I let you know that we shall contact Laue [?], only after careful consideration a[nd] in spite of the [...] remaining reservations that we expressed, e.g., concerning participation in the operation of the laboratory, to which we were formerly able to commit Einstein and later also Debye (undated, DMA NL 089: 411)." In Zurich the Government Council met on 16 July 1912 to discuss the nomination by the Philosophical Faculty, Section II, to appoint the private lecturer at the University of Munich, Dr. M. Laue, to the vacant academic chair in Theoretical Physics.

> At Munich he has conducted epoch-making investigations that have caused a sensation among specialist circles. The number of his scientific publications is already quite considerable, and furthermore, their intrinsic value is of major importance. There are also favorable evaluations of his lecturing activities. However, the two delegates found that there were still some deficiencies in his style of delivery, namely, his way of speaking softly, quickly, and unclearly, which the proposed candidate would have to overcome. It is noted, however, that Laue appeared at the course thoroughly prepared, his lecture elucidated the most complicated problems and held the students in its sway (*Regierungsrat*, UZA, 116.3, 18 Jul. 1912).

On 22 August 1912, Laue's "petition for dismissal" from his post as private lecturer at the Ludwig Maximilian University was granted (LMUA, No. 2403). Perhaps the couple was still on holiday on the island of Wangerooge at this time. From there, Laue had written to Edgar Meyer on 2 August 1912, asking him to excuse his delay in responding (DMA, NL 080: 97/1), before discussing earlier considerations by physicists about diffraction at cross-gratings arranged in a series. Regarding his preliminary results, he responded: "When the article will appear? Heaven knows! (Ibid.)." He was hoping for publication in September, and reported on experiments he intended to continue in Zurich, and the necessary funding. The Solvay Foundation had granted 1,000 francs. But, he said, an X-ray tube costing 130 marks needed to be replaced every ten days. He indicated Röttelstrasse no. 15 as his future address in Zurich as of 1 October 1912.

References

Authier, André. 2013. *Early Days of X-Ray Crystallography*. Oxford: Univ. Press

Barkla, Charles G. 1908. Der Stand der Forschung über die sekundäre Röntgenstrahlung. *Jb. Radioakt. Electr.* 5: 246–324

Boeters, Karl E., and Jost Lemmerich, eds. 1979. *Gedächtnisausstellung zum 100. Geburtstag von Albert Einstein, Otto Hahn, Max von Laue, Lise Meitner in der Staatsbibliothek Berlin Preußischer Kulturbesitz*. Exhibition catalogue. Berlin: Staatsbibliothek Preußischer Kulturbesitz

Born, Max. 1912. M. Laue. Das Relativitätsprinzip. *Phys. Z.* 13: 175 f. (book review)

Bragg, W. H. 1911. Corpuscular radiation. *Report of the British Association for the Advancement of Science*. 80th Meeting at Portsmouth, 31 Aug.–7 Sept. 1911: 340 f.

Budde, E. 1911. Zur Theorie des Michelsonschen Versuches. *Phys. Z.* 12: 979–991.

Eckert, Michael. 2012. Disputed discovery: The beginnings of X-ray diffraction in crystals in 1912 and its repercussions. *Z. Kristall.* 227: 29

Eckert, Michael. 2013. *Arnold Sommerfeld. Atomphysiker und Kulturbote 1868–1951. Eine Biographie*. Göttingen: Wallstein

Eckert, Michael, and W. Pricha. 1984. Die ersten Briefe Albert Einsteins an Arnold Sommerfeld. Erstveröffentlichung von bisher unbeachteten Briefen. *Phys. Bl.* 2: 29–34

Einstein, Albert. 1993. *The Collected Papers of Albert Einstein*. Vol. 5: *The Swiss Years, Correspondence 1902–1914*. Edited by Martin J. Klein, A.J. Kox, and Robert Schulmann. Princeton: Univ. Press

Ewald, Paul Peter. 1968. Erinnerungen an die Anfänge des Münchener Physikalischen Kolloquiums. *Phys. Bl.* 24: 538–542

Ewald, Paul Peter. 1979. Max v. Laue – Mensch und Werk. Memorial address held in Berlin on 2 March 1979. *Phys. Bl.* 8: 337–349

Fölsing, Albrecht. 2004. *Albert Einstein. Eine Biographie.* 6th ed. Frankfurt am Main: Suhrkamp

Friedrich, Walter. 1912. Räumliche Intensitätsverteilung der X-Strahlen, die von einer Platinantikathode ausgehen. *Ann. Phys.* 12: 377–430

Friedrich, Walter. 1922. Die Geschichte der Auffindung der Röntgenstrahlinterferenzen. In Zehn Jahre Laue-Diagramm. *Naturw.* 16: 363–366

Friedrich, W., P. Knipping u. M. Laue. 1912. Interferenz-Erscheinungen bei Röntgenstrahlen. *Sb. Bayer. Akad. Wiss.*, math.-phys. class 1912: 303–322, 363–373

Haga, Herman, and Cornelis H. Wind. 1903. Die Beugung der Röntgenstrahlen. *Ann. Phys.* 2: 305–312

Hartmann, Hans. 1952. *Schöpfer des neuen Weltbildes. Große Physiker unserer Zeit.* Bonn: Athenäum-Verlag

Hermann, Armin. 1994. *Einstein. Der Weltweise und sein Jahrhundert. Eine Biographie.* Munich, Zurich: Piper

Hildebrandt, Gerhardt, and Wilhelm Treue, eds. 1987. *Berlinische Lebensbilder.* Vol. 1: *Naturwissenschaftler.* Einzelveröffentlichungen der Historischen Kommission zu Berlin, no. 60. Berlin: Colloquium-Verlag

Ioffe, Abram F. 1967. *Begegnung mit Physikern.* Basel: Teubner

Klein, Felix, and Arnold Sommerfeld. 1965. *Über die Theorie des Kreisels.* 4 issues, 1897–1910. Reprinting of the original editions from 1897, 1898, 1903, and 1910. Stuttgart: Teubner

Kubbinga, Henk. 2012. Crystallography from Haüy to Laue: controversies on the molecular and atomistic nature of solids. *Z. Crystall.* 227: 1–26

Laue, Max. 1907. Die Mitführung des Lichtes durch bewegte Körper nach dem Relativitätsprinzip. *Ann. Phys.* 10: 989 f.

Laue, Max. 1909a. Zur Thermodynamik der Gitterbeugung. *Ann. Phys.* 12: 225–239

Laue, Max. 1909b. Thermodynamische Betrachtungen über die Beugung der Strahlung. *Phys. Z.* 32: 807

Laue, Max. 1910a. Zur Thermodynamik der Beugung. *Ann. Phys.* 3: 547–558

Laue, Max. 1910b. Die Wärmestrahlung in absorbierenden Körpern. *Ann. Phys.* 10: 1085–1094

Laue, Max. 1910c. Ist der Michelsonversuch beweisend? *Ann. Phys.* 11: 186–191

Laue, Max. 1911a. *Das Relativitätsprinzip.* Die Wissenschaft – Sammlung naturwissenschaftlicher und mathematischer Einzeldarstellungen, No. 38. Braunschweig: Vieweg; 2nd exp. edition 1913

Laue, Max. 1911b. Zur Diskussion über den starren Körper in der Relativitätstheorie. *Phys. Z.* 12: 85–87

Laue, Max. 1911c. Zur Dynamik der Relativitätstheorie. *Ann. Phys.* 8: 524–542

Laue, Max. 1911d. Über einen Versuch zur Optik bewegter Körper. *Sb. Math.-Phys. Kl. Bayer. Akad. Wiss.* 1911: 62, 405

Laue, Max. 1911e. Zwei Einwände gegen die Relativitätstheorie und ihre Widerlegung. *Phys. Z.* 13: 118

Laue, Max. 1911f. Bemerkungen zum Hebelgesetz in der Relativitätstheorie. *Phys. Z.* 12: 1008 ff.

Laue, Max. 1912a. Zur Theorie des Michelsonschen Versuches. *Phys. Z.* 13: 501–506. Budde's reply is on p. 825

Laue, Max. 1912b. Zur Theorie des Versuches von Trouton und Noble. *Ann. Phys.* 7: 370–384

Laue, Max. 1912c. Eine quantitative Prüfung der Theorie für die Interferenzerscheinungen bei Röntgenstrahlen. *Sb. Bayer. Akad. Wiss.*, math.-phys. class, 1912: 363–373

Laue, Max. 1913a. Wellenoptik. In Sommerfeld (1913): 362–373

Laue, Max. 1913b. Eine quantitative Prüfung der Theorie für die Interferenzerscheinungen bei Röntgenstrahlen. *Ann. Phys.* 10: 989–1002

Laue, Max. 1921. *Die Relativitätstheorie.* Vol. 1: *Das Relativitätsprinzip der Lorentztransformation.* 4th exp. Edition, Braunschweig: Vieweg

Laue, Max von. 1961a. *Gesammelte Schriften und Vorträge*. 3 vols, Braunschweig: Vieweg

Laue, Max von. 1961b. Mein physikalischer Werdegang. Eine Selbstdarstellung. In Laue 1961a, 3: V–XXXIV

Laue, Max, and F. F. Martens. 1907. Beiträge zur Metalloptik I. Über die polarisation der von glühenden Metallen seitlich emittierten Strahlung. *Dt. Phys. Ges.* 5: 522

Mehra, Jagdish. 1975. *The Solvay Conferences on Physics. Aspects of the Development of Physics since 1911*. Dordrecht, Boston: Reiche

Michelson, Albert Abraham. 1903. *Light Waves and Their Uses*. Chicago: Univ. of Chicago Press

Morley, Edward W., and Dayton C. Miller. 1905. On the theory of experiments to detect aberrations of the second degree. *Phil. Mag.* 9: 669–680

Planck, Max. 1937. Zur 25-jährigen Jubiläum der Entdeckung von W. Friedrich, P. Knipping und M. v. Laue. *Verh. Dt. Phys. Ges.* 18th ser., 3: 77

Röntgen, W. Conrad. 1895. Über eine neue Art von Strahlen. *Sitzungsberichte der Physikalisch-Medicinische Gesellschaft zu Würzburg* 1895: 132–141

Santesson, Carl Gustav. 1920. *Les Prix Nobel en 1914–1918*. Stockholm: Norstedt & Söner

Sommerfeld, Arnold. 1899. Theoretisches über die Beugung der Röntgenstrahlen. *Phys. Z.* 1: 105

Sommerfeld, Arnold. 1909. Über die Verteilung der Intensität bei der Emission der Röntgenstrahlen. *Phys. Z.* 19: 969

Sommerfeld, Arnold. 1912. Über die Beugung der Röntgenstrahlen. *Ann. Phys.* 8: 473–506

Sommerfeld, Arnold, ed. 1913. *Encyclopädie der mathematischen Wissenschaften mit Einschluss ihrer Anwendungen*. Vol. 5 in 3 parts: *Physik*. Leipzig: Teubner

Sommerfeld, Arnold. 2000. *Arnold Sommerfeld. Wissenschaftlicher Briefwechsel 1892 bis 1918*, Vol. 1. Edited by Michael Eckert and Karl Märker. Berlin: Verlag für Geschichte der Naturwissenschaften und Technik

Staley, Richard. 2008. *Einstein's Generation. The Origins of the Relativity Revolution*. Chicago: Univ. of Chicago Press

Tobies, Renate. 2010. *Morgen möchte ich wieder 100 herrliche Sachen ausrechnen. Iris Runge bei Osram und Telefunken*. With a foreword by Helmut Neunzert. Stuttgart: Steiner

Trouton, Frederick T., and Henry R. Noble. 1903. The forces acting on a charged condenser moving through Space. *Proc. Roy. Soc.* 72: 132

Walter, Bernhard, and Robert W. Pohl. 1908. Zur Frage der Beugung der Röntgenstrahlen. *Ann Phys.* 4: 715–724

Wolfke, Mieczyslaw. 1912. Über die Abbildung eines durchlässigen Gitters. *Ann. Phys.* 4: 797–811

Chapter 6
Professorship in Zurich

When Max Laue was appointed to the University of Zurich, Albert Einstein was back in Switzerland after an interlude at the German University in Prague. But this time he was on the faculty of the Swiss Federal Polytechnic (ETH) in Zurich (Fig. 6.1). This allowed Einstein and Laue to meet at the various academic events and also privately.[1] Einstein was busy studying the problems of space and time, the general theory of relativity. He received the news of Laue's appointment with skepticism. On 20 May 1912 he wrote to his friend Heinrich Zangger from Prague: "It's Laue who is coming to Zurich University now. I am looking forward to the exchanges with him. He will probably soon be called away again, but that'll be less of a pity than it is with Debye. For, in my opinion, Laue possesses only formal talent and is no teacher. But he does have some things that I lack, so it might be useful for the two of us to work together. I almost think it would have been wiser to appoint Ehrenfest as extraordinarius. Don't tell anyone, though, about such useless statements (Schulmann 2012: 97)." For his current scientific interests Einstein found in Laue a competent listener but no collaborator. On mathematical problems Laue could, of course, offer Einstein valuable advice.[2] Einstein had arrived in Zurich with his assistant, Otto Stern. Born in 1888, Stern had studied chemistry and physical chemistry at Freiburg im Breisgau, Breslau, and Munich and was both an outstanding theoretician and an imaginative experimenter. A lifelong friendship with Laue developed soon after he joined them (Schmidt-Böcking and Reich 2011: 33; Trageser 2005).

---

[1] Unfortunately, no record of these meetings seem to exist.

[2] Fölsing (1995: 368): "Einstein was more in his element at a once-a-week colloquium, which Max von Laue came over from the university to attend with some students."

© Basilisken-Presse, Natur+Text GmbH 2022
J. Lemmerich, *Max von Laue*, Springer Biographies,
https://doi.org/10.1007/978-3-030-94699-9_6

Fig. 6.1 Albert Einstein (1879–1955), around 1910/11. Laue met him for the first time in the summer of 1906 in Bern. *Source* Archive of the Max Planck Society, Berlin-Dahlem

The move to Zurich cost Laue 735 francs. At his request he received a 500 franc grant from the University of Zurich with the proviso that he repay it if he resigned before the end of a three-year period.[3] His annual salary was 4500 francs, plus attendance fees. The teaching appointment stipulated four to six hours of lectures a week. The term of his employment was fixed at six years. In Zurich Laue received a letter from William Henry Bragg in England dated 10 October 1912, and his reply, sent just five days later, was lengthy: "Thank you very much for your kind letter of the 10th of this mo[nth]. The interest in my work expressed in it pleased me very much. You have obviously already thought much more than I have about the theory of photography with inclined crystals; throughout the last three months, urgent work of a different nature completely occupied me (for example, the 2nd edition of a text on the principle of relativity); I am not even in a position to check your formulas for the position of the spots with inclined crystals, but this will soon be rectified. For, Stark's paper in the last issue of the Phys. Zeitschr. will probably force me to do further calculations (RIGBA, MS WGB 4a: 8p.)." There followed a discussion about

---

[3] Minutes of the Regierungsrat, 24 Oct. 1912 and 18 Jul. 1912. UZA, Rektorats-Archiv, 116.3.

the dependence of the interference figure on the anticathode material of the X-ray tube and about the excitation of the atoms in the crystal by certain wavelengths.

Exactly one month later, on November 10th, Laue replied from Zurich to William Henry Bragg's next letter: "I must beg your pardon for having kept you waiting so long for my response, but I am being pressured by the publishing establishment with the correction proofs for the second edition of my book on the principle of relativity; for a while I was receiving a proof sheet daily. This, combined with the efforts attached to adjusting to the local working conditions, has kept me completely preoccupied for the past few weeks (RIGBA, MS WGB 4a: 9p.)."[4] In the same letter Laue asked Bragg to recommend a publisher for an English edition of his book on relativity because he had someone ready to translate it: "That the method you have given for calculating the position of the [interference] spots leads to good results, I readily believe. It is, after all, completely identical to mine (ibid.)." Laue also enclosed a table with his own calculated figures. The differences between Bragg's and his own values would not be empirically ascertainable, he continued. Regarding the complex problems of how the interference points form and how X-rays are spectrally composed, Laue explained:

> I believe for this reason that no physical conclusions can be drawn from your observations which contradict my views. Nonetheless, your remarks are extremely valuable to me. As, assuming that all possible wavelengths of a certain order of magnitude were present in the portion of the incident X-radiation responsible for these phenomena, interference points would have to occur in all directions in which atoms can bind to each other. These directions are obviously so close to each other that—under the present assumption—the whole plate would have to be blackened. The fact that only individual points are actually blackened proves to me that not pulses, but one or a few spectrally homogeneous types of radiation (of as yet unknown origin) play a role in this (ibid.).

He then also applauded the idea of performing experiments on different anti-cathode material that Friedrich had perhaps already conducted, and broached a publication by Stark about X-ray corpuscles. In another letter to Bragg on November 13th he returned to the comparison between their calculated figures and proved their good agreement (RIGBA, MS WGB 4a: 9 pp.).

On 14 December 1912 Laue held his inaugural lecture at Zurich on the topic 'The wave theory of X-rays' (Laue 1913a).[5] (Fig. 6.2) He started with a tribute to Röntgen's discovery and mentioned the two conceptions of the nature of radiation: the corpuscular and wave theories. A historical review of Newton's and Huygens's theories of light followed. Then moving on, he talked about experiments with X-rays, pointing out that their wavelength is so small that experiments with diffraction gratings are infeasible. Only in the last section did Laue discuss diffraction in the space lattice of crystals and the theory of X-ray diffraction, referring to the experiments performed by Friedrich and Knipping (ibid.).

---

[4] The demand for the book was so great that the publisher wanted to release a second edition. Since several important publications had appeared in the interim, Laue wanted to incorporate this material, including the paper by Planck's doctoral student Ernst Lamla (1912).

[5] Since the inaugural lecture did not take place until December 1912, Laue offered no courses during the winter semester 1912/13.

UNIVERSITÄT – ANSICHT VON NORDWESTEN

Fig. 6.2 The University of Zurich around 1900, view from the northwest. *Source* Photograph in the author's estate

Laue was impatient to start his own research on X-ray interference, for now there was competition from his colleagues in Germany, England, and France, who were expanding the field with their own investigations. Laue naturally felt obliged to respond to any criticism made against his theory (Laue 1913b, 1913c, 1913d). From 1913 onwards almost every issue of the *Physikalische Zeitschrift* contained several experimental and theoretical papers on X-ray diffraction. Paul Peter Ewald's comments about his paper on triple-symmetric images from regular crystals were particularly important to Laue because he had independently developed a theory of his own Ewald (1913).

At the end of 1912, Robert Pohl's *Die Physik der Röntgenstrahlen* was published by Vieweg & Sohn.[6] It was his habilitation thesis, reworked into a book of 156 pages. In the preface Pohl remarks: "This account adheres throughout to the electromagnetic conception of X-rays as short aetheric pulses, without, however—as I hope—in any way constraining the interpretation of the experimental data. After completing the manuscript I learned of the experiments by Messrs. Laue, Friedrich, and Knipping, and by the favor of these gentlemen I was able to add a chapter dedicated to interference phenomena as a postscript, which are of fundamental importance in many respects (Pohl 1912)."

At Zurich, Laue could not find the appropriate equipment for his experimental research. Through Sommerfeld he applied for a grant of 5000 francs from the Solvay

---

[6] The volume in the library of the Technical University of Berlin is stamped "M. Laue" and contains a handwritten correction of a formula, probably by Laue.

Foundation to purchase X-ray apparatus. He also wrote to the academy in Berlin: "to the physical–mathematical class of the Royal Prussian Academy of Sciences I submit my application for approval of a grant in the sum of perhaps 2000 M for the furtherance of my investigations on the interference phenomena of X-rays caused by the space lattice of crystals."[7] Mentioning the existing studies on regular crystals, he explained that the aim was to examine other systems and the influence of temperature on the thermal motion of the lattice components. The *Institut International de Physique Solvay* had already granted him 4000 marks, which he would use to purchase a high-voltage transformer and a rotating rectifier from Siemens & Halske. He mentioned having received 800 marks from the Swiss Ministry of Education. On 22 April 1913 the Berlin Academy granted Associate Professor Laue 1500 marks.

News about Laue's scientific achievement spread among physicists. Wilhelm Wien received an inquiry from the USA, from Michael Pupin at Columbia University in New York, asking about the possibility of luring Laue away to a professorship in New York. Laue replied to Wilhelm Wien on 9 May 1913: "Zurich, at any rate, would not hold me back; the petty spirit prevailing in the Ministry of Education, and to some extent also at the university, does not make the stay so very pleasant, in spite of the natural beauty everywhere. But there is, of course, a German university keen on appointing me which I would gladly go to! (DMA, NL 056: 409–452)." Planck also corresponded with Wilhelm Wien about Laue's intention to come back to Berlin, who had probably also alerted him about Laue's desire for a change of location. Thus Planck wrote to Wien on 22 June 1913, since he was acquainted with Columbia University from his lecture series in 1909:

> About Laue, I was at the ministry yesterday. They are doing everything they can for him. But nothing definite can be promised until the budget situation for the Extraordinariate for Theoretical Physics has passed the Ministry of Finance. The bills for the new budget (from April 1914) are going to be processed by the Ministry of Finance in the course of October. The gentlemen won't decide earlier because they are waiting to see what other submissions are filed. [...] Would it not be possible to postpone the negotiations with America until October, either by putting Mr. Pupin off until then, or by Laue himself asking for time to think things over?
>
> Strong as my hopes are, I do not wish to be the cause of his relinquishing anything as long as he doesn't have anything definite. I shall write to him presently to the same effect (DMA, NL 056).

A few weeks later, on July 31st, Planck wrote to Wien: "Our Academy elected Einstein as a member last week. He will move to Berlin in the spring at the latest. If, as I very much hope, Laue is then also appointed to the university, theoretical physics will be all set here (ibid.)." But it was too soon for that. On June 30th, Laue informed the Board of Education of the Canton of Zurich that he had declined the call to Columbia University.[8]

---

[7] Laue to Kgl. Preuß. Akad. Wiss., no location, undated. Hist. Abt. Wiss. Untern., Phys.-Math. Kl., II–VII, 1912–1914, fols. 64f., 115.

[8] Laue to the University of Zurich, 30 Jun. 1913, quoted from: Extract from the minutes of 2 Jul. 1913, Protokoll des Erziehungsrates des Kantons Zürich, UZA, Rektoratsarchiv 116.3.

In 1913, Kaiser Wilhelm II, King of Prussia, celebrated his 25th anniversary as regent. He took this as an opportunity to show his benevolence by awarding rises in rank, titles of nobility, promotions, commendations, and orders of merit in large numbers on the basis of recommendations. Max Laue's father had already been promoted several times in the meantime and held the rank of *Wirklicher Geheimer Kriegsrat*. In a letter dated 16 June 1913, the Office of Heraldry contacted him with the Kaiser's pronouncement: "I find Myself gracefully inclined on this day of My twenty-fifth anniversary in Regency to confer hereditary nobility upon the following personalities, namely: 26th to the Real Privy Councillor of War (with the rank of Councillor 1st Class) Laue, Military Intendant of the IIIrd Army Corps."[9] The letter did not bear the personal signature of the regent. Julius von Laue had to present his baptismal and confirmation certificates and state his marital status. A draft coat of arms was also requested. He wanted the Iron Cross, which had been awarded to him after the war of 1870/71, to be included in the coat of arms along with the motto "Intrepid and true" (*Furchtlos und treu*, see the frontispiece). This honor was not for free, though. Julius von Laue was expected to pay 4800 marks for it. To raise this sum, he had to sell securities from his savings. The number of recipients was large: over sixty were ennobled, and the enumeration of the lower distinctions filled many pages in the official notice. Henceforth his son Max used the prefix "*von*" on official letters, but also occasionally in letters to friends. Otherwise it altered nothing in his style of living.

Max von Laue offered a lecture course on Theoretical Optics with Experiments and Exercises during the summer semester of 1913. His scientific research made slow progress. One problem he studied was the shape of the blackenings of the interference spots on the photographic plate. They were not strictly circular, but somewhat elliptical, becoming rounder when the distance between the anticathode and the crystal was increased. Laue, together with his Swiss collaborator Franz Tank, studied the problem of the shape of the interference points in X-ray interference. The problem was successfully solved on a few pages of mathematical calculations following Fresnel. It pointed out in conclusion: "This fact is qualitatively very easy to understand, since the sharpness of the maxima depends essentially on there being exactly the same phase difference between each of the two interfering oscillations; but this condition is satisfied only at infinite distance from the light source (Laue and Tank 1913)."

In Laue's lecture, the question Planck had posed during the discussion following Laue's talk on the influence of the thermal motion of atoms before the German Physical Society was also mentioned. Experiments at low temperatures were planned to resolve it.

Several times Laue responded to papers by other physicists about interpretating the photograms. Sometimes he pointed out the errors quite bluntly and succinctly, sometimes he set forth his arguments at length. One example, is his rejection of a computation by the Dutchman Leonard S. Ornstein: "So far complete agreement

---

[9] Heroldsamt to Carl Julius Laue, Berlin, 29 Jul. 1913. GStA PK, HA, rep. 176, no. 5833. Since then, his son was authorized to sign his name as "Max von Laue."

Fig. 6.3 A view inside the auditorium during the 85th Assembly of German Natural Scientists and Physicians in Vienna in 1913, with the Laue couple on the right-hand side. *Source* The author's estate

exists between him and me, although Mr. Ornstein failed to realize it. But this ceases as soon as the third of the necessary summations is treated. How Mr. Ornstein's approach works I cannot gather from his paper (Laue 1913e)."

For the summer vacation Laue arranged to meet Wilhelm Wien at a ski resort in St. Anton and also intended to tell him something about X-ray interference, as he promised in a letter from Zurich dated July 7th (DMA, NL 056: 409–452). That month a paper appeared in the *Philosophical Magazine* by the Danish physicist Niels Bohr (1913) from Copenhagen 'On the constitution of atoms and molecules.' In it he presented his boldly designed model of the hydrogen atom and was able to derive from it the experimentally determined frequencies of its spectral lines according to the formula of his Swiss colleague Walter Ritz. Whether this was discussed and rejected during the ski trip has unfortunately not come down to us, but many physicists were unable to follow these considerations. Max von Laue wrote to Niels Bohr to discuss his mathematical formulations and show him what he thought was a more concise approach.

The 85th Assembly of German Natural Scientists and Physicians in Vienna from 21 to 28 September 1913 provided a good opportunity for Laue to present his ideas on X-ray interference to a larger audience (Laue 1913f) (Fig. 6.3). Albert Einstein was among the speakers, who discussed the current state of the gravitational problem. At the joint meeting of the sections for physics, chemistry, and mineralogy, Max von

Laue presented his talk on X-ray interference. By way of introduction, he discussed the work of crystallographers and the dimensions of space lattices, which can only be analysed with very short-wave radiation: X-rays. He acknowledged the diffraction experiments of Herman Haga and Cornelis Wind as well as Pohl, then discussed Friedrich's and Knipping's experiments and the conclusions resulting from them. With didactic skill he presented his theory of diffraction, starting from the diffraction of a ray at lattice points on a plane. A monochromatic ray was a secondary condition. He proceeded to the three-dimensional case, pointing out that X-rays have an extensive continuous spectrum, which makes the theory more complex. He mentioned the researches by the Braggs, father and son, and Moseley and Darwin on the spectral decomposition and isolation of a monochromatic X-ray. Results from various crystals were discussed in depth, such as the recent ideas on diamonds by Debye. Friedrich then followed with a talk about X-ray interference, noting that Laue did not initially assume a continuous X-ray spectrum. He described in detail the reflection method of W. H. Bragg and W. L. Bragg and Darwin's observation of five different characteristic X-ray lines. In the subsequent lively discussion Laue set Friedrich's contention right:

> My honorable collaborator, Mr. Friedrich, has somewhat maligned me here by attributing to me the view that the spectrum of primary X-rays were monochromatic. I never thought such a thing; and if I may not have been clearly explicit earlier about the question of how the monochromaticity of the diffracted rays arises, it was because I myself had grave misgivings about the only explicable possibility I saw. But this possibility consisted of attributing to the atoms strongly selective properties, causing them to respond only to a few wavelengths among the abundance present in the incident radiation. In the meantime, however, this view has been refuted; if it had been correct, the existing interference points would have had to disappear and others would have had to appear upon even small rotations of the crystal, corresponding to the selective reflection in Bragg's experiments. But there is no question of this; on the contrary, the existing interference points thereby wander with only gradually varying intensity according to the law of reflection (Friedrich 1913: 1084).

Just a month later, Max von Laue again had the opportunity to deliver a fundamental talk. The great success of the 1st Solvay Conference of 1911 in Brussels had encouraged a continuation of these meetings. Significant advances had been made in several fields as a result, especially in X-ray analysis. The members of the preparatory committee were: the Dutchman Hendrik Antoon Lorentz, Madame Marie Curie, Marcel Brillouin, Heike Kamerlingh Onnes, Martin Knudsen, Walther Nernst, Ernest Rutherford, and Emil Warburg. Among those invited from France were Louis Georges Gouy and Paul Langevin, and from Great Britain: William Barlow, William Henry Bragg, James Hopwood Jeans, William Jackson Pope, and Joseph John Thomsen. Germany was represented by Eduard Grüneisen, Heinrich Rubens, Arnold Sommerfeld, Woldemar Voigt, and Wilhelm Wien. Albert Einstein and Max von Laue were present for Switzerland. Robert Williams Wood came for the USA, and Robert B. Goldschmidt was invited for Belgium. The theme of this meeting was 'The structure of matter.' As the conference was during the semester—von Laue's lecture course during the winter semester of 1913/14 was on Thermodynamics with Exercises, and on Structural Theory—he had to apply to the Cantonal Board of

Education in Zurich for leave of absence, which was granted until the beginning of November.[10]

The scientific secretaries Maurice de Broglie and Frederick Alexander Lindemann took the minutes. The sessions were held from 27 to 31 October 1913 in an elegant hotel in Brussels (Mehra 1975: 75 f.).[11] It was not Max von Laue who was asked to open the meeting with a talk but Joseph John Thomson, who spoke about The Structure of the Atom. Because Niels Bohr was not attending, the discussion was limited to the often contradictory scattering experiments of α and β-rays. The topic of Laue's contribution was 'The interference of X-rays on three-dimensional crystal lattices.' As in Salzburg, his discovery was not mentioned at the outset; first Laue paid tribute to the research by crystallographers and passed on to the unexpected findings of the experiments conducted by Friedrich and Knipping. On the theory of diffraction on a three-dimensional lattice, he referred to Ewald's contribution as well as those by the two Braggs, and Moseley's and Darwin's work. Laue discussed the composition of the radiation emanating from the anticathode and its interaction with the atoms in the crystal, and mentioned Ewald's opinion about the resonance of the atoms with the radiation. He concluded with a discussion of how exactly one could determine the structure of the crystals. He also included Debye's theory about the influence of temperature on diffraction patterns.

Lorentz opened the discussion with a remark about real reflections off atoms according to Bragg and about von Laue's formula on the intensity of diffraction maximums. There followed discussions by Einstein, Lindemann, Lorentz, Nernst, and Wien about observations of zero-point energy. William Henry Bragg spoke about the reflection of X-rays and the spectrometer he had developed, and the first results with anticathodes made of osmium, platinum, and iridium. Absolute measurements had been taken of alkali halides. Bragg also gave details about the structures of these substances. In the discussion Sommerfeld made extensive corrective remarks about the formula given by Laue for the intensity of the interfering radiation. Barlow and Pope spoke about crystalline structure and chemical constitution. Unfortunately, there are no notes about private conversations among the participants (ibid.).

Further research into the structure of various crystals broadened in scope, and questions of principle were also waiting for a solution. In the visual range, the dependence of the refractive index on temperature had been tested experimentally by Hippolyte Fizeau in 1864, in order, as he wrote, to draw conclusions "for information about the structure of bodies and the theory of light." He had determined positive as well as surprisingly negative changes of n (Fizeau 1864).

---

[10] Erziehungsrat des Kanton Zürich 1720: "Upon receipt of an application by Prof. Dr. M. Laue from November 24th and an expert opinion from the Faculty of Philosophy Section II of 5 December 5 1912, the Education Directorate decrees: 1. Dr. M. Laue, Professor at the Faculty of Philosophy Section II, be granted leave of absence for the purpose of attending a congress of the Institut international de physique Solvay to be held in Brussels at the end of October 1913, for the period from the beginning of the winter semester 1913/14 to the beginning of November 1913." UZA, Rectorate Archives, 116.3.

[11] Hermann (1963) mentions that Lorentz had also invited W. C. Röntgen, but he could not come for health reasons. The atomic model of Niels Bohr was not discussed.

It was now possible to proceed with the research utilizing X-ray interferences. Peter Debye used the theoretical approach to address the problem of the thermal motion of the atoms in the lattice, which should be ascertainable from the structure of the blackenings in the X-ray interferences. In several publications from 1913 he tackled the problem of how the thermal motion of atoms alters X-ray interference patterns. At the beginning of the year, an overview appeared under the title 'Interference of X-rays and heat motion' with the following theses:

1. The thermal motion of atoms has an essential influence on the interference phenomena observable in X-rays.

2. The sharpness of the interference maxima is not influenced, but their intensity is, along with the spatial intensity distribution.

3. The inter-intensity decreases exponentially due to thermal motion (Debye 1913: 91).

He added some details about the last point. On the basis of Debye's contributions published up to mid-1913, Laue published a paper 'On the influence of temperature on the interference effects of X-rays.' The editors of the *Annalen der Physik* received it on October 18th, but it did not appear until after the Solvay Conference. Again it critically reviewed his own publications. First he acknowledged Debye's contribution to the problem. Then a mathematical treatment of the shape of interference spots followed:

> Well, we had formerly found that the elongation diminishes with increasing distance between the anticathode and the crystal far sooner than is intelligible from the formula, if the irradiated part of the crystal is taken as a uniform space lattice. At that time, we expressed the suspicion that the thermal motion disturbed the uniformity and thus explained this result. Now we see that this assumption is not confirmed. It seems, therefore, as if the crystal used at that time (fluorspar) was a conglomerate of many components not assembled with the necessary perfection. After all, this would not be surprising if one considered that errors in the composition of magnitude of 10^{-9} cm must already be considerably noticeable (Laue 1913g: 1566).

The detailed mathematical treatment led to a formula reflecting the influence of temperature. Conjectures already expressed by W. L. Bragg and Debye could be confirmed.

At the end of 1913 Laue worked on a contribution for a commemorative volume by the staff of Zurich University (1914). Once again the topic he chose for his contribution was: "The interference phenomena of X-rays, caused by the space lattice of crystals." After a renewed survey of other investigations and findings including his own, he treated the following points: I. The position of the interference points, II. The intensity and shape of the interference points (Laue 1914a). Concerning the second point Laue remarked that following Debye's reasoning the influence of the thermal motion is negligible, only a change in position would be noticeable. Laue arrived at the following conclusions: "The thermal motion thus suppresses the reflectivity of the less densely occupied planes of the network. Due to the thermal motion, the shorter wavelengths are disadvantaged during interference (ibid.)." In the section on rock salt and diamond Laue inserted drawings to illustrate their structures (Fig. 6.4).

(a)

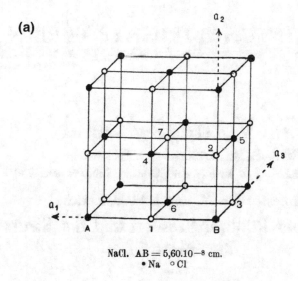

NaCl. AB = 5,60.10⁻⁸ cm.
• Na   ○ Cl

(b)

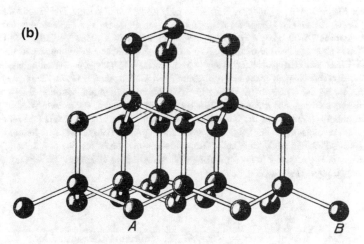

Fig. 6.4 The lattice structure of rock salt (top) and diamond (bottom). *Source* Laue (1914a, reprinted 1961, vol. 1: 316, 321)

In December 1913 Laue and J. Stephan van der Lingen were able to conclude their experiments on the Debye effect (Laue and van der Lingen 1914). X-ray diffraction images of rock salt were taken at 600 °C and in a Dewar vessel at the temperature of liquid air, both times in the same X-ray tube. The experiments verified Debye's theory, at least qualitatively. The authors gratefully acknowledged the financial assistance of the *Institut international de Physique Solvay* and the Royal Prussian Academy of Sciences.

WISSENSCHAFTLICHER VEREIN

37. Vortrags-Abend

Mittwoch, den 3. Dezember, abends 8 Uhr,
im Wissenschaftlichen Theater der „Urania", Taubenstraße 48-49

Professor Dr. M. von LAUE, Zürich,

Neues über Röntgenstrahlen und die Struktur der Materie.

❋

Den Mitgliedern ist zu diesem Vortrag je eine Karte zurückgelegt. Die Eintrittskarten liegen bis zum 3. Dezember, mittags 1 Uhr, an der Kasse der Gesellschaft Urania, **Taubenstraße 48-49,** gegen Vorzeigung der Mitgliedskarte für das Vereinsjahr 1913 14 zur Abholung bereit **(Kassenstunden 10—7 Uhr).** Schriftlichen Bestellungen ist ein mit Marke versehener Briefumschlag sowie die Mitgliedskarte beizufügen. Bestellungen durch den Fernsprecher können **nicht** entgegengenommen werden, ebenso ist eine Zurücklegung von Karten an der Kasse nicht angängig. **Wir bitten diejenigen Mitglieder, welche verhindert sein sollten, die abgeholten Billetts zu benutzen, dieselben wieder zurückzugeben, damit diese im Interesse des Vereins verwertet werden können.**

Für die Angehörigen der Mitglieder, sowie für eingeführte Gäste steht eine beschränkte Anzahl Eintrittskarten zur Verfügung, die zum Preise von 4 Mark für Sitze im 1. Rang und Parkett, zum Preise von 2 Mark für Sitze im 2. Rang in der Reihenfolge der einlaufenden Nachfragen abgegeben werden.

───

Der **nächste Vortrag** wird am **21. Januar 1914** stattfinden.

Fig. 6.5 Poster of a lecture given by Laue in the Berlin *Urania* in December 1913. *Source* Bleyer et al. (2013)

On December 3rd Laue was in Berlin again, probably together with his wife, because he had agreed to hold a lecture at the *Urania* (Fig. 6.5). The invitation to this lecture certainly brought back memories for Max von Laue of his first visit to the *Urania* as a pupil. One can assume that some of his Berlin colleagues attended the lecture. His father was probably also among the audience.

On 19 January 1914, Laue wrote to Johannes Stark from Zurich to offer to submit an article about X-ray interference to his journal *Jahrbuch der Radioaktivität und*

Elektronik (SB PK, Stark papers). Stark replied immediately, but when Laue heard that a paper by Bragg would also to be included, he had second thoughts. He wrote to Stark four days later that he considered it better that his "review not appear alongside Bragg's [paper] in your *Jahrbuch* (ibid.)." He had summarized extraordinarily much of Bragg's research and would otherwise have to make heavy revisions. In the end, though, both papers, Bragg (1914) and Laue (1914b), did appear in the same issue.

His lecture course, these external talks, his publications on X-ray interference, and his research on the theory fully taxed von Laue's workday. Nevertheless, he was still interested in questions concerning the special theory of relativity. How often Laue and Einstein discussed the latter's ideas about general relativity in Zurich has not come down to us, but it is very likely that it was quite frequent. At that time Einstein apparently had not yet succeeded in convincing Laue of his ideas about general relativity. Meanwhile, Planck's and Nernst's efforts to create a salaried position for Einstein at the Royal Prussian Academy of Sciences had made some progress.

The editors of the *Jahrbuch der Philosophie* asked Laue for a contribution. He gave his article the title 'The principle of relativity' (Laue 1913h). Written in the style of a lecture, the introduction tempts the reader to read on: "Processes which are the subject of physics take place in space and time. An event is established by determining the space and time of its occurrence. This determination is done scientifically by defining four numbers. One measures the time (t) from an agreed origin, the other three are the 'coordinates' of location. In what follows we always consider Cartesian coordinates (x, y, z), which measure the three perpendicular distances of a point on three mutually perpendicular planes (ibid.)." This was followed by an introduction to Newtonian mechanics, force = mass × acceleration, and the problem of an inertial frame: "But does at least the definition of the inertial system agree with just one frame of reference? With increasing proximity, must one eventually arrive at a particular system? Not at all (ibid.)." Right afterwards came Laue's treatment of the Galilean transformation with different velocities to establish the following: "Thus any velocity figure only makes sense if it is additionally specified, relative to which system among the above infinite manifold [of inertial systems] it is measured. Hence the theorem about this manifold of systems is called the principle of relativity of mechanics. [...] Now, perhaps all of this sounds very dry; and yet these considerations are among the most important findings we have arrived at in physics (ibid.)." For a full understanding of the subsequent sections, some mathematical knowledge was presumed, such as in § 4 'Einstein's kinematics.' At the end of § 7 'The invariants of the Lorentz transformation,' Laue made a critical remark about contemporary views:

> The historical development which has led from electrodynamics to the theory of relativity has caused some to set the new principle of relativity, being the one of electrodynamics, alongside the old principle of relativity of mechanics. This description could hardly be more erroneous—if only because it has occasionally given rise to the misunderstanding that the one could be valid in electrodynamics and the other in mechanics. Therefore it must be emphasized: If different principles of relativity existed in two different parts of physics, then one frame of reference would be justified according to both. This frame would then be distinguished by both at once; one would have to define all phenomena that belong to either area as absolute motions. Two principles of relativity would cancel each other out (ibid.).

He again paid tribute to Einstein, Poincaré, and Lorentz for their reflections. Just at the time he was writing his article for the *Jahrbuch der Philosophie* Laue heard of a paper by Schaposchnikow (1913) 'On the relative dynamics of the homogeneous body'—that is, prior to its publication by the editors of the *Annalen der Physik*, Wilhelm Wien and Max Planck. Its author contended that a contradiction existed between Planck's and Laue's researches on this problem. Laue conceded that there were small differences in designation between the two papers, but he managed to prove their point nevertheless: "We thus assert most resolutely that there is not the slightest difference between the two accounts of relativistic dynamics on any point (Laue 1913i)."

On 4 November 1913 Max von Laue sent a manuscript to Wilhelm Wien (DMA, NL 056/409–452) that Einstein had encouraged him to write, dealing with a regularity in the imaging of illuminated and luminous objects. His careful precision in identifying a problem and solving it in his own way is once again evident. The theory underlying microscopy developed by Ernst Abbe and extended by Leonid Mandelstam (1911) was his point of departure. Laue claimed it was only an approximation: "We mean by this mainly the assumption of a relatively coarse structure (i.e., compared with the wavelength of light). However, a strictly valid equivalence theorem exists in this field, which is valid for any bodies and any optical instruments of arbitrary design and adjustment setting. [...] In order to pose our problem at all, it must be stated how the body should radiate in the one case, i.e., with what intensities its different parts should shine in the different directions, and how in the other case the intensity of the extraneous light should depend on direction (Laue 1914c: 165)." The coherence relations between the radiation in different directions would also have to be considered. Laue solved this problem by providing a statistical approach. His proof was: "If we think of the body as well as its surroundings as radiating simultaneously and in the same manner, then with our instrument we are looking into cavity radiation which is equal on all sides and, as is known, we cannot discern any outline of its boundaries. [...] But this radiation is composed additively of the images which the body would produce solely by its own light and solely by the external light. And that means: Both images are complementary to each other (p. 166)." He discussed a talk that Henry Siedentopf had given in Munich about observing a completely transparent body. It surprised Laue that his peers did not know the proof. He then compared it with Mandelstam's finding (p. 168).

In Laue's next publication, on 'The degrees of freedom of beams of rays,' there is another reference to his friend under the subheading "According to a personal note by Mr. Einstein." Again the point of departure is a paper by someone else. The English physicist James Jeans had conducted researches with John William Rayleigh on the theory of radiation. Laue cited Jeans's 1905 publication on the degrees of freedom of radiation and his example of a cube with completely reflective walls: "It is easy to see that a beam of rays of finite length possesses a finite number of degrees of freedom. If we suppose that inside a cavity at a given moment there is a single beam of rays of a given length and no other process of oscillation, we can state which natural oscillations of the cavity are involved and of what strength (Laue 1914d: 1198)." Following on from the foregoing publication, Laue treated the problem mathematically to arrive at a formula yielding the number of degrees of freedom for a strictly monochromatic, linearly polarised beam. Section 3 was concerned with incorporating Einstein's theory of radiation fluctuation. It yielded an expression similar to Boltzmann–Gibbs statistics for the relative energy fluctuation of a body. Laue was also able to apply his considerations to elastic thermal oscillations (pp. 1207–1209).

At the end of March 1914, Einstein travelled to Berlin to take up his post at the Academy. His family were still in Zurich. The bantering shoptalk between the two friends could only be resumed several years later, when Laue himself moved back to Berlin.

Max von Laue got in trouble with the university administration in Zurich in the spring of 1914. He had announced a two-hour course on Radiation Theory for the summer semester but had already dealt with the topic in an unofficial course during the winter term. That was why he wanted to lecture on Statistical Mechanics instead, as he informed the dean in a letter from Bellagio on April 5th (UZA). But then, on May 11th, he changed the topic again to Hydrodynamics. He immediately received a response from the dean's office that subsequent changes to lectures could not be approved by the dean's office. According to § 8 of the new university regulations, he must rather apply to the University Commission.

But there were joyous events as well. Max von Laue received a fine distinction: the award of the Erich Ladenburg Prize from the University of Breslau (Wroclaw). He traveled there to attend the award ceremony.[12] In his speech on June 14th, he first thanked Otto Lummer, who had been appointed to the chair for experimental physics at the University of Breslau in 1904. Laue recalled Lummer's lecture course in Berlin and the experiments he had witnessed in 1903. This introduced the topic of optics, and he used optical imaging to address the limit of observability, the impossibility of seeing atoms, and the criticism about whether they existed. "One should ask those skeptics sometime if they think the Sun and the fixed stars are real, or if they also declare astronomy's claim that they are huge bodies unimaginably far away from us, to be but a working hypothesis (Laue 1914e: 758)." Laue then asked what one might possibly be able to observe using very short-wave radiation. That is how he moved on to X-ray diffraction analysis. A longer section was devoted to the question of optical resolving power. The concluding words revealed Laue's thoughts about scientific research and its progress:

Gentlemen, according to the statutes of the Ladenburg Foundation, anyone receiving the Erich Ladenburg Prize should also talk about the goals behind his research. [...] But the actual goals that every scientific researcher stakes out for himself are a delicate matter. Especially when they are set high, it is somewhat doubtful whether they can be achieved, indeed it is not even certain that they are not going to shift in the course of time with advances in science. That is why I beg the high Faculty to excuse me for not venturing into this area. One simply has to wait and see what unfolds. In any case, I can promise the Faculty what I

---

[12] Laue to the Dean, 9 Jun. 1914. UZA 116.3: "Please allow me to inform you that, by permission of the Rector, I shall be out of town from Thursday the 11th inst. to the 18th inst. The main purpose of the trip is the conferral of the Ladenburg Prize to me in Breslau."

already vowed according to the custom there at the time of my doctorate in Berlin, to conduct research [...] (Laue's speech ends with a recital of the oath in Latin. Ibid., pp. 759 f.).

On June 16th Laue informed the dean by letter from the address Meierottostr. no. 8 in Berlin: "On Sunday, the 14th inst., I received in Breslau an invitation by the Prussian Ministry of Culture to come to Berlin for a meeting on the 17th inst. I have therefore asked the Rector for an extension of my leave until the 24th inst. However, I will try to return to Zurich earlier (UZA, 116.3)." The dean's bureaucratic reply from June 18th read: "Highly esteemed Professor, Concerning your communication *in re.* extension of your leave of absence by eight days, I am compelled to inform you that it is not within the Rector's authority to grant such a long leave during the semester. You must address such a petition directly to the Education Directorate (ibid.)." Max von Laue reported to the dean on June 22nd: "I returned yesterday and am resuming my lectures today. The meeting with Privy Councillor Elster at the Prussian Ministry of Culture has not led to anything definitive. Since there is nothing in writing about an appointment, I cannot resign my position here yet (ibid.)."

A political event on 28 June 1914, the assassination of the Austrian heir to the throne and his wife, led to the outbreak of a world war a few weeks later, on August 1st. The Laue couple were on holiday in Switzerland at the time. They returned to Germany on August 28th. Laue had already contacted the dean by letter from Zurich on July 15th with the news: "Allow me to inform you and the Faculty that I intend to accept a professorship at a German university today, effective as of the coming semester (ibid.)." The government council took note of Laue's resignation at its meeting on 1 August 1914: "I. Dr. M. von Laue is granted the requested dismissal as Professor at the Philosophical Faculty II of the University of Zurich on October 15th, 1914, with thanks for the services rendered (Minutes of the Regierungsrat meeting, UZA)." He was also reminded to repay the relocation allowance from 1912. The *Neue Zürcher Zeitung* published a brief announcement about Max von Laue's departure.

References

Bleyer, Ulrich, Dieter B. Herrmann, and Otto Lührs, eds. 2013. *125 Jahre Urania Berlin.* Exhibition catalogue. Berlin: Westkreuz-Verlag

Bohr, Niels. 1913. On the constitution of atoms and molecules. *Phil. Mag.* 26: 476–502, 857

Bragg, W. L. 1914. The reflection of X-rays. *Jb. Radioakt. Elektronik* 11: 346–391

Debye, Peter. 1913. Interferenz von Röntgenstrahlen und Wärmebewegung. *Ann. Phys.* 348: 49–92

Ewald, Paul Peter. 1913. Zur Theorie der Interferenzen der Röntgenstrahlen in Kristallen. *Phys. Z.* 14: 465–472, 1038

Fizeau, Hippolyte Louis. 1864. Untersuchungen über die Ausdehnung und Doppelbrechung des erhitzten Bergkrystalls. *Ann. Phys.* 11: 515–526

Fölsing, Albrecht. 1995. *Albert Einstein. Eine Biographie.* Frankfurt am Main: Suhrkamp. English trans. by E. Osers. 1998. *Albert Einstein. A Biography.* London: Penguin

Friedrich, W. 1913. Röntgenstrahlinterferenzen. *Phys. Z.* 14: 1079–1084

Hermann, Armin. 1963. Das Jahr 1913 und der zweite Solvay-Kongreß. *Phys. Bl.* 10: 453–462

Lamla, Ernst. 1912. Über die Hydrodynamik des Relativitätsprinzips. Abridged version. *Ann. Phys.* 4: 772–796

Laue, Max. 1913a. Die Wellentheorie der Röntgenstrahlen. *Himmel und Erde* 25: 433–438

Laue, Max. 1913b. Kritische Bemerkungen zu den Deutungen der Photogramme von Friedrich und Knipping. *Phys. Z.* 14: 421 ff.

Laue, Max. 1913c. Die dreizählig-symmetrische Röntgenstrahlaufnahmen an regulären Kristallen. *Ann. Phys.* 12: 397–414

Laue, Max. 1913d. Berichtigung zu der Arbeit über die dreizählig-symmetrischen Röntgenstrahlaufnahmen an regulären Kristallen. *Ann. Phys.* 16: 1592

Laue, Max von. 1913e. Zur Optik der Raumgitter. *Phys. Z.* 14: 1040

Laue, Max von. 1913f. Röntgenstrahlinterferenzen. Vorträge und Diskussionen von der 85. Naturforscherversammlung zu Wien. *Phys. Z.* 14: 1075–1079

Laue, Max von. 1913g. Über den Temperatureinfluß bei den Interferenzerscheinungen an Röntgenstrahlen. *Ann. Phys.* 16: 1561–1571

Laue, Max. 1913h. Das Relativitätsprinzip. *Jahrbuch der Philosophie* 1: 99–129

Laue, Max von. 1913i. Zur Dynamik der Relativitätstheorie. Entgegnung an Hrn. Schaposchnikov. *Ann. Phys.* 16: 1575–1579

Laue, Max von. 1914a. Die Interferenzerscheinungen an Röntgenstrahlen, hervorgerufen durch das Raumgitter der Kristalle. In Universität Zürich 1914: 203–245 (reprinted in Laue 1923)

Laue, Max von. 1914b. Die Interferenzerscheinungen an Röntgenstrahlen, hervorgerufen durch das Raumgitter der Kristalle. *Jb. Radioakt. Elektronik* 11: 308 345

Laue, Max von. 1914c. Zur Theorie der optischen Abbildung. *Ann. Phys.* 1: 165–168

Laue, Max von. 1914d. Die Freiheitsgrade von Strahlenbündeln. *Ann. Phys.* 16: 1197–1212

Laue, Max von. 1914e. Über optische Abbildung, *Naturw.* 31: 757–760

Laue, Max von. 1923. *Die Interferenz der Röntgenstrahlen 1912–1914.* Ostwalds Klassiker der exakten Wissenschaften, no. 204. Edited by Friedrich Rinne and Ernst Schiebold. Leipzig: Akademische Verlagsgesellschaft

Laue, Max von. 1961. *Gesammelte Schriften und Vorträge.* Vols. I–III, Braunschweig: Vieweg

Laue, Max von, and J. Stephan van der Lingen. 1914. Experimentelle Untersuchungen über den Debyeeffekt. *Phys. Z.* 2: 75 ff.

Laue, Max, and Franz Tank. 1913. Die Gestalt der Interferenzpunkte bei den Röntgenstrahlinterferenzen. *Ann. Phys.* 10: 1003–1011

Mandelstam, Leonid. 1911. Zur Abbeschen Theorie der mikroskopischen Bilderzeugung. *Ann. Phys.* 10: 881–897

Mehra, Jagdish. 1975. *The Solvay Conferences on Physics. Aspects of the Development of Physics since 1911.* Dordrecht, Boston: Reiche

Pohl, Robert. 1912. *Die Physik der Röntgenstrahlen.* Braunschweig: F. Vieweg & Sohn

Shaposchnikov, K. 1913. Zur Relativdynamik des homogenen Körpers. *Ann. Phys.* 16: 1572 ff.

Schmidt-Böcking, Horst, and Karin Reich. 2011. *Otto Stern. Physiker, Querdenker, Nobelpreisträger.* Frankfurt am Main: Societäts-Verlag

Schulmann, Robert, ed. 2012. *Seelenverwandte. Der Briefwechsel zwischen Albert Einstein und Heinrich Zangger 1910–1947.* In collaboration with Ruth Jörg. Zurich: Verlag der Neuen Zürcher Zeitung

Trageser, Wolfgang, ed. 2005. *Stern-Stunden. Höhepunkte Frankfurter Physik.* Frankfurt am Main: Johann-Wolfgang-Goethe-Universität Fachbereich Physik

Universität Zürich, ed. 1914. *Festschrift des Regierungsrates Festgabe zur Einweihung der Neubauten, 18. April 1914.* Zürich: University of Zurich

Chapter 7
Professorship in Frankfurt am Main, World War I, and Nobel Prize in Physics

Max von Laue probably did not know about the founding history of the University of Frankfurt when he was first informed of his appointment there during his negotiations with the Ministry of Culture in Berlin (Wachsmuth 2005: 19 f., Beck 1989: 24).

This wealthy imperial residence was a free city of the Holy Roman Empire until 1866. A physical society had been actively promoting scientific life in Frankfurt am Main since 1824, the *Physikalischer Verein*. Towards the end of the nineteenth century, the idea of founding a college, academy, or university was enthusiastically discussed by several wealthy citizens as well as by leaders of the city administration, especially by the Mayor Franz Adickes; and it eventually materialized. It was Adickes who engaged the Oppenheimer family as its generous benefactors. Thus, in 1906, the Physikalischer Verein had its own large building with a lecture hall and laboratory facilities. Tough negotiations and further funding finally led to the foundation of the university in 1914. The physical society's building became part of the university. On 10 June 1914, the kaiser approved the founding of the university, but it had to settle its own finances. Its first rector was the physicist Richard Wachsmuth. After earning his doctorate he had worked under Helmholtz at the imperial bureau of standards, the *Physikalisch-Technische Reichsanstalt*. His further career had taken him via Göttingen, Rostock, and Freiberg to the Physikalischer Verein in Frankfurt am Main. Soon afterwards he was appointed to the local Academy as a lecturer in physics. With great energy Wachsmuth selected and appointed the first fifty professors for this new university. For the subject of theoretical physics, the usual list of three candidates were Peter Debye first, then Gustav Mie and, because of "his youth," Max von Laue in third place. The grounds stated: "Numerous longer and shorter publications [...] attest to his scientific achievements, which consistently stand out for their independent ideas, acute thinking, and precise exposition. But his name has become best known from the unexpected and top-ranking experimental discovery which he put forward last year (with the collaboration of Friedrich and Knipping, who carried his ideas to execution): The generation of interference phenomena in

© Basilisken-Presse, Natur+Text GmbH 2022
J. Lemmerich, *Max von Laue*, Springer Biographies,
https://doi.org/10.1007/978-3-030-94699-9_7

the passage of X-rays through crystals."[1] The consequences of the discovery for crystal physics and atomic theory, it continued, cannot be predicted. "Laue is a fine, modest, and amiable person. According to information from Professor Sommerfeld in Munich and from Switzerland, his teaching ability has developed extraordinarily, hence his reputation as a bad lecturer, originating from his first semesters, can be vigorously countered." In the final verdict Laue was placed ahead of Debye and Mie. (Debye was five years younger than Laue.)

The planned inauguration festivities of the university in the presence of the kaiser had been set for 18 October 1914. However, he was unable to come, as his telegram from the main headquarters informed them; World War I had already broken out on 1 August 1914.

The Ministry of Culture had sent Max von Laue the following advice on 7 August 1914: "Prof. Max von Laue—Zurich, is willing to accept a full professorship in the Faculty of Natural Sciences at the University of Frankfort-on-M[ain], starting with the winter semester 1914/15."[2] There followed details about the post as director of the Institute of Theoretical Physics, a salary of 6600 marks plus housing allowance, 1300 marks salary supplement, and for the management of the institute 1000 marks, an honorarium for lecture courses up to 3000 marks in full, above that 75%, examination fees 3900 marks annually.

With the onset of the war, there were many incidents of unrestrained nationalism. This was also evident in speeches held by professors at the University of Berlin and elsewhere. They flaunted superiority of the German character over that of other peoples, declared the war sacred, and proclaimed prophecies of a German victory. Because their opponents energetically defended themselves, a "war of intellects" in the truest sense of the word was unleashed. Inspired by the writers Ludwig Fulda and Hermann Sudermann, and with the collaboration of Georg Reicke, a related appeal was formulated and published bearing the signatures of 15 natural scientists. The names Max Planck and Walther Nernst were among them. But many of these signatories had not yet seen the exact wording of the final text. Each paragraph began with the words "It is not true […]." It came to be known as the "Manifesto of 93," after the total number of signatures (cf. Remane 2005: 399 ff.). Max von Laue's was not among them. Toward his Dutch colleague Hendrik Antoon Lorentz, Max Planck

---

[1] The Rector of the Academy (Wachsmuth) to Oberbürgermeister (Adickes), Frankfurt/M., 6. Jan. 1914. GSta PK I, HA, Rep. 76; Kultusministerium, Va, Sekt. 5, Tit. IV, no. 1, fol. 30. Max Planck was also involved. At the Faculty meeting at the Friedrich-Wilhelms-Universität in Berlin on 28 May 1914, a proposal to the Ministry was prepared: "… that Professor Max v. Laue in Zurich be appointed to the newly established permanent Extraordinariate for Theoretical Physics, with the condition of his appointment as personal Ordinarius." HUA, film 34, fol. 241. At the meeting on July 11th, the application on behalf of Laue was approved and its submission to the Minister resolved.

[2] Minister to Laue (copy), Berlin, 17 Sep. 1914. UAF, Laue papers, 1.10 as well as GSta PK I, HA, Rep. 76, Va, Sekt. 5, Tit. IV, no. 5, vol. 1. This material is very extensive.

was willing to defend the conduct of the German army on its march through neutral Belgium. Fritz Haber wrote to Svante Arrhenius in Stockholm on 23 September 1914: "But now we regard it as our moral duty to use all our powers to bring down the enemy and enforce a peace that will make the recurrence of a similar war impossible for generations to come and will provide a secure basis for the peaceful development of Western Europe (Szöllösi-Janze 1998)." Laue wrote to Wilhelm Wien in this regard on 27 November 1914 from Frankfurt am Main: "Recently I received a letter from H. A. Lorentz and P. Ehrenfest, in which it was supposedly establishable that the Belgians had not been guilty of any atrocities against the German wounded. But how should the two have procured authentic material? I hear from the local mineralogist Boeke, a Dutchman by birth, that Lorentz unfortunately has very little affinity for Germans and is becoming far more temperamental about this than ought otherwise to be expected of such a calm man. What a pity! (DMA, NL 056)"

Max von Laue officially assumed his duties as full professor of theoretical physics at the University of Frankfurt on 1 October 1914. On September 17th, the ministry had notified him "that His Majesty the Emperor and King, my most gracious Lord, has deigned to appoint you full professor in the Faculty of Natural Sciences of the Royal University of Frankfort-on M[ain] (on this call: GSta PK I, HA, Rep. 76, Va)." The Laue couple moved into an apartment in a large villa-style building at Beethovenstraße no. 33 not far from the avenue called Senckenberganlage. He had Otto Stern as a colleague, who had rehabilitated himself from Zurich. Stern's fields of work were molecular and quantum theory, and so they agreed between themselves on the lecture schedule. Stern had already been drafted at the beginning of the war, so Laue had to inform him of his thoughts by letter. Towards the end of 1914 Laue gave a rather popular lecture 'On interference phenomena' before the *Physikalischer Verein* in Frankfurt am Main (Laue 1914).

After a long vacation in Berchtesgaden, Laue succeeded in formulating a simple proof of the applicability of probability calculus to radiation theory which, he wrote to Stern from home on 24 September 1915, "Einstein can really be tickled about." (UAF, Otto Stern papers) But Einstein had other ideas. Laue asked whether it was permissible to consider the coefficients in a Fourier series intended to reproduce the oscillations of 'natural radiation' as statistically independent of each other. Einstein had spoken about this at the first Solvay Conference in 1911 dedicated to the topic of radiant energy and fluctuations (Einstein 1912: 419), using the main formula of thermodynamics for a reversible heat input, $dS = dQ/T$, the Boltzmann principle (1), as well as Planck's formula (2), and Einstein's own definition (3). Laue wrote under the title 'Einstein's energy fluctuations' directly after the introduction: "We want to show how the laws of fluctuation (1), (2) and (3), which obviously do express the same facts in different ways, can also be derived together, in a shorter, purely statistical way which completely avoids the concept of temperature. Since we do not thereby leave the foundations of Einstein's considerations, the results must, of course, coincide completely (Laue 1915a)." Laue succeeded in bringing the results into accord without qualification.

At this time Max von Laue was also working on a second fundamental response to a paper by Einstein and his collaborator Ludwig Hopf. He sent a draft of his text to Einstein, who found fault with it. In a long letter from Feldafing dated 27 May 1915, addressing him with the familiar *Du*, Laue tried to convince his pal of his views: "But now to your main objection. You say: Links exist between the coefficients in a Fourier series, one term (or several terms) of whose development range should reduce to zero. […] How much this speaks from my heart! […]. Otherwise, this proposition doesn't affect my considerations. As I *don't* demand that my Fourier series disappear in the invalid part of the development domain [of the radiation]; I don't demand anything of them at all. Rather, I am completely indifferent to them there, since they have nothing to do with reality (Einstein 1998: 131, orig. emphasis)." But he realized that the paper would have to be revised. So he corresponded with Wilhelm Wien about this, as we can gather from a letter dated 21 June 1915. Wien had sent him one of his publications about the "disorder" of radiation without further comment and Laue thanked him with the words: "I always had a vague feeling that I had read something like this before […]. I have come up with a probability hypothesis for monochromatic oscillations that satisfies all their stability requirements. I.e.: If you imagine resonators and radiation at thermal equilibrium inside a cavity, this order is demonstrably preserved. There is no mention of energy elements and other quantum wonders, everything proceeds naturally (DMA, NL 056)." He assured him of his complete confidence that Planck's "h" occurred in the fluctuation formula and was hoping that it would be Einstein's fluctuation law. "As you see, I am not exactly modest at the moment, but you must attribute that to my delight about this matter. I hope to satisfy your quite legitimate curiosity in a few weeks as well."

It was necessary to correct the publication by Einstein and Hopf 'On a theorem of the probability calculus and its application in the theory of radiation' from 1910, and Max von Laue did this with his characteristic straightforward tact. It concerned the mathematics behind the physical description, about "whether it is justified to regard the coefficients in a Fourier series intended to represent the oscillations of natural radiation as statistically independent of each other (Laue 1915b: 853)." The problem was, according to Laue, of significance to Planck's conception of natural radiation and to quantum theory. Point V provides the result of the detailed mathematical treatment: "The interaction of spatially disordered centers of vibration, however many they may be, cannot lend to radiation that degree of disorder necessary for the adjacent coefficients in Fourier series for radiation to be statistically independent. If in reality such disorderliness is present in the radiation, it is due to disorderliness in the oscillations of the individual resonators (p. 877)." Laue then discussed the objections Einstein had raised in writing and in person. The correction proof appends a quote from a lecture by Wilhelm Wien on the 'Aims and methods of theoretical physics' (Wien 1914). Einstein replied to Laue with a paper of his own: "In the cited work, Laue puts the mathematical basis of the statistics of radiation into a form that leaves nothing more to be desired in terms of concision and beauty. But as far as the application of this basis to the theory of radiation is concerned, it seems to me that he has fallen victim to a severe mistake […]. […] In my view, therefore, none of the cases considered by Laue is equivalent to natural radiation in terms of disorder, so nothing can be concluded about natural radiation from his results (Einstein 1915: 879 ff.)." He had found a mathematical escape from the objection that Laue had raised. Laue wrote from Frankfurt on 15 July 1915 to Wilhelm Wien (DMA, NL 056) about his paper in the *Annalen* and Einstein's reply, which he had deliberately not studied yet in order to avoid being prematurely influenced. In it he mentioned another conversation with Einstein in Göttingen. Laue was not entirely convinced by Einstein's solution and replied under the heading 'On the statistics of the Fourier coefficients of natural radiation.' His introductory remark was: "In a reply to our analyses of a theorem of the probability calculus and its application to the theory of radiation, Mr. Einstein raises two objections, which we should like to discuss here in detail, although—we must hasten to add—we do concede great legitimacy to the second. For, the state of affairs does not yet seem to us to be as clear-cut as would be desirable considering the importance of the issue (Laue 1915c: 668)." Laue started by posing a fundamental question: "Why ought radiation emanating from many 'naturally' radiating source points not be natural radiation? […] We therefore believe we should maintain that no grounds against a statistical coupling of different Fourier coefficients can be derived from radiation composed of many individual rays (p. 669)." Regarding the second objection, Laue argued that Einstein "is not at all concerned about the formation of radiation from individual emissions, but simply treats the Fourier expansion of an oscillation function $f(t)$, hence the oscillations of some linear form (ibid.)." Laue's colleague Ernst Hellinger and the Würzburg mathematician Emil Hilb had assisted him with the mathematical execution, and were duly thanked in the publication. The result obtained agreed with the one Einstein had reached. Laue noted, however: "So there is no probability law with coupling elements applicable to arbitrary time domains. Now it might well be intrinsically conceivable that preferred ranges of time existed in nature which one would be able to use here; for approximately monochromatic oscillations the period or the reciprocal value of the spectral width (measured in oscillation numbers) might provide a measure of such times. But such a theory seems to us unpromising (p. 680, the date of receipt by the publisher is 14 October 1915)."

In a letter to the "University President" on 11 November 1915, Max von Laue reported to the University of Frankfurt: "I take the liberty of announcing that the Swedish Academy of Sciences has today conferred upon me the Nobel Prize in Physics for 1914 (UAF, Laue papers, 1.11)." The university, through the chairman of its board of trustees, congratulated him on November 15th: "For your kind communication of the 11th inst. I extend to you, on behalf of the Board of Trustees and in my own name, most sincere congratulations on the award of the 1914 Nobel Prize in Physics. This award is not only very gratifying to our fledgling university, but under the prevailing circumstances it also represents a distinction of particularly high value (ibid.)." The first congratulations from abroad came from Otto Stern, who was on military duty in the East. The members of his own Faculty of Natural Sciences extended theirs on November 18th. The newspapers also carried a report about the youngest Nobel prizewinner, Laue having just turned 36. When Wilhelm Wien promptly saluted him, Max von Laue gratefully wrote back from Frankfurt on 14 November 1915:

> My most cordial thanks for your kind congratulations! The newspaper reports to which you refer are correct. When two editors called me the day before yesterday shortly after lunch to congratulate me on the basis of a telegram by Wolf, I did still at least have some doubts, especially since the telegraph office in response to my inquiry declared that it was not aware of any such telegram. But the dispatch from Stockholm arrived around 5 o'clock, signed by the secretary of the Academy, Aurivillius.[3] I have no idea yet what will happen with the trip to Stockholm. First of all, it seems to me questionable whether the Swedes will be sending out invitations to the occasion to international scholars during the war, or whether they wouldn't prefer to postpone it. I don't have the letter probably containing that information yet. And if they extend the invitation for next summer, I don't know whether I can travel there. For, according to a verbal communication by Prof. Pohl (from Berlin), your cousin Max [Wien] still has some intentions regarding me. But as soon as I am in uniform, I may not be able to get leave. I do admit, though, that I would regret that a little.
>
> As you have probably also read, the Nobel prize for 1915 was awarded by the Swedish Academy to Bragg and his son. I am very pleased about that, too, and I think that the Academy has thus delicately let the English know that the undoubtedly very great merits of their physicists are not infrequently preconditioned by German research results.
>
> Yesterday I expressed my thanks by telegram to the Stockholm Academy. But may I not also extend a little bit of this gratitude to you? From what I know about the statutes of the Nobel Foundation, your opinion had also been sought. And even if none of the parties involved is allowed to tell me anything about the course of these negotiations, it would be rather difficult for me to assume that you should be so wholly innocent in the conferral of the prize to me.
>
> With the request to relay best wishes from my wife and me to your wife, and with cordial greetings, yours very sincerely, Dr. M. Laue (DMA, NL 056/0423).

A letter written by Arrhenius in English to William Henry Bragg on 12 January 1916, reveals: "Professor von Laue wrote me and asked me to say [to] you and your son his very best compliments and congratulations to the Nobel Prize. As he says, it could not have fallen to more able and worthy scientists."[4]

The war prevented the award ceremony from taking place in Stockholm on 10 December 1915. The prize money of 156,661.05 kroner was transferred via the Deutsche Bank and had to be converted into German currency. The German embassy in Stockholm forwarded the sum, the certificate, and the medal to Laue in mid-1916

---

[3] Christopher Aurivillius was permanent secretary of the Royal Swedish Academy of Sciences from 1901 to 1923.

[4] Arrhenius to W. H. Bragg, Stockholm, 12 Jan. 1916. RIGBA, MS WHB, 106/3. Unfortunately, there are no letters with personal congratulations from Planck and Einstein among Laue's papers in the University Archives at Frankfurt am Main.

(Fig. 7.1). He divided the apparently nontaxable prize money fairly into thirds to share with Friedrich and Knipping. As a sign of true patriotism he invested his share in war bonds.[5] As a Nobel laureate, Laue was henceforth solicited annually by the *Nobelstiftelsen* for nominations of the next candidates. When asked in 1916, he proposed Max Planck, and the following year jointly to Albert Einstein in Berlin and Hendrik A. Lorentz in Leyden for postulating the theory of relativity.[6]

On 15 November 1915, Ferdinand Springer asked Laue if his publishing house could print his Nobel Lecture; Laue replied eleven days later from Frankfurt am Main to thank him for the offer, saying that his best friend had already commissioned a publisher in Karlsruhe (UAS, folder 1). It eventually appeared in 1920 with the C. F. Müllersche Hofbuchhandlung, after Laue had presented it personally in Stockholm (Fig. 7.2).

The battle on the Eastern front against Russian troops that had begun in the winter and the fierce fighting in France in the Champagne region continued into the spring of 1915. In an attempt to achieve a breakthrough along the frontline in France, a major assault with poison gas (chlorine) was carried out near Ypres on 22 April 1915, according to plans drawn up by Fritz Haber at the Kaiser Wilhelm Institute of Physical Chemistry and Electrochemistry. The horrors of the war continued to escalate. At the university, the lectures and exercises continued despite the war. However, many students were drafted or had volunteered for military service. So attendance was often poor. Laue offered courses on the Theory of Electricity and Statistical Mechanics with their associated practical sessions. Max von Laue was also fascinated by problems further afield. A colleague of his at Frankfurt, Richard Lorenz, sparked his interest in 'The migration of discontinuities in electrolytic solutions' (Laue 1915d). It involved ion migration and the further development of existing theories. Laue's publication appeared in the *Zeitschrift für anorganische und allgemeine Chemie* with the note that the manuscript had been completed in April 1915 but had not reached the editors until October 5th because it had initially been intended for a commemorative volume in honor of Julius Elster and Hans Geitel.

That autumn, on September 25th, Max von Laue was able to report to Wilhelm Wien from Frankfurt on the state of his war effort: "By the way, it is thanks to your cousin Max Wien that I have the prospect of being drafted into the military. The final decision is still pending, though (DMA, NL 056)."

The year 1915 ended with a surprise in science. The character of X-rays had been elucidated by the discovery of their interaction with crystalline structure. What kind of image would be obtained if a random arrangement of atoms were irradiated? The astonishing results were presented at a meeting of the Royal Scientific Society at Göttingen on 3 December 1915. These successful experiments by Debye and Paul Scherrer (1916) soon expanded the applications of X-ray diffraction investigations quite considerably.

---

[5] Laue to Ève Curie, [Berlin-] Zehlendorf, 5 Aug. 1938. Having read Ève Curie's biography of her mother, he told her that he too had spent his Nobel prize money on war bonds.

[6] Vetenskapsakademien to Laue, Stockholm, 31 Jan.1916, and Laue to Vetenskapsakademien, Würzburg, 12 Dec. 1917. UAF, Laue papers, 1.16.

Fig. 7.1 The certificate of the 1914 Nobel prize in physics awarded to Max von Laue, designed by Sofia Gisberg. *Source* Archive of the Max Planck Society, Berlin-Dahlem

Fig. 7.1 (continued)

M. v. Laue

Professor der theoretischen Physik an der Universität Berlin

Über die Auffindung der Röntgenstrahl-Interferenzen

Nobelvortrag, gehalten am 3. Juni 1920 in Stockholm.

16 Seiten broschiert Preis 2.50 Mk.

„Frankfurter Zeitung": Schon der Entdecker der Röntgenstrahlen hat nach Interferenz- oder Beugungs-Erscheinungen bei dieser Strahlungsart gesucht, um die Frage zu entscheiden, ob man es hier mit einem Wellenvorgang oder mit der Ausschleuderung kleinster Teilchen zu tun habe. Obgleich seine Untersuchungen erfolglos blieben, fand die Wellentheorie starke Anhängerschaft, denn nach Maxwell-Lorentz müssen elektromagnetische Wellen entstehen, wenn Elektrizitätsträger ihre Geschwindigkeit ändern. Die Vertreter der Korpuskulartheorie können freilich noch heute manche Erscheinungen anführen, die der Erklärung durch die Wellentheorie trotzen. Über diese Dinge, namentlich über die Vorarbeiten und selbständigen Untersuchungen, die der Nobelpreisträger für Physik, Professor Max v. L a u e (Berlin), früher Ordinarius an der Universität Frankfurt, ausgeführt hat und die zur Auffindung von Gitterspektren und damit von Interferenzen führten, gibt der „N o b e l v o r t r a g" Aufschluß, den Laue am 3. Juni 1920 in Stockholm vor der Kommission der Nobelstiftung gehalten hat.

Zu beziehen durch jede Buchhandlung od. unmittelbar vom Verlag.

Fig. 7.2 The advertisement by C. F. Müllersche Hofbuchhandlung of the published version of Laue's Nobel lecture. *Source* Scan in the author's estate

The correspondence about the award of his Nobel prize had taken up much of Laue's time. On 9 December 1915 he wrote to Otto Stern: "You can congratulate me again—namely, for not receiving anymore congratulations now. But it caused me three weeks of hard work […]. Well, I was very pleased at least to see how many long-lost acquaintances reappeared on this occasion (UAF, Otto Stern papers)."

Despite the increasingly vicious fighting on all fronts in the terrible war, life for many people continued largely unchanged. At the beginning of 1916, on February 10th, Laue wrote to Wilhelm Wien about a planned ski trip: "I'm looking forward to it extremely much, also to Mie's participation in it. Isn't Sommerfeld coming along? I would like to use the opportunity to try and improve relations with him a bit (DMA, NL 056/409–452)." He obviously wasn't aware of a letter Sommerfeld had written to Wilhelm Wien on that very day about the planned excursion: "I expect to finish up on the 3rd and shall then be ready to join you. I hope Mie is coming, and hopefully Laue isn't. I fear that my sense of ease would suffer much by his presence (DMA, NL 056)."

Laue had reason to be pleased again, because a very lauditory review appeared in the *Physikalische Zeitschrift* by Clemens Schäfer (1916) about his contribution on wave optics in the 'Encyclopaedia of the Mathematical Sciences Including Its Applications' (Laue and Epstein 1915). On 1 March 1916 Max Planck reported to Wilhelm Wien from Berlin: "I spoke with Laue last week and found him unexpectedly vivacious (SB PK, Wien papers)." Not a hint of any mental turmoil plaguing Laue. But not long afterwards, on April 11th, Planck replied to Wien's letter in consternation:

> What you write me about Laue saddened me deeply. As, I don't see it leading to any good. The unwisest thing he could do, of course, would be to simply drop his present position. But even if he were to be offered another that initially might satisfy him along all lines, I am firmly convinced that he would not be able to endure it for long, as has happened here: the way he was changing his mind and resolutions from one day to the next, and doing so right off with such passionate impetuosity as to bar the way to any word of reason. At the same time, that constant fear of intrigues by enemies secretly plotting against him. It is fortunate for him that he now occupies such a well-known position in science and in the world, making it easier to talk him out of such things. I have heard nothing yet about the project by the Kaiser Wilhelm Society and am too unfamiliar with the conditions there to have an opinion as to whether it would be suitable for Laue. But, as I said, I have become quite pessimistic. Perhaps now that he is with you in Mittenwald in splendid nature he will be distracted by other, calmer thoughts. Give him my best regards, and Sommerfeld, too, if he is there (SB PK, Wien papers).

Planck had guessed correctly. Laue regained his mental equilibrium from friendly conversations during that stay in Mittenwald. The birth of his son, christened Theodor, on June 22nd probably also played a part. Four days later he informed Wilhelm Wien from Frankfurt: "My wife and the boy are doing extremely well (DMA, NL 056)."

In the middle of 1916, Max von Laue was caught in a kind of appointment maelstrom. The University of Vienna lamented the loss of its physicist Friedrich Hasenöhrl. He was killed in action near Trento at the beginning of October 1915. Now various candidates for a successor were being contacted, including Laue. On July 24th Laue informed the chairman of the board of trustees of the University of Frankfurt of the inquiry from Vienna (UAF, Laue papers, 1.11). The University of Frankfurt tried to keep Laue. The board of trustees then wrote to the Minister of Intellectual and Educational Affairs on 28 July 1916: "The Board of Trustees would find it exceptionally regrettable, in view of the position which Professor von Laue occupies in the scientific world through his outstanding achievements, if the University of Frankfort-on-M[ain] were to lose his teaching capacity after such a short time."[7] The dean sent a telegram, to the ministry on 1 August 1916, requesting that "all steps be taken […] to obtain the retention of Professor v. Laue (UAF, Laue papers, 1.11)." From Frankfurt, Laue turned to Wilhelm Wien for advice on 26 July 1916:

> Highly esteemed Privy Councillor!
>
> I recently received an inquiry from the Viennese Ministry of Culture as to the conditions under which I would accept a call to Vienna, if such a call were to come.
>
> Now this is a very unpleasant thing. For I am not going there under any conditions (I remember too well how I longed for Germany in Zurich), but unfortunately I rashly answered

---

[7] Dekan der Universität Frankfurt/M. to Minister für geistliche Angelegenheiten. UAF, Laue papers, 1.11.

the Vienna faculty in January when they asked about it, that I was willing to negotiate about Vienna. I surely would not have done so if I had not firmly believed that my appointment to Berlin would have forestalled the whole Viennese business and settled it without my intervention; nor did I want to say no directly to the Viennese so as not to offend them. How should I now get out of this predicament? I would prefer to put it on the back burner for the time being, if only to avoid having to go to Berlin and Vienna in the August heat.

It would be the most amenable to me if I could tell the Viennese the truth and write: I hope to be called to Berlin, and consider Vienna a detour, if not a wrong track; therefore I prefer to stay in Frankfurt. But so much honesty is probably contrary to academic custom.

The only useful thing to be made of this call, in my opinion, would be to put pressure on the Berlin ministry to finally make up its mind now about the Berlin appointment (DMA, NL 056).

Wilhelm Wien gave him the good advice of stalling. On the same day, Laue wrote to Sommerfeld. First asking for some details about the periodic system for his lecture on X-ray spectra on the following Monday, he then continued: "I have now been asked by the Viennese Ministry for the conditions under which I would be prepared to go to Vienna. So I shall have to go to Vienna next week. Unfortunately, nothing seems to be coming of my enlistment, at least I have had no answer to date (DMA, NL 089)." However, on August 16th Laue was able to inform the Ministry of Culture that he had received word from Vienna that approval by the Minister of Finance had yet to be sought. Thus his trip to Vienna would make no sense.[8] On vacation in Tyrol, he reported to Wilhelm Wien on 19 August 1916 that his left knee was "in repair," adding: "The appointment matter will probably end well, provided I am granted all my wishes here, which I have arranged in a way that the University of Frankfurt gain some advantage even if I do leave (DMA, NL 056)." If the Viennese reopened negotiations, he wrote, he would decline. The management of the University of Frankfurt considered whether to establish an extraordinariate, in order to relieve Laue of the burden of teaching beginner courses, as encouragement for him to stay. In a letter from Laue in Würzburg to the mayor of Frankfurt dated 14 November 1916, he candidly communicated his plans for the future and appointment preferences: "I have today declined the appointment to Vienna. But I still ask you not to continue the negotiations about the establishment of an extraordinariate, at least not for my sake. For as soon as peace is restored, I shall carry out my old plan and go to Berlin to join my esteemed teacher Planck. It is all the same to me in which position; hopefully the call to Berlin will arrive by then."[9] Just five days later he wrote to the minister of culture that he had finally cancelled the appointment to Vienna, adding: "I am going to resign from my post latest at the end of this war in order to go to Berlin […], whose university I regard as my true home in scientific respects. […] Frankfurt could not have offered me a complete substitute for it in the long run, despite the many advantages of my position here and the extraordinary kindness I

---

[8] Laue to Kultusminister, 16 Aug. 1916. GSta PK 1, HA, Rep. 76 Va, Sekt. 5, Titel IV, no. 5, vol. 1.

[9] Laue to Oberbürgermeister von Frankfurt, 14 Nov. 1916. UAF, Laue papers, 1.10. The letter is available as a handwritten draft among his papers.

have always experienced from the Board of Trustees and my colleagues."[10] Laue continued to say that he intended to resign as of the first of April, 1917. "For the future I plan to seek employment at the Imperial Bureau of Standards [PTR]; I have already taken the first step in that direction (ibid.)." The letter bears a handwritten note from the Ministry: "I spoke with the director of the Phys. Tech. Reichsanstalt, Prof. Dr. Warburg, yesterday. He has received [no?] request from Prof. von Laue so far. […] Prof. v. Laue declared, while he was here on the 7th 12/16, that he was not going to give up his professorship in Frankfurt before the end of the war in any case (ibid.)."

Many a physicist who had joined the war effort suffered from a lack of scientific stimulation and debate. This was the case with Wilhelm Lenz, a native Frankfurter who had obtained an assistantship under Sommerfeld at the beginning of 1914. His leave time did not suffice to travel to Munich, but he could visit Laue in Frankfurt. Back in his barracks Lenz later wrote about it to Sommerfeld on September 25th, and his report reveals Laue's approach to problems in physics:

> We talked about Einstein's new relativity, of course, in whose interpretation Laue takes a curiously stubborn phenomenological standpoint. [… Laue] says that Einstein orientates the frame of reference by the bodies; according to his view, bodies should be orientated according to the frame of reference, as has always [happened] in the past, which we must take as something given. The perihelion motion of Mercury does not impress him at all, he says, and countless theories have already been made about it. Furthermore, he doesn't understand why one isn't satisfied with Nordström's theory.
>
> My own feeling leans more towards the Einsteinian stance, as only experience will tell us who is right.
>
> Quanta were the central focus of our conversation, of course. Laue […] had interesting, rather phenomenological points of view on that, too.[11]

From the end of October 1916 Laue was in Würzburg working on amplifier tubes at Wilhelm Wien's institute. Max von Laue felt obligated as a German to make his own contribution to the war effort. In 1909, after having participated in six military exercises, he had resigned as lieutenant at his own request. One reason for this decision had been an account he had heard in the mess hall: A first lieutenant who had served in South West Africa described the cruel methods they had used to execute Herero captives (Laue's letter to his son Theodore, Lemmerich 2011: 270). Now Laue's application was rejected. He was curtly refused readmission into the army by the XVIIth Army Corps in Frankfurt on 1 July 1915: "In reply to your letter addressed to the District Command II Frankfurt am M[ain] on 20 Jun. 15, in which you apply for deployment for the duration of the mobilization, the respectful notice that your wish unfortunately cannot met (UAF, Laue papers, 1.1)."

---

[10] Laue to Kultusminister, Frankfurt/M., 19 Nov. 1916. GSta PK 1, HA, Rep. 76 Va, Sekt. 5, Titel IV, no. 5, vol. 1.

[11] Lenz to Sommerfeld, Gr.[oßes] H.[aupt-]Qu.[artier], 25 Sep. 1916. DMA, NL 089/059. Lenz also mentioned considerations about Bohr's atomic model, the fourth quantum number. This letter is one of few sources describing a private conversation about science, obviously seen from Lenz's own perspective.

Laue then tried to find ways to make himself useful as a scientist in defense of his country. He probably discussed the possibilities with Wilhelm Wien, because on 15 September 1916 he sent a slightly ambiguous message from his holiday resort on Lake Starnberg to the dean in Frankfurt: "The Roy. Technical Signal Inspection Commission (of which, i.a., Prof. Max Wien of Jena is a member) has, at my request, proposed my re-enlistment into the Army for the Inspectorate of the Signal Corps and my deployment for the purposes of radio telegraphy. My registration can be expected in about four weeks (UAF, Laue papers, 1.11)."

Since he would not be able to give his lecture course then, Laue addressed the question of who might stand in for him. He suggested Max Abraham as his possible substitute. The faculty of science communicated their concerns to the Ministry on 18 October 1916 about maintaining these well-attended lectures in theoretical physics: "It believes, however, that during these difficult times, when everyone is putting his services at the disposal of the Fatherland, it should not stand in the way if such an outstanding researcher as Professor von Laue wanted to devote himself to militarily important research."[12] But these worries became moot, as Laue agreed to hold his lectures for two hours himself every Saturday morning. His other commitments also needed to be kept. On 11 November 1916, Laue delivered a talk at the German Chemical Society on 'Crystalline research with X-rays' (Laue 1917a; its 16 illustrations were probably projected as lantern slides). Laue had already demonstrated an interest in the history of the natural sciences in several publications and lectures. His talk first dealt with the conception of the structure of crystals at the turn of the nineteenth century by René Haüy, then by Paul Groth in Munich. This was followed by explanations of gratings and space lattices and imaging in the microscope. He said that when observing with only one wavelength, no structure could be discerned; only when three wavelengths were used was this possible. He remarked somewhat lightly about the current state of knowledge: "We have no lenses for X-rays and hence no microscope; for this reason alone we cannot simply put space lattices of crystals under a microscope. But would this become feasible for us if a good fairy granted our wish for an X-ray microscope? It is perhaps surprising if we should answer this question in the negative, and yet this is undoubtedly the correct response (ibid.)." Another historical survey covered the crystallographers Bravais, Wiener, Schucke, Sohncke, Fedorov, and Schoenflies from Frankfurt about the development of the classification of the symmetry properties of crystals, mirroring on a plane, rotation around an axis, and rotation with reflection. NaCl, KCl, $CaCl_2$ and diamond were given as examples. The war notwithstanding, Laue mentioned the researches of the Braggs in England.

The mental strain of the war and the problems of daily life in wartime worsened. Yet Max von Laue still found time during the 1916/17 term to write a detailed report about the gravitational theory by the Finnish physicist Gunnar Nordström for

---

[12] Faculty of Natural Sciences of the University of Frankfurt a. M. to Minister f. geistl. Angel. u. Unterricht, 18 Oct. 1916. GstA PK 1, HA, Rep. 76 Va, Sekt. 5, Titel IV, no. 5, vol. 1.

Stark's journal *Jahrbuch der Radioaktivität und Elektronik*.[13] Nordström had studied
at the University of Helsinki and received his doctorate there in 1908. He had been in
Göttingen two years before. Before returning to Helsinki, he had sojourned in Leyden
with Lorentz. Publications on his ideas about a general theory of relativity started to
appear in 1912. As was usual for such *Jahrbuch* articles, the literature covered is first
listed: eight papers by Nordström and the paper on Nordström's theory of gravitation
coauthored by Einstein and Adriaan Fokker, plus some secondary literature. Laue
now sided with Einstein, as the introduction shows:

> Among the theories of gravitation connected with the theory of relativity, Einstein's theory
> probably has the most followers at present. And it deserves this in a certain respect, because
> of the tremendous intellectual achievement of developing from the general principle: 'The
> laws of physics must be such that they be valid with respect to arbitrarily moving reference
> systems' a mathematical theory not only of gravitation but of physics in general, together with
> a completely new conception of space and time. Its mathematical elegance is also captivating.
> However, its greatest advertising point probably lies in the aforementioned principle itself,
> especially since it seems to close a considerable gap in the structure of our physical world
> view (Laue 1917b, introduction).

Then Laue cited the three important differences from Newton's theory: the deflection of light by gravitation, the perihelion rotation of Mercury, and the equivalence
of frames of reference. Then he led over to Nordström's theory, which he appreciated
by writing that it seemed "to have received less attention so far than it deserves [...]
(ibid.)." Laue posed five requirements that the Nordström theory had to satisfy. The
principle of relativity should be valid unchanged; for gravitational statics Newton's
law of attraction should be valid; and for all static and stationary systems the equivalence theorem $M = g \cdot m$ should be valid. In addition, there were two mathematical requirements. Laue then checked whether all these demands were satisfied
by Nordström's statements. Only in the case of gravitational statics did Laue doubt
Nordström's result and examine it more closely. As a versed thermodynamicist he
had to cover that chapter, too. Redshift by gravitation was treated, and Laue found
that Einstein's and Nordström's values even agree in exact result. In the section on
longitudinal gravitational waves he noted that the energy of such waves was too low
for detection. In § 6 'Approaches to further development of the theory,' Laue reproduced the gist of Nordström's point as, "how one can summarize the basic equations
of electrodynamics and gravitational theory by introducing a five-dimensional 'world
extension' in which, in addition to x, y, z, t, another constant w occurs (ibid., § 6)."
The only flaw, he said, was the "unexplained perihelion motion of Mercury (ibid.)."

Max von Laue was assigned to the Technical Department of Radio Engineers
"*Tafunk*" by the Inspectorate of the Signal Corps. On 24 March 1917 he informed
the minister that this service, which had been arranged only by verbal agreement until
then, should be put in writing in a formal contract.[14] With the development of high-
frequency technology for the transmission of messages, the problem of amplifying

---

[13] Laue 1917b. This can be viewed as in connection with his revisions of his book on special
relativity to incorporate general relativity.

[14] Laue to the Minister, Frankfurt/M., 24 Mar. 1917. GSta PK, Rep. 76 Va, Sekt. 5, Titel IV, no. 5,
vol. 1.

high-frequency signals had become acute. In 1906 the Austrian physicist Robert von Lieben and the American physicist Lee de Forest invented the electronic amplifier tube independently almost simultaneously. Von Lieben demonstrated his amplifier tube to the Berlin physicists and submitted his patent application that same year. In order to avoid conducting countless experiments to find improvements, one first had to understand how the electrons emitted by the incandescent filament behaved. In 1914, Wilhelm Wien in Würzburg and his cousin Max Wien in Jena recognized the military importance of improving the amplifier tube in communications engineering. Wilhelm Wien's physics institute at Würzburg and the chemistry institute with its many glassblowers manufactured amplifier tubes on a large scale. Max Wien was working very successfully at the University of Jena in the field of high frequency engineering. Max von Laue was engaged there to conduct basic research. Only towards the end of the war did his theory of "glow electrons" appear in three publications from mid-1918 to 1919. This was a new field for him. The place of publication was given as the Institute for Theoretical Physics at the University of Frankfurt am Main. But the research was very probably performed in Würzburg, because he thanked his friend Dr. M. Linnemann in Würzburg. Moreover, the Laue couple were living on Mergentheimer Straße in Würzburg at the time. His publication of over 50 pages length again opens with a critical commentary on the conceptions of his predecessors: "The purpose of the following investigation is to clarify the concepts about glow electrons. In the literature one often encounters the idea of an electron gas in otherwise empty space, which is treated like any other gas with respect to the equation of state and thermodynamic functions, just as if the electrons did not repel each other. Electron emission is occasionally considered thermodynamically as evaporation and treated with the Clausius-Clapeyron formula. All this is subject to certain doubts from the outset and does not in fact quite stand the test upon closer examination (Laue 1918a)." Laue treated the electron cloud mathematically. The publications by Walter Schottky and Owen Richardson and others were taken into account.[15] Graphical illustrations of the course of the potential supported the established findings: "As an essential result of the present investigation, we consider that the glow electrons in a state of equilibrium form a layer adjacent to the glowing electrodes; this layer, when external influences are not too strong, has an energy, entropy, and charge per unit area characteristic of the electrode material and temperature, and a negative surface tension, which results in a negative capillary pressure in the case of curved surfaces (Laue 1918a: 246)."

The second publication on glow electrons concerned the determination of their entropy constant (Laue 1918b). The third paper was closely related to practice and investigated the operating conditions of triodes (filament–control grid–anode) of various grid designs, a single wire, a plate, and a grid-shaped cylinder. The manuscript (Laue 1919) was not completed until the spring. Laue simulated mathematically the curves of the dependence of the anode current on the grid voltage, which Heinrich Barkhausen had plotted, for the three arrangements and was able to derive an optimized design with a grid-shaped cylindrical control grid (Kaiser 1996).

---

[15] Laue's views triggered a dispute with Walter Schottky. Cf. Kaiser in Hoffmann et al. 1996: 233.

The confidence many Germans had in victory in August 1914 seeped away as the heavy losses in battle continued to mount. Provisions for the general population deteriorated; that infamous "turnip winter" of 1916/17 showed this clearly. The gaps formed by shortages in the regular male labor force in the munitions factories and other production sites were filled by women. There is no documentary trace of whether the events of the time oppressed Max von Laue and his wife and how much so.

In Zurich, Alfred Kleiner, full professor of experimental physics, died unexpectedly in 1916. As was the academic custom, a memorial issue was published, for which Laue also wrote an essay on 'A failure of classical optics,' to which he added a preliminary communication in 1917. Again, Laue's readiness to doubt scientific findings and not necessarily regard them as definitive becomes apparent: "It seems to us that here we have experimental proof that where radiation theory and wave theory contradict each other in questions of fluctuation and the related thermodynamic statistical considerations, the wave theory is wrong. The path toward a modification of classical optics, taking this into account by considering Planck's constant h, does seem to us to be long (Laue 1917c: 21)."

In the spring of the fourth year of the war, the German troops in France attempted a major offensive to come closer to victory, but it did not lead to the hoped-for success. During this time, Max von Laue was working on finding a mathematical solution to X-ray interference phenomena in solid solutions (Laue 1918c). The impetus came from a publication in the *Physikalische Zeitschrift* in 1917 by Lars Vegard and Harald Schjelderup, who had analysed KCl and KBr. Laue wrote to Wilhelm Wien from Frankfurt on 5 May 1918, sending him his manuscript for the *Annalen der Physik*, which had caused him quite some labor. "The result that one can distinguish enantiomorphic crystals by their diffraction pattern precisely because the absence of a center of symmetry is never apparent in the diffraction pattern sounds very paradoxical. Yet it is undoubtedly true (Laue 1918c)."

On May 20th, Laue informed the board of trustees at Frankfurt University: "In order to bring nearer to fulfillment my old and most ardent wish to move to Berlin, I have written to His Excellency Naumann at the Ministry of Culture today asking for permission to exchange chairs with Professor Dr. Max Born in Berlin. I would thus become a personal full professor at the University of Berlin. I have previously obtained the consent of Professor Born and those members of the Frankfurt Faculty of Natural Sciences who are professionally interested in the filling of the post in theoretical physics."[16] Max von Laue had harbored the rather vague hope of getting funding for his scientific research at the Kaiser Wilhelm Institute (KWI) for Physics from a private donor. When this did not materialize, he asked Max Born if he would be willing to swap professorships with him once the salary issue had been resolved.

---

[16] Laue to Kuratorium der Universität Frankfurt/M., 20 May 1918. UAF, Laue papers, 1.10; Laue to Naumann, Frankfurt/M., 19 May 1918. GSta PK 1, HA, Rep. 76 Va, Sekt. 2, Titel IV, no. 68 D, vol. 1, fols. 205, 207, 209. Inquiry by the Ministry to the Philosophical Faculty of the Friedrich-Wilhelms University on 9 Aug. 1918 as to whether there were any reservations about the exchange of Born's and Laue's chairs. Max Planck asked the Ministry "that the appointee not be subsumed under *Ordinarius*, but rather should be independent."

Several months elapsed before this exchange of chairs, which was highly unusual in academia, could become a reality. Born thus became a fully tenured professor (*Ordinarius*), Laue came away with a personal associate professorship (*Extraordinariat*) and thus was still obliged to offer four hours of lectures per week, after all. Laue reported this to Einstein in a letter from Würzburg on 30 January 1918, who had not yet recovered from a stomach illness, adding: "To talk about something completely different now, I'd like to ask you whether you know of any place on the currently accessible part of the Earth's surface that mightn't be more suitable than Berlin for curing a sick stomach. I don't want to meddle into your doctors' business but would like to think that now in wartime, just about *any* other place is better for the purpose (Einstein 1998: 621)."

The next letter, dated February 18th, mentioned the challenges he was facing of writing a lecture on thermodynamics in honor of Planck's 60th birthday on 23 April 1919, without actually spelling them out (Einstein 1998: 654–655).

An important advance in theoretical physics had been made in 1913 by the young Danish physicist Niels Bohr. He treated the atom mathematically as a kind of planetary model. According to it electrons circled around the atomic nucleus along fixed orbits. He interpreted the absorption of radiation as an absorption of energy together with a transition to a larger orbit; radiation was the electron's return to its original orbit with a release of that energy. Max von Laue did not become active in this field, but he did study Bohr's publications, as a letter to Bohr from Frankfurt dated 28 October 1918 shows: "Esteemed colleague! While reading your beautiful paper on the quantum theory of spectral lines (part I), I encountered a difficulty. It concerns the derivation of your Equation (6). The equation is undoubtedly correct, but one does not quite understand why you may omit the sum […], which is in the preceding formula."[17] Laue's calculation followed with the suggestion: "If these remarks convince you, you might want to incorporate them into one of your subsequent parts in the service of the cause, which all of us are eagerly awaiting (Bohr 1918, part 1)." Bohr answered Laue's letter with one in another hand—that of his wife or secretary—on November 11th. The reply demonstrates the way Bohr typically tackled problems. First he discussed the mathematics, referring among others to Boltzmann: "I would regret it if the concise form in which the calculations are given in this and other parts of my treatise should cause any difficulty to the reader. The excuse for this brevity in the calculation is in the endeavor to divert the reader's attention as little as possible from the more philosophical aspect of the considerations, on which the chief emphasis of the work lies (ibid.)." Laue replied from Würzburg on November 20th: "Esteemed Colleague! I thank you most sincerely for your letter of the 11th inst.; I have thoroughly reconsidered the situation in detail, and must now admit that for a more symmetrical proof taking both integral limits equally into account, it can indeed be said that the expressions in parentheses are equal to each other for both limits. But I must deny that anyone who has nothing but your

---

[17] Laue to Bohr, Frankfurt/M., 28 Oct. 1918 and 5 Dec. 1918. NBA; see also Bohr 1918: 1. Laue highly appreciated Bohr's work on the atomic model. He proposed him several times for the Nobel Prize, which Bohr eventually received in 1922.

paper before him would see this, on the basis of my own experience and that of other thoroughly competent specialists (e.g., Dr. P. S. Epstein) (NBA)." Laue suggested that he cite the passage in Boltzmann's work. It was not long before both scientists appreciated each other not only as fellow professionals, but also personally.

In August 1918, the German Supreme Army Command had already admitted that the military situation on the frontlines forebade any continuation of the war. On 9 November 1918, Wilhelm II abdicated the throne as German Emperor and King of Prussia and left Germany forever, finding refuge in the Netherlands. Philipp Scheidemann proclaimed the first German Republic. On 11 November 1918 the armistice was signed. There were uprisings led by politically radical groups, and civilian militias tried to keep order.

The new state did not disempower the administration of the Wilhelmine state, but new authorities were created. Rubber-stamped designations replaced the letterheads of the former ministries. For instance, the former Ministry of Intellectual and Educational Affairs, was henceforth called the Ministry of Art and Science. Laue received notification by that minister in Berlin bearing the date 28 December 1918: "I have transferred the full professor Dr. Max von Laue to the Faculty of Philosophy of the University of Berlin in the same capacity as of April 1st, 1919."[18]

References

Beck, Friedrich. 1989. Max von Laue 1879–1960. In *Physiker und Astronomen in Frankfurt. Geschichte der Johann Wolfgang Goethe-Universität Frankfurt/Main.* Eds. Klaus Bethge and Horst Klein, p. 24. Frankfurt: Metzner

Bohr, Niels. 1918. On the quantum theory of line spectra. Part 1: On the general theory. *Det Kongelige Dansk Videnskabernes Selskab Skrifter* 8, no. 4.1: 3–36

Debye, Peter, and Paul Scherrer. 1916. Interferenzen an regellos orientierten Teilchen in Röntgen-licht. *Nachrichten der Königlichen Gesellschaft der Wissenschaften zu Göttingen* 1916: 1–26; also in *Phys. Z.* 17: 277–283

Einstein, Albert. 1912. L'État actuel du Problème des Chaleurs Spécifiques. *La Théorie du Rayonnement et les Quanta. Rapports et Discussions de la Réunion Tenue à Bruxelles, du 30 Octobre au 3 Novembre 1911. Sous les Auspices de M. E. Solvay.* Ed. by Paul Langevin and Maurice de Broglie, 409–435. Paris: Gauthier-Villars

Einstein, Albert. 1915. Antwort auf eine Abhandlung M. v. Laues 'Ein Satz der Wahrscheinlichkeitsrechnung und seine Anwendung auf die Strahlungstheorie.' *Ann. Phys.* 15: 879–885

Einstein, Albert. 1998. *The Collected Papers of Albert Einstein.* Vol. 8: *The Berlin Years, Correspondence 1914–1918.* Princeton: Univ. Press

Einstein, Albert, and Ludwig Hopf. 1910. Über einen Satz der Wahrscheinlichkeitsrechnung und seine Anwendung in der Strahlungstheorie. *Ann. Phys.* 33: 1096–1104

Kaiser, Walter. 1996. Electron Gas Theory of Metals and Its Application in Technology. In *The Emergence of Modern Physics. Proceedings of the Conference Commemorating a Century of Physics Berlin 22–24 March 1995.* Edited by Dieter Hoffmann, Fabio Bevilaqua, and Roger Stuewer. Pavia: Università degli studi di Pavia

Laue, Max von. 1914. Über Interferenzerscheinungen. *Frankfurter Universitäts-Zeitung,* 4 Dec

---

[18] Ministerium für Kunst und Wissenschaft to Laue, Berlin, 28 Dec. 1918. UAF, Laue papers, 1.10, see also the files in GSta PK, Rep.76.

Laue, Max von. 1915a. Die Einsteinschen Energieschwankungen. *Verh. DPG* 10: 198

Laue, Max von. 1915b. Ein Satz der Wahrscheinlichkeitsrechnung und seine Anwendung auf die Strahlungstheorie. *Ann. Phys.* 15: 853–878

Laue, Max von. 1915c. Zur Statistik der Fourierkoeffizienten der natürlichen Strahlung. *Ann. Phys.* 21: 668–680

Laue, Max von. 1915d. Die Wanderung von Unstetigkeiten in elektrolytischen Lösungen. *Zeitschrift für anorganische und allgemeine Chemie* 93: 329–341

Laue, Max von. 1917a. Krystallforschung mit Röntgenstrahlen. *Berichte der Deutschen Chemischen Gesellschaft* 50: 8–20

Laue, Max von. 1917b. Die Nordströmsche Gravitationsstheorie. *Jb. Radioakt. Elektronik* 14: 263–313

Laue, Max von. 1917c. Ein Versagen der klassischen Optik. *Verh. DPG* 19: 19 ff

Laue, Max von. 1918a. Glühelektronen. *Jb. Radioakt. Elektronik* 15: 205–256

Laue, Max von. 1918b. Die Entropiekonstante der Glühelektronen. *Jb. Radioakt. u. Elektronik* 15: 257–270

Laue, Max von. 1918c. Röntgenstrahlinterferenz und Mischkristalle. *Ann. Phys.* 15: 497–506

Laue, Max von. 1919. Über die Wirkungsweise der Verstärkerröhren. *Ann. Phys.* 13: 465–492

Laue, Max von, and Paul Sophus Epstein. 1915. Wellenoptik. In *Encyklopädie der mathematischen Wissenschaften mit Einschluß ihrer Anwendungen*, vol. 5, part 3: *Physik*, 359–487; Epstein, P.S.: Spezielle Beugungsprobleme, 488–526. Leipzig: Teubner

Lemmerich, Jost, Ed. 2011. *Mein lieber Sohn! Die Briefe Max von Laues an seinen Sohn Theodor in den Vereinigten Staaten von Amerika 1937–1946*. With 2 contributions by Christian Matthaei. Berliner Beiträge zur Geschichte der Naturwissenschaften und der Technik no. 33. Berlin, Liebenwalde: ERS-Verlag

Remane, Horst. 2005. Der Chemiker und Nobelpreisträger Emil Fischer und der "Krieg der Geister." *Acta Historica Leopoldina* 45: 399–412

Schäfer, Clemens. 1916. Book review on: *Encyklopädie der mathematischen Wissenschaften. Phys. Z.* 17: 215

Szöllösi-Janze, Margit. 1998. *Fritz Haber 1886–1934. Eine Biographie*. Munich: C.H. Beck

Vegard, Lars, and Harald Schjelderup. 1917. The constitution of mixed crystals. *Phys. Z.* 18: 93–96

Wachsmuth, Richard. 2005. Die Errichtung der Akademie und ihre Entwicklung zur Universität. In *Stern-Stunden. Höhepunkte Frankfurter Physik*. Ed. Wolfgang Trageser, 19 f. Frankfurt: Johann-Wolfgang-Goethe-Universität, Fachbereich Physik

Wien, Wilhelm. 1914. *Ziele und Methoden der Theoretischen Physik. Festrede zur Feier des 332-jährigen Bestehens der Julius-Maximilian-Universität in Würzburg am 11. März 1914*. Würzburg: Stürtz

Chapter 8
Berlin—General and Special Theories of Relativity

Scientists and, in particular, physicists in defeated Germany perceived little of the fundamental changes to political life in Germany's first democracy. Although the "war of intellects" had severed contacts with physicists in England, France, and the USA, the scientific debate resumed in Germany in 1919, at times more ironically and bluntly, but nonetheless along the usual academic channels. Laue returned to Berlin into an academic environment he knew well. That was to change very soon. Over the next 25 years hateful racist attacks were repeatedly made on scientific theories and experimental findings. So-called philosophers and prophets like Oswald Spengler with his book 'The Decline of the West' and Rudolf Steiner with his ideas of religious renewal added their bit to this atmosphere. Max Planck at the Prussian Academy of Sciences, and most notably and unflinchingly also Max von Laue, took on the task of opposing this trend, setting the record straight, and defending proven science.[1]

Albert Einstein had been in Berlin since 1914 as successor to the deceased member of the Royal Prussian Academy, Jacobus Henricus van't Hoff in April. Einstein received a personal salary and was allowed to offer lectures at the university if he wished. He had first arrived in the city on his own. When his wife joined him from Switzerland with their two sons, the marriage rapidly deteriorated and Mileva left again for Zurich taking the children with her. Einstein initially lived as a bachelor and later moved in with his cousin Elsa Einstein. Scientifically, he continued to work intensely on the general theory of relativity (Fig. 8.1). It was important to find confirmation of the redshift in the spectrum and also the deflection of light by the mass of the Sun. But, for other remote reasons, earlier measurements taken were too imprecise. Einstein was able to interest the astronomer Erwin Freundlich in his problems. A solar eclipse expedition to the Crimea was organized to take advantage of the favorable observing conditions there in August 1914. The outbreak of the World War aborted this enterprise.

---

[1] For instance, festivities on Leibniz Day in 1922 at the Preußische Akademie der Wissenschaften and Planck's speech on that occasion; cf. Heilbron (1986): 122 f.

© Basilisken-Presse, Natur+Text GmbH 2022
J. Lemmerich, *Max von Laue*, Springer Biographies,
https://doi.org/10.1007/978-3-030-94699-9_8

Fig. 8.1 Albert Einstein c. 1920. *Source* Photograph in the author's estate

In Berlin, Einstein soon befriended the theoretical physicist Max Born and his wife Hedwig. Born had been appointed to help reduce Planck's workload, but no joint research developed out of this arrangement.

While the administrative issues for the chair swap between von Laue and Born were still being worked out, Einstein wrote a letter on 8 February 1918 to Mrs. Born, as Max Born himself was away, to tell them his thoughts about it.

> Laue wants to come here. Some time ago, when he had the prospect of obtaining by private endowment a sort of research post here without teaching duties, he justified his strivings to come to Berlin by his dislike of teaching. Now that this plan is not going to materialize, it seems, he is thinking of exchanging jobs with your husband. Primary wish, therefore: 'Onwards to Berlin'; motivation: (his wife's?) ambition. Planck knows about it; the Ministry, hardly likely. I haven't spoken to Planck about it yet. I imagine his ambition is, to become his successor. Poor man! Nervous vacillation. Striving for a goal that is hostile to his natural need for a peaceful life without complicated human relations. Please read Andersen's pretty little fairy tale about the snails. The objective possibility of Laue's plan coming about depends on two conditions:

1.   Sufficient endowment of your post for Laue,

2.   Your husband's inclination to exchange jobs.

Let us suppose that 1. is fulfilled, then the question arises as to whether you should consent; that, of course, is the question which is tormenting you already now. My opinion is: Definitely do accept.[2]

Einstein then went on to discuss the opportunities available to Born at Frankfurt to develop his own potential (Born 1969: 22 f.). Planck was able to reassure Laue about his appointment. On his 60th birthday, his Berlin colleagues at the German Physical Society paid tribute to Max Planck, making a number of addresses. The speeches by Max von Laue, Arnold Sommerfeld, Emil Warburg, and Albert Einstein later appeared in print (Warburg and Planck 1918, 1958). Warburg presented the volume to Planck, who later thanked Laue on 19 July 1918 for the "dear memories," telling him: "The matter of your transfer to Berlin is in good progress. The minister has inquired of our faculty, as he has already done of the Frankfurt faculty, on the position it takes in this matter, and I do not doubt that it will repeat the wish it has often expressed in the past once again (AMPG, V rep. 11 no. 889)". Planck also mentioned Born's achievements in Berlin (ibid).

No sooner had Laue arrived in Berlin, in advance of his family, than he began— perhaps partly at the instigation of Planck—to lobby for the conferral of the title of professor to Lise Meitner. He submitted an application to the Ministry of Science, Art, and Culture on March 3rd, explaining the importance of her scientific researches. Meitner was awarded the title in July 1919.[3] Laue then also wrote the faculty report for Meitner's habilitation in 1922 at Berlin University.

A scientific event distracted attention away from everyday worries: The English solar eclipse expedition in Funchal managed to take several good images of starlight. By autumn, the evaluations were almost complete. They showed the postulated deflection due to the mass of the Sun and the redshift calculated by Einstein! This was an important confirmation of Einstein's considerations.

It was the time of the signing of the peace treaty at Versailles on 28 June 1919, imposing most punishing conditions, such as the cession of territory, reparation payments, usufructuary rights to property of the German state, and the surrender of rolling stock of the railways and all kinds of industrial goods. Once again many Germans felt that their national pride had been hurt, and once again people began looking for culprits to blame for this catastrophe. Laue wrote to Wilhelm Wien from Charlottenberg in Berlin on 21 June 1919 that he was in favor of signing the surrender: "Otherwise the Entente armies will devastate the whole country, destroy the crops, and dismember the Empire into individual states (DMA, NL 056)". Laue was also of the opinion that even with that signing, the situation was rotten. It all looked so

---

[2] Born (1969): 22 f. (our translation). Einstein wrote to the Ministry of Culture on 25 Nov. 1918, to inform Minister Heinrich Becker somewhat ambiguously that Born had told him about the exchange. He had said that Laue's appointment to Berlin had brought "a certain uncertainty about my employment here." He had then elaborated that he had received a call to Zurich. GSta PK, Rep. 76 Vc, sec. 2.

[3] GSta PK, Ministerium für Wissenschaft, Kunst und Volksbildung, U I, K 7453, fol. 326.

bleak. A few weeks later, in a letter from July 18th, he told Wilhelm Wien that he was not yet teaching any courses and had not found an apartment either. It was doubtful whether his wife would be able to move (ibid.). In a letter written several days later, on July 31st, he reported about a possible appointment to the polytechnic in Zurich, which delighted him: "The situation here is such that my present salary doesn't suffice in the long run, and one can hardly rely permanently on interest income (ibid.)". His wife stayed in Würzburg and Laue frequently took the train from Berlin to see her. In Berlin he probably agreed to join a vigilante group to maintain public order. There were repeated armed clashes between radical political groups in the city after the capitulation, especially in the so-called newspaper district. On 31 August 1919 he wrote to Wilhelm Wien: "Tomorrow afternoon I have my first watch at Red City Hall. I have already taken part in the machine-gun course on Friday. (Ibid The city hall in Berlin was known as *Rotes Rathaus* after the color of its brickwork.)" He was apparently spared a mission. Despite all this mental strain, Laue continued his scientific research undiminished. The basis of the special theory of relativity, the constancy of the speed of light in moving systems, was still not sufficiently experimentally secured for some physicists, fourteen years after its publication by Einstein. At the end of 1919 Laue submitted a longer paper to the editors of the *Annalen der Physik* bearing the title 'On the experiment by Harress' (Laue 1920a). It was another one of his corrections. When around 1909 Laue had published his idea of an experiment with revolving mirrors, Georges Sagnac had published a similar arrangement. Franz Harress had also addressed the problem in his dissertation with an improved apparatus that was later built by Carl Zeiss in Jena. The translatory motion was converted into rotation on glass bodies.

Harress was killed in action during the World War before he could complete his dissertation. As Otto Knopf (1920) reported, he felt obliged to finish Harress's work. Laue noted briefly a number of self-evident facts about the physics underlying the experiments by Harress and Knopf, but when it came to fundamental findings in the work, he was scrupulously precise. Paul Harzer had already made some initial corrections to Haress's publication. His conclusion had been that curvature did occur in the light beam during the rotation, but it would be too small to be observable. Laue deemed clarification of Knopf's results important enough to ask Einstein for his opinion in two lengthy letters from Würzburg explaining the details on October 18th and 27th (UAF, Laue papers, 9.1; Einstein 1998: 207, 219). He singled out a change in the number of oscillations of the light upon reflection as well as the interference of the beams at the point of convergence. The last section, § 8, began with the question: "What can Harress's experiment teach us, once it has been carried out to full attainable perfection?" Laue urged,

> The optics of bodies moving without acceleration is such a homogeneous and densely woven theory that it can only be confirmed or refuted as a whole. Any confirmation of any one of its statements always applies to the whole. The premise of the theory so far is that the accelerations associated with rotation do not affect the speed of light in any perceptible way.

The general theory of relativity supports this assumption. Whether it is correct can only be shown by a repetition of this experiment, which would be very welcome.[4]

Max von Laue and Max Wien later wrote to the Helmholtz Society along these lines on 28 July 1921 from Berlin about the experiments by Harress on the propagation of light in a moving glass body, the translational motion in Fizeau's well-known interference experiment. Similar experiments had already been performed by Pieter Zeeman and Auda Snethlage on quartz and glass. In order to strengthen the experimental basis of the theory of relativity, they proposed, further experiments should be done in this direction; the honor of German physics was at stake (DMA, NL 056/6090).

Happy family events banished the day-to-day worries during this time of upheaval. On 1 October 1919 Laue was able to inform Wilhelm Wien: "We baptized our daughter Hildegard on September 19th." (DMA, NL 056) But peace inside Germany was not stable. On 13 March 1920 a group affiliated with the Pan-German (*Alldeutscher*) Wolfgang Kapp attempted to seize power with a brigade under the command of General Walther von Lüttwitz. The government fled, but the passive resistance of the population and the authorities caused the coup to collapse. Peace had not settled the "war of intellects." The ties with fellow British and French scientists that the World War and the 'Appeal to the civilized world' by German professors had severed were not yet mended. So Niels Bohr's visit to Berlin from Copenhagen in April 1920 signified a very welcome reopening of the scientific dialogue. Bohr held a lecture at the university on the extension of his atomic model. His soft voice and heavy Danish accent made it difficult for the younger members of the audience to understand it, purely from a linguistic point of view; they were, of course, seated on the back benches. So they invited him to hold a separate colloquium free of "*Bonzen*" for them, which he gladly did. Laue already numbered among those shunned "big-wig" faculty (Fig. 8.2).

Max von Laue was to be admitted to the Prussian Academy in order to draw him into the closer circle of Berlin physicists. According to the statutes, a vacancy in its membership had to exist. On 11 May 1920, at the instigation of Max Planck, the members of the Physical–Mathematical Class: Albert Einstein, Theodor Liebscher, Walther Nernst, Max Planck, Heinrich Rubens, Erhard Schmidt, and Emil Warburg, submitted such a nomination along with a laudation by Planck (BBAW, PAW 1812–1945, II-II-38, fol. 12). As a recognized scientist and Nobel laureate, Laue's unanimous election at the meeting of the Academy on June 24th was rather more a formality. The Ministry of Culture still had to approve and did so on August 14th. Laue, henceforth receiving a token annual salary of 300 marks (by comparison, Einstein received 36,000 marks), expressed his gratitude on September 2nd. He was the first Nobel laureate in this section and the youngest among its members.

The regular speech held on Leibniz Day, celebrated every year at the Academy, was conventionally delivered by a recent electee as his inaugural address, the content of which ranged from superficiality to profundity, depending on the speaker's character.

---

[4] Laue held a talk at the meeting of the Deutsche Physikalische Gesellschaft on 23 Jan. 1920 under the title 'About recent experiments on the optics of moving bodies' (Laue 1920b).

Fig. 8.2 The participants at the "bonzenfreies Colloquium" in Berlin 1920. *From left to right*: Otto Stern, Wilhelm Lenz, James Franck, Rudolf Ladenburg, Paul Knipping, Niels Bohr (laughing), Ernst Wagner, Otto von Bayer, Otto Hahn (looking to the left), Lise Meitner, Georg von Hevesy, Wilhelm Westphal, Hans Geiger, Gustav Hertz smoking a pipe, Peter Pringsheim *Source* Archive of the Max Planck Society, Berlin-Dahlem

Then came a rejoinder by the secretary of the class, in this case by Max Planck. What research, which reflections did Laue consider important enough to deserve mention on this occasion? Without a trace of vanity, he reported:

> My scientific activity began with researches on optics—namely, on its connection with the principles of statistics and thermodynamics. I used Planck's definition of 'natural radiation' to investigate how such radiation propagates in a dispersive body. This with Planck's active encouragement was followed by work here in Berlin on the thermodynamics of interference phenomena. It involved the problem of whether it is permissible to calculate the entropy in a system of several light rays without regard for the issue of their coherence, as had hitherto probably always been tacitly assumed. A consideration of interference phenomena, however, shows that this approach leads to a conflict with the second law; and Boltzmann's reduction of entropy to probability does indeed demand a special stance for coherent rays (ibid and Laue 1921a: 479).

He then discussed Einstein's theory of relativity: "this new and powerful chain of reasoning [soon] enthralled me. The question of whether Fizeau's interference experiment with flowing water fitted into it was the subject of my first small study on it. Further individual applications of the new principle followed, by means of which I gradually familiarized myself with this field (p. 479)." Laue mentioned his own misgivings about the theory and hinted at the current criticism of relativity. He had written his book about the theory of relativity with great enthusiasm, he said, and the second volume on the general theory would be appearing in the coming weeks.

My move to Munich in 1909 brought me to a university where, thanks to Röntgen's and von Groth's work, there was much talk about X-rays, on the one hand, and about the space lattice structure of crystals, on the other. In the case of X-rays, it was still up in the air at that time as to whether they represented electromagnetic waves of short wavelength, as Stokes and Wiechert, W. Wien and Sommerfeld wanted, or corpuscular radiation, as Bragg and others thought (p. 480).

There followed a passage about interference phenomena and space lattices of crystals. "But the spatial lattices that Nature presents us ready-made in crystals make it feasible to imitate the well-known grating spectra from light also for X-rays (ibid)." But no details were offered about the discovery of X-ray interference, no names mentioned of those involved, no date. He only briefly alluded to his service during the war and his research on electron tubes.

Seven years ago, on the appropriate occasion, Einstein spoke here about the two tasks of theoretical physics; namely, first, to find out from observations the fundamental laws governing them, and then, where such laws are known and have proved themselves, to draw from them conclusions which complete the knowledge of the field in question and, if possible, lead over to new experiments and observations. I have not been able to contribute to the solution of the first, higher task. If I have succeeded, it has always been because I dared to draw quite far-reaching conclusions from existing principles and to apply them to things, for the interpretation of which they were not initially established. I took the courage to do this once from the deeply felt need to expand and complete the physical world view in the sense of its unity, and from the joy of being able to control Nature by means of thought (ibid and Laue 1921a: 480–481).

Planck's rejoinder followed directly on the heels of Laue's speech and discussed his method of working and its accomplishments:

You spoke of the two tasks set before the theoretical physicist and to which he devotes himself, to the degree that his school of thought guides him. Now, if the art of coaxing out the fundamental laws from certain experiences that cannot be reconciled with the usual views comes from that divine vision which knows how to distill the essential from the inessential, the necessary from the accidental, the real from the conventional; doesn't the first great scientific throw which you succeeded in making belong among those accomplishments of a theoretical physicist which you are probably justified in calling higher?—The discovery of the limits to the validity of the additive theorem of entropy, which had hitherto always been unquestioningly applied, and the completion of the fusion of the concepts of entropy and probability arising therefrom? (Planck 1921: 481).

Planck mentioned Laue's determined courage to stand by a conviction once gained. Concerning the discovery of X-ray interference, he thought that it had been a call of Laue's scientific conscience to search for resolution not only of the question of the nature of X-rays but also of crystalline structure. The conclusion contained an indirect request: "Thus, to our delight, you have entered this circle as the youngest scion in the family of academic physicists, who are united in close alliance by a regular, trusting exchange of ideas. Of course, to be quite frank, I must not conceal that even now I still hesitate to give myself completely over to the reassuring feeling of secure possession, since the fear seems not to be excluded that sooner or later your agile mind might once again be drawn and lured away by foreign peals (p. 482)."

Max von Laue took his duties as member of the academy very seriously and frequently gave talks about the latest results of his own and other people's research. He missed very few meetings. Other obligations also had to be met in addition to his duties at the academy. In 1845 a small group of young scientists in Berlin had founded a "Physical Society." Initially it operated only locally, but gradually physicists from all parts of the Reich joined as well as foreigners. It did not pursue any particular political or educational policy. The predominance of Berlin members soon led to the formation of separate societies in other university towns, regional associations called *Gau-Vereine*, which held their own local events. Laue had become a member already during his first stay in Berlin through Planck. He resumed his active partic- ipation at its meetings currently known as the *Deutsche Physikalische Gesellschaft* (abbreviated as DPG). At the end of 1921 the Berlin branch elected as members of its board: "Chairman: Laue, deputy chairman: Goldstein, secretary: Scheel, treasurer: Pirani, panel members: Berliner, Boas, Gehlhoff, Henning, Kohlschütter, Lamla, Nernst, Planck, and Rubens." From 1918 to 1920 Arnold Sommerfeld was elected chairman of the umbrella association of German physical societies, succeeded later by Wilhelm Wien and Franz Himstedt. In 1919, the society for technicians in the field was founded, the *Deutsche Gesellschaft für Technische Physik* (DPGA, 21,216).

Just a few weeks after his election as member of the Berlin Academy, Laue and his wife left together to take part in the Nobel award ceremony in Stockholm at the beginning of June 1920. During the World War, some of the awards of the prize had been temporarily postponed, and all the ceremonies cancelled. Now that the war had ended, it was time to catch up. That summer of 1920 Laue attended the awards together with the other German Nobel laureates: Max Planck, Fritz Haber, Johannes Stark, and Richard Willstätter (Figs. 8.3–8.7).

Max von Laue's acceptance speech was a condensed account of the discovery.

> When other Nobel laureates express their gratitude for this high distinction by telling the story of their discovery, they can relate how they began with a lofty goal that seemed attainable, how they strove for it along various paths that at first mostly proved to be erroneous, and how they finally reached their goal after many years of exhausting labor. And in my eyes their merit grows with the difficulties they overcame. What I have to tell you here deviates somewhat from this. Certainly, I did know about the problem of producing X-ray interference for a long time before it could be solved. But I never believed that I could contribute anything to it myself, and therefore I did not make any further efforts until I suddenly saw the path which then immediately proved to be the shortest to the goal. Therefore, as there is but little I can tell you about my own preliminary work, I shall rather describe to you the situation in scientific and personal terms in which the idea came to me (Laue 1920c).[5]

In introduction Laue mentioned the researches by Wilhelm Conrad Röntgen on interference or diffraction phenomena and the assumption of wave radiation by George Stokes and Emil Wiechert. He also reported that Wilhelm Wien had made an estimation of the wavelength using the notion of quanta. The experiments to observe

---

[5] The reedition of this paper published by the Müllersche Hofbuchhandlung in 1920 differs from the Nobel lecture: Laue (1920c).

diffraction phenomena were also mentioned as well as the ideas by William Henry Bragg about the corpuscular character of X-rays. This was followed by a biographical passage about the influence that the lectures by Voigt, Planck, and Lummer had on his interest in wave theory. "It was a great stroke of luck that Sommerfeld assigned me the article 'Wave optics' for the Encyclopaedia of Mathematical Sciences. Because then I had to look for a mathematical description of the theory of gratings, which—without being exactly new—could be transferred quite easily to crossed lattices. At that time I already thought about how to transfer them onto space lattices; but I did not pursue the matter, because such lattices don't figure at all in optics proper (p. 4; Laue 1913a)." In the next section he told about his early days in Munich, about Röntgen and Sommerfeld and the crystallographers. An account of the discovery history followed:

> This was how things stood when one evening in February 1912, P.P. Ewald came to see me. He was working at Sommerfeld's instigation on the mathematical analysis of the way long electromagnetic waves behave in a space lattice, and later published a theory of crystal optics based on it as his dissertation. At that time, though, he was still facing difficulties and asked me for advice. I was not able to help him either, of course. But during the conversation a related idea occurred to me prompting me to ask about the behavior of waves that are short against the lattice constants of the space lattice. And my optical sense immediately told me: Then lattice spectra must occur (Laue 1920c: 5).

As he then recollected, he told Ewald that, given crystals and X-rays, he would expect interference phenomena to occur. He mentioned Friedrich's willingness to do the corresponding experiments and alluded to doubts among the "acknowledged masters of our science"—Sommerfeld—about getting anywhere with it. Laue then

Fig. 8.3 Max von Laue (1879–1960), Nobel prize in physics 1914. *Source* Archive of the Max Planck Society, Berlin-Dahlem

Fig. 8.4 Richard Willstätter
(1872–1942), Nobel prize in
chemistry 1915. *Source*
Eidgenössische Technische
Hochschule, Zürich,
University Library

Fig. 8.5 Max Planck
(1858–1947), Nobel prize in
physics 1918. *Source*
Archive of the Max Planck
Society, Berlin-Dahlem

Fig. 8.6 Fritz Haber (1868–1934), Nobel prize in chemistry 1918. *Source* Archive of the Max Planck Society, Berlin-Dahlem

Fig. 8.7 Johannes Stark (1874–1957), Nobel prize in physics 1919 *Source* Archive of the Max Planck Society, Berlin-Dahlem

gave a detailed report about the successful recordings and the response to the initial results as well as about the various papers by colleagues at home and abroad (pp. 7 f.). Back in Zehlendorf on the outskirts of Berlin, he wrote enthusiastically to Wilhelm Wien about the journey and enclosed the Baedecker guidebook on Sweden that Wien had actually lent to Max Planck, "but which we, in fact, used. We experienced the trip as a fairy tale, and it will be difficult for us to find our way back to reality. First the departure with our 4 German colleagues, then a week of uninterrupted and such fine festivities that one could not get tired of them, finally another week in which we enjoyed the scenic beauty of Sweden by ourselves, namely, on the east coast—there is much for our memories to feed on. Our return trip from Stockholm to Malmö was by ship, which I'd recommend to anyone to do likewise."[6]

Most impatiently Laue was still waiting for confirmation of his appointment to the Friedrich Wilhelm University in Berlin. In a letter to Einstein from 27 March 1920 he candidly revealed his wishes, at the beginning hinting at the price rises from inflation, which was slowly gathering momentum:

> I abducted a box of matches of yours yesterday; which is as wicked now as if some one had stolen a silver spoon before. That's why I'm returning the box to you as full as it was.
>
> Since the signs that an appointment for me is imminent are multiplying—whither I have not yet been able to establish for certain—I want to tell you right away what I'm going to be demanding of the local case; for, you're going to be asked for an opinion in any case and can thus better prepare yourself for the shock you'll receive.
>
> I want to try to get rid of the lecture obligations. Reason: I lack the nervous stamina to hold lectures to the same extent as before and to conduct scientific research at the same time. If our honorable fellow men want me to achieve something in science again—and that would be quite reasonable, especially since nothing really useful springs from my lectures—they will have to grant me the request, for better or for worse. If we had the economic conditions of the prewar period, the matter would be quite simple. Then I could appoint myself as private scholar, which has always been my ideal. Now, unfortunately, that's not possible anymore, and so I must try to create a similar liberty for myself by negotiation, although then one can easily get into a position in which one feels indebted to one's fellows. But, as things stand, I must dare to take that risk (UAF, Laue papers, 9.1).

Laue's wish was not fulfilled despite all his endeavors: He had to continue lecturing. But on April 1st the family was able to move into a single-family home with a garden at Albertinenstraße no. 17 in Berlin-Zehlendorf (Fig. 8.8).

The friendship that soon developed between Lise Meitner and the Laue family became very important to her. It was not just the ability to talk about everyday worries with them; she also very much enjoyed the excursions with the couple into the beautiful environs of Berlin. With Max von Laue she could discuss the problems that academia regularly posed.

On 3 August 1920, Laue wrote to Sommerfeld from Berlin:

> Your letter of July 31st pleased me greatly; for it strikes a note that I can trust. Let me immediately add that I would not have this feeling if it didn't also contain a few reproaches against me. […] When I wrote that letter to W. Wien in January (or was it February?), I was under the influence of certain very unpleasant experiences here, making my judgment of my fellow men not exactly fair.
>
> Besides, my Nobel lecture may tell you how I assess the influence of the intellectual atmosphere at Munich on the discovery of X-ray interference; it could well be appearing any day now. If I did sometimes think back on the Munich period with a certain bitterness, the reasons for that did not lie within the range of scientific stimulation; I could scarcely have appreciated that so richly anywhere else.

---

[6] Laue to Wien, Berlin, 17 Feb. 1920. DMA, NL 056. The date is obviously incorrect.

Fig. 8.8 The Laue family home at Albertinenstraße no. 17 in Berlin-Zehlendorf. *Source* Archive of the Max Planck Society, Berlin-Dahlem

I certainly have had some unpleasant personal experiences there. I just want to address one point, which isn't the worst among them. Why did you exclude me when you celebrated the discovery of X-ray interference with Friedrich and Knipping and our other younger physicists?

Well, you can, of course, rightly point out that I haven't always behaved correctly towards you, particularly shortly after my move to Munich. But you did know the state of mind in which I joined you. If you had granted me 'extenuating circumstances,' you could at any rate have improved our personal relations very considerably.

But let us let bygones be bygones; let's simply say, 'forget it!' It has always pained me deeply not to be on the best of terms with a professional colleague whose achievements I must rate so highly. It will be a great relief to me if that changes now. And that's why I am most especially looking forward to our meeting in Nauheim (DMA, NL 089).

This allusion was to the forthcoming convention of German Natural Scientists and Physicians in Bad Nauheim from 19 to 25 September 1920. A public politically racist attack on relativity theory happened even before that convention took place. Life in Berlin during the Weimar Republic was unfortunately not nearly as pleasant as the Stockholm interlude (Fig. 8.9). There were few joint efforts by the political parties to bear the burdens of the peace treaty in a meaningful way. Max Planck soon became a member of the *Deutsche Volkspartei* (DVP) and Max von Laue also joined its patriotic ranks. Both, however, remained inactive in party politics. Slogans from the right began to be heard, such as "The Jews are our bane" or "The Jews are to blame for everything"; and misinterpretations of the special theory of relativity and, more so of the general theory, became a pretext to agitate against Einstein. The many supportive and explanatory publications appeared in vain against this sentiment. An

Fig. 8.9 Otto Hahn (1879–1968) and Max von Laue, in the early 1920s. Hahn, the "father of nuclear chemistry," was awarded the Nobel prize in chemistry for 1944. *Source* Archive of the Max Planck Society, Berlin-Dahlem

antirelativity association was formed in Berlin, with Paul Weyland, who was not a physicist, and Ernst Gehrcke as its leaders. They organized a "major rally" on 24 August 1920 by their 'Syndicate of German Scientists for the Preservation of Pure Science' in Philharmonic Hall in Berlin.[7] There wasn't much harmony to speak of at that event, only impertinent personal attacks on Einstein, who was present. Max

---

[7] Boehlich (1965): 7 f. For further literature on Weyland, see Kleinert (1993); Grundmann (1967): 1623 f.; Hermann (1979): 54; Grundmann (2005): 108–109. The first slogan: "Die Juden sind unser Unglück" comes from the historian Prof. Heinrich von Treitschke (1834–1896).

von Laue, Walther Nernst, and Heinrich Rubens were also in the audience. Gehrcke was one of the speakers. Laue immediately wrote Sommerfeld the next day from Zehlendorf to tell him what had happened:

> The article and yesterday's speech by Mr. Weyland were on the same level: Einstein as a plagiarist, any follower of relativity theory as a propagandist, the theory itself as Dadaism (this word really did occur!). And most unbelievably, people with a scientific reputation like Lenard and Wolf in Heidelberg let themselves be seen in such company as speakers. Yesterday Gehrcke spoke after Weyland; although he rehashed the same old story, his calm, matter-of-fact manner of speaking was a respite after Weyland, who can measure up to the most unscrupulous demagogues. It is scandalous that such a thing can happen.
>
> Today Nernst, Rubens, and I sent out a short but energetic statement to all the major Berlin newspapers against these activities. Whether all the papers will print it remains to be seen; it is not to be taken for granted in view of the embroilment in anti-Semitic policy, which already manifested itself in the distribution of inflammatory leaflets in the foyer of the hall. Be that as it may, more must be done (DMA, NL 089).

Laue asked Sommerfeld, to bring about a counter-resolution as chairman of the German Physical Society in Nauheim. "It is and remains incomprehensible, though, how demoralizing a revolution is even among our circles (ibid)." The editors of the paper *Berliner Tageblatt* received several letters from physicists expressing their emphatic opposition to the way Weyland had presented himself on August 24th. The editors reached the decision: "We are publishing just one of these letters, but one which by virtue of the notable signatures it bears weighs as heavily as many (Bochlich 1976: 7 f.)." Its coauthors, Max von Laue, Walther Nernst, and Heinrich Rubens pointed out Einstein's significant contributions to physics, but also his aversion to publicity. They concluded with the statement: "It seems to us a matter of justice to lend prompt expression to this conviction, the more so as no opportunity to do so was afforded us yesterday evening (ibid)."

A few days later, on 27 August 1920 Laue's next letter to Sommerfeld left Zehlendorf: "If anything else is likely to excite your zeal, it is certainly the news that Einstein and his wife seem to be determined to leave Berlin and Germany in general at the next possible opportunity because of those hostilities. Then, on top of all the other misfortunes, we shall also live to see would-be nationalistic circles expelling a man who is the pride of Germany as very few others (DMA, NL 089)." Laue mentioned that his wife, wanting to deliver something to Einstein but not sure if it was the right building had asked a gentleman entering with her if Einstein lived there. The reply had been: "He still does, unfortunately." Laue then asked Sommerfeld to contact Planck, who was at Grundner Hof near Gmund. "One really ought to stand firm for once in Nauheim against this conflation of anti-Semitic agitation and scientific animosity. Where will this end otherwise? (Ibid)."

The protesters at the rally had also attacked the priority of Einstein's ideas about Mercury's perihelion motion. The problem and the solution allegedly had already been known before Einstein's publication. Laue felt called upon to come to his friend's defense. In a paper entitled 'Historical critique on the perihelion motion of Mercury' he referred to Paul Gerber's work from 1898 and other subsequent articles in which the formula and magnitude of the perihelion motion had already been published (on Gerber see Hentschel 1996: 3). They agreed with the magnitude that Einstein had derived from general relativity even though he was unaware of Gerber's paper at the time. Laue then interpreted the results: "As easy as it was [for Gerber] to make this discovery, it could have been a significant one if Gerber had managed to derive his proposition from rational physical conceptions in a mathematically sound way. Then his explanation would have had the merit of relating the perihelion motion to certain other facts—no other kinds of explanation exist in physics at all (Laue 1920d: 736)."

The German Physical Society informed its members by letter about their upcoming convention in Bad Nauheim that autumn. A participant ticket cost 100 marks, the ladies' ticket 50 marks. An exhibition on industrial science and technology was to be shown on the Exhibition Hall grounds. Professors Ambronn, Siedentopf, and Zsigmondy additionally offered a demonstration course on scientific microscopy in cooperation with the optical firm Carl Zeiss and the Institute of Pathology at the university. The journal *Physikalische Zeitschrift* reproduced its program, a broad spectrum of talks held at this Nauheim convention from 19 to 25 September 1920 for the field of physics. There were presentations on atomism, spectra, crystal physics, and radioactivity, with a contribution by Otto Hahn on joint experiments with Lise Meitner. Various meetings of the society were scheduled for September 17th.

> A joint meeting with the Mathematics Section is planned, in which questions of relativity will be mainly treated. The Physics Section will hold its meetings together with the German Physical Society; the latter will also present for detailed discussion at an administrative meeting its internal issues, especially the reorganization of its periodical publications.
>
> The membership fee for those attending the meeting is 40 marks, and accommodations are provided free of charge for all attendees. Breakfast, lunch, and dinner will cost a total of 20 M, or 28 M, or 34 M, breakfast in general 3 M. In addition, rooms in city quarters are to be made available with the right to eat in certain inns at midday and in the evening at a discount. Negotiations are underway as to the granting of fare reductions on the railway (DPGA, 21,218).

There were six presentations about relativity, which were also published in the *Physikalische Zeitschrift*: Hermann Weyl: 'Electricity and gravitation'; Gustav Mie: 'The electric field of a charged particle rotating about a center of gravitation'; Max von Laue: 'Theoretical issues on recent optical observations on relativity theory'; Leonard Grebe: 'On the gravitational redshift of Fraunhofer lines'; Hugo Dingler: 'Critical remarks about the foundation of relativity theory'; Franz Paul Liesegang: 'A diagram to illustrate the time–space relations in the special theory of relativity.' Laue's talk discussed the deflection of light by the Sun. He started from the theorem that the deflection of light according to general relativity is represented by the geodesic zero lines of the world. "This theorem has not yet been proved. [...] Of course, it must be possible—as it is in the restricted theory—to prove this theorem for the general theory of relativity from the electrodynamics of empty space. And one can indeed transfer the familiar way to do this completely (Laue 1920e)." His mathematical proof worked. In the discussion that followed, Georg Hamel asked, "Is there an exact proof that redshift really occurs?" Albert Einstein stepped in to respond: "It is a logical weakness of relativity in its present state that it must introduce scales and clocks separately, instead of being able to incorporate them as solutions to differential equations. But as regards the reliability of the consequences with respect to their relation to the empirical basis of the theory, the consequences concerning the behavior of rigid bodies and clocks are the most secure among them. Since the emitting atom should be regarded as a 'clock' in the theory's meaning, the redshift is among the most securely established findings of the theory."[8]

The planned joint meeting with mathematicians about the theory of relativity under the leadership of Planck on 23 September 1920 turned into an emotional quarrel between Philipp Lenard and Albert Einstein. Max Born, Oskar Kraus, Gustav Mie, Melchior Palágyi, and H. Rudolph also intervened. Lenard defended the aether and gravitational fields, and Einstein sought to clarify definitions of the concepts. This debate was reproduced in *Physikalische Zeitschrift* (21: 666 f.), almost entirely filling two printed pages.

In the context of the international boycott of Germany, nothing was mentioned about the attacks on Einstein in the English journal *Nature*, for example, but some publications did appear on the theory of relativity, including a short one by Einstein (1921). Hermann Weyl wrote a detailed commentary on the scientific issues involved in the debate about relativity. On the dispute between Lenard and Einstein he remarked: "The last and most dramatic part, the general discussion on relativity theory, essentially turned into a duel between Einstein and Lenard. Planck performed his duties as chairman with great skill, rigor, and impartiality; it was not least thanks to him that this 'Nauheim relativity dialogue,' in which opposing epistemological views on the fundamentals of science clashed, took a dignified course (Weyl 1922: 51)."

Some philosophers incorporated the theory of relativity into their considerations on space and time. There, too, misunderstandings arose. Laue tried to help by intervening with an article in *Kant-Studien* under the heading 'The Lorentz contraction' (Laue 1921b). In it he also criticized the way Hermann Minkowski expressed himself:

> In the many, very welcome attempts by professional philosophers to assume a stance on Einstein's theory of relativity, one almost always encounters a strange misunderstanding regarding the role of the Lorentz contraction. It is explained by the way in which the theory arrives at this conclusion about this foreshortening, and finds its particular support in the wording of a famous talk by Minkowski; for according to this talk, the Lorentz contraction in the theory of relativity is 'purely a gift from on high.' We would have expected a much stricter assessment of the theory particularly about the philosophical aspects, if not even a downright rejection of it for want of scientificness. For what else is 'a gift from on high' than a new expression for 'a miracle'? (p. 91).

Taking some simple examples Laue tried to explain the physics of the Lorentz contraction, pointing out that physicists do not always follow a causal chain rigorously: "On the contrary, in science the value of the most comprehensive laws of nature lies in that they relieve us of the trouble of pursuing every single detail of the causal series in nature, which are often not very clear (p. 92)." Laue mentioned the

---

[8] DPGA, 21,218. One can take Einstein's remark about the emitting atom as a foreshadowing of the later true "atomic clock." I thank Prof. Dr. Lübbig for this point.

formation of water from hydrogen and oxygen, for instance. The chemical formula says nothing about the alterations in the atoms. As further examples he gave the laws of planetary motion and the law of gravitation. To explain what relativity theory achieved and the Lorentz contraction, he considered two point charges at rest exerting their Coulomb force on each other within a given system. "We set them at a common velocity, invariable as to direction and magnitude, against the same system at an unchanged distance from each other (p. 93)." Although the distance between them stays the same, the force between them changes. This is an alteration in their relations to the frame of reference. In the last section Laue briefly discussed the general theory of relativity. He was still working on extending his book on special relativity (which he also referred to as the "constrained theory") to include the general theory that Einstein had subsequently developed.

At the university in Berlin, the subject of physics was excellently staffed. In the winter semester of 1920/21, the curriculum included, among others, lectures by Heinrich Rubens: Experimental Physics II, Electricity and Optics; by Max Planck: Theory of Heat and Mathematical-Physical Exercises; by Max von Laue: General Relativity and Non-Euclidean Geometry; by Albert Einstein: Various Topics in Theoretical Physics. Besides his load of lecturing and writing for his volume on general relativity, Laue still found time in the fall of 1920 to review a paper by Walter Schottky (1920), who had earned his doctorate under Planck in 1912 with a thesis 'On relativistic theory of energetics and dynamics.' Again, Laue pointed out a correction (involving the necessity to include intrinsic fields for particles) that showed how intensely he was occupied with the mathematical problems of general relativity (Laue 1921c).

Towards the end of 1920, physicists celebrated the 25th anniversary of the discovery of X-rays. *Die Naturwissenschaften* published the talks held during the festivities. Laue chose as his topic 'In what sense can one speak of 'microscoping' the fine structure of crystals by means of X-rays?' He started with a very visual explanation of the processes in a light-optical microscope and showed diffraction images of lattices. As the distances between the grating elements diminish, imaging with light eventually becomes impossible. He then moved on to discuss a space lattice with symmetry properties and mentioned the experiments by Friedrich and Knipping, leaving out the initial difficulties. A mention of the researches by several of his fellow physicists followed. His paper ends with the words: "In this issue we are celebrating the twenty-fifth anniversary of Röntgen's discovery. It is an infinite blessing for the sick and ailing among humanity. But, in my opinion, it signifies even more to pure science. It has opened a new epoch, so to speak, an epoch in which brilliant advances in physics are continuing to accumulate as never before (Laue 1925a: 971)." He gave the quotation: "Earn what you have inherited from your fathers, in order to possess it (ibid)."

Political conditions in some European countries, such as the Soviet Union and Hungary, were often very unfavorable for racial and religious minorities, and students from these countries were lured away by the quality of education at German universities and polytechnics. For example, Dennis Gábor, Leó Szilárd, and Eugene Paul Wigner came to Berlin from Hungary. Although Gábor and Wigner were studying at the local polytechnic, the *Technische Hochschule* (TH) in Berlin, they also attended lectures by Planck and by Einstein on the theory of relativity at the university. Szilárd audited Laue's lecture on General Relativity at Friedrich Wilhelm University in 1921 and also participated in the introductory seminar. Upon asking Laue for a topic for his dissertation, Laue therefore suggested something about relativity. But Szilárd was not comfortable with the issue and found a completely different problem that he was able to solve. He—probably somewhat hesitantly—then asked Laue if he would read his first draft. Laue replied the very next day to agree, as Szilárd had chosen a problem in thermodynamics and had treated it in light of what he had gathered as an auditor of Laue's course (Szilard 1980, 1925).

It was natural for Laue to suggest a topic from relativity for Szilárd, because he had just completed the volume on the 'The General Theory of Relativity' that spring (1921d). His preface to it struck a confident note, far removed from the previous heated debates: "Much admired and much reproached—this is how the general theory of relativity stands now. The most strident voices on either side have in common that they know very little about it. And yet, if there ever was a branch of science acquirable not by forms of speech but by hard work alone, then it is this theory. We dedicate our book to those who want to work their way toward its apprehension (p. V)." There followed a list of books by Albert Einstein, Paul Kirchberger, Hermann Weyl, and Ernst Cassirer on the subject. He was blunt about some fellow physicists and their incomprehension:

> But no strictly scientific book of any completeness has yet been written about the subject by a physicist; and we should like to think that only a physicist can fully perceive and at least attempt to remove the difficulties which at present still keep the great majority of his fellow professionals away from the general theory of relativity. These difficulties lie in the intellectual foundations of the new theory for a group of men, some of them very eminent, who have found in Lenard their spokesman. They confront it with feelings which have been compared, not inappropriately so, to Goethe's aversion to Newtonian optics. Most, however, do not know how to secure a stance on this theory, because for lack of sufficient acquaintance with non-Euclidean geometry and the associated tensor calculus, they are unable to familiarize themselves enough with it (pp. V–VI).

Laue thanked David Hilbert for making his treatise on the foundations of physics available to him and also thanked Ludwig Bieberbach, Georg Hamel, and Emil Hilb: "In Mr. Max Gut (now in Zurich) I found a proofreader who subjected not only the proofs but also the exposition as a whole to his utmost perusal of every detail (p. VIII)." The concluding sentence of the preface read: "This book isn't any promotional text; and yet let's not deny the wish that the present version of the theory, too, might exert something of that persuasive power upon the reader (p. VIII)." The book was a success, for a second edition appeared only 19 months later with a reworking of the tensor calculus and a change of sign. Laue thanked Walter Gordon for his advice on the revision. More recent redshift data were added and Einstein's cosmological ideas were expanded upon. Laue divided the material into eight chapters of various length:

I.   Our empirical knowledge and previous theories about gravity
II.   The general theory of relativity
III.   World vectors and tensors in Riemannian spaces

Unfortunately, there is no written message by Einstein to Laue about his book, since they met regularly at the meetings of the Prussian Academy and were on casual terms. The Göttingen mathematician Felix Klein and Wolfgang Pauli were not satisfied with the exposition; they found it too complicated (Klein to Pauli, 28 Apr. 1921, in Pauli 1993: 29).

Since July 1917, Albert Einstein had taken over the directorship of the Kaiser Wilhelm Institute of Physics in addition to his post at the academy.[9] There was still no institute building, and Einstein could only distribute research funding for work to be performed at other institutes. As a result of the lost war, state funding for research had been cut back in many cases or had ceased altogether. With the founding of the Emergency Association of German Science—*Notgemeinschaft der Deutschen Wissenschaft*—in 1920, a new advisory panel was created under the chairmanship of Friedrich Schmidt-Ott, which acquired scientific equipment on the basis of a request by a researcher and then made it available to the applicant for a limited period of time. Afterwards they were loaned to the next researcher. Max von Laue supported the work of this Emergency Association. He was chairman of the panel for questions concerning the promotion of physics and was regularly reelected until 1933. For a short while Johannes Stark was also a member of this panel, but when an election procedure was introduced, he was not reelected.[10]

In May 1921, the 34th General Assembly of the Association of German Chemists was held in Stuttgart. The *Zeitschrift für angewandte Chemie* published a report on it in its 34th issue dated 27 May 1921 (pp. 209–230). About the external proceedings it judged that the joint meetings of specialist groups bore witness to the fact "that German chemists are not at all willing to lay down their weapons of technical and scientific know-how yet in this economic contest, which the Entente countries are waging up to our complete demise (p. 209)." In the general session, the chairman Friedrich Quincke discussed the historical development of the views on atoms, for which Planck and Laue among all German physicists had provided the best grounds. Max Planck was awarded the Liebig Memorial Medal and Laue the Adolf Baeyer Memorial Medal. In the laudatory address Laue was described as: "the pathfinder in

---

[9] Einstein's separate engagement by the Kaiser Wilhelm Society may have had financial reasons. Einstein wanted to marry for the second time. The question of a widow's pension had to be settled. As a member of the academy in government service, he was entitled to a pension, a percentage of which would be paid to his widow after his death. In the case of a second marriage, however, by law only the first wife, Mileva, was entitled to this benefit. Thus Elsa would not have received a widow's pension. Healthwise, Einstein was not well; he had just survived a serious gastric condition. The employment with the KWG thus secured Elsa a widow's pension. See the indications among the Krüß papers, SB PK, HA.

[10] Wien to Präsidium der Notgemeinschaft, Munich, 21 Jan. 1922. DPGA, 21,218. See also Hammerstein (1999) and Zeitz (2006): 47.

crystal structure who, by ingeniously linking crystalline structure with the wave-like nature of X-rays, opened up new avenues for research into the internal structure of inorganic and organic substances and thereby profoundly influenced the fundamental views of chemistry (p. 209)." Max von Laue then also spoke on behalf of Planck, who could not come, giving a historical outline of the theories of radiation (pp. 209 f.).

Laue's book on the special theory of relativity reappeared in 1921 already in its fourth edition. He had changed much of the text, which was not usual, and the reviewer, Rudolf Seeliger, remarked: "One could perhaps put the situation somewhat crudely thus, that the first edition was written by a mathematician, but this one by a physicist, and that we can now put the book down quite satisfied. [...] All its readers will probably look forward to the second volume in happy suspense, which will hopefully appear quite soon (Seeliger 1921)."

Several universities organized speaking events before the start of the winter semester. Max von Laue was invited by the University of Kiel in 1921. He chose as his topic 'The physical world view.' His introduction clearly showed how strongly political events moved him, but also that he thought beyond political boundaries:

> We are living in a time of political upheaval, and the city in which we are gathered is a particularly vivid reminder of historical events [meaning the sailors' uprising in 1918] which every German who loves his fatherland can only think of in the deepest sadness and shame. May the Kiel Autumn Week take you away somewhat from the less-than-pleasant political present, and this speech in particular intends to lead you out into the field of natural knowledge, in which one can take pure delight in the achievements that have come about through the cooperation of all civilized nations and are still coming about every day. Indeed, we are also experiencing a profound upheaval in our knowledge of inanimate nature—in this sense I ask you today to understand the word 'physics.' Further progress, which always constitutes living science, has since the beginning of this century reached a speed as probably never before; old things which seemed to stand rock-solid have fallen, or at least have been shaken; new and unheard-of things, which have not yet always reached the elucidation of the old, have taken their places (Laue 1921e).

Several times Laue emphasized the importance of the laws of conservation. He appreciated Einstein's bold ideas, which had been necessary to make the nature of gravity understandable. The end of the aether theories was also mentioned as well as radioactivity, the decay of atoms. "And the quantities of energy released in radioactive atomic decay do, in fact, show that a source of energy of incomparable power is situated there. If we could release at one time the energy that one gram of radium gives off over the course of its half-life of about 1800 years—we have no means of doing so, of course—we would have an explosive against which all the known ones must seem like child's play (ibid)." Bohr's atomic model was acknowledged and Planck's quantum of action. In section IV he summarized his statements: "Every science is a system of judgments, and in physics these judgments relate to facts of physical reality. In the system of this science we must understand Truth as a kind of agreement between these judgments and facts, and this agreement can only be reached because from each judgment a fact clearly emerges, and from each fact, a judgment (ibid, sec. IV)." This assessment by Laue is repeatedly evident in his critical analyses of the work of other physicists.

Fig. 8.10 Philipp Lenard (1862–1947), Nobel laureate in physics for 1905, opponent of modern physics and proponent of an "Aryan physics," on a photograph taken around 1932. *Source* Archive of the Max Planck Society, Berlin-Dahlem

Philipp Lenard (Fig. 8.10) continued to try to weaken the priority of Einstein's statements about the deflection of light by adding related headnotes and annotations to a reprinting of a paper by Johann Soldner from 1801: 'On the deflection of a ray of light from its rectilinear motion through the attraction of a body in space which it closely bypasses' (Lenard 1921). Max von Laue's reply was immediate. He started with the statement: "On the main subject we may be brief and, without entering into a discussion about the empirical amount of a light deflection at the Sun, which is at present futile, let us ask the question: Shouldn't some advance beyond Soldner's calculations have already been made merely by virtue of the general theory of relativity having linked this deflection with the electromagnetic wave theory instead of with the emission theory of light? (Laue 1921f: 283)." The inertia of electromagnetic energy, Laue explained, was already known before the theory of relativity. He mentioned Poincaré's, Abraham's, and Hasenöhrl's important contributions to electromagnetic momentum. Regarding Lenard's proposal to name a mass based on energy after a particular author, Laue remarked: "[That] would be entirely super-fluous, because in our opinion, which Mr. Lenard also seems to hold, no other kinds of masses exist at all (ibid)."

In the spring of 1922 Max von Laue received a fine tribute. The journal *Die Naturwissenschaften* presented its issue no. 16 under the heading 'Ten years of Laue diagrams' with contributions on the history of their discovery and their current scientific significance:

Walter Friedrich	The history of the discovery of X-ray interference
Paul Knipping	Ten years of X-ray spectroscopy
Gregor Wentzel	Report about recent results in X-ray spectroscopy
Lise Meitner	On the wavelength of γ-rays
Peter Debye	Laue interferences and atomic structure
Paul Niggli	The importance of the Laue diagram to crystallography
Ernst Schiebold	Contributions to the evaluation of Laue diagrams
Michael Polanyi	Radiographic determination of crystal arrangements

The eight articles printed in German in the special issue of *Naturwissenschaften* under the heading 'Ten years of Laue diagrams' (source: 10th ser., no. 16, on 21 Apr. 1922: 363–416).

Owing to the World War, actually less than ten years of research had been conducted in the field of X-rays. Even so, fundamental progress had been made in many areas that had previously not been experimentally accessible. Few discoveries could boast such a broad array of internationally acclaimed successes.

Laue did not always concentrate on elementary questions. Together with his assistant Walter Gordon, Planck's last doctoral student, he tried in early 1922 to determine theoretically the thermal conductivity of a glowing metal filament in an incandescent lamp when operated under alternating voltage. The influence of some parameters—such as the wire ends—had to be neglected in order to arrive at a result (Laue and Gorden 1922). A year later, Laue, together with Richard Bär and Edgar Meyer, studied the physics of the electric arc in helium (Laue et al. 1923).

The collaboration and discussions among the physicists in Berlin remained an important path to secure results, even if this did not always happen without minor tensions. Einstein occasionally speculated about physical effects—based on the theories of relativity—including on radiation, in the field in which Laue's opinion was authoritative. On vacation in Italy, Einstein wrote to his Swiss friend Michele Besso on 20 October 1921: "Interesting experiment on light emission in progress in Berlin. Light beam bent according to undulation theory." (Speziali 1972: 170) (Fig. 8.11).

The experiment was carried out at the bureau of standards, the PTR, by Hans Geiger and Walther Bothe. Einstein wrote to Max Born about it on 30 December 1921: "The light emission of a moving canal-ray particle is strictly monochromatic, whereas according to the undulation theory the color of the elementary emission

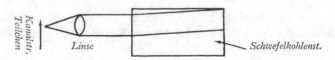

Fig. 8.11 Canal-ray particles are led through a lens into carbon bisulphide. Redrawing of a sketch in Einstein's letter to Besso. *Source* Speziali 1972: 170

should be different in different directions. This certainly proves that the undulation field does not really exist and that Bohr's emission is an instantaneous process in the true sense. It is the most powerful scientific experience I have had for years (Born 1969: 96)." However, Born and Franck had doubts, and to this Einstein replied: "The gist is this: The canal-ray particle, according to the wave theory, emits cont[inuously] variable color in different directions. Such a wave propagates in dispersive media at a velocity that is a function of location. From this should follow a bending of the wave surface, as in terrestrial refraction. But the experimental finding is negative (ibid)." Einstein relayed Laue's opinion to Born on 18 January 1922. It shows the way they acted towards each other: "Laue fiercely combats my experiment, resp. my interpretation of it. He maintains that the undulation theory does not require any ray bending at all. He proposed a pretty experiment to examine the possible undulatory ray bending with capillary waves, which do show strong dispersion, as a substitute for a theory which is so difficult to get a grasp on with the necessary rigor (p. 100)." Einstein later had to concede that he had "pulled a boner" and had been on the wrong track (Einstein 1922: 18), and corrected his statement with thanks to Laue, Geiger, and Bothe.

The unscientific attacks on the theory of relativity were not a singular phenomenon during the 1920s. The anthroposophists developed ideas in the field of natural science that were far removed from established knowledge. They posited that the exact sciences were to blame for the general moral crisis and were seeking to influence society. Rudolf Steiner had founded the Anthroposophical Society in 1913. On the occasion of Steiner's 60th birthday, Hans Wohlbold published the article 'Rudolf Steiner and the natural sciences' in a book edited by Friedrich Rittelmeyer, in which he formulated charges against scientific knowledge: "Modern life and thought are bloated full of scientific findings and impulses. On the surface floats what one likes to celebrate as the great achievements of the time and which through technology have an effect on all areas of practical living. [...] Everything that is great in our physical lives we thank to natural science, and at the same time we are indebted to it for all the mental and physical misery of recent years. Natural science impoverishes man when he strives beyond superficial existence toward an inner life [...] (Wohlbold 1921)." Max von Laue could not accept this and wrote a rebuttal. It shows very clearly his own view of the general importance of the natural sciences and religion, to which he felt attached. "It is right that the influence of technology on the course of world history, which has never been missing, has made itself felt more strongly than ever before since the end of the eighteenth century [...]. But are the natural sciences really solely responsible for the course of history, or even only partly? Are they the source of the driving forces that have determined the destinies of nations? Is it at all conceivable that a science, i.e., purely a system of knowledge, should set purposes for human actions and abstentions in practical life, hence above all, towards other people? (Laue 1922a)." Laue mentioned Wohlbold's term "scientific worldview" and asked again if there was such a thing. On one point he did agree with Wohlbold: "Only once the desire is aroused to use these [scientific] achievements selfishly, once this desire passes the limits of justifiability, once—spurred on by the new possibilities of gratification, hedonism, or hunger for power—the better instincts of the soul are stifled, then the dire consequences which Wohlbold describes not without eloquence are understood. Therefore, it is not a scientific worldview, but a materialistic worldview that has brought the present calamity upon humankind (ibid)." Quoting Wohlbold again: "Natural science has despiritualized the world; that is why Europe is now a heap of ruins and humanity is bleeding itself to death. But as it has dethroned religion, it must put something else in its place, spiritual knowledge according to its own principles (ibid)." Laue attacked this as a half-truth. But he also noted that the childish faith one used to draw one's moral strength from could no longer be retained: "But is one thus allowed to pass the blame onto the natural sciences? Isn't it rather that the churches were not able to separate the genuinely religious content of their teachings from mythical elements which purport to reveal all sorts of things about the course of natural and human history in the world? (Ibid)." A philosophically educated person would have an easier time, Laue thought, because Kantian philosophy helps him to grasp the scope of scientific knowledge: "To be sure, his longing also remains unquaked, and he must renounce—quite as Schiller does in that distich: 'What religion do I profess? None of all those/Which you name. And why none? By religion."[11] In parts II and III of his contribution Laue dealt with Steiner's works, especially his books about "our Atlantean ancestors," the legendary continent Atlantis, and his outline of "secret science," proving him guilty of many errors on physical questions, for Steiner claimed, among other things, that heat is a substance (Laue 1922a: 41—Steiner is not known to have replied).

There were still open questions about relativity theory which Laue felt called upon to offer an opinion, probably out of an inner drive to get a grasp of it. At the meeting of the Physical and Mathematical Class of the Prussian Academy on 20 April 1922, he presented his ideas on 'The significance of the zero cone in the general theory of relativity' (Laue 1922b). This was a follow-up on his publication in the *Physikalische Zeitschrift* from two years before (Laue 1920e). In 1908 Hermann Minkowski had published a graphic representation of the four dimensions—a double cone—which made it easier to understand special relativity. After presenting the mathematical formulation for the special theory of relativity, Laue asked: "Can this formula be

---

[11] ibid This mention of Immanuel Kant shows that Laue takes him as an authority. As I see it, this does not give reason to call Laue a Neo-Kantian, however, because his comments on Kant's views are too casual. The relativity theory was his motivation to analyse the concept of space–time. See also Scheibe (2006). Max von Laue took an active part in church life in Berlin. He regularly attended the Sunday service in Zehlendorf and later even became a church warden. His personal relationship with Protestantism, with faith, emerges from some of his letters to his son. See Lemmerich (2011b): 88.

preserved somehow in the general theory of relativity? Whereby, of course, we cannot expect it to keep the simple form of the integrand. The only essential thing is the reference to the pre-cone. We shall, in fact, be able to answer this question in the affirmative [...] (Laue 1922b: 118)."

The domestic policy of the Weimar Republic did not lead to any stabilization. Unfortunately, the opposite occurred. The assassins of Foreign Minister Walther Rathenau, who shot him in his car outside his home on 24 June 1922, came from among the ranks of the Pan-German association, founded in 1894, with its extreme nationalistic and anti-Semitic attitude. The state declared an official day of mourning.[12] A few days later, other assassination attempts were made on former Reich Chancellor Philipp Scheidemann and the journalist Maximilian Harden. Einstein was scared of becoming a target as well. He was scheduled to speak at the centennial meeting of the Society of German Natural Scientists and Physicians in Leipzig from the17th to the 23rd of September. He wrote to Max Planck on 6 July 1922 from Kiel about his worries (AMPG, V rep. 2 no. 26), who promptly forwarded it with a letter of his own from his home in Grunewald near Berlin to his "Dear Colleague," Max von Laue, two days later:

> The enclosed letter strikes me like a bolt from the blue. So this is how far the scoundrels have gone, wanting to foil an event of German science of historical importance. Einstein will therefore not be delivering the talk on 'The theory of relativity in physics' announced for the first plenary session of the Natural Scientists' Assembly on September 18th, and the importance of this meeting is thus most keenly jeopardized. Moreover, time is extremely pressing; for the printing of the programs must start in the coming days.
>
> In this predicament I know of only one salvation: that you should take over the talk, and I thus, in the name of the Board and in the name of the Society as a whole, address to you the most humble request that we may substitute you as speaker for Einstein on the above-mentioned topic. I am well aware of the sacrifice we are asking you to make. For, you will lose the complete relaxation of vacation after the burden of the terminating semester that you deserve as much as anyone, and are probably anyway still troubled with many other things. But I console myself a little with the thought that you would probably have gone to the meeting in Leipzig anyway, and found that you are right in the midst of the topic of the talk as few others. Among all the theoretical physicists in Germany I cannot think of anyone else I could approach with such a request with equal confidence." (UAF, Laue papers).

Planck's plea was not in vain; Laue took on the difficult task. Planck wrote back to Laue from home on July 9th in relief: "Dear Colleague: Your letter from yesterday made me very happy, and although I suppose that by complying you mean to make this sacrifice for the sake of the good cause, may I now also add that by it you are doing me a personal favor as a friend which I shall not forget (ibid)." Towards the end, Planck elaborated: "Besides, I do believe that the present substitution has its merits from a purely factual point of view. In particular, it makes evident that the theory of relativity is not just practiced by Jews, and it takes the wind out of the sails of those who always think that the whole principle of relativity is nothing but artificial

---

[12] Philipp Lenard ignored the ordinance to observe a public day of mourning with dropped flags. Work continued as usual at his Institute. This was noticed by members of the Socialist Student League. Banding together with laborers they demanded a face-to-face discussion with Lenard but were refused, which led to an escalation. See Schönbeck (2000): 36 f.

publicity for Einstein (ibid)." Max von Laue wrote to Einstein on July 8th that he had accepted, but was still hoping that the situation would change after all. The postscript reads: "Please write to me sometime about how you would have roughly arranged your speech. Perhaps I can use it (UAF, Laue papers, 9.1)." Einstein thanked his friend on 12 July 1922:

> Dear Laue:
>
> I am very grateful to you for being so willing to taken on the address in Leipzig and thus myself definitively and trustingly place it into your hands. Now I have one more request, the fulfillment of which I hope will suit you. I'm going away in October, you know, for God knows how long, and it is necessary that someone else take my place as Director of the K. W. Institute during that time. Now I do wish that you would accept this substitution from October 1st for an indefinite period. Of course, I'll leave to you the income associated with this position.
>
> Since I have, in a sense, already officially left Berlin, I won't be coming to the meeting tomorrow and ask you, provided you agree, to submit the above-mentioned to the Directorate as my proposal. Please do also declare that I approve of the granting of 4000 M to Messrs. Kallmann and Knipping for the purchase of a Hoffmann electrometer.
>
> With my best wishes for the holidays,
>
> Yours, A. Einstein (ibid; see also Goenner 2005: 190, Schönbeck 2000: 37).

There was no mention in the *Physikalische Zeitschrift* of any disruptions during Laue's address.[13]

In July, the professor of experimental physics, Heinrich Rubens, died unexpectedly. He had successfully led the famous Berlin "Wednesday Colloquium" for many years. Max von Laue was now assigned this important task. He skillfully managed to maintain its high level right up to the 1940s, despite the political pressures and uncertain environment after 1933! Walther Nernst assumed the chair of experimental physics after the negotiations with Wilhelm Wien ended unsuccessfully.

Physicists had several periodicals in which to publish their current results. But Stark's *Jahrbuch der Radioaktivität und Elektronik* was the only one that focussed on summary accounts. The publisher Ferdinand Springer recognized a gap in the scientific literature, as the *Jahrbuch* certainly did not cover the whole field. Springer started releasing the first volume of *Ergebnisse der exakten Naturwissenschaft* ('Results in the Exact Natural Sciences') in the fall of 1922, writing in his editorial preface to the first issue: "The 'Results of the Exact Natural Sciences' represent what has 'come about,' the momentary state of knowledge in the individual fields, i.e.: they should provide a survey of a topic, not of its publications." Max von Laue wrote the eleventh contribution to it, on X-ray spectroscopy, divided into the three sections 'The measurement methods,' 'The continuous spectrum,' and 'The law of absorption.' He did not cite any of his own papers, but praised Manne Siegbahn's publications and also included ones about γ-rays. An extensive bibliography concluded the essay and demonstrates the care with which Laue assessed what he considered to be essential for research (Laue 1922c).

---

[13] In his autobiography Werner Heisenberg (1972: 66 f.) recalled his trip to Leipzig. At the entrance to the lecture hall he was handed a leaflet by the Einstein opponents; he even claimed to have heard Einstein's lecture there, which obviously hadn't been the case.

Fig. 8.12 In front of Blaueishütte during the ascent of Hochkalter in the Berchtesgaden Alps. *From left to right*: Heinrich Barkhausen, Max von Laue, Otto Hahn with Felix Bobek (*seated*), and an unknown person, 1923. *Source* Ernst Berninger: Otto Hahn. Eine Bilddokumentation. Munich: Moos, (1965)

During the academic term Laue could not get any major writing done; part of his vacations between semesters had to be spent on that. He otherwise went skiing in wintertime and hiking in the mountains in summertime with his closer colleagues (Fig. 8.12).

In October 1922, Laue gave a talk on 'Our current notions about the nature of light' at the 10th annual meeting of the German Society for Lighting Engineers in the lecture hall of the State Museum of Art. A printed version appeared in December. The speaker did not make it easy for his audience; without knowledge of current physics they could follow him only at the beginning. Laue started from Kirchhoff's radiation law and Wien's displacement law. These two laws "offered the two guidelines for the production of light sources involving thermal radiation; one should try to approach the temperature of the Sun as closely as possible and use as emitters bodies that have the greatest possible absorption capacity for the visible light that one wants to produce, and the smallest possible absorption capacity for all other radiation (Laue 1922d: 547)."[14] Then he mentioned Planck's analysis of cavity radiation and remarked: "In connection with this, he was forced to introduce a new universal constant, the

---

[14] He spoke before the Deutsche Beleuchtungstechnische Gesellschaft; see also Ewald (1960): 151, there cited as "Lichttechnik." I am indebted to Frau Fischer at SB PK for locating the correct source.

elementary quantum of action (ibid)." He said that Einstein's work from 1905 and also the Zeeman effect of 1896 were essential for understanding emission and absorption. Laue then gave an outline of atomic theories, mentioning Joseph J. Thomson and Niels Bohr as well as Sommerfeld's application of relativity to the orbital motion of electrons around the nucleus, the Stark effect, and the electron collision experiments by James Franck and Gustav Hertz (ibid).

Laue did not go into the "duality" of radiation—as wave or particle—although this topic was quite current at the time. It had caused a bit of a sensation in physics. The American physicist Arthur Compton had discovered by a very simple experiment that when X-rays are scattered by elements of low atomic weight, there is a reduction in the frequency of the scattered radiation. He was able to deduce this from the law of energy and momentum (Compton 1923: 483). This process was interpreted by Peter Debye (1923: 161) independently as well as by Niels Bohr and his collaborators Hendrik Anthony Kramers and John C. Slater (1924), who thought that the law of conservation of energy and momentum had no validity for the isolated process. Hans Geiger and Walther Bothe performed a coincidence experiment at the PTR in 1925 that clearly showed the opposite. The broad implication was: "One can apply the collision laws of mechanics to particles of light." The "wave-particle" property of radiation had been experimentally proven. The first application of it was by Einstein to explain the photoelectric effect. Some physicists did not accept it initially, including Max Planck (Geiger and Bothe 1925a, 1925b). In his obituary of Hans Geiger, Laue wrote: "Physics was saved from going astray." (Laue 1950) Bothe and Laue were neighbors during Bothe's period in Berlin until 1930. They went on excursions together into the surrounding March of Brandenburg and, being highly musical, Bothe recalled a memorable experience listening to a radio broadcast at the Laues' home with a device constructed out of amplifier tubes (Bothe to Laue, no location, UAF, Laue papers, 3.1).

On 9 November 1922, Max von Laue sent the president of the Kaiser Wilhelm Society, Adolf von Harnack, a handwritten note on small letterhead of the Prussian Academy: "Your Excellency, I permit myself to impart the information that my colleague Einstein has assigned to me the management of the K.W. Institute of Physics and the associated compensation for the period from 1 Oct. 22 up to his return. May I ask your Your Excellency to have this compensation transferred to my postal check account 60651 at the Berlin Postal Check Office. In great respect to Your Excellency, most sincerely, M. v. Laue (AMPG, III rep. 1A, Laue staff file 1920–1925, fols. 25, 9, 10)." The letter bears Harnack's signature, and Friedrich Glum, the general secretary of the KWG, issued the appropriate instructions. A problem arose, however, as Einstein had already received his salary for the final quarter of that year, totaling 15,000 marks. Thus Harnack had to inform Laue of the fact that he could not expect any remuneration until January 1st: "Regarding this salary arrears, I humbly urge you, E[steemed] S[ir], to contact Prof. Dr. Einstein directly." The beginning inflation apparently caused Laue's transfer at the beginning of 1923 to total 13,500 marks (ibid).

In his new position as Einstein's representative, Laue had to distribute the institute's funds as research grants (Fig. 8.13). In May 1923, Laue asked the board of

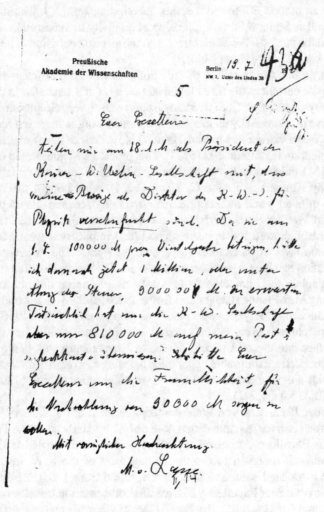

Fig. 8.13 A letter from Laue to the Prussian Academy of Sciences, illustrating his battle with finances. *Source* Archive of the Max Planck Society, Berlin-Dahlem, I rep. 1A, no. 1659

trustees of the Kaiser Wilhelm Institute of Physics to transfer two million marks to Professor Ludendorff, the director of the Astrophysical Observatory in Potsdam, for a solar eclipse expedition (AMPG, I rep. 1A no. 1659, fols. 29, 36b). Arnold Sommerfeld needed a million marks in mid-1923 for staff compensation. He applied to Laue as administrator of the KWI of Physics to receive this sum.[15] Laue retained this position even after Einstein had returned from his world tour. Since they often

---

[15] Laue to the Board of Trustees, Zehlendorf, 17 May and 23 Jun. 1923. Memorandum, fol. 44.

met at the university and at the academy and were able to discuss their views on the funding of various projects in person, there are very few letters between them about this collaboration. A letter from Einstein to Laue dated 3 October 1925 reveals the trust between them: "I send you herewith an application that in-and-of-itself is certainly justified, but cannot come into consideration for us to grant. But perhaps you can give the man some advice. I told you—I think—that Tomaschek intends to 'Miller' on the Zugspitze peak. Piccard (Swiss physicist, Brussels), whom you know, likewise intends (?) to 'Miller,' namely, by balloon. What could possibly be done to bring some system into this plague? It would be a pity if too much money were spent on this foul matter. Best regards, Yours, Einstein".[16]

As chairman of the Berlin regional branch of the German Physical Society (*Gauverein der DPG*), Laue held the obituary speech on Max Abraham on 8 December 1922, paying tribute to his contributions to electrodynamics, in particular his book 'On the Dynamics of the Electron.' He suggested that Abraham's intelligent yet critical attitude towards relativity might perhaps have been one reason why he never received tenureship (see the DPG proceedings for 1923, 3: 22).

An important visitor from the USA also stopped by in Berlin while visiting Europe: Robert Millikan, president of the California Institute of Technology and director of the Norman Bridge Laboratory of Physics in Pasadena—but not yet a Nobel laureate, which he was to receive for determining the elementary charge of an electron in 1923. He was an important rapporteur in the USA about who might be eligible to receive financial support for research in Europe (Fig. 8.14).

The economic situation deteriorated rapidly at the beginning of 1923, because rampant inflation not only destroyed all deposits in banks and savings associations, it also led to catastrophic unemployment. In the proceedings of the DPG a notice appeared under the heading 'Job placement for physicists with a doctorate and for student aids in the physical and physico-technical fields, offered by the Professional Association for German University Teachers in Physics." Only cautious allusion is made to economic hardship: "It has been shown that these [student aids] often render excellent services, and stand in for at least one certified laboratory technician, so their vacation employment has very frequently led to permanent later employment, or even, if unfavorable financial circumstances do not permit further study, to a permanent position beforehand (DPG proceedings for 1923: 47)."

On 10 February 1923 Wilhelm Conrad Röntgen died after a few days of suffering in Munich. As chairman of the Gauverein of the DPG, Laue held the funeral oration for its honorary member at the meeting in Berlin on February 23rd. Laue had experienced this very reserved man as a student and private lecturer at Munich. Thus he was able to report about Röntgen's excessive anxiety about publishing final results and also about Röntgen's own assessment of his great discovery, which he regarded as equivalent

---

[16] Einstein to Laue, 3 Oct. 1925. UAF, Laue papers, 9.1. "*Millern*" alludes to yet another repetition of the Michelson experiment of 1881 by Dayton C. Miller in 1904 in a futile attempt to prove at higher precision the existence of an aether, Morley and Miller (1904): 753 f. Auguste Piccard ascended by balloon for this purpose and was even to set a record of 15,785 m altitude in 1931, Piccard and Stahel (1927): 140.

Fig. 8.14 The American physicist Robert Millikan (1868–1953, *2nd from right*) among his Berlin colleagues Walther Nernst, Albert Einstein, Max Planck, and Max von Laue, probably in Laue's home in Berlin-Zehlendorf, 1928 or 1931. *Source* Archive of the Max Planck Society, Berlin-Dahlem

to his experimental proof that a Maxwellian displacement current arises in moving bodies (DPG proceedings for 1923: 8).

Scientific life continued normally despite the oppressive worries during this period. At the meeting of the physical and mathematical class of the academy on 15 February 1924, Einstein had Planck present a paper on general relativity, and Max von Laue spoke about 'The solutions to the field equations of gravity by Schwarzschild, Einstein, and Trefftz and their unification.' According to Einstein, the whole cosmos was uniformly filled with nebulous matter. Under certain conditions, stars could form. Then Laue compared his statement with Trefftz's solution and concluded: "thus we see in this way the possibility of constructing a spherically closed cosmos […] under mathematically rigorous satisfaction of the field equations and the boundary conditions (Laue 1923a: 27)."

On 23 March 1923, the board of trustees of the KWI of Physics instructed the banking house Mendelssohn & Co. to transfer to Laue 90,000 marks for the quarter. On September 21st, this same sum corresponded to 70,500,000 marks. This inflation also devoured Laue's savings. In June, he informed the board of trustees of the KWI of Physics that he would be hiring a secretary, Fräulein Hildegard Bathe, as of 1 July 1923. She was to receive 225,000 marks for a ¾-year period. Laue had spent

8000 marks on paper supplies and 6000 marks on postage, for which he requested reimbursement (AMPG, I rep. 1A no. 1659/1–3).

The stream of grant applications for research in physics continued to flow onto his desk. Given the rampant devaluation, in order to be able to calculate, grant amounts were counted in 'gold marks.' Most of the approved applications could still be met. On 13 October 1923 the 'Rentenmark' was introduced together with a decreed devaluation of $1:10^{12}$. The economic recovery from inflation was sluggish.

A letter from Planck to Laue dated 19 November 1923 documents their close exchange of ideas. It is written on the academy's letterhead. They had very probably been discussing a problem after a meeting of the academy, together Walther Nernst, who is also mentioned.

> Dear Colleague:
>
> I have carefully considered your reasoning and would like to give you my opinion briefly on it here. From your way of looking at things, it is clear that, according to the theorems of general thermodynamics, it is impossible for the two metal bodies of the same material but of different states of aggregation, which are in equilibrium with each other, to exhibit opposite free charges. Nernst's objection with the two independent components cannot be upheld.
>
> But a potential difference between the two bodies is also possible without free charges, namely when the two bodies carry electrical double layers not only at their contact planes but also at their surfaces (contact planes in vacuum). Of course, these must depend on the state of aggregation, namely in such a way that the three corresponding potential differences satisfy the law of voltage rest.
>
> How this is in reality, I leave completely open. I just wanted to say that in my opinion nothing can be deduced for this from thermodynamics alone. However, we shall probably continue to discuss this point at length (AMPG, V rep. 11 no. 881).

On 6 December 1923, the physical and mathematical class of the academy convened again for another plenary meeting. Max Planck spoke about the nature of thermal radiation and submitted two more theoretical papers. Albert Einstein also submitted a paper, 'Does field theory offer possibilities for solving the quantum problem?' Max von Laue delivered 'On the theory of positive ions and electrons emitted by glowing metals.' This topic picked up on his wartime research with electron tubes and a publication by Schottky. The introduction listed the limiting parameters that all charge carriers were electrons and all positively charged monatomic ones were charged with one elementary quantum of electricity. Laue used thought experiments in an attempt to visualize the processes. The publication had longer passages partly containing speculative considerations about the behavior of the charge carriers also at the interfaces (Laue 1923b).

Laue was still interested in the mathematical problems of relativity, as can be seen from a publication in the *Annalen der Physik* under the title 'G. A. Schott's form of relativistic dynamics and the quantum conditions' from spring (Laue 1924a). He was disturbed by the occurrence of complicated root computations when calculating the energy of a moving point of mass. In a footnote he alluded to a discussion about this with Walter Gordon.

Fig. 8.15 A rare family portrait depicting Magda von Laue with her two children, Theodore and Hildegard, ca. 1925. *Source* Universitätsarchiv Frankfurt am Main

Unfortunately, even ten years after the discovery of X-ray interference, its exact circumstances were still the subject of differences of opinion, about who had said what and when. Laue once firmly negated Ewald's contention that a conversation about it had taken place at the time between Ewald, Sommerfeld, and himself. Ewald lamented about this in a letter to Sommerfeld from Stuttgart on 28 April 1924: "My defective historical sense renders me almost worthless as a witness in this doctorate issue. I remember distinctly only my tremendous astonishment, while we were on a walk in the Engl[ish] garden, that Laue should reveal such an absolute cluelessness about the notion of a space-lattice structure for crystals. I believe that the idea about the interferences first occurred to Laue during a consultation that took place between him and me in the evening in his apartment, in the matter of my d[octoral] thesis (Sommerfeld 2004: 160)."

On the other hand, Laue could be pleased about three articles that appeared in the 16th issue of *Naturwissenschaften* in 1922 discussing extensions to his fundamental discovery (Debye 1922): Peter Debye's paper on 'Laue interferences and atomic structure,' Paul Niggli's (1922) on 'The importance of the Laue diagram for crystallography' and Ernst Schiebold's (1922) 'Contributions to evaluating Laue diagrams'. By then Laue's relations with Sommerfeld had become completely amicable again. When Sommerfeld came to Berlin to give a lecture on 16 February 1924 he even stayed with the Laues in Zehlendorf, as Sommerfeld's letter to Planck on 11 Dec. 1923 indicates (AMPG, V rep. 13 no. 886) (Fig. 8.15).

Laue had a variety of obligations to fulfill in addition to his teaching activities that were not always scientific. He was on the board of directors of the Kaiser Wilhelm Institute of Physics—that research establishment for a long time without a building to call its own—together with Einstein, Haber, Nernst, Planck, and Warburg. It was endowed with a budget and, comparable to the *Notgemeinschaft*, was able to procure equipment for individual researchers and professors at university institutes upon application and to award smaller grants. This directorate often met right before or after the meetings of the Academy of Sciences.

Several invitations reached Laue by scientific associations to present a talk. On 7 May 1924, he spoke before the Berlin district branch of the Association of German Engineers on 'Atomic structure and atomic fission.' He evidently reused his manuscript from the lecture at the Kiel Autumn Week and referred again to the large amount of energy released during radioactive decay. The example he gave was the helium atom at atomic weight 4.000, which could be imagined as constructed of four hydrogen nuclei at atomic weight 1.008. This left a surplus mass of 4.032 against 4, which according to the theory of relativity was a large amount of energy occurring when the helium nucleus formed. This large amount of energy would also have to be applied to split the helium nucleus, which was not possible by current means. He critically discussed Bohr's atomic model with its contradictory assumption of curved electron orbits without any release of energy (Laue 1924b).

His speaking obligations did not diminish. At the 42nd meeting of the Berlin chapter of the German Physical Society on 27 June 1924, Laue held the congratulatory address in honor of his colleague Walther Nernst on his 60th birthday, alluding to his many scientific achievements. Laue diplomatically circumvented Nernst's occasionally somewhat opinionated contributions to discussions. "Since thermodynamics leads in so many ways to quantum theory, Mr. Nernst has also commented on these. We know only too well that everything is still in flux in this field; hardly two physicists are in complete agreement on it. Thus, Nernst's views have also been subject to some criticism and change, also his own, in the course of time. But no one can fail to see how stimulating his views have always been for research, because they have always concerned essential points which have encouraged further progress (*Verh. DPG* 1924 no. 3)." The new building of the Institute of Physical Chemistry, which Nernst headed as director, was officially inaugurated for occupancy on that very day. During his time as president of the *Physikalisch-Technische Reichsanstalt*, Nernst had pointed out to the Ministry of the Interior that the PTR needed a theoretical physicist as an advisor. He proposed Max von Laue for this post, but the poor financial situation prevented its immediate materialization. Nernst's successor, Friedrich Paschen, however, did succeed in arranging it, and approval of Laue for this part-time position was granted.[17] Most of the time, Laue went to the PTR in Charlottenburg once a week, on one hand, to offer his expertise, on the other hand, to find inspiration for his own inquiries. A few years had to elapse before he had found a fruitful topic arising out of Walther Meissner's experiments: the theory of superconductivity.

If one compares Laue's publications against different areas of theoretical physics and the disparate, sometimes remote problems he addressed, the range is astonishing. Most theoretical physicists, such as Max Planck, worked intensely on a delimited area. Was Laue driven to this by an inner restlessness, which then manifested itself in a desire to change the environment of his activity in order to escape everything?

---

[17] Paschen to Ministry of the Interior, Berlin-Charlottenburg, 16 Mar. 1925. HUA, L 51, Laue staff file, vol. 1.

Or was it the unsuccessful search for a new, as yet unsolved important physical problem? To have become involved with Bohr's model of the atom would probably have led to disagreements with Sommerfeld, who had already published important contributions to it. Nor did he join the effort to construct a new theory, as some of his colleagues were doing. This was also the case with the quantum mechanics presented by Werner Heisenberg in July 1925 and then formulated together with Max Born and Pascual Jordan. It did not move Max von Laue to try his own hand at this new field, neither did the theories of quantum mechanics formulated by Erwin Schrödinger or Paul Dirac. But he read their publications with a critical eye. When Wolfgang Pauli earned his doctorate under Sommerfeld and published part of the thesis in the *Annalen der Physik*, Laue disagreed with a remark about a problem in thermodynamics and wrote to Sommerfeld from Zehlendorf on 12 Nov. 1924 to ask for a copy of the dissertation (DMA, NL 089).

In general, though, Laue remained, with few exceptions, "true" to X-ray interference and the theory of relativity, and later also to superconductivity. Nuclear physics was still a predominantly experimental science. In Berlin, it was Lise Meitner at the Kaiser Wilhelm Institute for Chemistry who was researching this field with success. Her close friendship with the Laues certainly gave plenty of occasion for the two physicists to talk about current and fundamental problems in their science.

During this period Laue managed by verbal agreement to convince the Ministry into exempting him from the obligation of offering a major lecture course at the university.[18] Reading the publications by his fellows in the field, Laue occasionally felt called upon to publish additional notes or remarks. The physicist Kurt Zuber at the University of Zurich examined the delay times that occur when a spark generates an electrical discharge (1925). He carefully measured the times. Laue may have known him personally or else had heard of his work through Edgar Meyer. His "remarks" about the results were of a mathematical nature and concluded with the statement that "in a spark gap of the kind he used, chance plays a decisive role, based on the disorder of molecular processes." (Laue 1925b).

At this time Laue completed two extensive theoretical papers on problems of X-ray interference (Laue 1925c, 1926a). In both instances it was simultaneously an opinion on research by another European physicist, offering a different interpretation and a different theory. On 29 June 1926 he wrote to Paul Peter Ewald to discuss the influence of temperature on the lattice vibrations (UAF, Laue papers, 3.2). Laue doubted the calculation by his colleague Debye. The method by Ivar Waller, he contended, yielded the same values as his calculation of the lattice vibrations. Very soon afterwards, on July 17th, he wrote Ewald that Ralph Walter Graystone Wyckoff, an American crystallographer, had found complete agreement (ibid). Then a meeting was planned for the coming vacation to discuss the problem. This time it was not in the high mountains because Mrs. Laue suffered from asthma, but in Holzhausen, where Ewald's mother lived.

In July, Laue submitted the manuscript of his study on crystals supposedly composed of two kinds of atoms: 'On X-ray interference in solid solutions' (Laue

---

[18] Laue to Generalverwaltung KWG, 18 Apr. 1924. AMPG, III rep. 1A, Laue staff file, fol. 13.

1925c). At the beginning of the paper he discussed the views of the Göttingen professor Gustav Tammann about the arrangement of the atoms and their lawful distribution in a space lattice (Tammann 1919). According to Tammann even a slight deviation from this distribution in the X-ray diagrams should cause an appearance of complete disorder. The X-ray diagrams permitted one to conclude, however, that the space lattices were uniformly occupied at all points. This, Laue argued, he had already shown in his theory from 1918, and he reiterated the mathematical proof. He chose NaCl as his example, with the qualification that irregularities could occur to the point of complete disorder. He was able to refute Tammann's objections (Laue 1925c: 167). In the second part Laue analysed the complete disorder, pointing out that it was possible that the disorder could never be complete, and in the vicinity of an atom of one species several of such atoms accumulate. His result was a broadening of all maximums. In the third part he examined the probability of deviations from the mean value calculated in parts I and II occurring in individual cases. But the X-ray diagrams of different crystals of the same composition did not yield any deviations. For this he also cited the results obtained by the method employed by Debye and Scherrer in their examination of powdered substances (ibid).

In July of the following year Laue (1926a) delivered the manuscript entitled 'The influence of temperature on X-Ray interferences' to the editors of the *Annalen der Physik*, Wilhelm Wien and Max Planck. There was an introduction with an appreciation of foregoing publications on the problem. Max Born and Erwin Schrödinger were especially praised, but the Swede Ivar Waller from Uppsala was also mentioned. A discussion of other papers was followed by the passage:

> If, on account of this lack of agreement with experiment, one looks for points in the calculation which are not quite certainly justified, one may perhaps form a suspicion against the fundamental idea upon which Debye bases the consideration, and which all his successors retain. Because the thermal oscillations have a very small oscillation number compared to X-rays, Debye first averages the electromagnetic energy for the calculation of the intensity over a time in which the atoms move just a little away from their instantaneous position, but which nevertheless constitutes many X-[ray] periods, in such a way as if the atoms were at rest. Only the subsequent averaging over all amplitudes and phases of the heat oscillations eliminates what in his result at first points to the mentioned location of the atoms. This procedure has never been objected to; but in any case this averaging in two steps is unusual, and we felt the need to substitute it with a calculation more closely adapted to other procedures in optics for verification (Laue 1926a).

Laue noted that he had also used this method. Then he analysed the following mathematically: the influence of the Doppler effect of the atom oscillating about the rest position; the assumption that the atoms of the space lattice oscillate independently of each other about the rest position; the natural oscillations of the crystal lattice and their influence at the interference points; and the diffuse scattered radiation. The conclusion he arrived at was: "The modification of the method carried out here to calculate the interference effect for thermally moved atoms in the space lattice according to Doppler's principle thus radically changes the course of the calculation; on the other hand, it changes nothing in the results by the older method, which thus yields full confirmation (ibid)."

No scientific cooperation existed yet between the Friedrich Wilhelm University in Berlin and the suburban polytechnic in Charlottenburg.[19] Even so Max von Laue made the acquaintance of the professor of applied mathematics, Richard Edler von Mises, who had come to the *Technische Hochschule* in Berlin-Charlottenburg via Strasbourg and Dresden. He was the founder of the journal for applied mathematics and mechanics, *Zeitschrift für Angewandte Mathematik und Mechanik*, and was interested in crystallography. Laue and he planned to publish a collection of stereoscopic images and drawings of crystal lattices. They contacted the publisher Ferdinand Springer in the 1920s and told him that they wanted their edition also to include texts in English. The drawings were made by Elisabeth Verständig, the photographs were probably taken by Clara von Simson. A contract with the publisher was signed on 26 January 1926. The book appeared in 1926 under the title 'Stereoscopic Images of Crystal Lattices' (Laue and Mises 1926–1936).[20] The author's honorarium of 1,500 Reichsmark was divided up: 500 Reichsmark went to Clara von Simson, 500 Reichsmark to the teacher trainee Elisabeth Verständig, 500 Reichsmark to the university treasury (as specified in von Mises's letter to the publisher from 13 November 1923). Copies of the book went out to the two female collaborators, to Professors Max Planck, Walther Nernst, Arthur Wehnelt, Arnold Sommerfeld, Paul Peter Ewald, and Paul von Groth, as well as to the Englishman Norman N. Greenwood.

At the academy meeting on 18 February 1926, Laue presented a paper on 'The scattering of inhomogeneous X-rays by microcrystalline bodies,' a copublication with Hermann Mark. As often before, the introduction was a historical survey of the current issues in physics around 1911. Laue and Mark (1926) noted a discrepancy between theory and observation in an inaccurate assumption about the coupling of electrons in atoms. The previous scattering experiments, they reported, had examined microcrystalline materials with the broad continuous spectrum and characteristic lines. The coauthors subjected to mathematical treatment the scattering of the bremsstrahlung on powdered crystal and could obtain sufficient agreement between the theoretical values and observed data from experiments with polycrystalline aluminium and diamond powder (Laue and Mark 1926: 586). Under the heading 'Lorentz factor and intensity distribution in Debye–Scherrer rings' Laue expanded on his joint publication with Mark, as Debye's and Scherrer's arrangement for examining the structure was also suitable for powdery substances. The introduction stated the topics of this two-part paper: "The first part thereby assumes, as did Lorentz in his day, the interference maxima furnished by a single ideal crystal in strictly monochromatic X-ray illumination to be infinitely sharp; its purpose was merely to clarify the theoretical issues. In Part II we drop this assumption in order to discuss the influence of particle size and form on the intensity profile in a Debye–Scherrer ring, and to allow us to observe their determinations (Laue 1926b)." One conclusion from the

---

[19] No joint research existed between the Friedrich-Wilhelms-Universität and the *Technische Hochschule Berlin-Charlottenburg* in the fields of physics and/or mathematics until 1945, not even when Gustav Hertz became full professor at the polytechnic in 1928.

[20] See also ZLB, Disp. Springer-Verlag. The surviving files, which have not been indexed, are very incomplete. It is unclear when the work began. Mises was forced into exile after 1933 and went to Istanbul.

calculations was: "It should accordingly be possible to determine the general shape of the crystal particles from the width of the interference rings (ibid)."[21]

A turning point was reached in the subject of theoretical physics at the Friedrich Wilhelm University when Max Planck resigned as institute director on 1 April 1926. But he continued to offer lecture courses, as he wrote to Wilhelm Wien on 12 October 1926 (DMA, NL 056; copy in AMPG, V rep. 11 no. 943). The faculty drew up a list of possible successors, naming Arnold Sommerfeld, Erwin Schrödinger, and Max Born. Einstein had no interest in the position, nor would his many travels have made it possible. Werner Heisenberg was missing from this list. Was he, at the age of 26, too young for such a chair?

A glance at Max von Laue's previous publications reveals no other revolutionary experimental or theoretical advancement in another fundamental field of physics— namely in atomic theory. Niels Bohr's model of the atom provided some help in understanding the system of spectra. Bohr was able to explain much about the structure of the periodic system of the chemical elements. But some things remained unclear, such as why electrons orbiting around the atomic nucleus emit no radiation. Werner Heisenberg, a graduate student of Sommerfeld, transferred to Göttingen to continue his studies under the theoretical physicist Max Born. In 1923 they together attempted to find a mathematical description of the helium atom, as Niels Bohr had done for the hydrogen atom, but failed. After taking many detours, including ones set by other physicists, Heisenberg concluded in 1925 that a theory of the atom could only be constructed on the basis of observable quantities such as the frequency of spectral lines. He worked out this theory together with Max Born and Pascual Jordan by means of the matrix calculus (Born et al. 1926). Very soon, other physicists, such as Wolfgang Pauli, made important supplementary contributions. Max Planck and Max von Laue watched critically from a distance. Only when Erwin Schrödinger described the atom and its behavior with a well-known mathematical tool, differential equations, did they accept this advance. Max Born pointed out the problem of causal relations in physics in his publication 'On the quantum mechanics of collision processes': "One gets no answer to the question 'What is the state after a collision?', only to the question 'How probable is a given effect of the collision?' The whole problem of determinism arises here. [...] I myself am inclined to give up determinism in the atomic world. But this is a philosophical question, for which physical arguments alone are not decisive. Practically, at any rate, indeterminism remains standing for both the experimental physicist and the theoretician (Born 1926)."

In March 1927, Heisenberg's paper 'On the visualizable content of quantum theoretical kinematics and mechanics' appeared in *Zeitschrift für Physik*, in which he developed the concept of "indeterminacy" in quantum theory (Heisenberg 1927). He started from very simple considerations. When determining the location of a particle with radiation, its wavelength determines the accuracy of the location information.

---

[21] The paper appeared in *Zeitschrift für Kristallographie*, founded in 1877 by Paul Heinrich Ritter von Groth. Groth became professor in Munich in 1883 and was director of the important collection Mineralogische Staatssammlung. He lived to see Laue's discovery as a septuagenarian. In 1920 he retired as editor of the journal, which was continued by his Swiss colleague Paul Niggli in collaboration with Max von Laue, Paul Peter Ewald, and Kasimir Fajans.

As the wavelength decreases, the accuracy of the location determination increases, but so does the energy of the radiation, making the particle no longer remain at that location. Elsewhere he treated the problem of determining the energy, stating: "But in the precise formulation of the law of causation: 'If we know the present accurately, we can calculate the future,' it is not the secondary clause but the presupposition that is false. We cannot, in principle, get to know all the determinants of the present. [...] Physics is only supposed to describe formally the relation to perceptions. The true state of affairs can much better be characterized thus: Because all experiments are subject to the laws of quantum mechanics and therefore to the equation of the uncertainty relation,[22] thus the invalidity of the law of causation is definitively established by quantum mechanics (Heisenberg 1975)." This blunt declaration about the invalidity of the law of causality was perhaps reason enough for the appointment commission not to place Heisenberg on the list of candidates. Planck as well as Laue and some philosophers later took a stand on the problem of causality.

During this period, political attempts were made to suspend laws. Joseph Goebbels, with a doctoral degree in German studies, became a regional head (*Gauleiter*) of the National Socialist German Workers Party (NSDAP) in Berlin-Brandenburg in 1926. Under his leadership an anti-Semitic smear campaign of the worst kind was launched against the vicepresident of the Berlin police force, Bernhard Weiss. When Goebbels publically encouraged that the editors of critical newspapers become targets of brutal assault, Weiss responded by banning the Nazi party from Berlin and the whole province of Brandenburg the very next day, on 5 May 1927. A large group of Nazis defiantly demonstrated against the ban on Wilhelm-Platz (now called Richard-Wagner-Platz) on May 12th. There were almost daily reports in the papers about bloody brawls. Pubs visited by communists, amongst other regular apolitical customers, were attacked (Reichhardt and Bering 1983: 87 f.). Max Planck and Max von Laue, as well as Lise Meitner, experienced none of this unrest in their neighborhoods, and there were no disturbances in their lectures either. Again it looked as if Laue might get his own KWI of X-ray Physics. Planck was very optimistic for a time. But on 22 December 1926, Laue wrote from Zehlendorf to Lise Meitner: "I just went to see Planck this afternoon to discuss the question of the KWI of X-ray Physics again. There wasn't much left of that joyful enthusiasm with which he had spoken about it the other day at the Kaiser Wilhelm Society. I'm afraid nothing will come of it, and my last shred of hope of getting out of teaching at last is disappearing. It's so disappointing (CAC, MTNR 5/27)." In hasty scrawl revealing his emotionality, Laue wrote a letter to Wilhelm Wien dated 5 March 1927 to inform him about his professional situation. He did not believe a KWI for X-ray Physics would ever materialize, but he had to get out of academic teaching. He asked whether a position at the bureau of standards (PTR) as director might not be an option for him, as he "had the impression that they were also thinking of the possibility of getting me there at the Reichsanstalt." (DMA, NL 056) Like Planck, Einstein also knew of Laue's dissatisfaction with the burden of lecturing that his teaching position imposed upon

---

[22] The equation mentioned is this: $\Delta p \cdot \Delta q \sim h$, where p is the momentum, q is the uncertainty, and h is Planck's quantum of action.

him. He tried to reassure him in a letter from May 12th with plans for the KWI of Physics: "I'd like it best, for your sake, if you got the K.W. Institute. You'd not only be able to give inspiration as a leader, but also create a good, friendly atmosphere. I'd also gladly go there, and often as well, to participate to the best of my abilities (UAF, Laue papers, 9.1)."

The endeavor to appoint Sommerfeld as Planck's successor collapsed. He declined the call to Berlin because the Bavarian state government assured him that funding for Munich would be improved. Sommerfeld informed Laue of this step as early as June 8th, who had been campaigning for his appointment (DMA, NL 089). He wrote again on 14 June 1927 to say that despite feeling guilty he would not be coming to "unruly Berlin," preferring rather to decline thus by letter, and appending some personal advice:

Dear Laue:

Please accept my sincere thanks for your letter and also for not chiding me. My letter to [the official responsible for the appointment at the Ministry of Culture] Richter already went out last Saturday. It would not have been easy for me to persuade him, against my conviction, to arrange Planck's succession in the way you intended, as I already told you in Berlin that morning in the hotel.

Because I believe that given the situation of the Berlin affairs you are fully occupied with scientific and organisational matters and are indispensable for that. So why would you want to burden yourself with an activity that does not suit you? Why don't you take up an exceptional post among the German professoriate, just like Einstein? (DMA, NL 089; Eckert 2013: 186 f.).

The next candidate for the Berlin chair was the almost forty-year old Erwin Schrödinger (Fig. 8.16). He already had a professorship in Zurich. A presentation by Prince Louis de Broglie, in which the young French physicist boldly ascribed to matter the character of a wave, inspired Schrödinger to construct such a wave theory for matter. Instead of using the Heisenberg-Born-Jordan matrix calculus, Schrödinger preferred to use the differential equations familiar to physicists. This, too, was a contributing factor in the decision to place Schrödinger on the candidate list for the Berlin post. He accepted the call and took over the main lecture courses in theoretical physics.[23] Planck wrote to Laue from Grunewald on 3 July 1927 in connection with Schrödinger's appointment. After applauding an extension to Paul Peter Ewald's fellowship, he continued:

Now, as I read in the newspaper, Schrödinger's appointment really has gone out. I assume that Schrödinger will be coming to Berlin in the near future to take a closer look at things here; I have invited him to stay over here with me again. Of course, it is of primary importance that he contact you, as the nearest colleague involved, and I would like to emphasize right now already that in your negotiations with him the consideration of the final state (i.e., when I myself am no longer there) solely should be decisive. In any case, you must in any event have a frank talk with him, one on one. Hopefully something positive really will come out of this soon (AMPG, V rep. 11 no. 964).

---

[23] Erwin Schrödinger did not always follow the usual dress code in the lecture hall. During the summer he was known to appear in his tennis outfit.

Fig. 8.16 Erwin
Schrödinger (1887–1961),
successor to Einstein and
Laue in the chair for
theoretical physics in Zurich.
He received the Nobel prize
in physics in 1933 jointly
with Paul Dirac
(1902–1984). *Source* The
author's estate

Erwin Schrödinger

For Laue, nothing changed, as he continued to be in close contact with Planck at the university and the academy. They discussed the possibility of procuring a paid assistant for Einstein. For a short while it once again looked as if a professional turnaround might be feasible for Laue after all. The Kaiser Wilhelm Society never had enough funds to found new institutes. Sometimes a generous sponsor was found, sometimes an industrial interest group provided the financial means. In the spring of 1927, the possibility of founding a KWI of X-ray Physics in Munich briefly arose, probably at the instigation of Wilhelm Wien; but Harnack was skeptical. The society's funds were needed for other projects.

Scientific life in Berlin was very diverse. Laue's friendship with Lise Meitner brought not only excursions into the vicinity of Berlin and scientific conversations. It also led to a coauthored paper on 'Calculating the scattering range from Wilson exposures.' A doctoral student of Ms. Meitner's, Kurt Freitag, used a cloud chamber to investigate the range of α-rays emanating from a ThC + C´ preparation and entering the chamber through a narrow slit. Differences in the path length were found owing to density fluctuations of the gas during expansion. Using the observed range and gas density, the problem could be satisfactorily solved by a series expansion (Laue and Meitner 1927).

In the foregoing year, invitations had already gone out to an International Congress of Physicists in honor of Alessandro Volta († 1827) in Como in September 1927. In Italy, a fascist regime ruled under Mussolini. Many German physicists who were invited feared a political rally, perhaps even a personal appearance by Mussolini, especially since the South Tyrol conflict was being exacerbated on the Italian side. Laue corresponded with Sommerfeld about this on 27 July 1926: "Last night Planck, J. Franck, and I talked about the invitation to Como that we all had received. Our opinion tended toward accepting the invitation. If the Italians do something tactless, it falls back on themselves (DMA, NL 089)." He also mentioned that his wife was staying in Schwarzeck in the Bavarian Forest and feeling variously unwell. "She is very grateful for your wife's kind solicitude during the Munich days (ibid)."

In the end, the participants at the congress in Como from Germany were: Max Born, James Franck, Walther Gerlach, Eduard Grüneisen, Max von Laue, Lise Meitner, Friedrich Paschen, Max Planck, Arnold Sommerfeld, Otto Stern, and Karl W. Wagner. The topic of Laue's talk was: 'On the Doppler effect in the scattering of X-rays by thermally moved atoms.' Heisenberg was listed under Denmark, even though he was already back in Leipzig. Several participants took this opportunity to travel through Italy afterwards.[24] In Germany, too, political right-wing pressure was growing among students. The German student association had merged with the strongly right-wing Austrian student association. With the help of the German People's Party, an attempt was made in the Prussian Parliament to revise the legal regulations on students. The Minister of Culture, Carl Heinrich Becker, recognized the danger. He chose very clear words to point it out, not without granting young people a special "state of mind":

> Besides, it is a law for every youth to prove manly pride before royal thrones and ministerial seats. […].
>
> And then the political masterminds come along to exploit this state of mind.
>
> 'You're in favor of Greater Germany, aren't you?' Certainly.
>
> 'You're pro academic freedom, aren't you?' Of course.
>
> 'The government is jeopardizing this, so you vote No. But then our self-government will be lost, after all. Our Austrian brothers would take that as a betrayal of the Greater German cause (GstA PK, C.H. Becker papers, Sachakten 1038, 1041).

Then Becker indicted senior academics. "However, there is much resentment against the new state here, a reactionary tendency in Germany, not at universities, not among the professoriate, but among many senior academics in the country. You have to know that most of the major associations explicitly or unexplicitly uphold the anti-Semitic principle (ibid)." The great majority of physics students and of the German Physical Society's membership did not hold such views. Lenard and Stark and their few followers were of a different opinion.

---

[24] James Franck spelled out such political concerns about attending the conference, see Lemmerich (2011a): 151.

It was probably again the proximity to Lise Meitner that inspired Laue to publish a paper about nuclear physics, an otherwise foreign field for him. Hans Geiger and John M. Nuttall (1911) had determined a formulaic relation between the half-life and energy of α-particles. It was unclear how low-energy particles could escape from the field of the atomic nucleus. The Russian physicist George Gamov published a theory in early 1928 about how to use Erwin Schrödinger's wave mechanics to understand this (Gamov 1928). Laue responded to this with a very typical remark: "The quantum theory of the atomic nucleus as outlined by Gurney and Condon and elaborated by Gamov builds the first bridge between the physics of the atomic electron shell and the physics of the nucleus. Their basic idea, that wave mechanics in principle permits one to overcome even the highest potential thresholds, is so plausible, and their derivation of the hitherto quite mysterious Geiger-Nuttall relation is so convincing, that these researches may well number among the most beautiful successes of recent physics. But the mathematics in Gamov's analysis is not quite right. Lest erroneous conclusions become entrenched in the literature, do permit the following correction (Laue 1928: 726)." Gamov's theory was concerned with the reflection of Schrödinger waves in a potential well and the possible escape of waves: tunneling (ibid).

On 20 February 1928 Max von Laue wrote to Wilhelm Wien from his home in Zehlendorf about complications in finishing off his article for the 'Handbook on Experimental Physics' (Wien and Harms 1928), about reproductions of diffraction phenomena, etc., and the publication of a photographic image of standing waves. About his professional situation he noted:

Please do apply your full influence toward placing me in a position that I finally be able to work properly again! For 16 years I have been suffering under the misery that this professorship means to me. If I still had my fortune, I would have turned it in long ago and assumed the position that suits me best, that of a man of independent means. Unfortunately, that's impossible for me, and I have to watch the best years of my working life pass by without being able to use my capacities, which certainly are still there. I am, by nature, simply not tuned to teaching and administrative tasks. I never wanted to be anything other than a researcher. And those monstrous motives behind my being denied another position so far—despite all the endeavors by Planck and others! I have had to listen to the most unworthy, mendacious vituperations about my character! Well, I am certainly no angel, but I would like to think that in a moral respect I am sky-high above my enemies, indeed that this very fact makes them my enemies. I, at least, can only call individuals who collude with detectives and have not stopped short of arranging that my home be broken into in order to pillage my desk for material against me, by names which I must unfortunately refrain from writing down.

But enough of this! You won't hold this emotional outburst against me. After all, it's not the first time that you have heard me give vent to my feelings, and you will appreciate that as a sign of trust in you.

I remain with warm regards, ever sincerely yours,

M. Laue (DMA, NL 056).

Max Planck replied to a letter from Wien on February 26th:

Your communication about Laue's letter is very distressing to me, because I had assumed that he is less tormented now by his fixations than he was last year. He reveals absolutely none of this towards me, nor do I have any reason to ask him about it and thereby draw his attention to it. As far as I can judge, the best thing would be for you to console him briefly with the prospect of seeing you here soon and speaking with him. I haven't the least idea what he means by the position he is aspiring toward; after that plan about the X-ray institute evaporated, he never gave me the slightest hint as to the ideas guiding him in that direction. I do urgently hope that this depressive condition will gradually calm down this time again, and that he will be able to recuperate during the coming vacation (DMA, NL 056, fol. 37).

Whether Wien was able to do anything on von Laue's behalf is unclear. The only researcher position at the Berlin academy was occupied by Einstein. The prevailing economic hardship made it unrealistic to hope that the ministry would approve a second one.

The above-mentioned handbook article was intended for the 18th volume of *Handbuch der Experimentalphysik* edited by Wilhelm Wien and Friedrich Harms, which was published in 1928. Laue had committed himself to writing three contributions. In 'The optics of moving bodies' (pp. 39–103)—actually about sources of light—he mentioned Maxwell and Einstein in a historical introduction and then reviewed the work by Fizeau, Doppler, Michelson, and others. The second article (pp. 125–207) was on 'Reflection and refraction of light at the boundary between two isotropic bodies.' It was treated according to Maxwell's theory. The third article (pp. 211–361) was on 'Interference and diffraction of electromagnetic waves (apart from X-rays).' Laue also began each of these contributions with a historical introduction before proceeding with a very extensive integration of the research of other scientists, incorporating some briefer publications.

An important anniversary was approaching: Max Planck's 70th birthday. Physicists had already issued an appeal for donations the previous year. Max Born had the idea of awarding Planck a gold medal for his services to theoretical physics bearing his portrait on the front. Such a medal would then continue to be awarded annually to eminent physicists. At a joint meeting of the *Physikalische Gesellschaft zu Berlin* and the *Deutsche Gesellschaft für Technische Physik* on June 28th, the medal was presented to Planck, who, after expressing his thanks, awarded a second one to Albert Einstein. The post-meeting celebration took place on the premises of the "*Deutsche Gesellschaft 1914*" on Schadowstraße (*Verh. DPG*, 1928, III, no. 1, vol. 15; 1929, no. 2, vol. 26).

Max von Laue had the great privilege as a young man of having become close friends with his considerably senior colleagues Max Planck and Wilhelm Wien, as the letters from their estates attest. So the sudden death of Wien on 30 August 1928 hit Laue personally very hard. He sent a telegram in September to Planck, who replied from Winterthur: "Dear Colleague: I am deeply shaken by the contents of your telegram, after being so delighted with your nice greetings from Brand. I immediately expressed my condolences to Mrs. Wien and will visit her in Munich at the end of this mo[nth] (AMPG, V rep. 11 no. 990)." At the funeral service in Munich, Max von Laue represented the Prussian Academy of Sciences.[25] Together with Eduard Rüchardt from Munich, he wrote a detailed obituary that paid tribute to Wien's versatile and successful scientific activities. About his character they wrote:

---

[25] Planck to Laue, Brand, 8 Sep. 1928. AMPG, V rep. 11 no. 991.

One only became fully acquainted with Wien's personality, and also his scientific ways, from direct dealings with him. Be it through working in his institute, be it through participating in those winter mountain tours, to which he invited a number of his students, coworkers, and friends among his colleagues every year before the war.

Especially then, under the open sky, he radiated such enthusiasm, such a delight in research that swept up everything around him. No literature can document how many papers he inspired in this way, how often he encouraged younger colleagues in the field, or steered them onto the right path in a way perhaps only noticeable to them; one would have to ask each individual who had experienced the unforgettable joy of being allowed to go skiing with Willy Wien.[26]

The many routine duties that Laue had to perform must have distracted him from his grief. Although he had a secretary, he typed much of his correspondence himself, as his handwriting remained difficult to decipher. His mostly handwritten manuscripts were difficult to read and the copy needed to be retyped again after proofing by Laue.

The Berlin physicists did not abandon their plans for a Kaiser Wilhelm Institute of Physics even though the economic situation continued to deteriorate. Thus it was virtually hopeless to find financing for the institute from the KWG, from industry, or from the state (Zierold 1968: 108). Nevertheless, schemes continued to be drafted, as we see from a letter by Einstein in Berlin to his friend from 30 January 1929. "Dear Laue: You can count on me in every respect. I also consider it objectively right and beneficial that you be relieved of all teaching duties. I'm convinced that it would have been the right thing to do to have you become the director of the K.W. Institute, as had been planned. I found it incomprehensible at the time that you didn't stay on. This could still be rectified (UAF, Laue papers, 9.1)." Then he tried to assuage Laue: "That amongst our circles there might be persons not looking favorably upon you or not holding your scientific merits in high esteem is an unfortunate illusion that haunts your mind from time to time (ibid)." On 5 March 1929, the president of the KWG, Adolf von Harnack, was handed an important document. It was the almost poetic 'Proposal: Establishment of an Institute of Theoretical Physics as an extension of the Kaiser Wilhelm Institute of Physics.' In introduction, reference was made to the activities in July 1917 toward founding an Institute of Physical Research, which "came to life" on October 1st under the direction of Albert Einstein. Laue's substitution for Einstein was mentioned as well as that a planned small building for the institute had not yet materialized. "In the 12 years that have elapsed, the importance of such a Kaiser Wilhelm Institute has not diminished but considerably increased. The development that has occurred in theoretical physics has no equal in the history of the field and stands so mightily and successfully before us that we are almost at a loss to name its equal in the history of science (AMPG, I rep. 1A, no. 1650)." Then, in a kind of self-praise, the accomplishments in providing financial support for external research projects were described. "For this reason, one can expect great success from an institute of physics that not only distributes funds, but unites

---

[26] Laue and Rüchardt (1929). Wilhelm Wien was called "Willy" by his friends. The article appeared many months after Wien's death. It cannot be ruled out that Laue, under the impression of the loss of this patron and friend, was unable to collaborate on the obituary at first. The statement about "the unforgettable joy," I think, indicates Laue's return to equanimity. See also Wien (1930).

outstandingly qualified people in common activity. This applies the more so, as in the German Empire there is no research institute exclusively for theoretical physics, and this most creative and fruitful branch of the 'exact' natural sciences in fact has no site here, where the professional atmosphere condenses into the exhalations of a cooperating circle of suitable personalities (ibid)."

The proposed location for the prospective institute was midway between the two already existing institutes in Dahlem on the outskirts of Berlin: the Institute of Physical Chemistry and Electrochemistry under Fritz Haber, and probably the KWI of Chemistry. The proposal was cosigned by Einstein, Haber, Laue, Nernst, Paschen, Planck, and E. Warburg.[27]

'On Albert Einstein's fiftieth birthday' was the title with which Max von Laue congratulated his friend on his birthday on 14 March 1929. After an introduction, it read:

> For we become truly perplexed when we are supposed to say which of his ideas has had the most lasting influence on research. Neither can we imagine physics today without the theory of relativity of 1905, which, by eliminating the old prejudice of an 'absolute' time, solved at one stroke the partly century-old puzzles of the luminous aether, of electrodynamics, and of the optics of moving bodies, which made venerable Newtonian mechanics more precise and generally brought closure to and crowned a whole epoch of physics. Nor can we imagine thermodynamics today without the quantum theory of specific heat, nor atomic theory without the 'hv-relation,' which has maintained its position unaltered in the face of all changes to this theory and, as a direct expression of fundamental experience, is likely to maintain its position in the future as well. All this we owe to the deeds of the young Einstein.
>
> Then we have witnessed how the more mature mind, slowly, gropingly, but unflinchingly striving towards its goal, created the general theory of relativity, thus fitting the doctrine of gravity into mechanics and teaching us how to understand the equivalence of ponderous and inertial mass (Laue 1929a: 173).

Some more details followed, along with this congratulatory note: "On a birthday, for lack of anything better, one offers good wishes for the future. Einstein is entering the decade called a man's best years, not only in jest. We act as brash egotists in wishing him further success, especially fulfillment of the hope he attaches to his new unified field theory (ibid, not reprinted in Laue 1961)."

The city of Berlin wanted to give Einstein a very special birthday present, a plot of land by one of the many bodies of water in the Berlin environs. But nothing came of it. National Socialist deputies and administrative clumsiness prevented the donation from taking place. A family among Einstein's friends then took matters into their own hands and Einstein acquired a plot of land in Caputh at the confluence of Lake Templin and Lake Schwielow. A young architect, Konrad Wachsmann, designed a spacious, modern two-story wooden house and supervised its construction through

---

[27] ibid A second document with very similar content was cosigned on 6 March 1929 by H. Freundlich, R. Ladenburg, ?, M. Polanyi, Wilhelm Eitel, Otto Hahn, Lise Meitner, ?, (Otto) Warburg and Carl Neuberg. The following important note was appended: "There is no field of the exact natural sciences in which we are more blessed with extraordinary young staff. The direction of the existing Kaiser Wilhelm Institute of Physics offers us the guarantee, however, for the best possible management of such an institution".

to completion that fall. The architect donated a guest book, and Laue was the first to sign it, on 4 May 1930, with the verse:

„Einer muss der erste sein
   Der sich in dies Buch trägt ein
   Darum ohne viel zu grübeln
   Ob es Würd'gere mir verübeln
   Und trotz der mir eigenen Klaue
   setze ich an diese Stelle
   meinen Namen hin
   M. Laue"

Somebody's got to be the first
   to sign this book.
   That's why without much thought
   whether more worthy people might take it amiss
   and despite my own scrawl
   I put in this place
   my name,
   M. Laue.

Einstein could dock his sailboat there in Caputh, and the premises became a venue for Planck and other physicists whom Laue delivered by car. There they could converse and safely crack their private jokes about themselves and others (Grüning 1990: 218, 222; Grundmann 2011: 225 ff.). Laue had bought an automobile, a "Steyr," that same year (Fig. 8.17). He had been riding a motorcycle for some time by then. He also drove it to the university; that was certainly exceptional by the standards of the time. Occasionally, when Mrs. Laue's health did not allow her to join them, Lise Meitner had taken her seat on the pillion and let herself be driven out into the Berlin countryside on two wheels.

Very soon a touring book was started. Those who had been taken along on an excursion by car were asked to write something in it. On 6 May 1929 Ernest Rutherford, Lise Meitner, and Otto Hahn signed it. On May 10th, Kasimir Fajans, Hermann Mark, Hans Pelzer, and Alexander von Brill recorded their trip to Schildhorn in Grunewald.[28]

Max von Laue had been a member of the senate of the KWG since 1925, but unfortunately no statements by him about this position have been preserved. When a new body, the Scientific Council (*Wissenschaftlicher Rat*), was founded at the initiative of Fritz Haber, there was a change in the statutes that led to Laue's resignation from the senate. The president, Adolf von Harnack, thanked him for his work in June 1929. Laue then became a member of the Scientific Council.

Time and again, Max von Laue found occasion to add his own theoretical reflections about research by fellow physicists. Josef Hengstenberg and Hermann Mark

---

[28] AMPG, III rep. 50, suppl. 1, Autoreisebuch (car touring book).

Fig. 8.17 Max von Laue in his Steyr, a photograph by Otto Hahn, 1932. *Source* Archive of the Max Planck Society, Berlin-Dahlem

(1928) described an X-ray method to determine the size and shape of submicroscopic crystalline particles. Laue reported on this to his colleagues at the meeting of the academy's physics and mathematical class on 16 May 1929. He had already dealt with a similar topic in May 1926. The purpose now was to explain the method and the advance to an audience whose field of expertise was not necessarily physics. "It takes up where the well-known experimental method by Debye and Scherrer left off in investigating crystal structures, i.e., one irradiates with a homogeneous X-ray beam many such preparations containing particles wholly or partially disordered in direction, and photographs the interference cones thus formed […] (Laue 1929b: 227 f.)." It is then not the position and structure which is evaluated, but the broadening of the rings and the influence of the thickness of the preparation. Satisfactory agreement was achieved (ibid).

Laue assigned the title 'On the electrostatics of space lattices' (1930a) to a paper that appeared in the proceedings of the physical and mathematical class of the Prussian Academy of Sciences. This topic was revised again in 1940, with Laue mentioning recent publications by Hans Bethe, but adding that they were irrelevant to his considerations. The arrangement of the lattice components and the geometry of the boundary layers were described in detail.

On 28 June 1929 the golden anniversary of Max Planck's doctorate was celebrated at the Friedrich Wilhelms University. Max von Laue picked up the Plancks from Wangenheimstraße by car. The Prussian Academy (1929) dispatched an official 'Address' containing the highest praise: "As a radiant crown jewel of immeasurable value draws the eye away from all the other pearls and gems and holds it spell-bound, thus we are captivated by the deed by which, on the eve of the century, you inaugurated a new physics, the physics of the Planckian quantum of action."

In September 1929, the Laue couple drove away to Prague in their Steyr, taking Walther Bothe and Mr. and Mrs. Orlich along with them as passengers. On 9 October 1929, Max von Laue celebrated his fiftieth birthday. Max Planck published his congratulations in the journal *Die Naturwissenschaften*. He began by mentioning the connection between age and successful research noting, however, that the more recent advances had mainly come from younger researchers: "at any rate, we are confronted by the peculiar fact that the major decisive steps of recent times have been taken predominantly by younger aged researchers; and among these researchers Max von Laue belongs first and foremost (Planck 1929: 787 f.; cf. Hartmann 1936: 96)." He emphasized that Laue's discovery of X-ray interference had not been a random inspiration, "but the necessary result of a consequential intermingling of ideas, which matured in him earlier than in any other physicist because it was closely related to the problems wholly preoccupying his scientific thought (Planck 1929)." Planck spoke of Laue's urge to deepen scientific knowledge, of his "delight at imposing order and neatness in a theory; in a word, his endeavor to think every thought about physics through to the finish [...] (ibid)." After alluding to the various scientific fields that Laue had enriched, he paid tribute to the man:

> But one would not do justice to Laue's character if one did not want to regard him from another aspect: as the kind person, the loyal friend, and—that which characterizes him even more—as the helpful promoter of the rising generation. How many young persons has he guided onto the right path through difficulties of a scientific, social, or economic nature, by offering advice or a helping hand with untiring patience? It is not recorded in any journal or any report; it lives on only in the grateful hearts of those who had dealings with him in this respect (ibid).

On 25 October 1929 the global economy collapsed; it came as a complete surprise to the unsuspecting general public. Millions of people were plunged into poverty on what came to be known as "black Friday" and that accelerated the growth of radical movements.

On one of his weekly visits to the *Physikalisch-Technische Reichsanstalt*, Laue learned of a technical problem that was otherwise far removed from his scientific interests: birefringence occurring when a crystal glass pane was cut for optical purposes. Could a theoretical explanation be found for this? Could it be prevented? A local optical company, Bernhard Halle, supplied plates whose birefringence suggested a completely radial symmetry of residual stresses. Laue investigated how the residual stresses could be deduced from the birefringence after cutting with a diamond saw. In his paper he compiled the existing knowledge about the problem, adding the apology: "It will perhaps be said that few things in the first paragraphs are not already available elsewhere in the literature. But since they are rather scattered there, and therefore are probably less well known to opticians who might now find them important, the detail with which we present them here may well be forgivable

(Laue 1930b)." He summarized the result of his analysis as follows: "Thus, while residual stresses decrease to within a few percent of the initial ones, changes in thickness of 1 to 2 optical wavelengths occur. The latter might perhaps be compensated for by giving the strip, before it is cut out, a thickness variation of δ by grinding, which is oppositely equal to the expected change in thickness according to the calculation (ibid)." He expressed thanks to the PTR employees Erich Einsporn and Otto Schönrock (ibid).

The KWI of Theoretical Physics was still only real on paper. On 15 December 1929, notwithstanding the economic crisis, after consulting with President Harnack personally and by letter, Max Planck decided to submit another official reminder about the founding of a KWI of Theoretical Physics. It mentioned that Laue would work out a draft, which he sent to Harnack on 4 February 1930, who signed it off on February 7th (AMPG, I rep. 1A, no. 1650). Of course, he referred to the elaboration they had worked out together on 5 March 1929. Laue drew clear distinctions between the prospective institute and the PTR and corresponding university institutes. He also mentioned the differences to the existing Kaiser Wilhelm Institutes. Coinciding points would only exist with Haber's institute, through whose initiative "certain fields of work belonging to physics" were also being researched. "As soon as there is a physics institute, these will surely be able to be transplanted into it by friendly agreement between all parties concerned (ibid)." Laue then discussed the fields of research of the new institute and mentioned in the first place the diffraction phenomena of electronic and molecular beams. The research on this was only in its infancy and was of interest to the theory of crystalline structure. In second place he mentioned research on molecular magnetism, which had been set on a new theoretical basis by Schrödinger's wave mechanics. The new institute would also serve purely theoretical research. "To accomplish these purposes, I imagine the Institute as having two directors generally on equal footing, one scientific member, and about 4 assistants (ibid)" Laue had already taken action at his own initiative and had asked an architect for a nonbinding cost estimate. And he mentioned the sum of 1½ million Reichsmarks. In closing he quoted from an address by Ernest Rutherford to the Royal Society about the importance of research (ibid).

References

Berninger, Ernst. 1965. *Otto Hahn. Eine Bilddokumentation.* Munich: Moos
Boehlich, Walter, Ed. 1965. *Der Berliner Antisemitismusstreit.* Frankfurt am Main: Insel-Verlag
Bohr, Niels, H. A. Kramers, and J. C. Slater. 1924. The quantum theory of radiation. *Philosophical Magazine* 47: 785–802
Born, Max. 1926. Zur Quantenmechanik der Stoßvorgänge. *Z. Phys.* 37: 863–867
Born, Max, Ed. 1969. *Albert Einstein, Hedwig und Max Born. Briefwechsel 1916–1955.* With commentary by Max Born. Munich: Nymphenburger Verlagshandlung
Born, Max, Werner Heisenberg, and Pascual Jordan. 1926. Zur Quantenmechanik II. *Z. Phys.* 35: 557–615

Compton, Arthur H. 1923. A quantum theory of the scattering of X-rays by light elements. *Physical Review* 21: 483

Debye, Peter. 1922. Laue-Interferenzen und Atombau. *Naturw.* 16: 384

Debye, Peter. 1923. Zerstreuung von Röntgenstrahlen und Quantentheorie. *Z. Phys.* 24: 161

Eckert, Michael. 2013. *Arnold Sommerfeld. Atomphysiker und Kulturbote 1868–1951. Eine Biographie.* Göttingen: Wallstein

Einstein, Albert. 1921. A brief outline of the development of the theory of relativity. *Nature*, Feb. 1921: 782

Einstein, Albert. 1922. Zur Theorie der Lichtfortpflanzung in dispergierenden Medien. *Sb. Preuss. Akad. Wiss.*, Phys.-Math. Kl. 1922: 18–22

Einstein, Albert. 1998. *The Collected Papers of Albert Einstein.* Vol. 8: *The Berlin Years, Correspondence 1914–1918.* Edited by R. Schulmann, A.J. Kox, Michel Janssen, and József Illy. Princeton: Univ. Press

Ewald, Paul Peter. 1960. Max von Laue. *Biographical Memoirs of the Royal Society.* London, 6: 135–156

Gamov, George. 1928. Zur Quantentheorie des Atomkerns. *Z. Phys.* 51: 204–212

Geiger, Hans, and Walther Bothe. 1925a. Experimentelles zur Theorie von Bohr, Kramers und Slater. *Naturw.* 20: 440 f

Geiger, Hans, and Walther Bothe. 1925b. Über das Wesen des Comptoneffekts. Ein experimenteller Beitrag zur Theorie der Strahlung. *Z. Phys.* 32: 639–663

Geiger, Hans, and J. M. Nuttall. 1911. The ranges of the α particles from various radioactive substances and a relation between range and period of transformation. *Philosophical Magazine*, ser. 6, 22, no. 130: 613–621

Goenner, Hubert. 2005. *Einstein in Berlin (1914–1933).* Munich: Wiley-VCH

Grundmann, Siegfried. 1967. Das moralische Antlitz der Anti-Einstein Liga. *Wissenschaftliche Zeitschrift der Technischen Universität Dresden* 16: 1623 f

Grundmann, Siegfried. 2005. *The Einstein Dossiers. Science and Politics – Einstein's Berlin Period with an Appendix on Einstein's FBI File* (trans. Ann M. Hentschel). Heidelberg: Springer

Hammerstein, Notker. 1999. *Die Deutsche Forschungsgesellschaft in der Weimarer Republik und im Dritten Reich.* Berlin: Beck

Hartmann, Hans. 1936. *Max Planck als Mensch und Denker.* Berlin: Springer

Heilbron, John L. 1986. *The Dilemmas of an Upright Man. Max Planck as Spokesman for German Science.* Berkeley, Los Angeles, London: Univ. of California Press

Heisenberg, Werner. 1927. Über den anschaulichen Inhalt der quantentheoretischen Kinematik und Mechanik. *Z. Phys.* 43: 172–198

Heisenberg, Werner. 1975. Bemerkungen über die Entstehung der Unbestimmtheitsrelation. *Phys. Bl.* 5: 193–196

Hengstenberg, Josef, and Hermann Mark. 1928. Über die Form und die Größe der Mizelle von Zellulose und Kautschuk. *Z. Kristallogr.* 69: 271–284

Hentschel, Klaus, and Ann M. Hentschel, Eds. 1996. *Physics and National Socialism. An Anthology of Primary Sources.* Rev. reprint 2010. Basel, Boston, Berlin: Birkhäuser

Hermann, Armin. 1979. *Die neue Physik. Der Weg in das Atomzeitalter.* Munich: Moos

Kleinert, Andreas. 1993. Paul Weyland, der Berliner Einstein-Töter. In *Naturwissenschaft und Technik in der Geschichte. 25 Jahre Lehrstuhl für Geschichte der Naturwissenschaften und Technik am Historischen Institut der Universität Stuttgart.* Edited by Helmuth Albrecht, 199–232. Stuttgart: GNT-Verlag

Knopf, Otto. 1920. Die Versuche von F. Harreß über die Geschwindigkeit des Lichtes in bewegten Körpern. *Ann. Phys.* 13: 389–447

Laue, Max. 1913. Wellenoptik. In *Encyklopädie der mathematischen Wissenschaften mit Einschluss ihrer Anwendungen.* Ed. by Arnold Sommerfeld. Vol. 5, part 3: *Physik*, 359–487. Leipzig: Teubner

Laue, Max von. 1920a. Zum Versuch von Harreß. *Ann. Phys.* 13: 448–463

Laue, Max von. 1920b. Über neuere Versuche zur Optik der bewegten Körper. *Verh. DPG* 3: 17

Laue, Max von. 1920c. Über die Auffindung der Röntgenstrahlinterferenzen. Nobel lecture held on 3 June1920 in Stockholm. 16 pp. In Carl Gustav Santesson: *Les Prix Nobel en 1914–1918.* Stockholm: Norstedt & Söner; rev. separate, Karlsruhe: C.F. Müllersche Hofbuchhandlung

Laue, Max von. 1920d. Historisch-Kritisches über die Perihelbewegung des Merkur. *Naturw.* 8: 735–736

Laue, Max von. 1920e. Theoretisches über neue optische Beobachtungen zur Relativitätstheorie. *Phys. Z.* 21: 659–662

Laue, Max von. 1921a. Antrittsrede. *Sb. Preuss. Akad. Wiss.* 1921: 479–481

Laue, Max von. 1921b. Die Lorentz-Kontraktion. *Kant-Studien* 26: 91–95

Laue, Max von. 1921c. Zu Schottkys Gleichgewichtssätzen für die elektromagnetisch aufgebaute Materie. *Z. Phys.* 22: 46 ff

Laue, Max von. 1921d. *Die Relativitätstheorie.* Vol. 1: *Das Relativitätsprinzip der Lorentztransformation.* 4th exp. ed. with new Vol. 2: *Die Allgemeine Relativitätstheorie und Einsteins Lehre der Schwerkraft.* Braunschweig: Vieweg

Laue, Max von. 1921e. *Das physikalische Weltbild. Vortrag gehalten auf der Kieler Herbstwoche 1921.* Karlsruhe: Müller

Laue, Max von. 1921f. Erwiderung auf Hrn. Lenards Vorbemerkungen zur Soldnerschen Arbeit von 1801. *Ann. Phys.* 20: 283 f

Laue, Max von. 1922a. Steiner und die Naturwissenschaften. *Deutsches Revue* 47: 41–49

Laue, Max von. 1922b. Die Bedeutung des Nullkegels in der allgemeinen Relativitätstheorie. *Sb. Preuss. Akad. Wiss.* 1922: 118–126

Laue, Max von. 1922c. Röntgenstrahl-Spektroskopie. *Ergebnisse der exakten Naturwissenschaften* 1: 256–269

Laue, Max von. 1922d. Unsere jetzigen Vorstellungen von der Natur des Lichtes. *Licht und Lampe Rundschau* 26: 547

Laue, Max von. 1923a. Die Lösungen der Feldgleichungen der Schwere von Schwarzschild, Einstein und Trefftz und ihre Vereinigung. *Sb. Preuss. Akad. Wiss.* 1923: 27

Laue, Max von. 1923b. Zur Theorie der von glühenden Metallen ausgesandten positiven Ionen und Elektronen. *Sb. Preuss. Akad. d. Wiss.*, 13 Dec. 1923, 33: 349

Laue, Max von. 1924a. G.A. Schotts Form der relativistischen Dynamik und die Quantenbedingungen. *Ann. Phys.* 3/4: 190–194

Laue, Max von. 1924b. Atomaufbau und Atomzertrümmerung. *Zeitschrift des Vereins Deutscher Ingenieure* 68, no. 30: 769

Laue, Max von. 1925a. In welchem Sinne kann man von einem 'Mikroskopieren' des Feinbaus der Kristalle mittels der Röntgenstrahlen reden? *Naturw.* 8: 968–971

Laue, Max von. 1925b. Bemerkungen zu K. Zubers Messungen der Verzögerungszeit bei der Funkenentladung. *Ann. Phys.* 2-3: 261–265

Laue, Max von. 1925c. Über Röntgenstrahlinterferenzen an Mischkristallen. *Ann Phys.* 18: 167–176

Laue, Max von. 1926a. Der Einfluß der Temperatur auf die Röntgenstrahlinterferenzen. *Ann. Phys.* 25: 877–905

Laue, Max von. 1926b. Lorentz-Faktor und Intensitätsverteilung in Debye-Scherrer-Ringen. *Z. Kristallogr.* 64: 115–142

Laue, Max von. 1928. Notiz zur Quantentheorie des Atomkerns. *Z. Phys.* 52: 726–734

Laue, Max von. 1929a. Zu Albert Einsteins fünfzigsten Geburtstag. *Naturw.* 17: 173

Laue, Max von. 1929b. Eine röntgenographische Methode, Größe und Form submikroskopischer krystalliner Teilchen zu bestimmen. *Sb. Preuß. Akad. Wiss., Phys.-Math. Kl.* 1929: 227 f

Laue, Max von. 1930a. Zur Elektrostatik der Raumgitter. *Sb. Preuß. Akad. Wiss., Phys.-Math. Kl.* 1930: 26–41. Revised version in *Z. Kristallogr.* (1940) 103, no. 1: 54–70

Laue, Max von. 1930b. Über die Eigenspannung in planparallelen Glasplatten und ihre Änderung beim Zerschneiden. *Z. techn. Phys.* 11: 385–394

Laue, Max von. 1950. Nachruf auf Hans Geiger. *Jahrbuch der Deutschen Akademie der Wissenschaften* Berlin 1946–1949. Berlin 1950: 150 f

Laue, Max von. 1961. *Gesammelte Schriften und Vorträge.* 3 Vols. Braunschweig: Vieweg

Laue, Max von, Richard Bär, and Edgar Meyer. 1923. Über den niedervoltigen Lichtbogen in Helium. *Z. Phys.* 20: 83–95

Laue, Max von, and Walter Gordon. 1922. Ein Verfahren zur Bestimmung der Wärmeleitfähigkeit bei Glühtemperaturen. *Sb. Preuss. Akad. Wiss.*, Phys. Math. Kl. V, 20 Apr. 1922: 112–117

Laue, Max von, and Hermann Mark. 1926. Die Zerstreuung inhomogener Röntgenstrahlen an mikrokristallinen Körpern. *Sb. Preuss. Akad. Wiss.* 58: 586

Laue, Max von, and Lise Meitner. 1927. Die Berechnung der Reichweitestreuung aus Wilson-Aufnahmen. *Z. Phys.* 41: 397–406

Laue, Max von, and Richard von Mises, Eds. 1926–1936. *Stereoskopbilder von Kristallgittern.* Berlin: Springer

Laue, Max von, and Eduard Rüchardt. 1929. Willy Wien. *Naturw.* 17: 675–680

Lemmerich, Jost. 2011a. *Science and Conscience. The Life of James Franck.* Trans. by Ann M. Hentschel. Stanford: Univ. Press

Lemmerich, Jost, Ed. 2011b. *Mein lieber Sohn! Die Briefe Max von Laues an seinen Sohn Theodor in den Vereinigten Staaten von Amerika 1937–1946.* With two contributions by Christian Matthaei. Berliner Beiträge zur Geschichte der Naturwissenschaften und der Technik no. 33. Berlin, Liebenwalde: ERS-Verlag

Lenard, Philipp, Ed. 1921. Reprint of Soldner (1801) with a prefatory note by P. Lenard. *Ann. Phys.* 15: 593–604

Morley, E. W., and D. C. Miller. 1904. Extract from a letter dated Cleveland, Ohio, August 5th 1904 in *Phil. Mag.* 6th ser, 8: 753–754

Niggli, Paul. 1922. Die Bedeutung des Lauediagrammes für die Kristallographie. *Naturw.* 16: 391

Pauli, Wolfgang. 1993. *Wissenschaftlicher Briefwechsel mit Bohr, Einstein, Heisenberg et al.* Vol. 1. New York, Heidelberg, Berlin: Springer

Piccard, Auguste, and E. Stahel. 1927. *Neue Resultate des Michelson-Experimentes. Naturw. 15: 140*

Planck, Max. 1921. Erwiderung des Sekretärs. *Sb. Preuss. Akad. Wiss.* 1921: 481 f. (reply to Laue 1921a)

Planck, Max. 1929. Max von Laue zum 9. Oktober 1929. *Naturw.* 17: 787 f

Planck, Max. 1958. *Physikalische Abhandlungen und Vorträge. Aus Anlass seines 100. Geburtstages (23. April 1958).* Edited by Verband Deutscher Physikalischer Gesellschaften and Max-Planck-Gesellschaft zur Förderung der Wissenschaften. 3 vols. Braunschweig: Vieweg

Preußische Akademie der Wissenschaften. 1929. Adresse der Preußischen Akademie der Wissenschaften an Max Planck zum fünfzigjährigen Doktorjubiläum am 28. Juni 1929. *Sb. Preuß. Akad. d. Wiss.*, Phys.-Math. Kl. 1929: 341 f

Reichhardt, Hans J., and Dietz Bering, Eds. 1983. *Von der Notwendigkeit politischer Beleidigungsprozesse. Der Beginn der Auseinandersetzungen zwischen Polizeipräsident Bernhard Weiß und der NSDAP.* Berlin in Geschichte und Gegenwart. Jahrbuch 1983 Berlin: Landesarchiv

Scheibe, Erhard. 2006. *Die Philosophie der Physiker.* Munich: Beck

Schiebold, Ernst. 1922. Beiträge zur Auswertung der Laue-Diagramme. *Naturw.* 16: 399

Schönbeck, Charlotte. 2000. Albert Einstein und Philipp Lenard. *Schriften der Heidelberger Akademie der Wissenschaften*, Math.-Naturw. Kl. 8: 1–42

Schottky, Walter. 1920. Gleichgewichtssätze für die elektromagnetisch aufgebaute Materie. *Z. Phys.* 21: 232–241

Seeliger, Rudolf. 1921. Die spezielle Relativitätstheorie. *Jb. Radioakt. Elektronik* 17: 301

Soldner, Johann Georg. 1801. Über die Ablenkung eines Lichtstrahls von seiner geradlinigen Bewegung, durch die Attraktion eines Weltkörpers, an welchem er nahe vorbeigeht (1801). *Astronomisches Jahrbuch für das Jahr 1804:* 161–172

Sommerfeld, Arnold. 2004. *Wissenschaftlicher Briefwechsel 1919–1951.* Vol. 2. Edited by Michael Eckert and Karl Märker. Berlin, Diepholz: Verlag für Geschichte der Naturwissenschaften und Technik

Speziali, Pierre, Ed. 1972. *Albert Einstein – Michele Besso. Correspondence 1903–1955.* Paris: Hermann

Szilárd, Leó. 1925. Über die Ausdehnung der phänomenologischen Thermodynamik auf die Schwankungserscheinungen. *Z. Phys.* 32: 753–788

Szilárd, Leó. 1980. His Version of the Facts. In *Leó Szilárd, His Version of the Facts: Selected Recollections and Correspondence*. Edited by Spencer R. Weart, and Gertrud Weiss-Szilárd, 9–10. Cambridge: MIT Press

Tammann, G. 1919. Zum Gedächtnis der Entdeckung des Isomorphismus vor 100 Jahren. Die chemischen und galvanischen Eigenschaften von Mischkristallreihen und ihre Atomverteilung. *Z. anorg. allg. Chemie* 107: 1–239

Warburg, Emil, and Max Planck. 1918. *Zu Max Plancks sechzigstem Geburtstag. Ansprachen gehalten am 26. April 1918 in der Deutschen Physikalischen Gesellschaft*. Karlsruhe: C.F. Müller

Weyl, Hermann. 1922. Die Relativitätstheorie auf der Naturforscherversammlung in Bad Nauheim. *Jahrbuch der Deutschen-Mathematischen-Vereinigung. Jahresbericht* 31: 51–63

Wien, Wilhelm. 1930. *Aus dem Leben und Wirken eines Physikers. Autobiographie*. Leipzig: Barth

Wien, Wilhelm, and Friedrich Harms, Ed. 1928. *Handbuch der Experimentalphysik*. Vol. 18. Leipzig: Akademische Verlags-Gesellschaft

Wohlbold, Hans. 1921. Rudolf Steiner und die Naturwissenschàften. In *Vom Lebenswerk Rudolf Steiners. Eine Hoffnung neuer Kultur*. Edited by Friedrich Rittelmeyer. (2nd ed. 1922) Munich: C. Kaiser

Zeitz, Katharina. 2006. *Max von Laue (1879–1960). Seine Bedeutung für den Wiederaufbau der deutschen Wissenschaft nach dem Zweiten Weltkrieg*. Stuttgart: Steiner

Zierold, Kurt. 1968. *Forschungsförderung in drei Epochen*. Wiesbaden: Steiner

Zuber, Kurt. 1925. Über die Verzögerungszeit bei der Funkenentladung. *Ann. Phys.* 2/3: 231–260

Chapter 9
Physics and Politics in Berlin During the 1930s

On 16 January 1930, at the meeting of the physical and mathematical class of the Prussian Academy of Sciences, Laue presented a report about electron diffraction, starting with experimental data published by various authors. Hermann Mark and Raimund Wierl had obtained diffraction patterns from "fast-moving electrons" reflected off a thin layer of various metals and compared them against X-ray diffraction patterns. Emil Rupp at the research department of the electric company AEG had reflected electron beams off a surface of NaCl, KCl, or KBr. The theory of interference could explain the peaks of the curves obtained. Laue then presented a paper of his own: 'On the electrostatics of space lattices' and discussed his considerations of the problem. In order to arrive at a solution, he had to assume some simplifications, such as an ideal undisturbed lattice structure of cells and a dipole moment. He concluded that there is no constant mean potential in the crystal and that a potential jump occurs at the interfaces. Using a continuum theory, he then investigated the potential distribution at the corners and edges. Optimistically and self-critically, he remarked at the end of the publication: "But now that electron diffraction provides a new means of investigating the boundary layers and especially the electric fields inside them and in their vicinity, let us reiterate that these considerations constitute a first apparently necessary step toward drawing conclusions from the data (Laue 1930: 41)." Therefore, the problem was yet to be solved. Rupp additionally experimented with lead halides and tellurium halide crystals. In experiments with diamonds, the results showed no agreement with the theory.[1] Laue surely knew about the experiments by Clinton J. Davisson and Lester H. Germer (1927) on the reflection of electron beams from nickel monocrystals, which "confirmed the wave character of electrons." It is unclear why he did not mention them.

---

[1] Some of Emil Rupp's experiments at the AEG research laboratory later proved to have been "unscientifically processed" and Rupp lost his job.

© Basilisken-Presse, Natur+Text GmbH 2022
J. Lemmerich, *Max von Laue*, Springer Biographies,
https://doi.org/10.1007/978-3-030-94699-9_9

Max von Laue still longed to be liberated from lecturing duties and perhaps also to be able to head his own institute of physics within the framework of the Kaiser Wilhelm Society (KWG). He started looking for a partner experimental physicist for the envisaged institute. He knew Otto Stern from his days in Zurich and Frankfurt and the molecular beam technique he had developed. Together with Walther Gerlach at the University of Frankfurt am Main in 1921/22, Stern had been able to demonstrate the directional quantization of silver atoms in an inhomogeneous magnetic field. Laue succeeded in winning Stern over to his project. They discussed their plans with the architect Lewicki in June 1929, but in the end Stern decided that he preferred to stay at the university in Hamburg after all.[2]

On 10 June 1930 Adolf von Harnack unexpectedly died, and the Kaiser Wilhelm Society was forced to find a new president. The ministry had its own ideas about the succession, wanting to increase its influence and control over the society. However, the senate of the KWG succeeded in persuading Max Planck to accept the presidency. His election took place on July 18th. Retired State Minister Carl Becker from the Prussian Ministry of Culture became the third vice-president.[3] Planck had always expressed great interest in founding a Kaiser Wilhelm Institute (KWI) of Physics. But even now, as president, he did not see himself in a position to raise the necessary funding, owing to the general financial situation not only of the society but also of its regular donors. Although the proposals for a Kaiser Wilhelm Institute of Physics had been "set in motion," they made no progress.

Thanks to Otto Warburg, its materialization drew closer along a roundabout route (Fig. 9.1). Otto Warburg had studied chemistry at Berlin University under Emil Fischer. He was the only son of Emil Warburg, the physicist and full professor at the University of Berlin and president of the bureau of standards, the *Physikalisch-Technische Reichsanstalt* (PTR) from 1905 to 1922. Otto Warburg worked with his father as an apprentice at the PTR, where he learned about the importance of physical metrology. He then volunteered for service during World War I. When a letter from Einstein compellingly persuaded him of the nonsensicalness of his war deployment, he found a position at the Kaiser Wilhelm Institute of Biology on Boltzmannstraße in Berlin-Dahlem in 1918. His research combined metrology in physics with analytics in cell physiology. His results soon caused a stir, and in the fall of 1929 he accepted an invitation by the Rockefeller Foundation to speak about his research in Baltimore, USA. Warburg realized that for further advances to be made in cell research, closer ties with chemistry and physics were crucial. This motivated him to appeal for the creation of an institute of physics within the KWG, with Max von Laue as director.

---

[2] AMPG, I rep. 1A, no. 1671/1–5. Minutes of the meeting between Stern, Laue, and the architect Lewicki on 4 Jun. 1929 about a new building for the Kaiser Wilhelm Institute of Physics; Stern to Laue, Berkeley, USA, 26 Jan. 1930, UAF; according to Schmidt-Böcking and Reich (2011): 109. I am grateful to the authors of this biography, esp. Prof. Dr. Schmidt-Böcking for their reference to the Otto Stern papers at the Frankfurt University Library.

[3] AMPG, I rep. 1A, no. 55 fols. 47, 73, 84; AMPG, III rep. 47 no. 1408; AMPG, I, Planck staff file.

Fig. 9.1 Otto Warburg
(1883–1970) on the terrace
of the Kaiser Wilhelm
Institute of Cell Physiology,
late 1930s. *Source* Archive
of the Max Planck Society,
Berlin-Dahlem

Warburg's father Emil had been the one to approach Laue about serving as advisor on theoretical questions at the PTR. But they did not manage to raise the necessary funding. The Rockefeller Foundation appreciated Otto Warburg's ideas and at the beginning of 1930 not only offered to the KWG to construct an institute for Warburg, but was also prepared to cofinance the construction of a KWI of Physics.[4]

A high point for Berlin physicists occurred in the summer of 1930 when Niels Bohr came to Berlin to receive the Planck Medal. The award ceremony took place in the university's main auditorium. Bohr's speech was 'On the quantum of action,' which played a key role in his atomic model. A number of private invitations were attached to Bohr's visit, of course. Laue took Bohr and Planck on a tour by car. Following the physicists' convention in Königsberg, the Laue couple spent their autumn holiday in East Prussia on the Curonian Spit in Nida together with a cousin, Rudolf Laue.[5]

---

[4] Elisabeth Warburg to Einstein, 21 Mar. 1918 and Einstein to Warburg, 23 Mar. 1918. AMPG, Va, rep. 2, Einstein collection no. 12. See also: Krebs (1979): 25. On the history of the *Max-Planck-Institut für Zellbiologie*, see Jahrbuch der Max-Planck-Gesellschaft II, (1961): 816; Macrakis (1986): 348.

[5] Königsberg (Kaliningrad) and Nida (in Lithuania) are both on the Baltic; AMPG, III rep. 50 suppl. 1, car touring book. Planck to Laue, Berlin, 19 Aug. 1930. AMPG, Va, rep. 11 no. 1023.

Fig. 9.2 Rudolf Ladenburg
(1882–1952), scientific
member of the Kaiser
Wilhelm Institute of Physical
Chemistry and
Electrochemistry. *Source*
Archive of the Max Planck
Society, Berlin-Dahlem

The Weimar Republic still seemed to be stable enough to withstand the attacks from the left and the right. As a result of the World War and the economic difficulties of the interwar period, no new buildings for university institutes could be built in Germany, including one for physics. In planning a modern institute it was necessary to travel to the USA, where numerous new institutes had been built. Max von Laue had been designated several times now as the prospective director of the new institute for the KWG. So it was a matter of course that he be the one to take the voyage to America. Laue originally wanted to travel with his wife, but medical reasons spoke against this; she was asthmatic. So he chose the physicist Rudolf Ladenburg as his travel companion, who had joined the KWI of Physical Chemistry and Electrochemistry from the University of Breslau in 1924. Ladenburg had worked intensely on the theory of anomalous dispersion (Steinhauser et al. 2011: 48 f., 56) (Fig. 9.2).

They left by steamer for New York on 16 September 1930, arriving on the 27th. In total they visited 26 institutes, first in and around New York, including Bell Laboratories. Then they went to Princeton, and to the National Bureau of Standards and the Smithsonian Institution in "beautiful Washington." The other stops were Columbus and Chicago with its university. Traveling onwards to Colorado they saw Glenwood Springs and the Great Salt Lake and drove through the Sierra Nevada. Continuing on into November they arrived in Pasadena, where they met Robert A. Millikan and also visited the observatory. The next stop was the Grand Canyon before returning northeastwards, where they attended the physics convention in Chicago on November 28th. Their tour had also included the West Coast with its institutes; and the Niagara Falls had also been on the agenda, of course. The last stop was back in New York. Max von Laue had hoped to meet Einstein there, but Einstein's time was so scarce that Laue decided to write to him from there on December 13th:

> Dear Einstein:
>
> Since I couldn't manage to see poor you for a proper conversation this morning, I'd like to tell you something about the fundraising for the K.W.I. along this route. The Research Corporation, or more precisely its president, told Ladenburg that it had often donated money for scientific purposes in Germany on your advice. If they should ask you again, do perhaps think of the K.W.I. The situation at present is that the Kaiser-Wilhelm-Gesellschaft has an offer from the Rockefeller Foundation to finance the building and the initial equipment of the Institute; the only condition is that the operating budget covering a number of years be guaranteed elsewhere. This condition, as far as I know, has not yet been met. So please think of the K.W.I. when a suitable opportunity arises. In such a case it would probably be best if you only called attention to this institute and left all negotiations about the details to the Kaiser-Wilhelm-Gesellschaft.
>
> With cordial regards and warmest good wishes for your trip,
>
> Yours, M. Laue.
>
> Enclosed is a letter to your wife! (UAF, Laue papers, 9.1).

They had embarked on their return journey at the beginning of December.[6]

In February 1931, their comprehensive official report about their trip was ready. Its discussion ranged from architectural details such as room layout, to artificial ventilation and the institute workshops. Numerous important technical details were mentioned, for instance, how the various supply lines had been laid, the cabling for direct and alternating current, or equipment for the liquefaction of hydrogen and helium. Optical gratings were manufactured at three of the visited sites: in Baltimore, Chicago, and Pasadena. Ladenburg and Laue attached to their report a longer draft for the prospective institute of the KWG, and listed many points that they deemed important:

> 1) The vibrations from machinery should be kept away from the laboratories as best as possible; on the other hand, the engine house and large workshop should be located as centrally as possible and close to the laboratories, and the workshop should be illuminated with bright daylight but no direct sunlight.
>
> 2) The laboratories should be oriented mostly to the north and form a longer row in order to be able to accommodate a long light path [for optical experiments]; they should also be as equal in size as possible and where possible, pairs of laboratories should be convertible into single double laboratories.
>
> 3) Some of the laboratories should be as deep down into the ground as possible, specifically the grating room and a large thermoconstant room.

---

[6] Laue, Max von: Amerikareise. UAF, Laue papers, 11.1. Numerous postcards and small photographs of his own illustrate the text. Laue to Born, postcard, Berkeley, 29 Oct. 1930. Private collection. Laue enthusiastically described the many stops along the way, the "almost" ascent of Pikes Peak, the drive across the Great Salt Lake, and the Sierra Nevada. AMPG, I rep. 1A, no. 1650.

The next item concerned the engine room again. It continued:

4) The following laboratories should also be designed for special purposes:

1.   High-voltage rooms for X-ray and other purposes, in such a dimension as to also be suitable for extremely high voltages
2.   a magnetic room, preferably away from the other laboratories
3.   a larger chemical laboratory
4.   an open-air laboratory
5.   a vertical deep shaft in which sunlight can be directed downward and at the same time into the suite of laboratories (AMPG, I rep. 1A, no. 1650).

The authors then described the building in detail, apparently after consulting the architect at the Laternser company, who worked for the KWG. "A 3-room apartment with a separate entrance and private cellar is provided for building attendants and mechanics at the southwest corner in a special wing; above it are 6 rooms for assistants to live in (Senatsprotokoll, 15 May 1930. AMPG, I rep. 1A, no. 1650, 1245–1247)." They quoted a total cost of 930,000 Reichsmarks. Max Planck signed the report and had copies sent to the committee members Heinrich Konen, Fritz Haber, and Walther Nernst (ibid.).

The general director of the KWG, Friedrich Glum (Fig. 9.3), felt that by involving the Laternser firm without first consulting him his authority had been ignored, and pointed this out quite clearly in his letter to Laue dated 12 May 1931: "The general administration of the Kaiser Wilhelm Society cannot, of course, object to your privately contacting a firm for this purpose. I would just like to point out that, quite apart from the fact that experience has shown that the design of a building plan depends very largely on who is entrusted with the scientific management of an institute and its individual departments, any ties to an individual building firm could have unpleasant consequences for the Kaiser Wilhelm Society (AMPG, I rep. 1A, no. 1671)."

Laue's impulse to maintain order and insist upon exact exposition in the physical fields within his expertise by offering meticulous critique was not always appreciated, and he was fully aware of this. Inspired by one paper by Hans Bethe and another by the American, Philip M. Morse, on the influence of external and internal electric fields on electron diffraction in crystals, Laue published a very critical remark about their method. The paper appeared in English in the *Physical Review* in 1931: "They treat the triply periodic internal field as though it ceases suddenly at a certain plane. This is in complete contradiction to electrostatics, as one may not consider this plane as charged without coming into conflict with the atomistic foundations of the whole theory (Laue 1931a: 53)." This was followed by a mathematical treatment and solution to the problem using the Schrödinger equation. At the end, Laue cited the comparable researches by Otto Stern and his coworkers (p. 59).

Findings in astrophysics about star formation attracted widespread interest. Lise Meitner from the KWI of Chemistry presented a talk on 'Interrelations between matter and energy' at the 20th general assembly of the KWG in Berlin on June 1st. She had probably discussed this with Laue as, besides her visits and excursions with

Fig. 9.3 Friedrich Glum
(1891–1974), since 1920
general secretary, and from
1922 to 1937 general
director of the Kaiser
Wilhelm Society. *Source*
Archive of the Max Planck
Society, Berlin-Dahlem

the Laues into the environs of Berlin, they also saw each other regularly at meet-ings of the Physical Society and at the colloquium that Laue chaired. A reporter for *Vossische Zeitung* described Meitner's talk in a newspaper article dated 2 June 1931: "The speaker, one of the few female physicists of stature, showed in her exem-plary lucid expositions how literally one must take the revolutionary theorem of the new physics that mass is just a special form of energy." He reviewed Lise Meitner's explanation of the evolution of energy by the process of nuclear fusion in stars. A subsequent astrophysical paper by Laue entitled 'The propagation of light in spaces with temporally changing curvature according to the general theory of relativity,' was presented by its author at the meeting of the physical and mathematical class of the Prussian Academy. It elaborated on the recent results of the astronomers Hubble, de Sitter, and Tolman. Edwin P. Hubble had been employed at the Mount Wilson Observatory since 1919. In the late twenties, he studied the redshift of galactic radi-ation together with Milton L. Humason. They succeeded in estimating the velocity of the expansion of spherical space, and came to the conclusion that the velocity was proportional to the distance. In his introduction Laue remarked: "Hubble uses the usual Doppler formula specifically to determine the velocities, and the quadratic law of the decrease in intensity with increasing distance, to determine the distance.

We shall investigate the extent to which these laws can be applied to a space of the kind mentioned. [...] Today we can probably only say that a uniform distribution [of masses], which de Sitter and Tolman take into account in their cited papers, is not fully legitimized by observations (Laue 1931b)." Richard C. Tolman presupposed that the sinusoidal oscillation of light would remain as a sinusoidal oscillation during propagation, but this assumption was not certain. So Laue started from Maxwell's equations. Expanding space R should "remain similar," Laue assumed. He tested the legitimacy of this assumption to determine the rate of expansion of the universe from the redshift of the spectral lines and established the relation between the energy of a light wave and the rate of change of space. Laue showed that as each ray of light progressed, its oscillation frequency v would change, but Rv would remain constant. He concluded by taking the example of a nebula to point out that a long transit time for the light has an influence on its observed brightness as a consequence of the great distance. In closing he commented: "We leave aside whether this already has consequences on current astronomy (ibid.)."[7] In connection with these considerations Laue published another critical statement in *Die Naturwissenschaften* under the heading 'Formation of the elements and cosmic radiation,' which concluded: "If, then, the formation of an element is associated with emissions of a certain oscillation frequency v_0, and if this process takes place as frequently everywhere in space, then according to that hypothesis this radiation certainly does not fall monochromatically onto the Earth's atmosphere from without; rather, the farther it has traveled, the more it migrates to the lower oscillation frequencies (Laue 1931c)."

Such advancements made in confirming the theory of relativity did not diminish the number of polemical attacks on Einstein, though. Under the title '100 authors against Einstein,' a collection of rather strange pamphlets was published at Voigtländer in Leipzig in 1931 by the editors Hans Israel, Erich Ruckhaber, and Rudolf Weinmann, to which Philipp Lenard also contributed. The series also included completely unpolemical papers dealing with purely scientific questions (Israel et al. 1931; SB PK, 202 060 R).

The unstable political situation is not apparent in Max von Laue's correspondence, even though it gave cause enough for serious concern, as the radicalization of left-wing and right-wing elements continued unabated. Laue managed to find the time and the mental focus to pursue theory. Nineteen years after the discovery of X-ray interference and Laue's first publication on the theory, and fourteen years after Paul Peter Ewald had presented the first version of his theory, Laue's paper on 'The dynamical theory of X-ray interference in a new form' appeared in the periodical dedicated to the "results of the exact natural sciences." The introduction showed that Laue was also quite capable of giving praise where praise was due: "Ewald's dynamic theory of X-ray interference is, in our opinion, one of the masterpieces of mathematical physics for all time. It mastered by brilliant methods what was at first an almost insoluble problem; it actually first justified the elementary theory based purely on phase composition by teaching how to estimate its shortcomings quantitatively;

---

[7] Wolfgang Pauli praised this paper by Laue in a letter to Eugene Wigner, Princeton, 29 Feb. 1936. In Pauli (1985): 784.

it demonstrated its inner connection with the optical theory of dispersion and in so doing predicted deviations from the results of the older theory, which a few years later became indispensable for a complete interpretation of Hjalmar's observations as well as for exact measurement of the wavelengths (Laue 1931d: 133)." Then Laue pointed out that Ewald and the Swede Ivar Waller also set out from the idea of dipoles at the lattice points of the cells. The space in between was supposed to be empty. William Lawrence Bragg and others showed, however, that the space was filled almost gaplessly with electron charges. As a prerequisite Laue set: "Let us think of the positive charges as distributed in such a way that they compensate for the negative charges everywhere exactly, provided there is no interfering field (p. 134)." For the further treatment of the problem of the propagation of monochromatic X-rays, Laue employed the Maxwell equations. An important statement concluded this first section on "the basic dynamic equation": "Ewald's basic equation states physically that each of his dipoles receives from the field the excitation it needs to perform the oscillation which defines the field. We have taken this statement into account by introducing a dielectric constant. Since, the proportionality between polarization and field strength expressed by its assistance does contain the entire dynamics of the charge carriers (p. 140)." The second section dealt with "approximations in solving the basic equation," where Laue initially set out from Ewald's results with the excitation sphere. Then in Sect. 3 he gave an example of nonabsorbent crystals and arrived at the assertion: "One probably sees here best how the rigorous interference condition of the elementary theory, while not exactly invalidated, is nevertheless mitigated in a physical sense (p. 145)." The mathematical treatment for two rays followed. The fourth section dealt with the "inquiry into the reality of dispersion surfaces," which he was able to prove for non-absorbent crystals. Finally, Sect. 5 examined the "reflection of X-rays at the boundary for non-absorbent crystals." Laue discussed the "internal waves" and the "boundary waves" as well as the processes at the interface, thus returning to his interest in the analyses by Mark, Wierl, and Rupp. He was unable to present a final solution because, he concluded: "We must not do this [simplification which Ewald et al. applied to the interfaces], because a mathematical plane would generally separate off parts of the continuous charge distribution which belong to an atom, the main part of which would still be lying on the inner side of the plane. Such a demarcation would certainly not correspond to reality (p. 153)."

Paul Peter Ewald received an offprint with the dedication: "Presented by M. Laue to his dear P. P. Ewald as a belated response to a question back in February 1912"; and on the first page of the text he wrote:

Who wouldn't laud Klopstock?

But is everyone a reader? No.

Let's be less exalted.

And – more diligently read.

G. E. Lessing.[8]

---

[8] Paul P. Ewald Papers, #4586 Div. of Rare Ms. and Collections, CUL. Gotthold Ephraim Lessing (1729–1781): "Wer wird nicht den Klopstock loben? Doch wird ihn jeder lesen? Nein. Wir wollen weniger erhoben/Und – fleißiger gelesen sein".

This publication about a "new form" (Laue 1931d) also required great concentration and creative solutions. And Laue's teenage children suffered the burden it cost. When their father sat introvertedly in deep thought at the dining table, then strict silence was manditory! But when such ponderings had borne fruit and the publication was out, they could all leave on light-hearted outings. Both their parents were good swimmers, and so the lakes in the vicinity of Berlin were frequent destinations. Mrs. Laue still missed the mountains of her youth, though. The couple enjoyed many of the diverse musical offerings in Berlin. Magda von Laue also extended coffee invitations to Lise Meitner, Edith Hahn, and Ellen Hertz, the wives of Otto and Gustav, along with other guests in their home in Zehlendorf. The relations between Mrs. Laue and her own relatives were very close.[9] In September, the Laues drove away in the Steyr to Bad Elster, where the annual meeting of the German Physical Society was scheduled to take place. On the way there they visited the retired headmaster of the Strasbourg grammar school, Dr. Veil. At the conference they met up with the Millikan couple and also took them out by car.

Max von Laue had sent an offprint of his "dynamic theory" to Sir William Lawrence Bragg, who thanked him for it by letter on November 3rd and pointed out an earlier work by Charles Darwin that Laue had not mentioned (RIGBA, Bragg papers, 77 E/89).

It was a worrisome time politically and economically. What were the prospects for the future? The national parliamentary elections on September 14th had resulted in a major gain in Nazi representation at the Reichstag; the NSDAP mandates jumped from 12 to 106. Brutal clashes, especially with the Communists, continued to escalate. In the middle of 1931 there was another banking crisis and the government yet again had to declare a state of emergency and govern by decree.

Laue was still waiting impatiently for a reaction to the proposals for the construction of a building for the KWI of Physics that Ladenburg and he had submitted to the directorate of the KWG; in June 1931 there was still no response.[10] When it finally did get discussed, again no solution could be found. This became clear at the meeting of the board of trustees of the KWI of Physics on 14 July 1931, which Planck had convened as president of the society. The illustrious assembly of Albert Einstein, James Franck from Göttingen, Fritz Haber, Heinrich Konen from Bonn, Max von Laue, Walther Nernst, Max Planck, Friedrich Schmidt-Ott, Erwin Schrödinger, representatives of the KWG as well as of government ministries in Berlin discussed the building proposal. The minutes paraphrase Planck's introductory words as follows: "Great projects had often been erected in difficult times. One should not only think of the present, but should also look to the future. The importance of this Institute and its necessity were acknowledged, he contended. [...] For a variety of reasons it would be desirable to draw the new institute into a close relationship with the University of Berlin. He was thinking of an extensive cooperation, which could most certainly be brought about by appointing the director of the university institute as the director

---

[9] Personal communication to the author by von Laue's daughter, Mrs. Hildegard Lemcke, and his son Theodore.

[10] Laue to Lauda, Rockefeller Foundation, Berlin, 10 Jun. 1931. AMPG, I rep. 1A, no. 1650.

of the planned KW Institute, alongside Mr. v. Laue (AMPG, I rep. 1A, nos. 1650, 1664)." This was discussed, but no details were logged, no dates were set (ibid.).

Occasionally in science it happens that two researchers or two groups of researchers come up with something almost simultaneously (Fig. 9.4). This is what happened with the double-crystal spectrometer for X-rays to determine line width. In America it was Bergen Davis and Harris Purks and in Germany Werner Ehrenberg and Hermann Mark who invented this arrangement in 1927. This interested Laue, and he remarked about the arrangement: "The idea of the double spectrometer is to determine these defects [the mosaic structure of the crystal], or rather the dependence of the reflectivity on the angle of incidence, first independently of the shape of the line, and then to use the results thus obtained to improve the actual spectrometer measurement computationally. [... The calculation performed led to the] determination of the reflection curve for the individual crystal only if it is symmetrical. If this is the case, and one knows the point of symmetry as a function of wavelength, one can determine the intensity distribution (Laue 1931e: 472)."

From their almost daily meetings at the university and the academy, Max Planck clearly sensed Max von Laue's impatience, his restlessness. He wrote to him anxiously on November 10th. This letter also offers some insight into how difficult it was for the KWG to reach decisions and the position and influence that its president had on such decisions.

Dear Colleague:

Before we meet tomorrow, I would like to say a few things to you in response to the letter you began last night. Above all, I beg you now, when much is still in the balance not only regarding the K.W. Institute of Physics, not to take any step by which you commit yourself personally without gaining any advantage out of it. Specifically the question of whether you are willing to take over the management of a department as director at the Physics Institute, you can give your yes response once the respective offer is extended to you. I very much hope so, and will do everything I can to have this position made as palatable for you as possible. The attempt to liberate you once and for all from the pressure of the obligations oppressing you in your present university post really must work at last (AMPG, Va, rep. 11 no. 1039).

Planck added that the attitude at the ministry was positive (ibid.). But his letter on November 30th had reason to be even more personally supportive and consolatory.

Dear Colleague:

I read your letter with attention and deep sympathy. You certainly did suffer severely during your time in the military. It is completely new to me now and totally unexpected that you are still suffering from it, though. Perhaps you would like to talk to me about these things some more sometime, if that can give you some relief. Because I would very much like to do everything I can to support you.

With warm regards,

Yours, Planck.[11]

---

[11] Planck to Laue, Berlin-Grunewald, 30 Nov. 1931. AMPG, Va, rep. 11 no. 1042. There is no reference to such a conversation or the past events. Planck himself had never served in the military. Whether he knew of comparable experiences in the armed forces from his two sons is not known either. Nothing can be gathered about this from Erwin Planck's biography (cf. Pufendorf 2006).

150

Zur Theorie des Doppelkristallspektrometers.

Von **M. von Laue** in Berlin-Zehlendorf.

Mit 1 Abbildung. (Eingegangen am 9. August 1931.)

Die Theorie des Instruments wird hier von beschränkenden Annahmen befreit.

Bei Bestimmungen von Linienbreiten im Röntgengebiet mit dem einfachen Spektrometer geht in die Messungsergebnisse die Mosaikstruktur des Kristalls ein. Die Idee des Doppelspektrometers besteht darin, diese

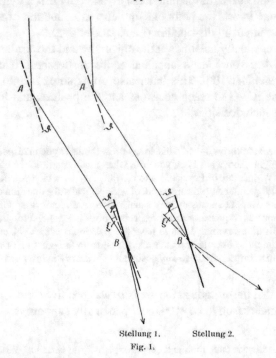

Stellung 1.   Stellung 2.

Fig. 1.

Fehler, oder besser die Abhängigkeit des Reflexionsvermögens vom Einfallswinkel, zunächst unabhängig von der Linienform festzustellen und dann die so gewonnenen Ergebnisse zur rechnerischen Verbesserung der eigentlichen Spektrometermessung zu benutzen. Deswegen kann der zweite der beiden Kristalle, B in Fig. 1, sowohl um die Parallelstellung (Stellung 1) ein wenig geschwenkt werden, als um jene andere „Antiparallelstellung"

Fig. 9.4 A sketch of the two positions plotted by the double-crystal spectrometer for computational improvement of a spectrometer measurement, 1931. *Source* Laue (1961a), II: 150

A new scientific problem preoccupied Laue's thoughts for almost three decades: Superconductivity. He published sixteen articles and one book on the topic in several editions until 1952. Various physicists, whether experimentalists or theoreticians, played a part in "reconnoitering" this effect. Thus, unfortunately, unscientific disputes arose again. The discovered phenomena were too complex, the experiments at the temperature of liquid helium, too difficult. The experimental advances in low temperature physics in the twentieth century initially were only made in London, at the Royal Institution of Great Britain under the direction of James Dewar, and in Leyden, at the University under Heike Kamerlingh Onnes. In 1906 Kamerlingh Onnes succeeded in liquefying hydrogen, and in 1908 helium. Measurements of the conductivity of metals as a function of temperature was carried out with a Wheatstone bridge. Good contact was achieved by leads consisting of mercury in thin glass tubing. Gilles Holst discovered superconductivity in mercury in 1911 while experimenting. In the original publication Kamerlingh Onnes thanked Holst and another collaborator. A current once induced in a superconducting ring flowed without losses. Some physicists, including Einstein, attempted to postulate a theory on the phenomenon, but only partly succeeded. A technical application for superconductivity was also considered, but that was still much too optimistic.

The president of the bureau of standards (PTR) in Berlin, Friedrich Paschen, secured the financial means to work in the field of very low temperatures there during the 1920s. Thanks to the fortunate choice of Walther Meissner as head of this new field of research and his prudent engineering work, the Laboratory for Low Temperatures could be established (Fig. 9.5). Meissner had first studied mechanical engineering at the *Technische Hochschule* in Berlin, but shortly before taking his final examinations he switched his focus to physics and earned his doctorate under Max Planck. That is how Laue made his acquaintance. In 1908, Meissner joined the PTR. He also had performed military service during World War I. Only afterwards was he able to set up the cryogenic laboratory which, in 1927/28, became the third location where liquid helium experiments could be conducted. The second one was in Toronto, Canada. Meissner broadened the scale of superconducting elements to include the rare earths, such as niobium and niobium carbide, as well as alloys.

It took some time for the friendship between Max von Laue and Walther Meissner to deepen because, unlike Laue, Meissner generally had less free time, no Saturdays off and much shorter holidays. The Laue family had to plan their vacations according the children's school schedule, which did not coincide with the much longer university term breaks. In the spring of 1932 there was a trip by car to Riesa and at Easter an excursion to Magdeburg. Ernst Orlich, a professor at the polytechnic in Berlin-Charlottenburg, was another frequent passenger in their car. When the *Kaiser* granted permission to polytechnics like his the right to award doctorates, the universities

---

It is possible that Laue had once been incarcerated. A personal letter to him offers some hint of this. Laue made the following remark in his selfportrait (1961b: XXVI): "But inwardly my stance was in stark contrast to the whole essence of the military. Setting down which nonmilitary circumstances increasingly aggravated this until finally leading to grave illness goes beyond the scope here."

Fig. 9.5 Walther Meissner
(1882–1974), a friend and
colleague of Laue's who
often joined him on outings
by car. *Source* PTB, after a
photograph in the author's
estate

made a point of insisting that their doctoral titles be distinguished by an abbrevia-
tion in German script, in Sütterlin. Max von Laue never gave this kind of academic
discrimination much notice.

An important honor, the award of the Max Planck Medal by the German Physical
Society, was bestowed on Laue in the spring of 1932. The person thus honored
usually gave a speech within his field of research. Max von Laue chose neither X-ray
interference nor relativity. He gave some theoretical commentary on the dissertation
by Gerardus J. Sizoo at the University of Leyden on a problem of superconductivity,
to which he applied Maxwell's theory. Laue started by investigating theoretically
the different influences exerted by a longitudinal and transverse magnetic field on a
superconducting wire and a circular cylinder, respectively. He then analyzed Sizoo's
experiment with two coils of different wire thicknesses connected in parallel under
superconductivity. Sizoo could not detect any magnetic field. Laue stated: "Between
several superconductors connected in parallel, the current is distributed in such a way

that the magnetic energy becomes a minimum (Laue 1932a: 793 ff.)." This could only be explained theoretically—deviating from Kirchhoff's law (ibid.).[12]

However, the broad topic of "X-ray interference and matter" never let Laue go. Data obtained by others served as his basis for studying the interference on cross gratings. There were divergent results on this. He analysed the publication by Y. Sugiura from Japan on proton diffraction off a metal layer. The introductory mathematical part was illustrated by two drawings. In Sect. 3 he applied considerations from his 1926 paper to cross lattices. First he treated a completely transparent cross lattice, then an opaque one. The outcome was that Sugiura's results could only be interpreted qualitatively (Laue 1926, 1932b).

Heisenberg had already published his theory of uncertainty and the invalidity of the law of causality in (1927) under the title 'On the visualizable content of quantum-theoretical kinematics and mechanics.' Not only physicists, especially those at the meeting in Como in honor of Alessandro Volta, but also philosophers and many other people tried to outline the consequences of Heisenberg's postulate. But initially neither strict rejection nor approval of his conclusions brought clarification of the issue. The journal *Naturwissenschaften* honored its editor Arnold Berliner on his 70th birthday. Einstein wrote a laudatio and Max Bodenstein paid tribute to the jubilarian's accomplishments. The first article in the issue was entitled 'On the discussions about causality' and immediately presented to the reader von Laue's stance on what to him was a fundamental question: "In the philosophical literature and, more recently, also in physics, a debate is taking place about the epistemological foundations of physics commonly subsumed under the 'principle of causality'. Quantum theory has given cause for it to become such a live issue right now, which in its present forms leads to the Heisenberg uncertainty relations; but these relations must indeed give rise to skepticism about the unequivocal determinacy of natural processes. We shall not comment here on the philosophical portion of this debate (Laue 1932c: 915)."

A historical introduction followed on how Newton and Galileo treated the initial locality and momentum of a mass point. With the elimination of the possibility of simultaneously determining location and momentum with equal precision, the system of mechanics accordingly collapses. "And now has come to pass what was bound to happen sooner or later. Facts have emerged which refute the hypotheses made in mechanics. This is a process such as physics has often experienced; one need only think of all the transformations in optics, thermodynamics, or electrodynamics (ibid.)." The term "mass point" reappeared again in connection with the discovery of spin and Bohr's wave-particle duality. "We are confronted here by the quantum

---

[12] Laue had received Sizoo's dissertation from Walther Meissner. In his memoirs, twelve years later, he wrote: "It was known that a sufficiently strong magnetic field cancels it [superconductivity] out; but the measurements showed an incomprehensible dependence of the necessary field strength against the axis of the wire; only in this form were those metals examined at that time. Then it occurred to me that the superconducting wire itself distorts the field, in the sense that considerably greater field strengths occur at its surface than are measured at some distance away from it. My assumption that the same field strength was indeed always required to cancel superconductivity could now be easily grasped quantitatively and transferred to other body shapes, e.g., superconducting spheres." Laue (1961b): XXV, see also Gavroglu (1995): 114.

puzzle: what is a body? This problem makes the question that the indeterminacy relations answer moot; there is no such thing as a mass point, so one need not worry about its position and momentum. [...] The present forms of quantum theory are now attempting to salvage the 'mass point'. This then inevitably leads to the uncertainty relations precisely due to those wave processes; and from these relations, in turn, they conclude that physics must dispense with the causal interpretation of an individual process and confine itself to the establishment of statistical laws. We do not mean to censure this procedure at all; it probably represents the best way out *at present* (ibid.)." Max von Laue mentioned the successes of Newton's theory and Einstein's general theory of relativity, and in closing again stressed the significance of the evolution of physical concepts: "But these difficulties can compel no one to change his epistemological standpoint, whatever it may be; although, as with any profound physical question, they point again to the importance of epistemological considerations (ibid.)."[13]

In June 1932 Max Planck delivered a long lecture on 'Causality in nature' at the Physical Society of London. This Guthrie Lecture of his went into the philosophical foundations postulated by Kant in order to discuss various physical quantities and their measurement. Planck regarded quantum physics as a means of transgressing any limit set by the perceptible world by providing increasingly precise measurement methods.

> In contrast, the wave function of quantum mechanics initially offers no grounds at all for an immediate interpretation of the sensory world. The name 'wave', however descriptively and appropriately chosen, should not blind us to the fact that the meaning of this word in quantum physics is quite different from its former meaning in classical physics. There, a wave denotes a certain physical process, a sensually perceptible motion or an alternating electric field accessible to direct measurement. Here it only denotes, as it were, the probability of the existence of a certain state [...].
>
> Such considerations have prompted the indeterminists to renew their attack on the law of causation (Planck 1944: 223 f.).

Planck then discussed various physical phenomena and their assessment and explanation by determinists and indeterminists. He then commented on the importance of free will and the assumption of an ideal intelligence. He was critical of the view that the concept of probability should be made the ultimate basis of physics. "Just as the law of causation already immediately takes hold of the awakening mind of a child and puts into its mouth the interminable question 'Why?', thus it accompanies the researcher throughout his life and incessantly poses new problems to him (ibid.)."[14]

Obviously, nothing in Planck's talk hinted at the political situation in Germany. As president of the KWG, he was aware of the society's uncomfortable financial situation. But was science, his beloved physics, independent of politics? Were the anti-Semitic attacks on Einstein and the special as well as the general theories of

---

[13] Wolfgang Pauli did not agree at all with Laue's views, see Pauli (1979): 405, 2nd footnote.

[14] Planck also spoke at the Prussian Academy on the 'Causal concept in physics.' A shortened version appeared in its proceedings for (1932): 13.

relativity the only targets? At the Friedrich-Wilhelms-Universität, the Faculty of Philosophy had to find a successor to Walther Nernst. The preferred candidate, also for Planck, was the Nobel laureate James Franck from Göttingen. At the same time, Haber wanted him to become his successor at the KWI. A few months later, politics and anti-Semitism were to make that unfeasible. James Franck emigrated with his family to the USA (Lemmerich 2011a, 2011b: 182–189).

The Schrödingers hosted one of their popular "sausage evenings" on December 3rd, which Lise Meitner and Max Planck also attended. Max von Laue and Otto Hahn had declined.

At the end of May 1932, Reich President Hindenburg had accepted the resignation of Brüning's government, and a government he appointed with von Papen as chancellor remained in office only until early December. Who at the ministries supported funding the KWG? Planck's son, Erwin, had no influence on this. He was serving as a nonpartisan state secretary in the Reich Chancellery, and in the same position in the next government, which was installed by the Reich President—without elections—on 3 December 1932. It was led by General Kurt von Schleicher, a confidant of Hindenburg.

On 5 January 1933, a long article appeared on the first page of the morning edition of the *Rheinisch-Westfälische Zeitung*, signed "B J G A" under the headline 'About a novel idea for the state.' The author was very probably the general director of the KWG, Friedrich Glum. It was a critical view of the current political situation and the achievements of the parties: "What we need, but have lacked for a long time, is: To lay the foundation for an independent and active foreign policy which will lead us out of isolation, protect us against the impending threat from the East, grant support to endangered German culture beyond our borders, and make room for our people commensurate with its diligence and intelligence in the anticipated rivalry among nations for a somewhat enlarged scope in the global food market which is still deficient (BJGA 1933; cf. Lemmerich 2015: 80)." The author regarded a stable government as the key to realizing a purposeful foreign policy. In this context he named Hitler, Brüning, Papen, and Hugenberg. Then he discussed the possibility of stabilization: "We do not have this option in Germany right now, because the parties of the Right were unable to find each other just at the time when their great hour seemed to have struck (ibid.)." He briefly touched on the relationship between the civil service and the army which, he contended, were no longer real factors of power: "So, do we go back to the monarchy? This question is not properly posed. In history there can be no going back (ibid.)." And at the end of the paragraph he wrote: "The goal of this idea of the state can only be one thing, a goal that Stein and Fichte already had in mind: greatness for the nation (ibid.)." Then he elaborated on his thoughts: "Working towards achieving this calls for many ways and many endeavors. But this is the essence of practical politics. It seems to us that even in this Republic the national parties of the Right would still have to do much to develop the national idea of the State in such a way that it really be capable of establishing permanent authority (ibid.).

Politics aside, the KWG's efforts to attract sponsors by organizing interesting lectures at Harnack House continued. The general administration repeatedly tried to persuade the society's own scientists into presenting such lectures. In the summer of 1932, Max von Laue had agreed to offer a lecture at Harnack House on thermodynamic fluctuation phenomena.

Professor Henry Siedentopf from the Zeiss Works in Jena came with a special projector to demonstrate Brownian molecular motion and other phenomena to Laue's audience on 11 January 1933. Laue requested that a small table be set up near the lectern so that his secretary could take shorthand notes of the lecture. This text never appeared in print though. In Planck's absence from Berlin, His Excellency Schmidt-Ott agreed to introduce the speaker in his stead (AMPG, I rep. 1A, no. 808 fol. 73).

Parliament was convened on January 24th by Hermann Göring, a member of the Nazi party (NSDAP), as president of the *Reichstag*. Was another change of government imminent? On this day a "small dinner" was held at Harnack House under the chairmanship of President Max Planck, to which 40 gentlemen were invited: Reich Chancellor Kurt von Schleicher, Reich Minister Franz Bracht, several state secretaries including Erwin Planck, military officers, several chief editors and professors. There is no documentation of the subject of their discussions![15]

Von Schleicher's unelected government had been in office by appointment of Hindenburg since the beginning of December 1932. On January 28th, Kurt von Schleicher resigned because Hindenburg had refused him broad powers to deal with the crisis. The new government of "national concentration," again unelected, was supposed to overcome the crisis. Adolf Hitler was appointed Reich Chancellor by Hindenburg on January 30th. Only Ministers Wilhelm Frick and Hermann Göring were Nazis, two were members of the German National People's Party, and six ministers were non-partisan (Verhandlungen des Reichstags, vol. 457/58, 1933: 28 C). How long would Hitler's cabinet rule? Only a matter of months again? Did the KWG leadership have to conform to the new circumstances? The editorial in the morning edition of *Vossische Zeitung* of 31 January 1933 (p. 2) commented critically on the political developments: "Reich President von Hindenburg has assumed immense political responsibility by dismissing the Schleicher government, for which no objective cause is apparent to us, and even more so by appointing the Hitler cabinet with its incalculable risks (This soon came to be referred as the "*Machtübernahme*"— the takeover of power.)." On February 14th another "small dinner" was held for 44 people at Harnack House. The former Reich Chancellor Heinrich Brüning and His Excellency Schmidt-Ott were invited, two members of the NSDAP came as well as General Stülpnagel and Admiral Raeder. Erwin Planck, who had immediately resigned from his post at the change of government, did not partake in the meal.

On February 27th, parts of the *Reichstag* building went up in flames. The National Socialists accused the Communists of arson in order to persecute them without legal basis, to arrest and eliminate them politically—and execute them. Hitler and his government could not yet legitimize themselves through an election; that happened on March 5th. With a turnout of 89%, the National Socialists achieved 44% and

---

[15] An account of the political events as experienced by Erwin Planck can be found in Pufendorf (2006): 278–312.

together with the reactionaries of the "Black-White-Red Front" they barely reached 52%. A spectacle, the "Day of Potsdam", was staged to mark the historical 21st of March. Hindenburg and Hitler ceremoniously made their way to the coffin of Frederick the Great in Garrison Church; Hindenburg had stood there before as a young lieutenant in 1866. That solemn handshake sealed Hitler's legitimacy. A public flagging order had gone out. Lise Meitner wrote to Otto Hahn, who was in the USA, from Berlin on 21 March 1933: "Here, of course, everything and everyone is under the impression of the political upheaval. Today is the ceremonial opening of the Reichstag in Potsdam. Last week, the K.W.G. already instructed us to hoist the swastika flag next to the black-white-red one, at the K.W.G.'s expense (AMPG, III rep. 14 no. 4869; CAC, MTNR 2/15, diary of Lise Meitner)." Meitner and two friends of hers had listened to the radio broadcast of the ceremony. "It was quite harmonious and dignified. Hindenburg made a few short statements a[nd] gave the floor to Hitler, who spoke very moderately, tactfully, and personally. Hopefully it will continue in this spirit (ibid.)." Then she mentioned that periods of transition are the precondition for all sorts of blunders, "it's almost inevitable." (ibid.) On March 24th, the Enabling Act was promulgated, abolishing democracy, and perfecting the dictatorship. Hitler proclaimed at the *Reichstag*: "The irremovability of judges, on the one hand, must be set in relation to the elasticity of judgment for the purpose of the preservation of society. The individual cannot be the focus of legal concern, but the nation! Treason and national betrayal shall henceforth be gutted out with barbarous ruthlessness! The grounds of the existence of the judiciary can be none other than the grounds for the existence of the nation (Parliamentary proceedings, vol. 457/58/1933: 28 C)."

At the beginning of 1932, Einstein had discussed with the ministry, the academy, and Planck about how his salary should be treated, because he wanted to accept an invitation to collaborate at Abraham Flexner's new Institute for Advanced Study in Princeton for a few months each year. His salary was to be withheld for this period, and the academy reserved the right to negotiate with Einstein about how it be dispensed. Einstein was loosening his ties with Germany, with Europe, with his many friends and patrons. At the end of 1932 he and his wife left Berlin. His secretary Helen Dukas, and his collaborator, the mathematician Walther Mayer, also turned their backs on the city in October 1933. Einstein initially traveled to Pasadena. Thus, at first, he only became aware of what was happening in Germany through the reports he read in the papers, and said what he thought about it. He did not stay long in the United States and returned to Europe, first to Belgium, then to England. Many newspapers took advantage of his publicly expressed opinions to boost their sales. During his return voyage, still on board ship, Einstein informed the Prussian Academy of Sciences on 28 March 1933: "The current state of affairs in Germany compels me to resign herewith from my position at the Prussian Academy of Sciences. For 19 years, the Academy has given me the opportunity to devote my time to scientific research, free from all professional obligations. I know how very much I am obliged to her. It is with reluctance that I withdraw from this circle, also because of the intellectual stimulation and the fine human relationships I enjoyed throughout this long period as a member and have always valued highly. But under the present circumstances I consider my position's inherent dependence upon the Prussian government intolerable. With all due respect, Albert Einstein (Akad. der Wissenschaften der DDR 1979: 246; trans. Grundmann 2005: 281–283)." Einstein's letter to the academy arrived on March 30th and was read at the plenary session on the same day. The excerpt from the minutes stated: "The Chairman [von Ficker] then reported on a decree by the Ministry of March 29th of this year, in which the Academy is requested to examine whether the newspaper reports according to which Mr. Einstein is supposed to have taken part in the German agitation in America and France are true and, if necessary, to apply for disciplinary proceedings against Mr. Einstein. The Academy is of the opinion that Mr. Einstein's withdrawal has obviated further measures on its part. Report should be made to the Ministry to this effect (p. 282)." This occurred on the same day as well. The short notice was signed by the chairman and the lawyer Ernst Heymann, permanent secretary of the philosophical-historical class. Did this close the "Einstein case"? The members of the academy did not know what exactly Einstein had said. What had the reports altered? Heymann took action on April 1st, saying that the ministry expected the academy to issue a public statement about this case. Heymann's text, published in many German newspapers, read: "The Prussian Academy of Sciences was shocked to learn from newspaper reports about Albert Einstein's participation in the slander campaign in America and France. It has demanded an immediate explanation from him. Einstein has since given notice of his withdrawal from the Prussian Academy of Sciences on the grounds that under the current government he could no longer serve the Prussian state (p. 283)." Heymann then went on to mention Einstein's Swiss citizenship as well as his Prussian one, which he had assumed in 1913. "Einstein's agitatorial behavior abroad is particularly offensive to the Prussian Academy of Sciences, because it and its members have felt intimately attached to the Prussian state since times past; and for all its strict restraint in political matters, it has always emphasized and preserved the national idea. For this reason it has no cause to regret Einstein's resignation. For the Prussian Academy of Sciences, Heymann, Permanent Secretary (ibid.)." This independent action immediately elicited protests from its members Gottlieb Haberlandt and Max von Laue, who wrote to von Ficker demanding that a plenary session be convened on April 6th to "raise the issue of this statement." In a letter dated April 3rd the attempt was made to persuade Friedrich Paschen to undersign a joint declaration but even this statement distanced itself from unsubstantiated allegations attributed to Einstein: "after all [the Prussian Academy of Sciences] is not a political body but a scientific one and must therefore, while fully acknowledging the incompatibility of Einstein's political utterances with his former post, must nonetheless regret losing with him one of the most brilliant members that it has ever had (ibid.)." That meeting convened on April 6th, and the minutes read under point (5): "The Academy retrospectively approves Mr. Heymann's measure and expresses its gratitude to him for his appropriate actions. […] I. Addendum. At the request of Mr. von Laue it is stated that no member of the phys.-math. class had an opportunity to have a say on the Academy's statement of 1 April 1933 on the Einstein case (ibid.)."

On April 1st, the anti-Semitic terrorism manifested itself. A public strike was declared against Jewish shops; they became targets of destruction and devastation.

On April 7th, the "Law for the Restoration of the Professional Civil Service" was promulgated. Many newspapers, including the *Vossische Zeitung*, carried the text:

> (1) Civil servants who are not of Aryan descent are to be placed in retirement (§§ 8 et seq.); in the case of honorary officials, they are to be dismissed from office.

> (2) The above clause does not apply to officials who had already been in the service since the 1st of August, 1914, or who had fought in the World War on the frontline for the German Reich or for its allies, or whose fathers or sons had been casualties in the World War [...].[16]

Almost all universities and colleges had numerous teachers and faculty members who were Jews, as did the Friedrich Wilhelm University, the Prussian Academy of Sciences, and the Kaiser Wilhelm Society for the Advancement of Science. Einstein—who had lost his post as director of the KWI of Physics—was not the only one among Max von Laue's friends and acquaintances to fall under the purview of the law. Max Born, James Franck, Fritz London, Peter Pringsheim, and Otto Stern were others, as well as his doctoral student Leo Szilárd, to name but a few. Lise Meitner did not disclose that her family was Jewish until after 1933. Laue attacked the implementation of this law, he defended Einstein, he stood by his friends. These were turbulent weeks and months. Laue was probably not yet aware of the danger it posed to him personally. Max Planck was at this time on vacation in Taormina, Sicily. He was informed of the events in Berlin by the general director of the Kaiser Wilhelm Society, Friedrich Glum. On 18 April 1933, he wrote back to him to thank him for his telegram and that his message "was very reassuring and appreciated, because upon my arrival here I found together with it some other reports that were hardly encouraging. Mr. v. Laue even sent a telegram: 'Personal presence urgently desired.' However, I know my revered and dear colleague too well not to know that he can become very upset and then be somewhat unreasonable, and answered him by letter that I would only abort my journey if that were in the interest of the K. W. S[oc]., and in this respect I would be guided solely by the official reports from there. As, I am only assuming the chairmanship of the Academy on May 1st." He concluded, "Quite unpleasant things have been happening over there, though, on the occasion of Einstein's resignation; but if I should have had to bear the responsibility for that, I would not have embarked on this trip at all (BAK, N 1457, Glum papers)." He had already replied to Max von Laue on April 16th:

> My dear Colleague:
>
> Your telegram has naturally given me much to think about. On the one hand, I consider it a self-evident duty to make myself [ever] available for participation in all important decisions. On the other hand, it is an important practical duty of mine to use the vacation to gather my forces, which I am going to need for the straining weeks in May. Now, I do not know to which matter your telegram refers. And so I envisioned for myself the following. If the K.W. S[oc]. should somehow come into play, I would definitely cut my journey short and immediately come back to Berlin. Because that's where I bear the ultimate responsibility for everything that happens. But it is precisely there that I am confident, in that both His

---

[16] Gesetz zur Wiederherstellung des Berufsbeamtentums (7 April 1933), Reichsgesetzblatt (1933), part I: 175, reprinted in Meier-Benneckenstein (1935): 172–175; English trans. in Hentschel (1996): 22–23.

Ex. Schmidt-Ott as well as Glum and v[on] Cranach have expressly assured me that in this respect I can completely rely on being informed by them in time in case my departure from here should become necessary (BBAW, Laue papers, U VIII).

Max Planck then went on to discuss his duties in the Academy of Sciences. Max von Laue's position at the university did not allow him to exert any influence on the dismissals that were now pending. Paul Peter Ewald was rector of the Stuttgart Polytechnic in 1933. He resigned from his post on April 20th in protest against this defamation of Jews. But then—who would fill the post in his stead? (Lemmerich 2011a: 196).

In the 8th issue of the *Physikalische Zeitschrift* of 15 April 1933, Laue congratulated his revered teacher on his 75th birthday on April 23rd on behalf of the German Physical Society. He would have liked to have wished Planck "Otium cum dignitate"—leisure with dignity—, but it was unlikely that this would be granted him owing to the anxiety about the future of German science and Planck's offices as president of the K.W. Society and as secretary of the academy. His closing words were: "We wish him many more years. Because—we need him (Laue 1933a: 305; the manuscript was submitted before April 7th.)."

In the issue of April 21st, a laudation appeared by Fritz Haber 'On the seventy-fifth birthday of Max Planck, president of the Kaiser Wilhelm Society' in *Die Naturwissenschaften*. Haber paid tribute to Planck's commitment to the society, despite his advanced age. In his typical style it reads: "The effectiveness of a president of the Kaiser Wilhelm Society who performs his post is that of an eminent man in the position of an English king. His constitutional power is small. But in truth he is capable of effectuating all that is essential by virtue of his personality. […] In the three years of the global crisis during which he has held the office, it has naturally not been possible for President Planck, under the pressure of the economic emergency, to expand the scope of research by the Society through the construction of new institutes (Haber 1933: 293)." Then he praised how Planck had succeeded in assuring that: "the vigor and fruitfulness of the whole stand undiminished (ibid.)." The "Law for the Restoration of the Professional Civil Service" affected two department heads and sixteen assistants at Haber's Kaiser Wilhelm Institute. As a war veteran, Haber was not covered by the law and tried to keep the institute running smoothly and particularly to offer financial help to those who were affected by it. He agonized about whether he should bow to the demands of the National Socialists or whether he could resist them. Lise Meitner and Max von Laue stood by him in making this difficult decision. His friend and chosen successor, James Franck, had resigned his professorship in Göttingen in protest in April and was preparing to emigrate. Haber decided on April 30th to follow Franck's example and resigned his post in demonstrative protest against the Nazi stipulations. In a handwritten letter dated May 2nd, he asked Laue to replace him in office until Otto Hahn had returned from America, who should then step in. Fritz Haber emigrated to England.[17]

---

[17] DPGA 10011. Steinhauser et al. (2011): 102, with a copy of Haber's request; Stoltzenberg (1998); Szöllösi-Janze (1998); Laue to Meitner, Göttingen, 30 Jun. 1948 in Lemmerich, ed. (1998b): 519; Haber to Laue, 2 May 1933, DMA, (1964), 6 (30–34).

On May 10th, the flames from the book-burning campaign, mainly in university towns, were set ablaze. Students and other participants ransacked libraries. The works of Jewish authors, but also of non-Jews were thrown on the bonfires. In Berlin, this spectacle took place directly across the street from the university. Einstein's books were also tossed into the fire (see also Laitko et al. 1986: 32). On May 14th, Laue wrote to his friend, foreseeing the omnipotence of Nazi policy:

Dear Einstein,

I take this opportunity to send you a greeting, after an offprint with suitable dedication that I sent you has not, as far as I know, reached you. I hardly need to assure you that I am very sad about the events here: the worst is the complete powerlessness to do anything about it. Whatever would cause a stir only makes the situation worse.

With your departure from here, Berlin has for a large part become deserted to me, despite Planck, Schrödinger, and some others. But why did you have to be so politically prominent about it?! Far be it from me to reproach you for your statements.

I only think that a scholar should show restraint. The political contest requires other methods and other interests than does scientific research. As a rule, a scholar gets caught up under the wheels. That's what happened to you, too. What once was cannot be put back together again from the shattered pieces. But even if we should henceforth only rarely see each other, I think we shall both keep each other in good memory.

In old friendship.

Yours, M. Laue (Hoffmann and Walker 2007: 531, citing AEA, no. 16 088).

Max Planck became painfully aware of how close to reality Laue's remark to Einstein was. Adolf Hitler granted Planck's request as president of the Kaiser Wilhelm Society for an audience, on May 16th. It was evidently completely futile. When Planck tried to intercede on Fritz Haber's behalf, Hitler became abusive and railed at all Jews as communists. Planck retreated.[18]

Max von Laue tried to find positions abroad for his dismissed colleagues; in this matter he wrote on 21 May 1933: "Dear Sommerfeld, I thank you for your two letters; I will report the names of those affected by the 'Civil Service Law' next week when several lists have been compiled, firstly to P. Ehrenfest at Leyden, secondly to the Hamburg banker Dr. C. Melchior (Hamburg 1, no. 75 Ferdinand St.). The former is conveying these names to a Dutch organization, which will probably soon merge with the international relief organization for expelled scholars, the latter is conveying them to the relief organization of German Jews. What will happen then, I cannot say at present (DMA, NL 089)."

The dialogue about whether an academic be politically active continued. Einstein was still at Oxford and wrote on 26 May 1933:

Dear Laue,

I have received both copies of your beautiful summary work and thank you very much for it. I can imagine how you feel. For, these things go far beyond the personal in their significance. It is like an exodus from below, a trampling of cultivation by crudity.

---

[18] Planck (1947): 143, trans. in Hentschel, Ed. (1996): 359–361; see also Albrecht (1993): 41–63 and Henning (2004): 69 f. with a critical assessment of Planck's account of the visit.

I do not share your view that a scientist should remain silent in political, i.e., human affairs in the broader sense. You can just see from the conditions in Germany where such self-restraint leads.

It means leaving the leadership to the blind and irresponsible without resistance (UAF, Laue papers, 9.1.).

Einstein offered Giordano Bruno, Spinoza, Voltaire, and Humboldt as examples: "I don't regret a single word of what I said, and believe I have thereby done a service to people. Do you think I regret not being able to stay in your country under such circumstances? This would have been impossible for me even if I had been wrapped up in cotton wool. My feelings of warm friendship for you and a few others there remains. I hope we shall see each other again in better times (ibid.)." Then he mentioned a paper he coauthored with his assistant, Mayer, on a theory by Dirac (Einstein 1975: 234). Einstein's letter was delivered rapidly. Laue replied soon afterwards, on May 30th, setting forth his notions about the politically-minded scholar:

I maintain, moreover, that the likes of us should not actively interfere in politics. You cite Giordano Bruno, Spinoza, Voltaire, and Humboldt as examples of scholars who did. In doing so, you overlook, as far as I can tell, some rather significant differences. To what extent Giordano Bruno was a scholar I cannot say, owing to my complete lack of education concerning this man. Voltaire certainly was not, even if he was supposed to have written some truly scholarly essays. With Spinoza and Humboldt, on the other hand, I don't know how much they were 'active' in politics. But none of these were representatives of any kind of exact science! Can you name me one mathematician, one physicist, one chemist of repute, who had any success in concerning himself with politics? These scientists—whether one regrets it or not—are simply otherworldly; they make any intrinsically broad mind who deals with them professionally become otherworldly. Whoever engages in history, ethics, law, is so close to the workings of the world that he can take care of it well enough. He doesn't need to readjust himself intellectually. But the likes of us should keep our hands off it.

I cannot explain these differences to you better than by Kant and Fichte, two men who were quite close to each other intellectually. But the one was *quite* the scholar, like us. It never occurred to him to want to venture beyond his philosophy; indeed, he would probably have smiled wryly if someone had advised him to do so. The other was a man of tremendous ethical passion—he loved to hold speeches before the German nation.

There's a lot of unrest here, you don't get much work done.

With warm regards

Yours, M. Laue (Hoffmann and Walker 2007: 531 citing AEA, no. 16–091).

From Oxford, Einstein wrote to Laue on 7 June 1933 to ask him to arrange for the removal of his name from membership rolls of other German societies: "Dear Laue, I have learned that my unsettled relations with such German bodies, in whose membership rolls my name still appears, could cause discomfort for some of my friends in Germany. Therefore I ask you please to see to it sometime that my name be removed from the lists of those bodies. They include, for example, the German Physical Society, the Society of the Order pour le mérite. I expressly authorize you to arrange this for me. This ought to be the proper way, as it avoids new theatrical effects. Amicable greetings to you from your A. Einstein. P.S. I like your booklet on wave mechanics extraordinarily much (AMPG, III rep. 50)." In the sixth volume of the *Handbuch der Radiologie* Laue had published a contribution on "Corpuscular and wave theory" (1933b) which he had sent to Einstein. Einstein severed all ties to Germany—but the strongest one could not get completely undone—the one to his mother tongue, German.

At the beginning of June, politics briefly dropped into the background. The Laues left on a car trip with relatives to the Black Forest via Maulbronn. As early as June 14th, Laue was able to report to Einstein that he had been able to settle his request to some extent. "Upon my return yesterday from a wonderful drive to the southern Black Forest, I found your letter of Jun. 7th. I thereupon informed the managing director of the German Physical Society of your withdrawal and passed your letter on to Planck. As grateful as I am for your striving to make the situation as easy for us as possible, I could not do either without the deepest sorrow. I hope that in a not-too-long time from now the spirits will have settled down again and that the German Physical Society will be able to reestablish its ties with you in one form or another (Hoffmann and Walker 2007: 533 citing AEA no. 16–094)." The "civil service law" also applied to Laue's doctoral student Leo Szilárd and he left for England, where he registered a secret patent for an atomic bomb and then emigrated to the USA.

Despite these political reprisals: physics came first for a scientist like Laue, as the letter of June 26th to Einstein shows.

Dear Einstein,

I am making use of a special occasion to send you this letter. First of all, it should bring you warm greetings. Then the information that although Berlin physics is anticipating further painful losses, it is still pursuing the cause unswayed. Enthusiastic reports are still being presented at Schrödinger's seminar, and we do not need to hesitate at all there from mentioning even relativity theory.

Nothing happened even to me when at my first lecture of this semester I referred to the theory of relativity as being 'known to have been translated from Hebrew.' You probably gather the allusion to the 'famous' appeal by the so-called student body 'against the un-German spirit.' Or have you not yet set eyes on this gem of unintentional humor? (Hoffmann and Walker 2007: 533–534).

Max von Laue then asked about Einstein's location and indirectly asked him not to take anymore political measures, because then "German scholarship as a whole or thereabouts would be held responsible for it." He mentioned the foreign aid ventures and the difficulties in finding a successor for Walther Nernst (ibid.).

The German population had no previous experience of life in a lawless dictatorship. It was unknown how far one could dare to go with criticism, let alone resistance, without punishment. At the beginning it was also unclear whether Hitler could even outlast his predecessors. Was this yet another brief transitional government? A terrible example one year later would reveal the lawlessness.

That autumn Planck left on holiday again with his wife, this time to Fures, and he kept in touch with Laue, who informed him that Erwin Schrödinger had resigned his professorship for health reasons. On September 11th, Planck replied: "Thanks for your letter of the 9th inst. I feel Schrödinger's resignation as a new deep wound inflicted on our Berlin physics and which we must endure with all the energy we can muster. As far as I can see, though, this misfortune is unavoidable because if the decisive reasons do lie in the sphere of health and the ability to work, one cannot argue against that. I hope the ill effects of this affair have not taxed your own physical wellbeing." Whether political reasons weren't rather Schrödinger's main motivation was evidently not confided by letter. Planck then mentioned the fate of Lise Meitner at the university. (She had been banned from lecturing and was forbidden from advising doctoral students.) Planck reported that he had spoken with a staff member at the ministry about the option of voluntary resignation by Ms. Meitner. He then added with reference to Laue's chairmanship of the German Physical Society: "On your leadership of the proceedings at the physicists' convention in Würzburg I wish you all the best: health and vigor. It is such luck for physics that this time you are at the head. If the majority really is so unreasonable as to elect Stark as chairman, they will soon have to suffer for it [...] (UAF, Laue papers, 8.8)."

The vast majority of societies and associations quickly conformed to the political demands of the National Socialists. The German Physical Society (DPG) had, of course, numerous Jewish members; Einstein had been its chairman from 1916 to 1918. In 1933, the members Philipp Lenard and Johannes Stark immediately began to lobby with the National Socialist leadership to further their own aims. By catapulting Stark into a dominant position they could use state support in their hateful campaign against Einstein and his friends in pursuit of their "Aryan physics" (*Deutsche Physik*) agenda. The physics convention in the autumn of 1933 was an opportune time for this. Stark wanted to assume the office of chairman which Max von Laue held. The conference was to be held in Würzburg and the topic was: atomic research and the limits of electrical measurements. The Laue couple drove to Würzburg in their Steyr in the company of the Grotrian couple. In Veitshöchheim they stopped to visit a former fellow student who was a teacher there.

On September 18th, before the conference proper began, Karl Mey gave the first welcoming address, followed by Max von Laue as chairman of the DPG. Laue skillfully wove historical retrospection into his speech to launch indirect as well as quite direct criticism about the current state of affairs: "When we come together tomorrow at the physics department of the local university, we will be standing at a historic site. At the end of 1895 in this building Wilhlem Conrad Röntgen discovered the rays that were named after him. It would be superfluous, not to mention in bad taste, to discuss the importance to physics and the whole gamut of applications of these rays. But we would like to commemorate the great achievements of Röntgen, who was the first to see consciously that which many before him had narrowly missed, and elevate it from shady speculation to transparent and reliable scientific knowledge (Laue 1933c: 889, trans. Hentschel 1996: 68–71)." He was probably insinuating Lenard here, who had claimed his own priority after Röntgen's discovery had become public. Lenard's laboratory journals clearly show, however, the trouble he had in replicating Röntgen's experiment. Laue continued: "Next we commemorate Röntgen's successor at this place: Willy Wien taught and researched in this building for 17 years. His classical theoretical writings on heat radiation, which have become an integral part of our science in the form of Wien's displacement law, originate from his early years, when the young physicist was working under Helmholtz at the Reich

Physical and Technical Institute"—the PTR (ibid.). The next section paid tribute to Wien's successful period in Würzburg and praised his war efforts during the four years of the World War in developing amplifier tubes together with Max Seddig.

On the 22nd of June of this year physics could celebrate a particularly special anniversary. 300 years ago on that day Galileo's trial ended before the Inquisition. The grounds for the trial were, as is well known, Copernicus's theory of the motion of the earth and the other planets around the sun, a theory that caused a similar sensation and stir to that of the relativity theory in our century for being contradictory to the traditional notions of the day. Galileo was not its only advocate, but its most successful one, because he was able to support it so convincingly with his wonderful discoveries of Jupiter's moons, the phases of Venus, and the rotation of the sun. The trial ended with a conviction. Galileo had to renounce the Copernican theory and was condemned to lifelong imprisonment (ibid.).

Laue then mentioned the lenient conditions of his confinement and Galileo's two major works.

There is a well-known myth attached to this condemnation that while giving his oath and signing his renunciation of the theory of motion, Galileo was supposed to have said: 'And yet, it moves!' This is a myth, historically unverifiable and intrinsically implausible—and yet it is ineradicable in common lore. What is its vitality based on? Surely on the fact that throughout the whole proceedings Galileo must have posed the unspoken question: 'What's the point of all this? Whether I or whether any person now believes it or not, whether political or religious forces support or are against it, the facts remain unchanged. Might can certainly delay knowledge of the *facts* for a while, but eventually it will prevail.' And this is indeed what happened. The victorious advance of the Copernican theory was irresistible. Even the Church that had condemned Galileo gave up all forms of opposition, albeit only 200 years later.

The times were not always favorable to science later either, such as in Prussia under the otherwise so praiseworthy king Friedrich Wilhelm I. But in the face of all the repression, its supporters could stand steadfast in the triumphant certainty that is expressed in the modest phrase: And yet, it moves! (Ibid., orig. emphasis; see also Hoffmann and Walker 2007: 535–537).

The third speaker was Johannes Stark, who of course heavily criticized Laue's speech, because he was making a bid for the position of chairman. But he was not successful, perhaps because he had said something about "using force" at the end of his speech. The chairman of the German Society for Technical Physics, Karl Mey, was elected. He was the same age as Laue and had studied physics in Berlin under Emil Warburg. Research in incandescent lamp light at Osram was under his command. There was no debate about relativity at that convention. Werner Heisenberg was actually supposed to be awarded the Planck Medal on that occasion but he was not in Germany at the time. Max Planck later handed it to him at a meeting of the Physikalische Gesellschaft zu Berlin on 3 November 1933.

In that politically turbulent and emotionally oppressive year, Laue was nevertheless scientifically active. He prepared the article on 'Statistical physics' for the second edition of the *Handwörterbuch der Naturwissenschaften* (1934a). Laue also copublished a paper with his former doctoral student Friedrich Möglich 'On the magnetic field in the vicinity of superconductors' (1933) in the academy's proceedings. The question to be resolved was how to understand the magnetic field in three different experiments—by different physicists—with superconducting rings, and what further experiments physicists should perform to clarify the issue. In the first experiment, a continuous current was generated in a superconducting ring with an external magnetic field; in the second, the ring was placed in a magnetic field above the superconductivity threshold and then cooled down to below the transition temperature; and in the third experimental setup, a current was applied to a superconducting ring from the outside. "The following discussions are intended to produce the possibility of experimental analysis of current distribution, by making statements about the field for the three types of experiments mentioned, which are accessible to measurement. That is, from the premise that the superconductor be a perfect conductor in the sense of Maxwell's theory, these discussions are intended to treat the magnetic field for some typical conductor shapes, namely, for the very elongated loop of wire, and for the circular ring (torus) (Laue and Möglich 1933: 544)." The coauthors cited a paper by Gabriel Lippmann (1919), who had applied Maxwell's theory to the problem. For the solution, Laue and Möglich restricted the calculation to the two-dimensional case. They succeeded in arriving at a plausible result by means of Maxwell's theory. The result showed that this classical theory was on the right track toward further progress and better descriptions of the phenomena of superconductivity.[19] But, as with many theories, a final solution could not yet be reached. Many years passed before a solution was found along a completely different route from completely new ideas and experiments on the conduction mechanism in metals and metal compounds.

Johannes Stark had ousted Friedrich Paschen as president of the *Physikalisch-Technische Reichsanstalt* on 1 May 1933. This led to Laue losing his post as scientific advisor, because Stark wrote to the Reich Minister of the Interior on November 15th "that a successful influence on the activities of the Reichsanstalt is not being made by the cursory collaboration of Mr. von Laue." (Zeitz 2006: 45) He asked the minister for permission to terminate the contract with Laue on 31 March 1934. His petition was granted. Stark organized the PTR according to the National Socialist model. He was, however, interested in the experiments on superconductivity at the PTR, as they, of course, were "non-Jewish."

At the beginning of 1933, Robert Ochsenfeld had joined Walther Meissner as a staff member. It was only a temporary contract, because the PTR's budget could not afford a more permanent position. Meissner and Ochsenfeld investigated, among other things, the behavior of superconducting metals such as lead and tin as monocrystals in the magnetic field. In mid-October, they submitted a manuscript for the "short original communications" section of *Die Naturwissenschaften* under the heading 'A new effect at the onset of superconductivity.'[20] They established that in normal conduction the magnetic field lines penetrated the metal, but in superconductors the field lines were displaced from the metal and the magnetic field lines surrounded the

---

[19] Friedrich Möglich was Laue's assistant at the time. After 1945, he was Laue's interlocutor in contacting the German Democratic Republic, esp. as editor of the *Annalen der Physik*; on Möglich see Müller-Enberga et al. (2000): 587 f. See also Matricon and Waysand (2003): 416.

[20] Meissner and Ochsenfeld (1933). The publication had no field-line diagrams. The paper by Laue and Möglich was cited but only with respect to its application of Maxwell's equations, not their proposals about the form and arrangement of superconduction.

superconductor. This discovery, known as the Meissner-Ochsenfeld effect, caused a great stir among physicists.

Ochsenfeld left the PTR in 1933. In 1934, Walther Meissner accepted an appointment to the polytechnic in Munich and continued the experiments there with his colleague Fr. Heidenreich. Meissner's scientific exchange with Laue on the subject therefore continued in writing.

On the last day of November, Max Planck, as "permanent secretary" of the academy, applied to the Ministry of Science, Art, and Culture to assign the "full-time" membership position that Einstein had vacated with his resignation, to the academy's "member, the Prof. of theoretical physics Max v. Laue" with the corresponding salary. Thorough grounds followed (AMPG, PA Akad. II–III, 75/2). But: A long silence ensued! In the end, Laue did not get the position.

Since many years, the general public and many scientists waited yearly in late autumn for news from Stockholm about who would be awarded the Nobel prizes in chemistry, physics, and medicine. After the Nazi "takeover" in 1933 this was particularly interesting. The research results of Werner Heisenberg (for 1932) as well as Erwin Schrödinger and Paul Dirac (for 1933) were honored with the Nobel prizes in physics. Max von Laue paid tribute to the three in a short essay. After an introduction mentioning Niels Bohr's work, it continued: "Werner Heisenberg, alongside whom one must not forget Max Born and Pascual Jordan however, drafted a mathematical atomic theory, initially starting from the formalism of Bohr's theory but then completely stripping it away, whose agenda reads in complete renunciation of intuitiveness: mention should only be made of quantities that can be found in possible experience, thus, for example, energy levels, not orbits of a particular form (1933d: 719)." The named researchers managed to compute the energy levels of the hydrogen spectrum correctly, and Heisenberg derived his "famous indeterminacy relation." Laue described Schrödinger's idea "to generalize de Broglie's ingenious idea of linking a wave process with the motion of any body to the motion in any potential field. Thus was formed a partial differential equation, the 'Schrödinger equation,' which yielded, certainly to the greatest surprise of its author, the well-known energy levels of the Balmer spectrum [i.e., of hydrogen] as eigenvalues (ibid.)." The quantum mechanics of Heisenberg and Born agreed with Schrödinger's theory, because they could be mathematically converted into each other. "But both theories stem from Newtonian mechanics and therefore did not satisfy relativity. It is to the great merit of Paul Adrien Maurice Dirac that he created a kind of relativistic Schrödinger equation (ibid.)." Max von Laue then went on to explain the fine structure of the hydrogen spectrum. "That Dirac was compelled to conclude from his theory the existence of a positive electron must not remain unmentioned. For a long time this constituted the most serious objection to it. How great was the astonishment then, when Anderson and others discovered the positive electron in their Wilson chamber! (Ibid.)" Laue concluded his appreciation on a critical note: "In the face of this brilliant development, one must not forget that all this notwithstanding, the 'quantum puzzle' remains. A completely satisfying, self-contained physics does not exist at present; the often-discussed rift between corpuscular and wave conceptions is still gaping (ibid.)."

At the end of the year, an essay entitled 'Matter and the filling of space' appeared by Laue in the Italian journal *Scientia*. It took as its starting point a lecture by Erwin Schrödinger on the question: 'Why are atoms so small?' and Paul Ehrenfest's address (1933) in Leyden on the question: 'Why are atoms so fat?' Laue explained why Ehrenfest's statement, that the "Pauli exclusion principle" did not permit a different atomic structure, by discussing quantum numbers, the constraint on the number of particles on the individual energy levels K, L, M, N, and the $\pm$ ½ spin. His conclusion was that the duality between the corpuscle and wave representations flatly contradicts conventional dynamics. The trajectory of a particle is then only imprecisely determinable. "From this it then follows that in general all natural processes are only predictable to a limited, sometimes quite low accuracy—hence the famous contradiction to any kind of causality. [...] The task now posed to physics is more problematic than ever before. In a certain sense one may even say: it is set back to a position far before Galileo, very far (Laue 1933e: 402)."

On November 11th (!!!) an 'Avowal to Adolf Hitler and the National Socialist state by professors at German universities and colleges' appeared bearing very many signatures, but not Max von Laue's.

After the Nazi "takeover" Johannes Stark immediately started to try to draw the professional periodicals in physics under his control as well. The Hirzel publishing house in Leipzig, which issued the journal *Physikalische Zeitschrift*, apparently did not dare to ignore Stark's influence. Max von Laue informed Arnold Sommerfeld on 10 January 1934 about a letter from the science editor Arnold Berliner in Zehlendorf on the outskirts of Berlin, who had written with reference to another editor, perhaps Rudolf Seeliger or Karl Scheel: "I have advised S. to get in touch with you in order to discuss with you how one should proceed against the conniving conduct of both Hirzel and Stark—who had not informed S., for example. Among physicists a little more determination against such attacks should be shown than has gen[erally] been the case among German scholars. Wouldn't you also like to interest Mr. Sommerfeld in this? (in Sommerfeld 2004)." Laue told Sommerfeld that he had contacted the chairman of the DPG, Dr. Mey, but since that society had no direct connection with the *Physikalische Zeitschrift*, he did not see any possibility to intervene. On February 15th, Laue informed Sommerfeld from Berlin: "The fact that the Ministry of Culture has rejected without justification the full-time position in the academy applied for on my behalf is something you probably already know (ibid.)."

The emigré Fritz Haber could not gain a permanent footing in England and decided to move onward to Palestine. During that journey, in Basel on 29 January 1934 he collapsed and died. In the seventh issue of *Naturwissenschaften* dated 16 February 1934, a short obituary of Haber appeared on the first page by von Laue. He started by recalling the celebration of Haber's 60th birthday *in absentia* held by his staff on 9 December 1928. They planted a lime tree in front of the institute entrance in his honor. Laue then paid tribute to the work in so many fields in which Haber had carried out successful research. As an example, he singled out the separation of ortho- and para-hydrogen by Karl Friedrich Bonhoeffer and Paul Harteck. By a historical comparison Laue condemned the defamation of his Jewish fellow citizen:

On the 2nd of May, 1933, Haber handed in his resignation.

Themistocles went down in history not as the pariah at the court of the Persian king, but as the victor of Salamis. Haber will go down in history as the genius inventor of the procedure of binding nitrogen with hydrogen, which underlies the synthetic extraction of nitrogen from the atmosphere. He will be remembered as the man who, in the words used at the award of his Nobel Prize, had created in this way 'an exceedingly important means towards promoting agriculture and human prosperity.' He will be remembered as the man who had made bread out of thin air and who triumphed 'in the service of his country and of the whole of humanity' (Laue 1934b: 97, trans. Hentschel 1996: 77–78).

He concluded by quoting a reminiscence by the botanist Margarete von Wrangell, who had worked with Haber for a while (ibid.). Johannes Stark protested by letter to the German Physical Society on 3 March 1934 from Berlin. Haber, he contended, had not been banished into exile by the National Socialists. "This irresponsible action by Mr. v. Laue is the more regrettable and harmful, since the English weekly 'Nature' is slandering National Socialist Germany almost exactly at the same time in a similar way (DPGA, 10011)." The Prussian Academy of Sciences celebrated their deceased member in the regular manner with a memorial session. The physical chemist Max Bodenstein held the memorial address. He also recalled Haber's involvment in the deployment of poison gas during the war.

Despite the uncertainty in science policymaking, Max Planck, as president of the KWG, continued his untiring attempts to make the planned KWI of Physics a reality, and wrote to the Ministry of the Interior on 5 March 1934, with this in mind: "for the direction of the institute a top-ranking scientific researcher would be available, the director of the Physics Institute at the University of Leipzig, Professor Peter Debye, now at the prime of his creative energies [...] (AMPG, I rep. IA, no. 1671)." Apparently Planck and Laue had come to an agreement on the question of filling the director post.

Max von Laue received permission from the ministry to accept an invitation from Frederick Lindemann to visit Oxford. A native Englishman, Lindemann had completed his doctorate under Nernst and had returned to England in 1914. He was a professor at Oxford University since 1919. Laue met him there in March as well as Schrödinger and Fritz London who had also developed a theory of superconductivity based on Maxwell's theory. This led to an intense exchange of letters in 1935. After the visit Laue wrote to Fritz London nine times and received seven replies. But still there was no explanation forthcoming of the behavior of electrons in the superconducting state. The theory by Felix Bloch of a dependence of conductivity on temperature presuming an "electron gas" could not be applied to superconductivity.[21]

The surveillance of individuals in Germany increased. Max von Laue must have sensed this, as a letter from Max Planck in Cannero across the Italian border, dated 22 March 1934 shows:

---

[21] HUA, UKP, staff file Laue, fol. 55; Gavroglu (2005): 137, notes for ch. 7. The conflicting ideas about a theory of superconductivity could not be bridged, as Gavroglu's biography makes clear; the ideas of those involved were too different (p. 125).

Dear Colleague,

The first part of your letter of the 20th inst. pleased me very much, but all the less so the second. In the meantime, you have probably received my telegram as the first answer to it, with the content you confided to me. I just cannot foresee whether and how action might be taken against you, since, as you know, I have no contacts at the influential places of authority. But I can tell you one thing with complete certainty: that in your position I would not hesitate for a moment not to let my return to Berlin be influenced in the least by the fact that someone from the Party had given a report somewhere. To the best of my experience, any precautionary measure taken by you would be interpreted as a sign of a bad conscience on your part, and those who might be ill-disposed towards you would, as soon as they thought they detected any insecurity on your part, immediately become bolder and thus gain ground. My maxim always is: consider every step beforehand, but then, if you think you can justify it, don't put up with anything. I only hope that the vacation will exert a stronger effect on you after all, and that you return home refreshed. We shall see what happens later. I shall be back in Berlin on the 10th of April (AMPG, Va, rep. 11 no. 1075).

The letter to Planck has not survived, and so it's anyone's guess exactly what suspicions Laue had reported to Planck. Planck's reply was intended to have a calming effect, but could it do so given the mounting terror? Max von Laue was clearly an opponent of the Hitler regime and therefore a target for attack. It was terrifying enough just to have a member of the Nazi Storm Detachments (*Sturmabteilung*, SA) ring the doorbell in uniform. In the beginning, SA-men would conduct unauthorized mailbox searches for any "suspicious mail."[22] Planck's next letter to Laue from March 31st, again written in Cannero, was about a signature campaign in support of the Göttingen mathematician Richard Courant, who had been dismissed on the basis of the "civil service law." Planck had some doubts about signing this declaration. He wanted to discuss it with Laue in Berlin and invited him to tea at 5 o'clock on April 9, adding that he would be pleased if Mrs. Laue came along (AMPG, Va rep. 11 no. 1077).

In collaboration with Eduard Justi, Meissner's later successor at the PTR, Laue addressed a publication by Paul Ehrenfest on "phase equilibria." The thermodynamic problem was appealing, because the specific heat of liquid helium shows a drastic rise at slightly above 2 °K, followed by a drop. Measurements on solid methane yielded a similar effect, as Klaus Clusius and Albert Perlick showed. Justi had taken measurements of oxygen and also observed such phenomena. Thus they arrived at the result: "We would like to infer from these facts that equilibria of the third kind really do occur. On the other hand, we consider equilibria of the second kind, if they exist at all, to be unobservable, because for them one of the two conversions would conflict with the second law (Justi and Laue 1934: 237; on Justi see Lautz 1979)."

---

[22] Domestic mail was only monitored in exceptional cases, see Leclere (1988): 120; BAB, R 47.01 Reichspostministerium 18,331: 102. The Defense Department of the Army High Command under Wilhelm Canaris was responsible for censorship of foreign mail. He became an opponent of the regime in 1938. In Laue's letters to Meitner, apparently only military details were deleted. A former employee of this postal surveillance confirmed this practice to the present author. It supposedly had been a direct order from Canaris. Other, critical references to abuses in the Third Reich in the letters were to be ignored.

Increasing numbers of qualified people were being ousted from their posts and replaced by National Socialists. The former state minister, Friedrich Schmidt-Ott, headed the Emergency Association of German Science (*Notgemeinschaft der Deutschen Wissenschaft*) together with Fritz Haber. Minister Bernhard Rust, at Adolf Hitler's request, replaced him with Johannes Stark. Schmidt-Ott was informed about this at a meeting on 23 June 1934, but the letter of dismissal from the same day read: "In accordance with your wish, I have relieved you of your office as president of the Emergency Association of G. Sci." (GSta PK, file Schmidt-Ott, Friedrich) Max von Laue wrote to Schmidt-Ott on 27 June 1934: "Your Excellency! It was with deep regret that I heard of your resignation from the presidency of the Emergency Association. The vast majority of German physicists, especially the members of the physics committee, share this regret. As, you have held your office for almost 15 years in a way that makes it difficult for any successor to equal you. Under the present circumstances, moreover, the change in the presidency will, I fear, be the prelude to hard times for German science, and physics will probably have to suffer the first and heaviest blow (ibid.)."

The lawlessness and brutality of the state soon became apparent. Hitler solved a problem of internal party politics for the National Socialists brutally and cruelly at the end of June 1934. In 1931 he had appointed Ernst Röhm, a former professional officer and early supporter of the Nazi party in Bavaria, as chief of staff of the SA, which had 4½ million members. It was notorious for its brutal attacks, especially on communists. Gradually Hitler and Röhm became rivals within the party. Three years later, on Hitler's orders, Röhm and other senior SA leaders were arrested and murdered on June 30th and July 1st. Ostensibly they had been involved in an attempted coup, in which Kurt von Schleicher was also alleged to have participated. He and his wife were shot in their home. The official figure of such executions was 77, but the real number was far higher. None of the killers connected with this Röhm putsch affair was brought to justice.[23]

The political circumstances did not leave Laue and his scientific work untouched, as a publication in the 26th issue of *Naturwissenschaften* reveals: 'On Heisenberg's uncertainty relations and their epistemological significance.' (Laue 1934c) Inner tension pervades the style; not much philosophical detachment is perceptible in it. Max von Laue followed up on his "On the discussions about causality" from two

---

[23] In the Reich Law Gazette I, no. 7: 529 of 1934 it was announced: The Reich Government has decided on July 3, 1934: "The measures carried out to suppress highly treasonable and traitorous attacks on June 30, July 1 and July 2, 1934 are legal as the self-defense of the State."

Schmitt (1940), 2nd release: "The Fuehrer protects the law from the worst abuses when, in the moment of danger, he, by virtue of his leadership as Supreme Governor, directly settles justice. [...] In this hour he was responsible for the fate of the German nation and thus of the supreme governance of the German people. [...] The true leader is always also a judge. Judgeship flows from leadership. Whoever wants to sever the two, or even oppose them, acts as a judge. [...] In truth, the Führer's deed was veritable jurisdiction. It is not subject to the judiciary but is itself supreme right. [...] In extreme emergency the supreme right stands firm and thus emerges the highest degree of righteous avenging execution of this justice." The jurist Carl Schmitt (1888–1985) was a Prussian state councillor and member of the NSDAP. See also: Poliakov and Wulf (1959): 22, 65 f., 227 ff., 473–476.

years earlier (Laue 1932c) and then examined the "repercussions of measurement," concluding: "Nevertheless, I do not believe that the above consideration signals an insurmountable limit to knowledge in general. For, this conclusion is based on the tacitly made presupposition: For the revelation of new measurement potential, new experimental tools are *necessary* (Laue 1934c: 439, orig. emphasis)." Only those who assume this could conclude further: "Now that we have arrived at the finest aids, the atoms themselves, we can never progress further. Well, is this premise correct? (Ibid.)" A historical interlude about Heinrich Hertz's experimental appa-ratus followed as well as an example from metrology, the testing and possible cali-bration of voltmeters. "In general, it seems questionable to draw too far-reaching epistemological conclusions from the present state of physical knowledge. Quite apart from the fundamental misgivings about abandoning the principle that Nature be accessible to exploration because it is not yet fully understood how to apply it; one would at least have to start from a basis that is logically sound in itself and contains no inner contradictions. Unfortunately, this cannot be said of present-day physics (p. 440)." Then he dealt critically with the problem of filling space: "The time-honored idea that matter fills space is incompatible with the assumption of tiniest particles; and yet in present-day physics the two stand side by side. [...] And thus there is already talk about particles smaller than the 'smallest' ones (ibid.)." Laue emphasized that his remarks were not directed against modern quantum and atomic theory. "We just object to the conclusion that even modified statements of the problem can *never* lead to a full, causal understanding of physical processes. *The indeterminacy relations set a limit—this is my view—to any corpuscular mechanics, but not to any further physical insight* (p. 441, orig. emphasis)." He finally asked: "When should causality actually be considered 'empirically proven'? When the last scientific problem has been completely solved, perhaps? (Ibid.)" In closing, he criti-cized cultural pessimism (ibid.). Even though moral and political "uncertainty" was reigning in Germany, physics at least needed a firm foundation.

A few months after Laue's publication, a commentary on it appeared in *Naturwis-senschaften* by Richard Edler von Mises in Istanbul, whither he had emigrated. He first complimented Laue's remarks and then proceeded to discuss individual points: "But v. Laue, if I understand him correctly, also takes the standpoint that one can remain true to the deterministic conception of physics in spite of the uncertainty relation: '...We just object to the conclusion that even modified statements of the problem can *never* lead to a full, causal understanding of physical processes.' To this I would only like to say that, in my view, Heisenberg's uncertainty relation can *only* be expressed *as a theorem of statistical physics*, for which there is no place or possibility within the framework of a causal conception of physics (Mises 1934: 822, orig. emphasis)."[24] There followed a kind of critical terminological analysis of "measurement error," "uncertainty," and "true value." The Heisenberg relation

---

[24] Heisenberg enclosed an offprint with a letter to Max Born on 3 Nov. 1936 with the remark: "There are actually only very vague thoughts in it, and I had to burst out laughing in Holland when I heard that the boys there (Casimir, etc.) always talk about 'Heisenberg's dream with reference to this paper.' So you see how unclear everything is." Pauli (1985): 405.

accordingly says nothing about the true value or the error. "Rather, it states that the product of the *spread* of the two sets of observations has a minimum value (ibid., orig. emphasis)." About the kind of averaging, Heisenberg's inequality accordingly only says "something about the *average behavior* of very many (theoretically: infinitely many) measurements. If, however, one abandons the view that measurements form a collective, then there is no distribution, no spread, and therefore no uncertainty relationship. I also very gladly agree with v. Laue's rejection of 'ignorabimus.' But I think one will have to tolerate that adopting statistical explanations instead of causal ones doesn't mean renunciating knowledge, but only means a different and perhaps more advanced kind of knowledge (ibid., orig. emphasis)."

Occasionally, foreign visitors took Laue's letters to Einstein along with themselves. That made it possible for him to write more freely without worries about censorship or worse. On 22 August 1934, such an opportunity arose. He sent Einstein his critique of Heisenberg's uncertainty relation and noted: "We are experiencing all kinds of things here, also pleasant things. Among these I count, for instance, the outcome of the vote in 4 German academies on the nomination of Robusti [i.e., J. Stark] as president of the Notgemeinschaft. They voted almost unanimously against it, only in Heidelberg does it seem to have gone otherwise. That's no use right now, of course. But, after all, things may take a turn yet (UAF, Laue papers, 9.1)." He also mentioned that his own position was at risk because Planck and he had rejected Stark's demand that they decline any award of the Nobel prize to Germans (ibid.).

Even in 1934 a conference of German physicists and mathematicians still met. It convened from the 10th to the 15th of September in Bad Pyrmont, Lower Saxony. The two main topics from experimental physics on the agenda were: "low temperatures" and "the physics of atoms and nuclei." Six summary talks were given on the first topic, followed by another six talks on individual topics. In the first part, Walther Meissner gave a 'Report on recent work on superconductivity' by the various national teams of researchers. The scale of superconducting elements and alloys had been extended. In his report on the connection between magnetic field and superconductivity, Meissner mentioned Laue's statement: "The largest tangential component of the magnetic field is the determining factor in the disappearance of superconductivity (Meissner 1934: 931 f.)."[25]

Meissner mentioned the positive verification of this finding by a group of researchers in Leyden including Wander Johannes de Haas, Jacob Voogd, and Josina M. Casimir-Jonker; but there were still deviations. They recorded transition curves in tin with a transverse magnetic field and obtained hysteresis (ibid.).

On October 22nd, Max von Laue as a civil servant was made bounden to the "*Führer*" Adolf Hitler. The political monitoring of lecturers at colleges and universities by students was an integral part of the Nazi state. Lists were compiled. Laue's assessment in the one dated 21 December 1934, for example, read: "Max von Laue's

---

[25] On p. 933 Meissner wrote the following in connection with the hysteresis measurements: "This is undoubtedly related to a change in the permeability of superconductors at the onset of superconductivity, which Meißner and Ochsenfeld have established. [...] Indeed, it was these hysteresis phenomena which first indicated to me the probability of alterations in permeability."

talent as a scientist is outstanding, as a teacher less so. Nothing is known about his political views."[26] The opinion by the Reich Ministry for Science, Education, and Culture was a completely different one! How much people were being watched and patronized soon became apparent. Max von Laue had written a longer article 'On the activities of the Kaiser Wilhelm Society' in the *Deutsche Medizinische Wochenschrift*, a medical weekly that appeared that September, initially mentioning a drop in popularity for the society compared to universities. Being a physicist, he asked his readers to forgive him for not covering the institutes in the biological sciences and the humanities. "Let us first list the institutes falling into consideration: There is a Kaiser Wilhelm Institute of Chemistry in Berlin-Dahlem, there was one for physical and electrochemistry there until 1933. In Göttingen there is a Kaiser Wilhelm Institute of Fluid Dynamics Research, connected with an Aerodynamics Design Testing Station [...] (Laue 1934d: 1376)." This institute count was interrupted by a lengthier section about applications of X-rays in a wide variety of fields of research, such as metal and fiber analysis, and the discovery of chemical elements, e.g., rhenium. He quoted Ludwig Boltzmann: "Nothing in all the world is more practical than a—correct—theory (Boltzmann 1905: 79)." The following passage, which can also be interpreted as Laue's maxim in life, strayed a little from the main subject:

> Earning one's daily bread—this understood in the broadest sense—is important; and yet a human being does not live on bread alone. Life must have substance. 'What's the use,' Boltzmann asks, 'of merely furthering life by gaining practical advantages at the expense of that which alone makes it worthwhile living: the cultivation of the ideal?' And science is *one* example of such substance. *In this* we scholars see its rank and dignity; *that's why* we so highly value such discoveries as the two spectrum line splittings found by J. Stark, for which there is not even a prospect of any practical application, just because they enrich the substance of our lives. Let it not be alleged that this be the business of a small circle of experts. What now occupies professional scientists gradually penetrates into the nation as a whole (Laue 1934d: 1376, orig. emphasis).

Max von Laue then continued to discuss the KWI of Physics, which had been founded in 1917—without making any mention of Einstein—but only administers a fund to support physical research at other institutions. He hoped that this institute would soon become a reality and a stronghold of theory (ibid.). Eight days after this newspaper had appeared, the Reich Ministry of Science, Education, and Culture wrote a letter addressed "To Professor Dr. M. v. Laue via the Rector of the University of Berlin; To the President of the Kaiser Wilhelm Society in Berlin (AMPG, I rep.1A, staff file Laue, on the quarrel about Laue's article.)." What had prompted this complaint was Laue's comment about the KWI of Physical Chemistry and Electrochemistry:

> The allegation concerning the Kaiser Wilhelm Institute of Physical Chemistry and Electrochemistry is objectively incorrect. It is also likely to create a false impression among the circles involved about the measures taken by me in the interest of the state.

---

[26] HUA, staff file Laue, fol. 115; BAB, R 21/838, fol. 1; Reichsministerium für Wissenschaft, Erziehung und Volksbildung. BDC, Wi A 473.

It cannot be unknown to you that the Institute is still in existence, just that the previous director has left and that his position is being administered provisionally by one of the three department heads.

As a public servant you have the duty to promote the actions of the state in every way and to avoid anything that may be detrimental to them.

I therefore ask you to report immediately about how the misrepresentation in your article could have come about. At the same time, I request that you immediately correct your information in the same journal and at the same place (ibid.).

The following demand was directed at the president of the Kaiser Wilhelm Society: "I request to be informed whether you (or the administration of the K.W.G.) had knowledge of Professor von Laue's essay, whether you approved of it, and whether you, for your part, also intend to protest against von Laue's incorrect representation in a suitable publication. Signed: Bachér (ibid.)."[27] Franz Bachér was a private lecturer in organic chemistry at Rostock and had already joined the National Socialists, their party, and their Defense Squadron (*Schutz-Staffel*, SS) in 1933. At the time of this letter to Planck, he was leader of the lectureship at the University of Rostock, then professor at the Charlottenburg polytechnic, and deputy head of the University Department of the Ministry of Education, Science, and Culture, later its head. So this was a dangerous attack. Planck's reply to the Reich Minister of October 2nd has been preserved as a handwritten draft: "I humbly reply that prior to appearance I usually […] have knowledge of [publications by] Prof. v. Laue, [such as] his article 'On the activities of the K.W. Soc.' and that I generally approved of its contents. However, the remark contained therein concerning the K.W. Institute of Physical Chemistry and Electrochemistry is factually false and can give cause for misinterpretation. I have therefore drawn this point to the attention of Mr. v. Laue, and Mr. v. Laue has given me the assurance that he will soon publish a corresponding correction at the same place (AMPG, I rep. 1A, staff file Laue, partly illegible)." On this draft letter, Friedrich Glum, the general director of KWG, noted that he and the general administration took no responsibility for Laue's text, hence they wanted to have nothing to do with it: "(I note that neither Mr. v. Cranach, nor Mr. Telschow, nor I saw the essay prior to its publication. Glum). (Ibid.)" On October 6th the ministry reprimanded Planck's stance: "I have taken note of your letter of October 2nd, 1934, in the matter of von Laue's article about the Kaiser Wilhelm Society, and in particular of your instruction to Professor von Laue. With your prior knowledge of the article, I should have expected of you, as President of the Kaiser Wilhelm Society, to have assumed a stance against the offensive statement from the outset and prevented its inclusion (ibid.)." On the same day, the ministry wrote to Laue: "I have taken note of your letter of 30th September 1934. The form of correction you have chosen is not yet sufficient to completely exclude any possible misinterpretation. I therefore request that you give it the wording: 'The ... *sentence should read correctly*: There is a Kaiser Wilhelm Institute of Chemistry in Berlin-Dahlem, as well as one for physical and

---

[27] Franz Bachér was professor of organic chemistry at the Charlottenburg polytechnic near Berlin until the end of World War II. He was dismissed because of his party affiliation in 1945, but in 1954 he was classified as "exonerated" and returned to a professorship at the Technische Universität in Berlin. See *Extra: TU intern*, April 2016: 10.

electrochemistry; the director position of the latter has been provisionally administered since November 1933.' Incidentally, I still miss an answer to my question as to how you arrived at your earlier misrepresentation (ibid., orig. emphasis)." Was this insistence meant to demonstrate the ministry's power? Was it meant to deter other authors from publishing similar things? Was this supposed to set an example to show how the National Socialists treated world-renowned scientists, Nobel laureates? Did the ministry think that this kind of tutelage would motivate cooperation not just by those concerned, and generally promote research in the "Third Reich"? Some authors tried to pander to the regime by attacking Einstein and contradicting him, and some scientific journals printed such contributions.[28] Arthur Wehnelt, a member of the NSDAP, was a lecturer of experimental physics at the Friedrich Wilhelm University. Erwin Schrödinger's successor had not yet been decided (Kant 1994: 115).

The Laue couple, just as other parents at the time, had to allow their children to join a National Socialist youth organization as prescribed, if they were not to lose access to higher education and the possibility of academic study. But politics were hardly ever discussed with them for safety reasons.[29]

The Kaiser Wilhelm Society, together with the Society of German Chemists and the German Physical Society, decided to organize a special memorial event at Harnack House on the first anniversary of Fritz Haber's death, as was customary on such occasions. Planck sought official permission, even though he had been informed by the rector of Berlin University that it would be "seen as a challenge to the National Socialist state." Planck approached Minister Bernhard Rust for this permission. He defended Haber's decision to write a letter of resignation and warned the minister that a ban on the memorial event could be interpreted by ill-wishers as a sign of weakness; he also referred to plans in England to hold such a celebration in Haber's honor. Planck emphasized that the KWG had shown its positive attitude towards the present state and its loyalty to the *Führer* by word and deed often enough. Nevertheless, all civil servants—which included professors—were forbidden from attending. Karl Friedrich Bonhoeffer did not dare to come from Leipzig. The speech Otto Hahn read out contained no political statements. Lise Meitner, Max Delbrück, and Fritz Strassmann attended the ceremony. Following the usual convention for such ceremonious acts of remembrance, the tragedy of death and the merits of the living were the topics of that day, not the political and racist reason for Haber's emigration.[30] After the event no public or official action was taken by the ministry against the speakers or participants. Max von Laue had not dared to go to the celebration, because

---

[28] Bericht über die Sitzung des Vorstandes der Deutschen Physikalischen Gesellschaft on 10 Sep. 1934. DPGA, 10011.

[29] Personal communication by Hildegard Lemcke, née von Laue, to the author. Looking back, she regretted this deficit.

[30] Lise Meitner, diary, 19 Jan. 1935. CAC, MTNR 2/12; Hahn (1975): 51; Deichmann (1996); Hoffmann and Walker (2007): 557–561. Letter from Max Planck to Rust, 18 Jan. 1935. For further documents, see DPGA, 10,011.

on his way to Harnack House he had caught sight of Ludwig August Sommer, who was not well-disposed towards him, and was afraid he would file a complaint.[31]

The materialization of a KWI of Physics building came closer. On 11 February 1935, Max Planck in Berlin was able to inform Peter Debye by letter about decisive progress: "I can give you today the happy news that the existing difficulties concerning the construction of the Kaiser Wilhelm Institute of Physics ought now to be solved. For, firstly, the Rockefeller Foundation has declared that it is prepared to pay the sum of $360,000.– which it had intended to make available at the time, and secondly, the conditions which it attached to these payments have now been satisfied (AMPG, I rep. 1A, no. 1651 fol. 48)." Carl Sattler was supposed to design the building according to Debye's wishes (ibid.). How Laue reacted to this is not documented.

Despite all this political stress, Laue's scientific research continued. In March, he submitted a lengthy manuscript to the *Annalen der Physik*. It was an attempt to elucidate theoretically the influence of a magnetic field on heat conduction and friction in paramagnetic gases. The starting point was measurements of the reduction of the thermal conductivity of oxygen and nitric oxide as well as mixtures of the two gases by a strong magnetic field. Hermann Senftleben and his collaborator, among other physicists, had studied this rather weak effect. Max von Laue remarked by way of introduction: "Nothing is known at the current time about the cause; it most probably is due to that the magnetic field influences the transfer of momentum and energy at a collision between two molecules. Although we are not able to clarify this most essential point, the symmetry considerations in part I permit statements about the dependence of heat conduction on the direction of the field in relation to the temperature gradient, as well as related statements about the orientation of the friction (Laue 1935a: 1)."[32] The result of the comprehensive calculations and considerations did not lead to the hoped-for success: "We thus arrive at the result already announced in the introduction, that heat conduction and friction depend on different coefficients of the said series, that therefore no simple relation between the magnetic influence of the two phenomena is to be expected (ibid.)."

The next manuscript was already submitted on May 22nd, this time to the *Naturwissenschaften* under the heading 'The optical reciprocity theorem applied to X-ray interferences.' He had presented it a day earlier before the German Physical Society. It involved the application of Maxwell's theory. Hendrik Antoon Lorentz had proved in 1905 that it is possible to interchange the light source and the point of incidence without altering the intensity. "Since the wave field in the space lattice of crystals is known from the dynamical theory of X-ray interference for the case of *irradiation*, the theorem provides statements about the *eradiation* emerging from the crystal, provided an atom is radiating (Laue 1935b: 373, orig. emphasis)." In the second case, therefore, the atom in the crystal is the source of radiation. The associated experiments, involving the irradiation of electrons on crystals, were studied by Laue in

---

[31] Laue to Meitner, Göttingen, 15 Jun. 1948, in Lemmerich, ed. (1998b): 516; Eckert (2012): 108, footnote 31.

[32] The effect is called the "Senftleben effect." I thank Prof. Hess for this reference.

later papers. In the theoretical consideration Laue treated an incident ray that generates a ray penetrating into the interior, and another ray emerging out of it again (the Bragg curve). However, the experiments to date did not show sufficient directional resolution. The author announced that a detailed publication would be appearing in *Annalen der Physik* (ibid.).

Laue was also working in a completely different field of physics, as another paper was submitted to the editors of the *Physikalische Zeitschrift* that June entitled 'On the theory of superconductivity.' It had three authors, Max von Laue from Berlin, and Fritz and Heinz London at Oxford. Fritz London had already published substantial research on spectra and quantum chemistry before he became engaged with superconductivity. He was born in Breslau in 1900 and, after completing his doctorate under Sommerfeld had worked as assistant to Paul Peter Ewald, followed by Erwin Schrödinger in Berlin from 1928 to 1933. He and his younger brother Heinz, another physicist, emigrated to England in 1933. Thus, Laue's joint publication with emigrés could count as yet another sign of "disobedience" against the Nazi regime. Fritz London's superconductivity theory was applied to solve the problem of heat production in a stationary current within a superconductor. The following was assumed: "The presence of an electric field in the superconductor determines the occurrence of a heat evolution corresponding to Joule's heat, at least no other form of energy is known that could be involved here (Laue et al. 1935)." The assumption that the sources of heat were located exactly on the surface of the superconductor did not lead to any results. Neither did their mathematical treatment of a tubular superconductor offer any clue. A four-dimensional consideration did not lead to any insights either. The authors confessed: "At the present time, our knowledge about the superconducting state is still too incomplete to allow a decision between these two formulations to be reached [...] (ibid.)."

During the semester break, on August 31st, the administrative director issued a notice to the university faculty to submit by September 20th more detailed information about their memberships in the German National Socialist Workers Party (NSDAP). Max von Laue never became a member. It is very likely that many outright or secret opponents of the regime suspected that their lectures were being monitored. Attendance at Laue's specialty lectures was always low and so it was difficult to introduce an informant inconspicuously. The party thus had to rely on the testimony of a regular auditor. An expert opinion from mid-1935 supports this conjecture: "According to the lecturers at the University of Berlin, there is no objection to having von Laue appointed. If need be, a change of location might be useful for his further advancement toward the National Socialist state. Although he is a distinguished and kind-hearted man, but of soft and hitherto democratic disposition, he does undoubtedly rank among the greatest discoverers of this century who, by proving the wave character of X-rays, has raised the 'structure of matter' to a firm view. As a result of his speech impairment, Prof. v. Laue is not particularly suitable for large fundamental lecture courses."[33]

---

[33] BAB, R21/838, fol. 1, Reichsminister für Wissenschaft, Erziehung und Volksbildung, Deutsche Dozentenschaft.

The scientist thus described was at that time still busily working on an extensive publication about fluorescent X-ray emission by single crystals. More recent experimental apparatus for irradiating crystals with electron beams of short wavelength had meanwhile considerably improved. Walther Kossel, a professor at Danzig, was able to detect this interference effect by aiming an electron beam at single crystals. The Japanese physicist Seishi Kikuchi had already discovered peculiar lines in electron diffraction patterns in 1928. Laue's lengthy paper in eight sections took this as its starting point, and the term definitions just for the calculations filled almost an entire page. Almost 40 pages later Laue's critical summary was:

> In fluorescence *roentgen*-radiation the atoms certainly act as independent, incoherent sources of radiation. Otherwise it would not be permissible for us to infer from the radiation from one atom to that of many, by means of the addition of *intensities*. Accordingly, for the analogy to be correct, the atoms producing the Kikuchi lines must act as independent, *incoherent* de Broglie-wave-emitting electron sources. A theory of electron scattering that simply describes the atom as a space of changed potential in the Schrödinger equation does not, in my opinion, make this point intelligible; according to it, one could only expect the normal interference phenomenon known for irradiation. In reality, if there is any truth to the analogy between the Kikuchi lines and Kossel's roentgen-ray observations, other scattering processes must be at work in addition to those describable by such a theory, which completely blur the phase relationship between the incident and scattered wave. To be precise, they may, if at all, only slightly reduce the energy of the electrons, because hitherto one has not noticed any increase in wavelength for the Kikuchi lines with respect to the incident waves; admittedly, it has probably not been looked for either (Laue 1935c: 746, emphasis added).[34]

Laue's high scientific reputation and upright political stance prompted an invitation from America to deliver lectures at the Institute for Advanced Study and at Princeton University. The main topics were to be the new theories of X-ray interference, atoms as radiation sources, and electron beam diffraction. Laue first had to obtain permission for the trip, of course, which at this time was fraught with uncertainty, as the Nazi encroachments on the scientific establishment continued. Arnold Berliner, the editor of the journal *Die Naturwissenschaften* and a close friend of Laue, had to give up his post at the publishing house J. Springer. On 16 August 1935 Laue informed Max Planck of this, who was on vacation in Sulden in South Tyrol and replied on the same day by letter: "The news of Dr. Berliner's dismissal is as unpleasant to me as it is unexpected; of course, I shall first and foremost persist with the publishing house of J. Springer and do my utmost to avoid further evil consequences of this event as best as possible (AMPG, I rep. Va 11 no. 1109)." Then he thanked Laue for his report and continued: "It is very fortunate that you have received ministerial approval for your American trip, and my best wishes accompany you on this journey. Please extend my cordial regards to all colleagues among my acquaintance, and arouse appreciation everywhere of the difficulties which we have to grapple with here, but also of the good will we are trying to muster to overcome them. After all, calmer and more normal times will come (ibid.)."

Arnold Berliner was able to join him on this voyage, and the vessel "Berlin" took them to New York. While on board, Laue wrote a postcard to his daughter

---

[34] See also Möllenstedt (1988): (14, unpaginated). Gerhard Borrmann mentions that Laue came to Danzig in 1935 to see the experiments. Borrmann later became a coworker of Laue's in Berlin.

Hildegard on September 21st, then he reported to his wife about the route around Ireland. They arrived in New York on September 30th, where they were met by Richard Courant, who by then was teaching at New York University. By October 4th they were already in Princeton, and on the 6th he sent a "night letter cable" home; in a letter he asked if that message had arrived: "[I] would like to write you a letter on this day, our silver wedding anniversary. If our court jeweler has not forgotten, you have also received the 4 silver serviette rings for both of us and the children. I was always missing them, especially since the break-in of 1929 (UAF, Laue papers, 3.13: America travels 1935)." In the following letters he mentioned his visits with Hermann Weyl and Eugene Wigner; also that he had a desk at the institute and would be delivering lectures. Einstein would not be attending. On the 19th he was able to report: "Prof. Pannowski [Erwin Panofsky] and I have also concocted a plan concerning Theo. What would you think of the boy coming to America in October 1936? (Ibid.)" Considerations followed as to whether his son Theodore should go to Princeton or Swarthmore, and about a scholarship for one year. The entrance exam wouldn't pose him any difficulties, Laue thought. He also wanted to talk to Abraham Flexner. He had gone to see Einstein the day before, where he met "Priece" [Maurice Pryce] who was also visiting.

The journey continued to Baltimore, Maryland, where Laue visited James Franck. The Capitol and the White House were on the travel agenda in Washington. Arrived in Pittsburgh, he gave a lecture on 'Thermodynamic fluctuation effects' and met the emigré psychologist Wolfgang Köhler as well as Otto Stern. The trip then continued to Niagara Falls and postcards were sent home and photos taken. The next stop was Ithaca, New York, where he visited Hans Bethe before traveling onwards to Cambridge, Massachusetts, in mid-November. "I'm *partly* looking forward to going home. But Theo *must* get out of that stifling air next year, into freedom (ibid., orig. emphasis)." On November 22nd, dead tired from all the sightseeing and experiences, he wrote that he had arrived two days earlier on board the ship for the homeward voyage. On December 5th he was back home again (ibid.). It is very likely that he had been offered a professorship in the United States but had declined. Later reports purported that he had said at the time that he hated Hitler so much that he must return to Germany to fight him.

The five lectures given in the USA were promptly published by Julius Springer in 1935. They provided an overview of the state of research and of the theory of X-ray and electron beam interference, including the scattering cones discovered by Kikuchi and his coworkers (Laue 1935d). Paul Peter Ewald reviewed this slim volume for *Naturwissenschaften*: "The five lectures delivered at Princeton University in the fall of 1935 introduce with astonishing freshness and concision a theory notoriously known to be difficult: the dynamical theory of X-ray and electron interference. The lecturer has, as is known from journals and academy reports, in recent years directed his attention and fruitful originality anew to the theory of these interference phenomena (Ewald 1936: 714)." Ewald praised Laue's treatment of the interference fields inside the crystal. "The fourth lecture is devoted to these novel phenomena, discovered just in the autumn of 1935, in which Laue demonstrates his masterly application of the *optical reciprocity theorem* toward their interpretation (p. 715)." Concerning Laue's treatment of diffraction phenomena with electrons in crystals, Ewald noted that thus far the Kikuchi lines were unexplained: "In the last lecture, Laue recognizes in these lines an effect analogous to the one obtained by Kossel with X-rays in the source exposures (ibid.)." The book review concluded with the remark that these lectures afford "much pleasure and profit": "Whether it is in order to become acquainted with the foreign ideas, or to proceed along the straightest course in the first three lectures toward understanding the phenomena treated in the last two (ibid.)."

Laue was evidently permitted to travel to Oxford at short notice, as the sender's location of his letter to Einstein dated 1 December 1935 reveals. In it he reports about the imminent emigration of Ewald with his family, with the comment: "Thus the question arises, whither with him? (UAF, Laue papers, 9.1)."

A letter from Laue, written to the publisher on 7 December 1935, evidences his political wariness. It involved the publication of the second volume of 'Stereoscopic Images of Crystalline Lattices' (1936) coauthored with Mises, who was already in exile in Istanbul: "I would like to suggest that on the title page of the second forthcoming volume no details be given about the universities at which Mr. v. Mises and I are employed, in order to evade any possible problems (ZLB, Disp. Springer-Verlag)." Mises had been informed about this suggestion.

The year closed with a physics National Socialist rally. On December 13th the Physics Institute at the University of Heidelberg was ceremoniously renamed the "Philipp Lenard Institute." Johannes Stark gave the laudatory address on Lenard's achievements. On the following day, talks were held about Lenard's conceptions of the aether.[35]

Already during the Weimar Republic, Berlin had been awarded the contract to host the Olympic Games for 1936. This was a significant step towards Germany's readmission into the international community. The famous "Twenties" in Berlin had demonstrated the cosmopolitanism of Berliners. However, the existing sports facilities were ill-suited for such a large event. The architect Werner March designed a spacious sports facility, which was erected between 1934 and 1936. The National Socialist dictatorship now had to pretend to be open to the world. At the beginning of the year, Rudolf Hess, the *Führer*'s deputy, instructed the regional heads to alter the signs forbidding Jews from entering parks, swimming pools, etc., into "Jews unwelcome" (*Juden unerwünscht*). Several emigrés used this opportunity to risk visiting Germany again, because they knew that they need not fear open reprisals that summer. Max Born also took this chance and came to Berlin to visit Max von

---

[35] Schönbeck (2006): 1107, further literature there. In the Archive of the Berlin-Brandenburg Academy of Sciences there is, among the material on Laue, a newspaper with the following article: "The Heidelberg student no. 1/1935 (?) presented a short acknowledgement by Lenard to the students: 'I want the Institute to be the standard bearer in the battle against the Asiatic spirit in the natural sciences. Our Führer has eradicated this same spirit in politics and national economics—where it is called Marxism. In the natural sciences—with its homage to Einstein—it is partly still very firmly entrenched. It is important to know that it is unworthy of a German person in every respect—and is only detrimental to him—to follow a Jew intellectually. Research of Nature in particular is an entirely Aryan foundation, and German persons will still continue to pursue their own paths into the unknown.'" BBAW, Laue file.

Laue, among others. In his letter to Rutherford on 17 October 1936 he wrote in English: "In Berlin I had a long talk with von Laue who asked me to send you his regards. He told me incredible things about the University life, and as he is one of the few who do not submit to every nonsense he was afraid to lose his position soon."[36]

The year 1936 finally brought Max von Laue the decision on an independent post with its own small department within the KWI of Physics. On January 10th a draft of the statutes of the KWI of Physics was submitted to the senate of the KWG. In a meeting with Peter Debye, Laue was able to agree on how to assign the available positions. He was to receive two assistants and a mechanic (AMPG, I rep. 1A, staff file Laue).

Max von Laue never hesitated to publish ideas that had yet to lead to a final solution. The next two publications are examples. In early January he sent a short manuscript to the editors of the *Annalen der Physik* entitled 'On the theory of Kikuchi lines.' He had found in Max Born's treatment of collision processes an explanation, "that electron waves scattered under energy loss from different atoms are incoherent with each other, even if the energy losses agree exactly. [...] According to this, Kikuchi lines are explained by 'inelastic' scattering with small energy changes hitherto unnoticed against the accelerating potentials (Laue 1936a: 569)." Laue also mentioned his own doubts several times, such as about the ability to localize the energetic transitions of electrons in specific atoms. An extension of the theory to the conduction electrons was necessary, but they did not yield any lines (ibid., p. 573).

At the end of February he already submitted the next manuscript to the *Annalen*: 'The external form of crystals as it influences the interference phenomena in space lattices.' In the introduction he immediately pointed out that no influence had yet been observed of any particular crystallite form on X-rays. "Electron beams are different. Since, with them even far smaller objects produce a considerable diffraction effect, they are used with much success to investigate surfaces and thin membranes, things whose diffraction effects would hardly be observable with X-rays (Laue 1936b: 55)." Laue cited research by Hans Lassen and Fritz Kirchner, who had found "strange star-shaped interference spots" in the electron irradiation of thin layers of silver, gold, and other metals. This, he said, was what had prompted his publication. He then set various boundary conditions for the path toward a solution. Using Fraunhofer's theory of diffraction phenomena as well as Abbe's theorem, and making symmetry considerations, he established several "theorems" to explain them. Following the mathematical treatment he discussed his results and those of Lassen and Kirchner and was able to explain from the crystal structure, from the octahedron and tetrahedron, the results of Lassen and Kirchner. In the end Laue remarked: "Electron diffraction at thin layers still exhibits some unexplained peculiarities. The interpretation given here certainly does not presume to give information about everything of relevance; it already has some value if it manages to explain correctly *one* conspicuous feature of such diffraction images (p. 68)."

---

[36] He also wrote: "During the vacation I have traveled in Germany and seen most of my relatives at different places. I thought that I could do so this safely during the time of the Olympian games." (original English) University Library, Cambridge, reprinted in Lemmerich and Hund (1982): 127.

In January, another distinction was bestowed on Laue; this time, the University of Manchester awarded him the title of "Doctor of Science."[37] He wanted to travel to Manchester in May to receive it. That meant, of course, that he had to ask the education ministry for permission. His colleague and dean of the university, the mathematician Ludwig Bieberbach, submitted his petition to the Reich Minister of Science, Art, and Culture on April 2nd with the note: "To my knowledge Manchester is an emigré base (HUA, UKP, L 51, vol. III, fol. 5)." Nevertheless, the trip was approved, and so Laue was able to report to his English colleagues about the true conditions inside the Third Reich.

Returned again, Laue submitted a paper to *Zeitschrift für Astrophysik* bearing the title 'Theoretical views on the brightness of remote nebulae.' It starts with a brief introductory review of various explanations for redshift by de Sitter, Hubble, and Tolman. According to de Sitter the redshift was caused by an as yet unexplored energy loss suffered by light quanta along their long trajectory. The other hypothesis assumed that redshift was caused by the general expansion of the universe. He then remarked slightly scoffingly: "Both parties use the notion of a light quantum. This notion was created for the processes of emission, absorption, and other interactions between light and matter. For the question of propagation in empty space, physicists probably rightly prefer Maxwell's theory. Therefore, in the present case, because redshift is supposed to occur without matter being influenced, it does seem to me safer to resort to this theory (Laue 1936c: 208)." Laue showed that the decrease in the electromagnetic energy is compensated by the increase in gravitational energy. In real observations of the brightness, he argued, a telescope is used, which participates in the general expansion. His result confirmed the views held by Hubble and Tolman (ibid.).

Besides pursing his research and fulfilling his duty of leading the physics colloquium to success—he had given up his general lecture course, with due permission—, Laue had to keep worrying about his future position within the KWI of Physics. His agreement with Debye in January had not settled all the issues, as can be gathered from Max Planck's letter to Friedrich Glum from Berlin on 26 June 1936: "Yesterday I spoke with Mr. v. Laue about the employment contract he wants. The matter is not as simple as I supposed. The main point to which he attaches importance is not an increase in his personal emoluments (although he would gratefully welcome such within modest limits), but rather the provision of a special apparatus fund over which he can dispose independently. He thinks that otherwise his department could not operate completely independently, and refers to the Chemical Institute, where Ms. Meitner also has full independent control of her apparatus (AMPG, I rep. 14, staff file Laue, 10 Jun. 1936)."[38] Planck added that he had asked Laue to put his wishes in writing. One would, of course, also have to discuss this with Debye (ibid.).

---

[37] UAF, Laue papers, 1–18, Honors. Max von Laue was already a two-time honorary doctor and honorary member of four German, and five foreign academies.

[38] The allusion to Laue being amenable to an increase in his compensation was understandable. In 1935 food prices had risen by 30% and continued to rise in 1936 by another 26.4%.

Laue wrote the requested letter on July 1st. Then he went to Copenhagen to see Niels Bohr. After he had returned, there was another conversation with Planck, on July 28th, who added a note in the file: "Today I had a conversation in my apartment with Mr. v. Laue, who has just returned from Copenhagen, and discussed with him the individual requests that he had summarized in his letter to me. As far as the remuneration for his department is concerned, he agreed that for the time being no specific amount would be fixed for this in his contract with the K.W. Soc., but that the amount of the remuneration would be left to a separate agreement with Mr. Debye. […] He further declared his agreement that the responsibility for the political reliability of his staff would be assumed by Mr. Debye."

That summer Laue found another opportunity to write to Einstein, on June 8th from Berlin. It was about a paper by a student of his colleague Rudolf Tomaschek on the interference of canal-ray light. He explained to Einstein: "The basis of my method for reconsidering interference phenomena with light from moving bodies is, to transform the light source into the resting state. Then a quite normal spherical wave goes out from it, and the problem solely consists in how a *moving* interference apparatus responds to it. This, however, can be answered by Maxwell's theory, in the manner according to which the theory of relativity applies to moving bodies (UAF, Laue papers, 9.1, orig. emphasis)." He asked Einstein to publish a short note about his consideration. In his reply to Einstein on July 29th about the latter's consideration, Laue pointed out to his friend the inconsistency of using the primitive quantum conception at one time and the wave theory at another in the same train of thought. The theoretical wave argument probably convinces every one, he added (ibid.).

In April, the president of the KWG, Max Planck, had asked the Reich and Prussian minister of science, education, and culture to appoint a representative for the board of trustees of the KWI of Physics. On August 26th, Planck was informed that Professor Rudolf Mentzel (Fig. 9.6) had been nominated for the post (AMPG, I rep. 1A, no. 1655). This effectively assured political monitoring of the institute's activities and, to a certain extent, also its research. It very probably caused additional nervous strain on Laue. The circumstances of this period destabilized him further, only encouraging paranoid thoughts.[39]

None of the ideas that Max von Laue had proposed earlier for the design of the KWI of Physics was adopted. The society's architect, Carl Sattler, executed all of Professor Debye's wishes, and that's how the main building of the institute was given its peculiar character with a turret. As was customary at the KWG, there was also a large residential building for the director on the premises. However, more than two years passed before everything was ready.

Laue always made an effort to make his very specialized research accessible to his fellow physicists. He reported about Kikuchi envelopes at the meeting of the regional chapters of the DPG for Thuringia, Saxony, and Silesia in June. He drove to Halle an der Saale by car. "One easily understands the phenomenon from the premise that

---

[39] Mentzel had studied chemistry in Göttingen and had been a member of the SA since 1922. He had agitated against James Franck's demonstrative resignation in April 1933, see *Göttinger Tageblatt*, 24 Apr. 1933; Hentschel (1996): XXXIX.

Fig. 9.6 Rudolf Mentzel
(1900–1987). From 1933 to
1935 he was head of
department and from 1938 to
1945 scientific member of
the Kaiser Wilhelm Institute
of Physical Chemistry and
Electrochemistry, from 1941
to 1945 vice-president of the
Kaiser Wilhelm Society.
Source Archive of the Max
Planck Society,
Berlin-Dahlem

each atom radiates independently of the others, therefore, the individual radiations
by the atoms are incoherent [...]. In a certain sense, one can also make a crystal
'self-luminescent' for electron irradiation (Laue 1936d: 544)." Then Laue treated
many details about the formation of the radiation cones. At the end, Laue pointed
out that Kikuchi envelopes might be a valuable tool for studying structural defects
(p. 547).

Philipp Lenard no longer held his tenured professorship at Heidelberg. In 1936/37
the emeritus presented his historical conceptions of physics in a four-volume work
on *Deutsche Physik* (1936)—in other words, 'Aryan Physics'! Max von Laue wrote
a book review about his account. After reading it Walther Nernst (Fig. 9.7) wrote
to Laue on October 16th: "I find it very appropriate that you say nothing about
the title '*Deutsche Physik*', but only point out the silence about particular German
physicists, such as Röntgen a[nd] Planck; nothing could have led the stupid main title
more effectively ad absurdum. [...] What a shame that you didn't become a theater
critic! (UAF, Laue papers, 3–5)."

This letter reveals the close relationship between these two often difficult personalities. They had become very good friends over the years. Nernst had bought a very
large estate in 1922, "Zibelle" in Upper Lusatia, where he was living as emeritus
since 1933. There were carp ponds on his estate and he would supply his colleagues
with fish for the December holidays. He also celebrated his 70th birthday there;
Max von Laue and his wife were also invited to this party with his closest friends

Fig. 9.7 The physicist and
chemist Walther Nernst
(1864–1941) was awarded
the Nobel prize in chemistry
in 1920 for his work on
thermochemistry. In August
1920 he, Laue, and other
physicists defended
Einstein's personality and
teachings in national
newspapers against the
anti-Semitic onslaught by
Weyland, Lenard, Stark, and
their followers. *Source*
Archive of the Max Planck
Society, Berlin-Dahlem

and family. On that occasion Laue presented him with the certificate of honorary
membership in the DPG, and Nernst's five grandchildren performed a play in his
honor, a "Festspiel" (Mendelssohn 1973: 213 f.; AMPG, III rep. 50 suppl. 1, car
touring book).

Laue chose a more public forum to release a kind of "corrective note" on the
ongoing debate about 'Experimental and theoretical physics,' discussing not only
Philipp Lenard's and Johannes Stark's relationship to theoretical physics but also
coming to the latter's defense, arguing for the "right of existence of theoretical and
mathematical physics." After lauding Faraday's unmathematical way of thinking
and Maxwell's reformulation and manipulations of equations, Laue also mentioned
misapplications of mathematics. He addressed Stark's assertion that theory had been
"uninvolved" in the discovery of the electron, of X-rays, and the influence of the
electric field on spectrum lines, and asked: "What is J. Stark demanding of theory
here? Surely not that it replace observations, I hope? (Laue 1936e: 165)" On Planck's
radiation law, which Laue characterized as one of the most glorious pages in the
history of theoretical physics, he asked: "What would be J. Stark's assessment if he
had taken [this step from classical to modern physics] without keeping experience in
view? (p. 166)" The importance of the theory of relativity was also mentioned, and
the tests toward its verification.

A brief outline of quantum mechanics led Laue to conclusions that he had already
published about the uncertainty of locality and change of place (ibid.; cf. Laue 1934c).

In July 1936 a longer paper appeared in the *Physikalische Zeitschrift* by Walther
Meissner and his coworker F. Heidenreich, 'On the alteration in the distribution of
current and in magnetic induction at the onset of superconductivity.' The first notice

about it had already appeared in 1933 in *Naturwissenschaften*.[40] Robert Ochsenfeld reminisced in 1986 in a brief letter about the reasons behind the measurements performed at that time: "The inspiration for the measurements of the loop with flowing current in the normal state and in the superconducting state came from M. v. Laue. At the beginning of the experiments I also discussed it with him; while the experiments were ongoing he only spoke to me once about them, but at the initial stage. I don't remember any further conversations. Johannes Stark was very enthusiastic about the outcome of the measurements but did not want to approve publication until he had seen the effect for himself. That is what happened, too."[41] The introduction to the 1936 paper by Meissner and Heidenreich read:

> Some of the measurements dealt with in the following have already been reported briefly by W. Meißner and R. Ochsenfeld. Further preliminary communications can be found in talks held by W. Meißner before the German Refrigeration Association [*Deutscher Kälteverein*], at the Physicists' Convention in Bad Pyrmont, at the London conference on superconductivity, and in papers presented by him. A detailed publication has been delayed by external circumstances […].
>
> From the measurements it was concluded already at that time that the distribution of current in the superconducting state cannot be the same as in the non-superconducting state. This was confirmed by measurements of the magnetic field near the surface of current-carrying superconducting loops, which were thereupon performed together with Ochsenfeld. Finally, as will be detailed below, these measurements afforded the opportunity to investigate the magnetic field distribution near noncurrent-carrying superconductors at the onset of superconductivity, where the new magnetic effect was found: altered magnetic induction (Meissner and Heidenreich 1936: 449).

Squabbles about who had first recognized certain correlations were not uncommon in science and still aren't. Although Meissner and Ochsenfeld both recognized and also described the effect that came to be known by their names, Meissner stopped mentioning Ochsenfeld in later publications.[42] An explanation for this is given in a letter by Meissner to Rudolf Brill: "For your orientation, I am also sending you the enclosed photocopy of a comment signed by Ochsenfeld concerning the discovery of the magnetic displacement effect, which I had written in 1956, because foreign papers always speak of 'Meissner' and not the 'Meissner-Ochsenfeld effect,' against which I should have had to protest if Ochsenfeld had had any part in the discovery."[43]

National Socialist views and actions increasingly determined life in Germany. At the end of any church service, a blessing for the *Führer* was mandatory. Laue, who was a churchwarden, made sure he was seated close enough to the church exit so that

---

[40] Laue's joint paper with Friedrich Möglich was mentioned several times in the "Kurze Originalmitteilung" section of the Berlin academy's reports: *Berichte der Berliner Akademie* 16 (1933): 544.

[41] Ochsenfeld to Lübbig, Hilchenbach, 28 May 1986. By kind permission of Prof. Dr. Heinz Lübbig. See also Hoffmann (1993).

[42] Cf. Lindner (2014): 61–70. Max von Laue did not publish anything about the Meissner-Ochsenfeld effect until years later.

[43] Meissner to Brill, 16 Aug. 1960, DMA, NL 045 006/1. See also Hoffmann (1999); Matricon and Waysand (2003): 57 f. A search in the Federal Archives Berlin, files of the "Berlin Document Center", did not yield any further material.

he could get away before it was pronounced. The German Physical Society, just like all other societies in the Third Reich, also had to submit to the orders of the regime. When Karl Scheel, the long-time managing director, died on 8 November 1936, a successor had to be found. Laue had repeatedly held leading positions in the society, so why not have him become managing director? The chairman at the time, Jonathan Zenneck, warned Karl Mey in a letter dated November 7th:

(1) He is very unpopular among circles we cannot bypass.

(2) He has a lot of time on his hands. But I'm more afraid of people who have lots of time than I am of evil people. They can sour life for one quite considerably. From my experience with von Laue up to now, I fear that there would be constant upset.

(3) Von Laue is definitely impetuous, so you don't know for sure whether he won't do something really inept at some point (DMA, NL 053/012).

Walter Grotrian, the managing director of the DPG's regional chapter for Berlin, thought otherwise. He wrote to Zenneck from Potsdam on November 23rd, advising him to consider the following: "You probably know Mr. v. Laue better from his time in Munich. At that time, he was surely a very introverted person, and one could have rightly referred to his otherworldliness. But nowadays, in my opinion, that is no longer justified; he is one of the people who give the strongest impetus to physical and scientific life here in Berlin and for years has shown special interest in the affairs of the Physical Society. The only thing that can be said is perhaps this, that he attaches somewhat less importance to appearances than some other people would deem necessary (DMA, NL 053/012)." He had talked at length with Walter Schottky, who considered this solution a good decision (ibid.).

There was apparently no pause in his scientific research during the Christmas season, because several manuscripts were submitted in early to mid-January 1937. Usually, as author, Laue indicated a location on his typescripts, most of the time at the university. But he also occasionally gave his private address: "Berlin-Zehlendorf, Albertinenstraße no. 17." He frequently varied the exact details.

Sometimes it was "Berlin," sometimes "Berlin, Institute of Theoretical Physics," sometimes "Berlin-Dahlem, Kaiser Wilhelm Institute of Physics." The mathematical formulas were emended by hand. This presented a major challenge for the typesetters. The Stürtz press in Würzburg was famous for its error-free typesetting. And Laue's manuscript 'Brightness variation along Kossel lines' from 17 January 1937 was nevertheless no exception. The text began with some praise:

In the fine exposures which Borrmann and Voges have recently taken of the Kossel effect, the most striking feature, which is also strongly emphasized by the auth[ors], is that some Kossel lines stand out partly darkly, partly brightly from the background. As soon as this was experimentally established, the question arose whether such a transition could be interpreted in the theory of the Kossel effect for the ideal single crystal, or whether this theory still required a physical supplement, such as the assumption of a mosaic structure, for these observations to become reconciled. At the beginning I was inclined (as the communications by Messrs. Borrmann and Voges suggest) to the view that the physical foundations to date were not sufficient. But upon further reflection the picture has shifted. And this is what I would like to discuss here (Laue 1937a).[44]

In this case, applying the "dynamic theory" failed, as Laue made clear. He picked up on a remark made by the authors about the different backgrounds on either side of a Kossel line and reached the conclusion that the mathematics of the dynamical theory needed improvement (ibid.).

Max von Laue once again felt called upon to set things straight in his science. He had to let his colleagues Fritz Kirchner and O. Rüdiger know what he thought about their 'Irrational interference points,' the title of his rebuttal (Laue 1937b). But praiseworthy things always cropped up as well. In England, at the Royal Institution of Great Britain, Kathleen Lonsdale was working on X-ray diffraction and crystallography. She was the first to elucidate the structure of an organic compound, hexachlorobenzene, using Fourier analysis. As a supplement to the *International Tables for the Determination of Crystal Structures*, appeared in 1935, she published her *Simplified Structure Factor and Electron Density Formulae for the 230 Space Groups of Mathematical Crystallography* (1936), based on her findings. Max von Laue's review for *Naturwissenschaften* commended this book very highly, because: "She provides, in addition, for each space group, in a form convenient for numerical calculation, the equation by which, in accordance with the fundamental investigations of English structuralists, one determines the electron distribution from intensity measurements of X-ray interferences (Laue 1937c)."

In mid-January 1937 the Zurich Physical Society celebrated its jubilee. The topic chosen for this meeting in Zurich was the "solid body." Thus the fundamental discoveries by Laue and Bragg, Jr., as well as by Debye and Scherrer were honored; and many reports were given about determining structure. Max von Laue spoke about Kossel and Kikuchi lines. His introduction was another tribute to the history of physics: "Twenty-five years have passed since the discovery of X-ray interference phenomena raised to certainty the old and very probable hypothesis of the space lattice structure of solid bodies and, moreover, secured beyond all doubt the wave nature of X-rays. And about 10 years have elapsed since, by means of the same space lattices, Davisson and Germer confirmed the existence of the de Broglie waves for electrons, thus completing what, in m[y] op[inion], is the greatest discovery we have witnessed (Laue 1938a)." There followed a brief historical retrospection appreciating the contributions by Paul Peter Ewald and Arthur H. Compton, and a discussion of some details, before coming to the subject at hand: the factor of crystal form in elementary theory and the Kossel effect, the reciprocity theorem, and the application of the dynamical theory to the Kossel effect. As this was a talk, text obviously predominated, and only a few figures and some formulas supplemented his explanation of the nature of Kikuchi lines. At the end, however, Laue pointed out that the underlying theory was more complicated (ibid.).

---

[44] Gerhard Borrmann was 29 years younger than Laue. He was working for Kossel in Danzig. In the 1950s he was to become a very successful collaborator of Laue's at the Fritz Haber Institute of the MPG in Berlin.

At the beginning of every year, the Nobel Committee wrote to its Nobel laureates looking for suggestions. In 1933 Max von Laue had proposed Arnold Sommerfeld and Otto Stern again, in 1935 Wolfgang Pauli, and in 1936 Carl David Anderson. In 1937 Enrico Fermi was under discussion among German physics circles as a worthy recipient. Laue, however, then proposed Otto Hahn and Lise Meitner for "the discovery of radioactive elements and the elucidation of decay processes." This decision, backed by Heisenberg, was also politically motivated, in an effort to provide Lise Meitner with some protection. Unfortunately, his proposals were unsuccessful. That fall, a strict ban was issued on German scientists from participating in the nomination process or even from accepting the prize. The reason was political. The publicist, pacifist, and editor of the weekly newspaper *Weltbühne*, Carl von Ossietzky, had been arrested by the Nazis, dispatched to a concentration camp, and grieviously maltreated. His friends nominated him for the Nobel peace prize. Seriously ill in 1936, he was released, and the prize for 1935 was retroactively awarded to him. He died in 1938 as a consequence of this incarceration and abuse.

Berlin, October 6th, 1937.

The Reich and Prussian Minister for Science, Education, and Culture.

Regarding: Involvement of German scientists in nominations for Nobel prizes.

In order to remove ambiguities, the following is stated: Apart from the fact that the acceptance, both of the Nobel peace prize and of the scientific Nobel prizes, is forbidden to Germans according to the Decree by the Fuehrer and Reich Chancellor on the occasion of the endowment of a German National Prize for Art and Science of January 30th, 1937 (RGB fol. 1 page 305), the participation of German scientists in the nomination and appraisal of Nobel prize candidates must also cease. This applies to the Nobel peace prize as well as to the scientific Nobel prizes.

I request that this decree be brought confidentially to the attention of the aforesaid scientists.

The decree shall not be published in the Ministerial Gazette (UAF, Laue papers, 1.16 Nobel Prize).

The scientific quest continued. In February, Laue submitted the lengthy manuscript 'The detection of submicroscopic crystal surfaces by electron diffraction,' including several diagrams and illustrations (Laue 1937d). This paper shows again how closely he studied the publications of other researchers. He had the assistance of Georg Menzer and Karl-Heinrich Riewe for specific details in his elaboration (Laue and Riewe 1936). "Cochrane's recently published diffraction images so surpass anything earlier of the kind in sharpness now, that discussing the theory again on the basis of these experiments is probably worthwhile (Laue 1937d: 211)." William Cochrane (1936) had taken measurements with electrons at 39,000 V with grazing incidence on very thin metal layers of 10^{-4} to 10^{-6} cm. Comparable experiments had been done by L. Brück (1936). The aim had been to analyse the diffraction at layers of rhombohedra and octahedra. "Here the effectiveness of even a single atomic layer was shown in the experiment. However, the possibility of making this distinction revealed an astonishing regularity in the structure of these crystal faces. I would not have believed this when, in the cited Annalen paper, I calculated the crystal

Fig. 9.8 The Kaiser Wilhelm Institute of Physics in Berlin-Dahlem around 1937: *on the left*, the building with the high-voltage facility for atomic disintegration; *in the middle*, the office space and smaller laboratories; *on the right*, before and after the entrance area with the turret, another laboratory wing for, among other things, low-temperature experiments. *Source* Archive of the Max Planck Society, Berlin-Dahlem

form factor, more precisely, the integral $E(A_0)$ of Eq. (7), by means of an approximation which replaced every step-shaped boundary layer by macroscopically smooth surfaces (Laue 1937d)."[45]

In April 1937 Peter Debye presented the not-yet-quite-finished institute, in his essay 'The Kaiser Wilhelm Institute of Physics' (1937: 260). The special structural features of the rooms and their equipment were explained in depth; and details were given of the extra-high voltage tower 15 m in height with the planned 2.8-million-volt system and separately located refrigeration laboratory plus a liquefaction plant for hydrogen and helium (Fig. 9.8). "The research program automatically emerges from the description of the institute's installations (ibid., trans. Hentschel 1996: 151)."

---

[45] Laue also presented a talk on this topic at the 13th German Physicists Convention in Bad Kreuznach in June 1937. The *Frankfurter Zeitung* reported on the conference regarding the second topic, nuclear physics: "The present state of experimental and theoretical nuclear research is as follows: By bombarding atomic nuclei with rapidly moving protons and neutrons, it is possible to convert the atomic nucleus of one element into that of another. In addition to the known 'stable' atomic nuclei, hitherto unknown, 'artificial radioactive' atomic nuclei are often produced, which usually change into known stable nuclei after a short time by the emission of elementary particles. In this way, it is even possible to create 'transuranic elements'—elements that have a nucleus of more complex structure than the heaviest element known up to now, uranium. All these transuranic elements are radioactive, so they decay back into simple nuclei in a short time." DMA, NL 053/37.6. Lise Meitner's evaluation of these analyses concurred, after experimenting with uranium at the KWI of Chemistry with Otto Hahn and Fritz Strassmann.

But nothing else was imparted. Then the financing by the KWG, the Rockefeller Foundation, and the Reich government was mentioned. "Here again, in a time of extreme economic strain, which justifiably causes many agencies to build upon the familiar and to fix their eyes on the nearest at hand, the highest authorities of the government leadership demonstrate wise and far-sighted interest in independent, unappropriated, and unrestricted research, which the latter indeed can and must claim for the future (ibid.)."

The name "Max Planck Institute" is studiously ignored in this publication, because the ministry did not approve of this designation. The department headed by von Laue remained unmentioned, likewise the institute's scientists.[46] The completion of the building with all its special facilities was an extraordinary achievement, because in Germany at that time not only building materials were in short supply but also nonferrous metals. Less prestigious buildings had to make do with aluminium instead of copper in their cabling and piping; door fittings were made of aluminium, and often there weren't even enough nails. One reason for this was the construction underway of the "Siegfried Line" and the country's heavy investment in armaments.

At the end of March 1937, the *Steglitzer Anzeiger* published on the 27th a brief news item under the headline 'Impending completion of the Kaiser Wilhelm Institute' including a mention of its name and namesake and its inauguration on Max Planck's 79th birthday. Apart from Debye, Laue and Hermann Schüler were also mentioned (AMPG, I rep. 1A, no. 1652).

The opposition to relativity theory and theoretical physics was still being led by Philipp Lenard and, primarily, by Johannes Stark who made use of every possible means to sway popular opinion. The inflammatory paper *Das Schwarze Korps* devoted a whole page to the subject in mid-July under the headline "White Jews in Science" (Anon 1937: 6). The term *Gesinnungsjuden* was used as a heading in an anonymous article: "Jews by Mentality. The vernacular has stamped such germ carriers with the label, 'White Jew,' which is extremely suitable, because it extends the concept of the Jew beyond race. In the same sense we could also speak of Jews in spirit [*Geistesjuden*], of Jews by mentality, or Jew types [*Charakterjuden*]." (ibid., trans. Hentschel 1996: 153). It goes without saying that Albert Einstein and the theory of relativity were attacked, but so was Werner Heisenberg, who had defended theoretical physics, Einstein, and the theory of relativity in the *Völkischer Beobachter* in 1936. He was called the "Ossietzky" of physics. Johannes Stark wrote a positively disposed sequel to this inflammatory article. So as not to become personally liable, Stark was careful to avoid mentioning names. "Though ethnically Jewish lecturers and teaching aids were forced to resign their positions in 1933 and though Aryan

---

[46] Reichs- und Preußischer Minister für Wissenschaft, Erziehung und Volksbildung to Generaldirektor der KWG, 11 May 1937. AMPG, I rep. 1A, no. 1652, including a memorandum for the file dated 27 May 1937 about the naming of the institute.

See also Kant (1996), Heisenberg (1971a). Laue is only alluded to as deputy director. Nothing about his involvement in the founding, nothing about his scientific research and his independent department was mentioned. See also Heisenberg (1971b). In the chapter on his activity as "director" at the Kaiser Wilhelm Institute of Physics and about his internment at Farm Hall, Heisenberg leaves completely aside Laue's activities or personality.

professors married to Jewesses are presently also being discharged, the majority of Aryan comrades and pupils of Jews, who had previously supported Jewish power in German science openly or covertly, have retained their positions and keep the influence of the Jewish spirit alive at German universities (ibid., Stark 1937)." The final section was headed "New Flood of Jews" and attacked the publishing house of Julius Springer in Berlin and Vienna (ibid.).

At the 26th ordinary plenary meeting of the KWG in June 1937, the National Socialist *Führerprinzip* was formally adopted, thereby also extending the power of that society's president. Max Planck had already been serving as president beyond the regular term, and now he was able to hand over the position to Carl Bosch (1874–1940). Carl Bosch was a nephew of the industrialist Robert Bosch. He had studied mechanical engineering and chemistry. As a collaborator of Fritz Haber at the chemical company BASF, he had succeeded around 1908 in conceiving an efficient catalytic high-pressure method for the synthesis of ammonia from nitrogen and hydrogen on an industrial scale. In 1931, he had been awarded the Nobel prize in chemistry together with Friedrich Bergius. Bosch rose to the rank of chairman of the supervisory board of the chemical trust *IG Farben-Fabriken*. As a member of the KWG senate, he had already been an integral part of the society for some time.

On June 10th, Laue presented another paper to his fellow members at the academy: 'The distribution of current in superconductors.' Citing his publication from five years earlier (1932a), he referred to two theorems about current distribution: Current flows as a surface current in such a way that no magnetic field prevails in the interior, and the distribution of the current is such that the magnetic energy assumes the smallest possible value. Laue proved that it is possible to derive the first theorem from the second. The calculation was performed using Maxwell equations. He discussed the problems caused by the nonsuperconducting leads. Fritz London, in exile in France, had developed a mathematical formulation without reference to Maxwell's theory, and Laue was able to show that the distributive theorem for the currents remained valid even according to London's theory (Laue 1937e: 240).

Despite the political pressure, some societies tried not to give way completely. At the opening of a joint meeting of the Berlin branch of the Physical Society and the Deutsche Gesellschaft für technische Physik on June 9th, Planck recalled Laue's presentation before their ranks on 14 June 1912: "As I am probably the eldest among you, I feel the need to recollect with you a memorable day that has imperishable significance both for our society and for physical science. Twenty-five years ago yesterday, in Munich, Mr. Sommerfeld submitted to the Bavarian Academy of Sciences a paper by Walter Friedrich, Paul Knipping, and Max v. Laue, bearing the title 'On interference phenomena in X-rays.'" (Planck 1937: 77, Laue 1912) Planck spoke very earnestly and forcefully about the necessity for the theory and went into the details of the discovery and its repercussions[47] (Fig. 9.9).

It was important for the head of the DPG to be a strong and diplomatic personality. This became a current issue in 1937. In a letter dated August 27th from Berlin, Max

---

[47] See Meitner, Lise, CAC, MTNR 3/2, diaries: "9th VI, even[ing]. The post-session took place in the 'Heidelberger.' The next day there was a large gathering for v. Laue with 26 invited persons."

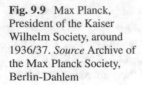

Fig. 9.9 Max Planck,
President of the Kaiser
Wilhelm Society, around
1936/37. *Source* Archive of
the Max Planck Society,
Berlin-Dahlem

von Laue answered the outgoing chairman Jonathan Zenneck: "I thank you kindly
for your letter of the 25th inst. I must share your misgivings about entrusting Planck
with the chairmanship. As far as Debye is concerned, he most definitely does have all
the prerequisite qualities for it now, except one, which I am not quite sure about. Does
his love for the German Physical Society suffice for him to assert his undoubtedly
considerable influence in its favor, if push came to shove? This seems to me to be
the decisive question (DMA, NL 053, no addressee)." Peter Debye was then elected
chairman (DPG proceedings for 1937, 18th ser., 3: 77).

Max von Laue's efforts to arrange that his son Theodore spend a longer period
of time in the USA were successful. After passing his school-leaving exams, Theo
first had to perform his obligatory labor service and a short period in the military in
order to become eligible for academic studies. In addition, as a student he had to be
a member of a National Socialist organization. So he joined the defense unit SA and
enrolled himself in history at Freiburg im Breisgau. Professor Wilhelm Westphal's
divorced wife lived there with their children. The Laue family knew her well, and
Theo settled in well in her home. Many of Theo's letters were quite critical reports
about his studies. He made good use of the freedoms of student life, with much

time spent on music making—he played the violin—and singing, and cultivating friendships. His father thought that a stay of one or two semesters in the U.S. would be an enrichment to his overall education. He succeeded in organizing an American scholarship for him. In the first days of September, Theo's parents accompanied him to Hamburg and brought him to the ship on the 4th. Max von Laue had been able to settle the details beforehand through friends in Baltimore.

The Laue parents returned to Berlin on September 7th and the father immediately sat down to write a letter to his son. Thus began an exchange of correspondence, conducted less intensively by the son than by the father. It provides insights into Max von Laue's experiences in life, his judgments about people and events otherwise not available for earlier periods of his biography.

> You should receive this letter immediately upon your arrival at Princeton, so that you don't feel abandoned there. Because, in my experience, this feeling comes very easily upon arriving in a new place, where one would only begin to feel at home gradually. At least I hope that the welcoming reception in Baltimore has already assured that you won't feel lonely; but what's best is simply best.
>
> Mama and I came home today from a lovely drive. We had splendid weather throughout and beautiful starry skies at night. I am enclosing a carbon copy of the travel report to be added to the car touring book. I can add as well that we also saw your barracks in Wik from the Hindenburg shore in Kiel. We were thinking of you often.
>
> Mama feels very emotional about it, now that you are gone, and as far as I can understand her, this manifests itself in a strange reluctance to pick up young people who are waiting on the roadside for someone to give them a lift by car. But I have still picked up several of them. Because it makes me think of you right then, and that some day you, too, might be in that position of having to secure your 'progress' that way (AMPG, III rep. 50 suppl. 6/7; Lemmerich, ed. 2011: 22).

Then he mentioned an air-raid drill lasting several days, at night everything had to be darkened. He would probably be going to Bad Kreuznach for the physicists' convention; whether his wife would be coming along was still uncertain (ibid.).

A 139-page booklet on 'Superconductivity' by Karl Steiner and Peter Grassmann had appeared in the "Vieweg Collection," which Laue reviewed for *Naturwissenschaften*. He started with the comment that various works on the phenomena had already appeared. "One cannot say that they therefore might appear less mysterious to us than at the outset; rather, the longer one occupies oneself with them, the more riddles they pose. The guiding uniform point of view is still entirely lacking when discussing them (Laue 1937f: 767)." Laue praised the way the material had been ordered and briefly went into the individual chapters discussing the experimental results. In chapter IV on the Meissner-Ochsenfeld effect, the theory predominated and the application of Maxwell's theory although it had not yet been successful. He recommended the work. Laue circumvented one important point, though, as one learns from a letter he wrote to Walther Meissner from Zehlendorf dated September 9th: "I enclose for you in first draft the review of the Steiner-Grassmann booklet for 'Naturwissenschaften.' Please tell me what you think. One could criticize some passages more sharply, namely also the preponderance of Starkian ideas in the final parts. But what's the use of drawing the authors' dilemma into the public eye? (DMA, NL 045/035)" Laue also reported that he had written an article against Stark: "You mustn't think, by the way, that I was endangering myself in doing so. I let Helmholtz do the speaking and copy it verbatim from his lectures—and Helmholtz never did get to know Stark, so he couldn't have given it his stamp either (ibid.)."

Philipp Lenard had, as already mentioned, written a multi-volume work on *Deutsche Physik*, the second volume of which, on 'Acoustics and Heat Theory,' had appeared in 1936. Laue wrote a critical review. "One could take up the present volume with considerable anticipation. One could reasonably expect that it would document Lenard's stature as a teacher, without the disturbing peculiarities of the first volume. Unfortunately, it is a disappointment (DMA, NL 053/040/3. Manuscript, apparently unpublished; cf. Laue 1936f)." Laue missed any mention of the advances in electroacoustics and sound. In thermodynamics, the absence of Nernst's heat theorem was noted, as well as Max Planck's radiation law. His renaming of Brownian motion as "tiny teemings" [*Klein-Gewimmel*] was also faulted. "One sees here, not for the first time, a good lecture course become a less good book when the author attempts *just* to reiterate it and, in particular, does not supplement his own views by careful study of the literature (ibid., orig. emphasis)."

That October, the survey article 'Twenty-five years of the Laue diagram. With a compilation of the present knowledge of atomic distances in crystals' by Fritz Laves appeared in *Naturwissenschaften*. The introduction read: "25 years have elapsed since M. v. Laue gave the world one of the finest ideas ever uttered. An idea that not only inspires those who are endowed with a reverential sense for pure knowledge in pretty form, but also one that forces acknowledgment upon those for whom the greatness of a thought can only be measured by success laid out in historical development (Laves 1937: 705)."

A new intervention by the state in university life occurred in November—namely, the exclusion of "non-Aryans" from participation at the colloquium; that also meant Lise Meitner. Max von Laue informed Max Planck about this by letter, who replied on November 17th with the comment that this "could not have been avoided in the long run (AMPG, II rep. 1A, staff file Laue)."

Laue had been collecting various publications for some time now in order to write a book about X-ray and electron diffraction. Publishing anything in science about general relativity during the Third Reich was very unusual; but Laue, whose interest in publications in this field was unabated, felt compelled to give his frank opinion to his physicist colleague Howard P. Robertson. He borrowed the latter's English title for his German reply in *Zeitschrift für Astrophysik*, 'The apparent luminosity of a receding nebula.' "The first point of contention between Mr. Robertson and me proceeds from the question whether, in the oft-discussed expanding cosmos, every one of its bodies, that is, including the astronomer's telescope, participates in the expansion (Laue 1938b: 160 f.)." Laue had already answered this affirmatively in his paper 'Theoretical views on the brightness of remote nebulae'(1936c). In his reply to Robertson, Laue (1938b) then referred to Einstein's field equations.

On January 12th, Laue wrote to his son from Berlin, remarking that he had learned from James Franck that he had been there to visit and that they had played some music together, before continuing: "I have not been able to take a Christmas vacation here, because I had to finish two manuscripts. On January 20th, I have to send a contribution to the Planck issue for the Annalen der Physik; you remember that Pl[anck] celebrates his 80th birthday on Apr. 23rd (remind Ladenburg and others to toast him then, too!). So, I had two topics, a safe and easy one to handle, and a more important one to me but more difficult. Thus, in any case, I wrote the easier one first. I fortunately just finished the other yesterday, and thereupon I took a long walk this morning from the big star in Grunewald via the Grunewald forester's lodge to Schildhorn and back via Teufelssee (Lemmerich, ed. 2011: 50)." What follows is a vivid account of the birdlife he had observed along the way. "Tomorrow evening Hahn and wife, Lise Meitner, as well as Mr. [Otto] and Mrs. v. Baeyer will be visiting us. […] On Saturday (Jan. 15th) I shall then be traveling to Copenhagen for 3 days to speak with Bohr (ibid.)." On the day after his arrival in Copenhagen, he reported about his train ride with four Danes in the compartment, one of whom had an anti-Semitic propaganda book and was leafing through it in such a way that Laue could see the pictures. Besides some politicians, Albert Einstein was also among those depicted. "I don't know whether those four Danes were spies sent to monitor me; at any rate, beware of such elements over there! (p. 51)".

Soon afterwards, in early February 1938, the next publication was sent off, this time to the editors of *Physikalische Zeitschrift*. Under the title 'The heating of the crystal in the Kossel effect,' Laue dealt with a technically important application. The anticathode of an X-ray tube is heated by the energy of the incident and penetrating electrons. Laue was able to show how to treat this mathematically and that the shape of the anticathode had little influence on the heating (Laue 1938c).

Max von Laue found another opportunity to indirectly criticize the hated regime. On 16 March 1638 Galileo's *Discorsi e dimostrazioni matematiche* had been published in Leyden. The Inquisition had not been able to prevent it! In the preface, Galileo had thanked Count Francesco di Noailles for his compassionate interest in his fate. Laue wrote an extensive review of this enriching work and its historical context for *Naturwissenschaften*. Galileo's 'discourses' on mechanics, the laws of projectiles and falling bodies, air resistance during fall, and acoustics were subjected to his scrutiny and won his appreciation. "This drive for truth moved many of his contemporaries, but probably none made such a powerful breakthrough as Galileo did—thereby exposing him to the *particular* hostility and persecutions of defenders of the old views. Therein lies one significance of Galileo that extends even beyond the field of natural science (1938d: 135, orig. emphasis)."

As a foreigner, Lise Meitner was not yet existentially affected by the racist restrictions imposed by the Nazi regime. Her Jewish colleagues with citizenship in the German Reich had all "given up" their posts in the KWG. Meitner still had her own department at the KWI of Chemistry, but she was not allowed to have any doctoral students. At Christmas 1937 she traveled to Vienna to visit her siblings as she so often did. She evidently did not see the mounting Nazism in Austria. Some of her Berlin friends—probably including the Laues—did not even know she was Jewish prior to 1933, because she had been baptized and attended mass in the village church in Dahlem. Meitner's diary entries in February 1938 indicate reciprocal visits with Arnold Berliner, Gustav Hertz, and the Hahns. On her way home from an invitation by Rausch von Traubenberg on March 11th, she fell and injured her ankle. That made her stay home on March 12th and she listened to the radio. Her diary reads: "12th III. Heard 12 h letter by Führer read out by Goebbels. Invasion of German troops in Austria (Meitner, Lise, CAC, MTNR 2/1, Diaries)." As a direct consequence of this annexation, the so-called *Anschluss* of Austria, Lise Meitner had thus become a Reich German, and the "Nuremberg Laws," the "Law for the Protection of German Blood and German Honor," now also applied to her. On April 15th the Laues left on a five-day excursion to the Upper Saale Valley, and Lise Meitner came along. She entered in the car touring book: "In the gloomy days of our lives God gave us solace for all sorrows, That our gaze may tend heavenwards, Toward sunshine, friendship, and beauty. For my recollection of this Goethean saying, I thank the dear Laue couple, who took me along on such a beautiful Easter trip to the Saale Valley."[48]

On 30 May 1938, the keys to the KWI for Physics were handed over to Peter Debye. But the institute's facilities were by no means complete; the helium liquefaction plant and the high-voltage plant were still missing (AMPG, I rep. 1A, no. 1652).

The year 1938 was a year of great festivities. The 80th birthday of Max Planck, honorary member of the German Physical Society (DPG), was celebrated on April 23rd at Harnack House in Berlin-Dahlem (Fig. 9.10). At the same time, the Planck Medal was awarded to Louis-Victor de Broglie, who could not come for health reasons, however, so the French ambassador accepted the award on his behalf. This was more than a birthday party, more than a medal award; it was a commitment to mutual understanding, to friendship with their French neighbors.

The chairman of the Berlin Physical Society, Carl Ramsauer, opened the festive meeting in the manner customary for the time, with a "Heil Hitler" salute, before welcoming the guests. Then he recalled the lecture held on 19 October 1900 by Planck explaining his radiation formula. This was followed by a laudatory address by the chief editor of the *Annalen der Physik*, Eduard Grüneisen. Max Planck then expressed his thanks with a historical retrospective on the development of the theory of radiation. Peter Debye conveyed the congratulations by the DPG and asked Planck to award the medal. In doing so, Planck looked back briefly on how he had made the acquaintance of de Broglie. "My great joy at the honor of being allowed to confer the medal is redoubled by a joy of a completely different kind. This is the circumstance that it thereby gives me the opportunity to address my professional colleague not only as a scholar, but also to address a member of a neighboring great nation as a German, that nation whose relationship with us is essentially determinative for an independent future (DPG 1938)." Planck then went into the self-evidence of love for one's fatherland and was convinced that there was an honest and ardent desire for a real, lasting peace, in Germany just as in France, before it was too late for Europe. With these thoughts in mind he presented the medal. His Excellency André François-Poncet conveyed the distinguished recipient's warmest thanks. He added: "For, science means cooperation and solidarity, and this is also how I understand the

---

[48] AMPG, III rep. 50 suppl., car touring book. "In unseres Lebens trüben Tagen / Gab uns ein Gott Ersatz für alle Plagen, / Daß unser Blick sich himmelwärts gewöhne, / Den Sonnenschein, die Freundschaft und das Schöne." Goethe.

Die Phyſikaliſche Geſellſchaft zu Berlin

beehrt ſich einzuladen

zu einer Feier des 80. Geburtstages ihres Ehrenmitgliedes

Herrn Geheimrat Profeſſor Dr. Max Planck

am Sonnabend, dem 23. April 1938,

im Harnack=Haus in Berlin=Dahlem, Eingang Ihneſtraße 20.

I. TEIL:

Feſtſitzung im Helmholtz-Saal des Harnackhauſes.

Beginn 18 Uhr pünktlich.

1. Begrüßungsanſprache des Vorſitzenden der Phyſikaliſchen Geſellſchaft zu Berlin, Profeſſor Dr. C. Ramsauer,

2. Anſprache des Schriftleiters der Annalen der Phyſik, Geheimrat Profeſſor Dr. E. Grüneiſen,

3. Anſprache des Vorſitzenden der Deutſchen Phyſikaliſchen Geſellſchaft, Profeſſor Dr. P. Debye,

4. Verleihung der Planck=Medaille durch Geheimrat Profeſſor Dr. M. Planck,

5. Wiſſenſchaftlicher Vortrag, Profeſſor Dr. M. von Laue:

Supraleitung und ihre Beeinfluſſung durch Magnetismus.

II. TEIL:

Feſteſſen mit Damen im Goethe-Saal des Harnackhauſes.

Beginn 20 Uhr.

1. Gemeinſames Abendeſſen,

2. Muſikaliſche Darbietungen des Kammerorcheſters der ſtaatlichen Hochſchule für Muſik in Berlin. Dirigent Profeſſor Dr. Fritz Stein,

3. Vorführung eines Filmes von Nordlichtaufnahmen.

C. Ramsauer

Vorſitzender der Phyſikaliſchen Geſellſchaft zu Berlin.

Anzug: Frack, dunkler Anzug oder Uniform.

Verkehrsverbindungen zum Harnackhaus: U=Bahnhof Thielplatz, Autobus M.

Fig. 9.10 Invitation by the Physikalische Gesellschaft zu Berlin on the occasion of Max Planck's 80th birthday. *Source Z. Phys.*1938, 4: 350

task of diplomacy. The German distinction made to Prof. Louis de Broglie today will be conceived and gratefully perceived in France as proof of this noble mentality."

Max von Laue then delivered a talk on Superconductivity and the Influence of Magnetism upon It. He reformulated the final publication in *Annalen der Physik* (Laue 1938e) somewhat for the assembled physicists. There followed a banquet in the Goethe Hall of Harnack House, which Lise Meitner also attended. Sommerfeld gave the dinner speech, introducing it with a poem by Gottfried Keller about scales, a rod, and a clock, and then continued with an anecdote by Kirchhoff, arriving at the topics of discovery and life for a scholar: "Genius is a gift of God. But other things belong to a great scholar: strict self-restraint, unerring objectivity, the 'seriousness that no effort can blanch,' the obstinacy to return again and again to the problem posed (DPG 1938)." Adriaan Fokker from the Netherlands brought good wishes for Planck: "because you have had such good friends among us and have always welcomed us into your friendship (ibid.)." The astronomer August Kopff then reported that a small planet had been named "Stella Planckia" in Planck's honor. At the end, Planck thanked all the speakers: "When one has attained 80 years of age, one has experienced much. I have always enjoyed my science, and also my pupils (ibid.)." He could have cited numerous names, but preferred rather to refrain from doing so. "But there is one name I should like to mention here, and that is Mr. Max von Laue, who from amongst my closest pupils has become not only a famous physicist, but also an intimate and loyal friend (ibid.)." Planck then also mentioned Moritz Schlick. In closing, he proposed a toast to the health of the German Physical Society and the Physical Society of Berlin. There was a comical conclusion. Walter Grotrian had written a one-act play about physics, called: 'The precision determination of Planck's action quantum' (Hoffmann et al. 2009). The *Annalen der Physik* dedicated two issues to Max Planck with a laudation by Laue (1938f) and contributions by his pupils and friends. Laue also wrote an address for the Proceedings of the Prussian Academy of Sciences, and likewise for the English-language journal *Research and Progress* (Laue 1938g, 1938h).

Laue's two papers, 'On the thermodynamics of superconductivity,' received January 21st, and 'The magnetic threshold for superconductivity,' received February 18th, belong together. The first deals with "the case of an arbitrarily shaped body in an arbitrary magnetic field (Laue 1938i)." In the normal conductor/superconductor transformation, Laue assumed that the space lattice coincided in both, which experiments by other physicists had proved. His thermodynamic observation showed that "along the whole shared surface, the difference in free energy per unit volume must be equal to the normal component of the force which the magnetic field exerts on the superconductor (ibid.)." He applied London's theory to the problem and, to test it, the data that the Englishman Rex Bush Pontius had obtained with thin wire out of lead. Laue often discussed Fritz London's theory when examining the mathematical treatment of a strongly curved body, i.e., a ring. The results of this mathematical treatment of the problem of a thin wire in a longitudinal magnetic field agreed well with the values measured by Pontius. The last section dealt with the process of transformation. "In the longitudinal field the transition certainly does not pass through states of equilibrium, but, once it has begun, proceeds tumultuously to the end (p. 84)." The second publication likewise dealt with a strongly curved superconductor according to London's theory. Laue assumed, as London also had assumed, that some magnetic lines of force in the superconductors penetrate into the surface of the superconductor. The calculation was carried out for a circular cylinder and a sphere (Laue 1938e) (Fig. 9.11).

At the beginning of May, Laue informed his son that a new car had been ordered, another Steyr, with 55 horsepower and a top speed of 120 km/h. In the interim, his son

Fig. 9.11 Max von Laue, late 1930s. *Source* Archive of the Max Planck Society, Berlin-Dahlem

had decided to stay in the U.S. for another year, which his father noted with pleasure.[49] Laue's concern that his letters were being read by "others" beforehand was not entirely unfounded. Prior to World War II, though, the monitoring of correspondence in Germany was probably limited to a small, specific group of persons.

The Laues took an excursion to Wiesenburg at the beginning of July together with Lise Meitner and Otto Hahn; Lise Meitner thanked them for the "wonderful trip" in their car touring book (AMPG, III rep 50 suppl.). For very many people, not just in Germany, it was still unthinkable that the discriminations and harassment of fellow Germans of Jewish origin, of Jewish faith, could be planned and carried out by the National Socialist state to the point of their complete eradication. Now Lise Meitner, too, was one of those ostracized. After it became clear that the leadership of the KWG would insist upon her resignation without being able to help her, Lise Meitner was forced to take the decision to emigrate. Through the mediation of Debye, Hahn, Laue, and Meitner's Dutch friends Fokker and Coster, she managed to escape by

---

[49] Laue to Theodore, 8 May and 19 May 1938. AMPG, III rep. 50 suppl. 6/7; also in Lemmerich, ed. (2011b): 59, 61.

train to Holland on July 13th, accompanied by Dirk Coster. Her final destination was Stockholm and a position at the Nobel Institute with Manne Siegbahn. Max von Laue was only informed about this "way out" after the escape had succeeded.[50]

On July 10th Laue submitted the theoretical paper: 'Diffraction of converging light waves' by Ilse Fränz-Gotthold and himself to the editors of the *Annalen*. It was dedicated to his Freiburg colleague Gustav Mie for his 70th birthday and referred to experiments by Ernst Lau on width reduction at the zero-order diffraction maximum. Using Sommerfeld's solution for diffraction at an edge and employing Debye's mathematical treatment of cylindrical waves, they first developed a general theory and then implemented a numerical application.[51]

The summer holidays were spent in South Tyrol, Italy, and so Laue was able to write more openly to his son that Lise Meitner had managed to "slip away" to Stockholm. From Wetterstein he wrote on July 20th: "But for me, of course, it is very sad that I can no longer speak with Lise Meitner; even more so for Otto Hahn, whose wife fell seriously ill after Pentecost. He has endured some difficult months (Lemmerich, ed. 2011b: 65)." On October 23rd, Laue sent Meitner his congratulations on her 60th birthday, regretting her fate, her flight to Stockholm, before continuing: "But be glad that it worked. I wrote you before, and I repeat it here: Many envy you for it. When it was looking like war in the final days of September, I told Hahn: 'Let's rejoice from the bottom of our hearts now, that Lise Meitner is gone away; *she*, at least, won't fall into the Styx.' And the risks of war are not the worst of it. It may well be so, that it's not so easy to get used to Stockholm. But always remember how you suffered here for 5½ years. Today it's worse though […]" (Lemmerich, ed. 1998b: 34, 36, orig. emphasis). And Magda von Laue added: "My dear Miss Meitner, I'd have much rather gathered you into my arms and whispered something heartwarming in your ear, now I'm afraid I must do it somewhat clumsily by pen (ibid.)."

Laue still was able to obtain the traveling permission to visit Niels Bohr. He used that opportunity to write some more letters without having to worry about postal surveillance, but he still chose his words carefully. The Prussian Academy like the others had to expel their non-Aryan members. On October 26th, Laue wrote a letter to Lise Meitner from Copenhagen: "Planck seems to be very sprightly but has a lot of work to do, since he has to draw up the new statutes as the Academy's presiding secretary. For, the Academy is now being brought into line. The three Jewish members, Norden, Goldschmidt, and Schur, have already left. Of course, Planck visited them personally (Lemmerich, ed. 1998b: 38)." Max von Laue knew that by reading the papers in Sweden, Lise Meitner was keeping herself informed about the political conditions in Germany. She could therefore also understand the rest of his letter: "Did he [Otto Hahn] write to you that he and I attended the marching

---

[50] The correspondence that began soon afterwards between Lise Meitner and Max von Laue contains many details about the scientific and private life of Max von Laue and his family in Berlin, see Lemmerich, ed. 1998. Meitner, Lise, CAC, MTNR 2/15, diary 1938. Details about Lise Meitner's escape are in Sime 1996 and 2001 resp., Lemmerich (2003): 86 f.

[51] Fränz-Gotthold and von Laue (1938). "Berlin-Dahlem. Max Planck Institute" appears as the location. This indicates that Ms. Fränz-Gotthold was working at the institute. She was also mentioned on other occasions. A personnel list confirms this.

parade in the Lustgarten on September 28th? He did not feel like going at all, but I persuaded him to do so, particularly by pointing out that if there was war, and if, as was to be expected, the SS took over the executive power, the reliability of all those persons would be called into question who had been too obviously opposed. It was worth having gone, too, because Mr. Graue was so astonished at our appearance, his two blackest sheep, that the mere sight of his face amply rewarded the effort (ibid.)."[52] Laue also wrote about other colleagues who had emigrated, and that Erwin Schrödinger had been dismissed without a pension. "At the Kaiser Wilhelm Institute of Physics, everything still is not finished. The liquid helium, for example, has been postponed until December. But the X-ray facilities are gradually getting under way. Dr. Menzer is busy at work on the final preparations (ibid. Georg Menzer was a staff researcher at the KWIP.)." About the children Laue remarked that Theo's second year in the U.S. was secured, and Hilde was driving the new car often and quite well. He also mentioned his activities to convert the laundry room in the basement into an air-raid shelter (ibid.).

That September, the political situation became threatening. More than three million Germans lived in the peripheral areas of Czechoslovakia, Bohemia, and Moravia. Their political representative, Konrad Henlein, was demanding cession and their "annexation" to the German Reich. On September 29th, Hitler met with Neville Chamberlain and Edouard Daladier in Munich and signed a treaty that ceded these territories, known as *Sudetenland*, to Germany. The public rally on September 28th was intended to increase tensions.[53]

On the night of the 9th to the 10th of November, the so-called "Night of Broken Glass" (*Reichskristallnacht*) occurred. Many synagogues were set on fire and destroyed, Jewish shops were looted and devastated, Jews were mistreated, arrested, even killed. The *Reichssteuerblatt* of November 15th published the "Ordinance on the Atonement of Jews of German Nationality" of 12 November 1938: "The hostile attitude of Jewry towards the German people and Reich, which does not even shrink back from commiting cowardly acts of murder, demands resolute defensive measures and harsh atonement. […] § 1 The Jews of German nationality in their entirety are ordered to pay a tribute of 1,000,000,000 Reichsmarks to the German Reich."[54]

On November 7th Lise Meitner celebrated her 60th birthday. No German scientific journal dared to publish any congratulations. Arnold Sommerfeld's 70th birthday was celebrated in Munich on December 5th with several scientific lectures held in his honor. The Laues drove to Munich by car, which took them nine hours, including the

---

[52] Dr. Georg Graue had taken his doctorate under Otto Hahn and was subsequently employed in the KWI of Physical Chemistry and Electrochemistry. He was a member of the NSDAP and a leader of the University Lecturers League.

[53] Laue wrote to his son Theodore on October 2nd from Berlin about the rally: "The procession, which moved along the (enormously widened) Charlottenburger Chaussee and through the Brandenburg Gate to the Lustgarten, consisted of people who were chatting nice and easily about all kinds of everyday things, who were talking about K.W. institutes, and probably also about science; but there was no talk anywhere of political matters, of war or peace. In Lemmerich, ed. 2011: 73.

[54] Verordnung über eine Sühneleistung der Juden deutscher Staatsangehörigkeit, in Dehlinger 1943: 1579. The ordinance was signed by Hermann Göring as the commissioner of the Four-Year Plan.

breaks. Some colleagues who would have wanted to be there were prohibited from participating because they were "not Aryan." Back in Berlin, Max von Laue attended the board meeting of the German Physical Society (DPG) on December 14th chaired by Carl Ramsauer. The minutes stated:

> The Chairman reports on the developments of the non-Aryan issue to date in the German Physical Society. Following the initiative expressed in a letter by Messrs. Stuart and Orthmann and after privately consulting with some of the members of the Board of the German Physical Society, and asking all the members of the Board in writing, the President of the German Physical Society, Mr. Debye, sent a letter addressed to all the German members of the German Physical Society, calling upon Jews in the sense of the Nuremburg Laws to withdraw membership in the Society. The second step to be taken is a change in the Society's statutes […].
>
>   In the discussion Mr. Orthmann points out that the first sentence of the letter addressed to the German members of the Society is formulated in such a way that the letter could be misunderstood.[55]

The letter read: "Under the compelling prevailing circumstances I am forced to view as no longer acceptable the continued membership in the German Physical Society of German Reich Jews in the sense of the Nuremberg Laws. I therefore request, in agreement with the Board of Trustees, that all members who fall within this provision inform me of their withdrawal from the Society. Heil Hitler! sig. P. Debye, Chairman" (ibid., trans. in Hentschel 1996: 181–182; cf. Wolff 2012: 50). Lise Meitner also received this letter, even though she was living abroad. Max von Laue wrote to her from Berlin on December 19th: "Debye has been attacked here because of the wording of his circular: In the first sentence 'regretfully' has been read between the lines, and the intention was to scare him with a reference to the 'Schwarze Korps,' which recently denounced some firms that had 'regretfully' dismissed Jewish employees. Debye's response was: 'I don't give a darn' (Lemmerich 1989: 42 f.)."

In his letter to Theo of December 11th, Laue discussed the nearing Christmas season and recommended that his son read the "Gospel according to St. Matthew" to contemplate about what people had done with this teaching. He criticized the abracadabra about the Holy Trinity and thought a new religious genius, a reformer, was needed. In the last letter of that year to his son, Laue wrote a passage about current politics: "Well, I also see something of a 'grand line.' It leads quite directly from the Reichstag fire, to the arrest of the parliamentarians, who according to the constitution that Hitler had sworn by were immune, to the 30th of June 1934. And if I try to follow it further into the future, it seems to me to lead to the complete destruction of all intellectual freedom, to the compulsory introduction of a common, prescriptive 'world view' convenient to the rulers, thereby destroying all that the

---

[55] BBAW, A II-I, 13 Tit.: Statutes 1937–1944, fols. 16 and 33; DPGA, 10013. Translation in Hentschel (1996): 182–183. On 8 Oct. 1938, Minister Rust requested that the Academy amend its statutes: "In this connection, I attach importance to the fact that the ordinary members of the Academy, insofar as they are non-Aryans, leave the Academy.".

reformers of the sixteenth century, and specifically the great scholars of the 17th, had fought and suffered for."[56]

A scientific event of globally historic impact occurred at the end of the year. Lise Meitner had persuaded Otto Hahn to repeat some experiments that the Italian, Enrico Fermi, had performed about the formation of new elements heavier than uranium upon the bombardment of uranium 239 with neutrons. In 1935 Hahn and Fritz Strassmann had started to conduct the extremely challenging experiments of chemical separation and detection. They always discussed their latest results with Ms. Meitner, but they were difficult to interpret. The search for transuranic elements continued on after her escape. At first it looked as if they had found radium. But then Hahn and Strassmann detected barium. On 19 December 1938 Hahn wrote a long letter to Lise Meitner about their analyses. The lines containing the most important scientific statements read: "Our Ra isotopes do not act like Ra but like Ba. [...] Perhaps you can come up with some sort of fantastic explanation. We know ourselves that it *can't* actually burst apart into Ba (Sime 1996: 233, orig. emphasis)." And in the next letter to Meitner, dated December 28th, Hahn was more specific: "Would it be possible that the uranium-239 breaks up into a Ba and a Ma? A Ba 138 and a Ma 101 [i.e., technetium] would give 239 (p. 239)." Lise Meitner spent Christmas at her friend's house in Kungälv. Her nephew, Otto Robert Frisch, joined her from Copenhagen. During that period between Christmas and the turn of the year, they together succeeded in explaining the process physically and anticipating the very large release of energy. Returned to Copenhagen, Frisch performed another very indicative experiment that demonstrated the major release of energy in the process of fission. (Meitner and Frisch 1939; Frisch 1967: 143–146; Lemmerich 1998a: 158 f., 180 f.) A few months later, there was talk of an atomic bomb and the first mentions of it appeared in writing.

References

Akademie der Wissenschaften der DDR, Ed. 1979. *Albert Einstein in Berlin 1913–1933*. Part 1: *Darstellungen und Dokumente*. Edited by Christa Kirsten and Hans-Jürgen Treder, with an introduction by Hans-Jürgen Treder. Berlin: Akademie-Verlag

Albrecht, Helmuth, Ed. 1993. *Naturwissenschaft und Technik in der Geschichte. 25 Jahre Lehrstuhl für Geschichte der Naturwissenschaften und Technik am Historischen Institut der Universität Stuttgart*. Stuttgart: GNT-Verlag

Anonymous. 1937. "Weiße Juden" in der Wissenschaft. *Das Schwarze Korps. Organ der Reichsführung SS*, 15 July 1937, no. 28: 6; in English, Anon: "White Jews" in science. In Hentschel 1996: 152–157

BJGA [Friedrich Glum]. 1933. Um eine Staatsidee. *Rheinisch-Westfälische Zeitung*, 5 Jan. 1933, morning ed. p. 1

Boltzmann, Ludwig. 1905. *Populäre Schriften*. Leipzig: Barth

Brück, L. 1936. The structure of thin metal films deposited on rock salt. *Ann. Phys.* 419: 233–257

---

[56] Laue to Theodore, Berlin, 28 Dec. 1938. AMPG, III rep. 50 suppl. 6/7; also in Lemmerich, ed. 2011: 82 f., 87.

Cochrane, William. 1936. The structure of some metallic deposits on a copper single crystal as determined by electron-diffraction. *Proc. Phys. Soc., London*, 48: 723–735

Davisson, Clinton J., and Lester H. Germer. 1927. Diffraction of electrons by a crystal of nickel. *Physical Review* 30: 705–740

Debye, Peter. 1937. Das Kaiser-Wilhelm-Institut für Physik. *Naturw.* 25: 257–260; translation in Hentschel 1996: 146–151

Dehlinger, Alfred 1943. *Systematische Übersicht über 76 Jahrgänge Reichsgesetzblatt (1867–1942)*. Stuttgart: Kohlhammer

Deichmann, Ute. 1996. Dem Vaterlande – solange es dies wünscht. Fritz Habers Rücktritt 1933, Tod 1934 und die Fritz-Haber-Gedächtnisfeier 1935. *Chemie in unserer Zeit* 3: 141–149

DPG. 1938. [Award of the Planck medal to L. de Broglie on the occasion of Planck's 80th birthday]. *Verh. Dt. Phys. Ges.* 19: 57–76

Eckert, Michael. 2012. The German Physical Society and "Aryan Physics." In Hoffmann and Walker 2007, trans. 2012: 96–125

Ehrenfest, Paul. 1933. Bijzondere Vergadering der Afdeling. *Natuurkunde*, 31 October 1933 (on his address "Warum sind Atome so dick?")

Einstein, Albert. 1975. *Über den Frieden. Weltordnung oder Weltuntergang?* Edited by Otto Nathan and Heinz Norden. Bern: Lang. Engl. orig.: *On Peace*. New York: Simon & Schuster, 1960.

Ewald, Paul Peter. 1936. M. v. Laue: Die Interferenzen von Röntgen- und Elektronenstrahlen. *Naturw.* 24: 714 f

Fränz-Gotthold, Ilse, and Max von Laue. 1938. Beugung konvergierender Lichtwellen. *Ann. Phys.* 425: 249–258

Frisch, Otto Robert. 1967. The interest is focussing on the atomic nucleus. In *Niels Bohr*. Edited by S. Rozental, 137–148. Amsterdam: North Holland Publishing

Gavroglu, Kostas. 1995. *Fritz London. A Scientific Biography*. Paperback ed. 2005. Cambridge: Univ. Press

Grundmann, Siegfried. 2005. *The Einstein Dossiers. Science and Politics – Einstein's Berlin Period*. Trans. by Ann M. Hentschel. Berlin, Heidelberg, New York: Springer

Haber, Fritz. 1933. Zur fünfundsiebenzigsten Geburtstage des Präsidenten der Kaiser Wilhelm-Gesellschaft Max Planck. *Naturw.* 21: 293

Hahn, Dietrich, Ed. 1975. *Otto Hahn. Erlebnisse und Erkenntnisse*. With an introduction by Prof. Dr. Karl-Erik Zimen. Düsseldorf, Vienna: Econ-Verlag

Heisenberg, Werner. 1927. "Über den anschaulichen Inhalt der quantentheoretischen Kinematik und Mechanik". In *Werner Heisenberg Collected Works*, Series A II. Ed. by W. Blum, H.-P. Dürr, and H. Rechenberg. Berlin, Heidelberg 1989. Springer. In *Z. Phys.* 43 (1927): 172–198

Heisenberg, Werner. 1971a. Das Kaiser-Wilhelm-Institut für Physik. Geschichte eines Instituts. *Jahrbuch der Max-Planck-Gesellschaft zur Förderung der Wissenschaften e.V.* 1971: 46–89

Heisenberg, Werner. 1971b. *Physics and Beyond*. New York

Henning, Eckhart. 2004. Max Planck – „ein armer Wirrkopf" als Kollaborateur der Nazis? In E. Henning: *Beiträge zur Wissenschaftsgeschichte Dahlems*. Veröffentlichungen aus dem Archiv zur Geschichte der Max-Planck-Gesellschaft no. 13, pp. 69 f. Berlin: Max-Planck-Gesellschaft

Hentschel, Klaus, and Ann M. Hentschel, eds. 1996. *Physics and National Socialism. An Anthology of Primary Sources*; revised reprint 2010. Basel, Boston, Berlin: Birkhäuser

Hoffmann, Dieter. 1993. Wendepunkt in der Geschichte der Supraleitung: Vor 60 Jahren wurde der Meissner-Ochsenfeld-Effekt entdeckt. *Phys. Bl.* 10: 899–901

Hoffmann, Dieter. 1999. Robert Ochsenfeld. *Neue Deutsche Biographie*, vol. 19: 412 f. Berlin: Duncker & Humblot

Hoffmann, Dieter, Hole Rößler, and Gerald Reuther. 2009. „Lachkabinett" und „großes Fest" der Physiker. Walter Grotrians „physikalischer Einakter" zu Max Plancks 80. Geburtstag. Max-Planck-Institut für Wissenschaftsgeschichte. Preprint no. 369, 2009. www.mpiwg-berlin.mpg.de/Preprints/P369.PDF [last accessed: 5 Oct. 2021]

Hoffmann, Dieter, and Mark Walker, eds. 2007. *Physiker zwischen Autonomie und Anpassung. Die Deutsche Physikalische Gesellschaft im Dritten Reich*. Weinheim: Wiley-VCH; trans. ed. by

Ann M. Hentschel (without the documentary appendix): *The German Physical Society in the Third Reich. Physicists between Autonomy and Accommodation.* Cambridge: Univ. Press, 2012

Israel, Hans, Erich Ruckhaber, and Rudolf Weinmann, eds. 1931. *Hundert Autoren gegen Einstein.* Leipzig: Voigtländer

Justi, Eduard, and Max von Laue. 1934. Neuartige Phasenumwandlungen bei einheitlichen Stoffen. *Sb. Preuß. Akad. Wiss.* 1934: 237

Kant, Horst. 1994. Arthur Wehnelt und die Elektronenphysik. *Physik in der Schule* 32: 115

Kant, Horst. 1996. Albert Einstein, Max von Laue, Peter Debye und das Kaiser-Wilhelm-Institut für Physik in Berlin (1917–1939). In *Das Harnack-Prinzip.* Berlin: de Gruyter, edited by Bernhard vom Brocke and Herbert Laitko, 227–244

Krebs, Hans. 1979. *Otto Warburg. Zellphysiologe, Biochemiker, Mediziner, 1883–1970.* Stuttgart: Wissenschaftliche Verlagsgesellschaft

Laitko, Hubert, Dieter Hoffmann, and Horst Kant, Eds. 1986. *Berlin – Hauptstadt der DDR*, Part III. Berlin: Präsidium der Urania

Laue, Max. 1912. Eine quantitative Prüfung der Theorie für die Interferenzerscheinungen bei Röntgenstrahlen. *Sb. Bayer. Akad. Wiss.*, math.-phys. class, 1912: 363–373

Laue, Max von. 1926. Lorentz-Faktor und Intensitätsverteilung in Debye-Scherrer-Ringen. *Z. Kristall.* 64: 115–142

Laue, Max von. 1930. Zur Elektrostatik der Raumgitter. *Sb. Preuß. Akad. Wiss.*, Phys.-Math. Kl., 16 Jan. 1930: 26–41

Laue, Max von. 1931a. The diffraction of an electron-wave at a single layer of atoms. *Physical Review* 1: 53–59

Laue, Max von. 1931b. Die Lichtfortpflanzung in Räumen mit zeitlich veränderlicher Krümmung nach der allgemeinen Relativitätstheorie. *Sb. Preuss. Akad. Wiss.*, Phys.-Math. Kl. 1931: 123–131

Laue, Max von. 1931c. Entstehung der Elemente und kosmische Strahlung. *Naturwiss.* 19: 530–531, 641

Laue, Max von. 1931d. Die dynamische Theorie der Röntgenstrahlinterferenzen in neuer Form. *Ergeb. d. exakten Naturw.* 10: 133–158

Laue, Max von. 1931e. Zur Theorie des Doppelkristallspektrometers. *Z. Phys.* 72: 472–477

Laue, Max von. 1932a. Zur Deutung einiger Versuche über Supraleitung. Lecture given at the award of the Planck Medal. *Phys. Z.* 33: 793 ff

Laue, Max von. 1932b. Kreuzgitterspektren. *Z. Kristall.* 82: 127–141

Laue, Max von. 1932c. Zu den Erörterungen über Kausalität. *Naturw.* 20: 915 f

Laue, Max von. 1933a. Zu Plancks 75. Geburtstag. *Phys. Z.* 34: 305

Laue, Max von. 1933b. Korpuskular- und Wellentheorie. In *Handbuch der Radiologie*, Vol. 6, Part 1: *Atome und Elektronen*, edited by Erich Marx, 2nd ed., offprint 114 pp. Leipzig: Akad. Verlagsanstalt

Laue, Max von. 1933c. Ansprache bei der Eröffnung der Physikertagung in Würzburg. *Phys. Z.* 34: 889–890

Laue, Max von. 1933d. Zur Verteilung der Nobelpreise für Physik am 10. Dezember. *Metallwirtschaft.* 12: 719

Laue, Max von. 1933. Materie und Raumerfüllung. *Scientia* 27: 402

Laue, Max von. 1934a. Statistische Physik. In *Handwörterbuch der Naturwissenschaften.* 2nd ed., vol. 9: 537–550. Jena: Fischer

Laue, Max von. 1934b. Fritz Haber gestorben. *Naturw.* 22: 97

Laue, Max von. 1934c. Über Heisenbergs Ungenauigkeitsbeziehungen und ihre erkenntnistheoretische Bedeutung. *Naturw.* 22: 439 ff

Laue, Max von. 1934d. Aus dem Wirken der Kaiser-Wilhelm-Gesellschaft. *Deutsche Medizinische Wochenschrift* 37: 1376

Laue, Max von. 1935a. Der Einfluß eines Magnetfeldes auf Wärmeleitung und Reibung in paramagnetischen Gasen. *Ann. Phys.* 415: 1–15

Laue, Max von. 1935b. Der optische Reziprozitätssatz in Anwendung auf die Röntgenstrahlinterferenzen. *Naturw.* 23: 373

Laue, Max von. 1935c. Die Fluoreszenzröntgenstrahlung von Einkristallen. (With an appendix on electron diffraction). *Ann. Phys.* 415: 705–746

Laue, Max von. 1935d. *Die Interferenzen von Röntgen- und Elektronenstrahlen. Fünf Vorträge.* Berlin: Springer

Laue, Max von. 1936a. Zur Theorie der Kikuchilinien. *Ann. Phys.* 417: 569–576

Laue, Max von. 1936b. Die äußere Form der Kristalle in ihrem Einfluss auf die Interferenzerscheinungen an Raumgittern. *Ann. Phys.* 418: 55–68

Laue, Max von. 1936c. Theoretisches über die Helligkeit ferner Nebel. *Zeitschrift für Astrophysik* 11: 208

Laue, Max von. 1936d. Kikuchi-Enveloppen. *Phys. Z.* 37: 544–547

Laue, Max von. 1936e. Experimentelle und theoretische Physik. *Ostdeutsche Tagespost*, nos. 89 and 91; also in *Z. phys. chem. Unterricht* 50 (1937): 164–167

Laue, Max von. 1936f. Deutsche Physik. *Frankfurter Zeitung*, Saturday, 29 February, vol. 80, nos. 110–111: 10. Book review of Lenard 1936, vol. 1; English trans. in Hentschel 1996: 127–129

Laue, Max von. 1937a. Helligkeitswechsel längs Kossellinien. *Ann. Phys.* 420: 528–532

Laue, Max von. 1937b. Irrationale Interferenzpunkte (Erwiderung an Kirchner und Rüdiger). *Ann. Phys.* 422: 616 ff

Laue, Max von. 1937c. Simplified structure factor and electron density formulae for the 230 space groups of mathematical crystallography. Book review of Lonsdale 1936. *Naturw.* 25: 140

Laue, Max, von. 1937d. Die Erkennung submikroskopischer Kristallflächen durch Elektronenbeugung (Nach Versuchen W. Cochranes und L. Brücks). *Ann Phys.* 421: 211–238

Laue, Max von. 1937e. Die Stromverteilung in Supraleitern. *Sb. Preuss. Akad. Wiss.*, Math.-Phys. Kl. 1937: 240

Laue, Max von. 1937f. Karl Steiner & Peter Grassmann: Supraleitung. Book review. *Naturw.* 25: 767

Laue, Max von. 1938a. Kossel- und Kikuchi-Linien. *Phys. Ges. Zürich* 1938: 131–154

Laue, Max von. 1938b. The apparent luminosity of a receding nebula. (A reply to Robertson) *Z. Astrophys.* 15: 160 f

Laue, Max von. 1938c. Die Erwärmung des Kristalls beim Kosseleffekt. *Phys. Z.* 39: 339

Laue, Max von. 1938d. Zum dreihundertsten Geburtstag des ersten Lehrbuches der Physik. *Naturw.* 26: 129–135

Laue, Max von. 1938e. Der magnetische Schwellwert für Supraleitung. *Ann. Phys.* 424: 253–258

Laue, Max von. 1938f. Zum 80. Geburtstag von Max Planck. *Ann. Phys.* 424: 2–4

Laue, Max von. 1938g. (Address to Max Planck). *Sb. Preuss. Akad. Wiss.*, Phil.-hist. Kl. 1938: 157

Laue, Max von. 1938h. (On Max Planck). *Research and Progress. Forschung und Fortschritte. Nachrichtenblatt der Deutschen Wissenschaft* 4: 103

Laue, Max von. 1938i. Zur Thermodynamik der Supraleitung. *Ann. Phys.* 424: 71–84

Laue, Max von. 1961a. *Gesammelte Schriften und Vorträge.* Vols. I–III. Braunschweig: Vieweg

Laue, Max von. 1961b. Mein physikalischer Werdegang. Eine Selbstdarstellung. In *Max von Laue Gesammelte Schriften und Vorträge.* Vol. 3: V–XXXIV. Braunschweig: Vieweg

Laue, Max von, and Friedrich Möglich. 1933. Über das magnetische Feld in der Umgebung von Supraleitern. *Sb. Preuss. Akad. Wiss.* 16: 544 f

Laue Max von, and Karl-Heinrich Riewe. 1936. Der Kristallformfaktor für das Oktaeder. *Z. Kristall.* 95: 408–420

Laue, Max von, and Richard von Mises, Eds. 1936. *Stereoskopbilder von Kristallgittern.* Vol. 2. Berlin: Springer

Laue, Max von, Fritz London, and Heinz London. 1935. Zur Theorie der Supraleitung. *Z. Phys.* 96: 359–364

Lautz, Günter. 1979. Von der Festkörperphysik zur Energieumwandlung. *Phys. Bl.* 9: 409–411

Laves, Fritz. 1937. Fünfundzwanzig Jahre Laue-Diagramm. Mit einer Zusammenstellung der heutigen Kenntnisse über Atomabstände in Kristallen. *Naturw.* 25: 705–708 and 721–733

Lemmerich, Jost. 1998a. *Die Geschichte der Entdeckung der Kernspaltung.* Exhibition catalogue. Berlin: Technische Univ., Universitätsbibliothek

Lemmerich, Jost, ed. 1998b. *Lise Meitner – Max von Laue. Briefwechsel 1938–1948*. Berliner Beiträge zur Geschichte der Naturwissenschaften und der Technik, no. 22. Berlin: ERS-Verlag

Lemmerich, Jost, ed. 2003. *Lise Meitner zum 125. Geburtstag*. Exhibition catalogue. Staatsbibliothek zu Berlin, Preußischer Kulturbesitz, 7 Nov.–13 Dec. 2003. Berlin: ERS-Verlag

Lemmerich, Jost. 2011a. *Science and Conscience. The Life of James Franck*. Trans. by Ann M. Hentschel. orig. German ed. 2007. Stanford: Univ. Press

Lemmerich, Jost, ed. 2011b. *Mein lieber Sohn! Die Briefe Max von Laues an seinen Sohn Theodor in den Vereinigten Staaten von Amerika 1937–1946*. With two contributions by Christian Matthaei. Berliner Beiträge zur Geschichte der Naturwissenschaften und der Technik, no. 33. Berlin, Liebenwalde: ERS-Verlag

Lemmerich, Jost. 2015. *Politik und Werbung für die Wissenschaft. Das Harnack-Haus der Kaiser-Wilhelm-Gesellschaft zur Förderung der Wissenschaften in Berlin-Dahlem*. Rangsdorf: Basilisken-Presse in Verlag Natur+Text

Lemmerich, Jost, and Friedrich Hund, eds. 1982. *Der Luxus des Gewissens: Max Born, James Franck. Physiker in ihrer Zeit. Ausstellung der Staatsbibliothek zu Berlin, Preußischer Kulturbesitz*. Exhibition catalogue. Wiesbaden: Reichert

Lenard, Philipp. 1936. *Deutsche Physik in vier Bänden*. Vol. 1: *Einleitung und Mechanik*; Vol. 2: *Akustik und Wärmelehre*. Munich: J.F. Lehmann. See Laue (1936f) for a review of vol. 1

Lindner, Sigrid Annemarie. 2014. *Walther Meißner (1882–1974). Physiker und Institutsgründer. Ressourcenmobilisierung in drei politischen Systemen*. Studien zur Geschichte der Mathematik und der Naturwissenschaften. Augsburg: Rauner

Lippmann, Gabriel. 1919. Sur les propriétés des circuits électriques dénués des résistances. *Comptes rendus de l'Académie des sciences* 168: 73–78

Lonsdale, Kathleen. 1936. *Simplified Structure Factor and Electron Density Formulae for the 230 Space Groups of Mathematical Crystallography*. London: G. Bell & Sons

Macrakis, Kristie. 1986. Wissenschaftsförderung durch die Rockefeller Foundation im "Dritten Reich." *Geschichte und Gesellschaft* 12: 348

Matricon, Jean, and George Waysand, eds. 2003. *The Cold Wars. A History of Superconductivity*. New Brunswick, NJ: Rutgers University

Meier-Benneckenstein, Paul, Ed. 1935. *Dokumente der deutschen Politik*. Vol. 1: *Die Nationalsozialistische Revolution 1933*. Berlin 1935. Junker & Dünnhaupt

Meissner, Walther. 1934. Bericht über neuere Arbeiten zur Supraleitfähigkeit. *Phys. Z.* 35: 931 f

Meissner, Walther, and F. Heidenreich. 1936. Über die Änderung der Stromverteilung und der magnetischen Induktion beim Eintritt der Supraleitung. *Phys. Z.* 37: 449–470

Meissner, Walther, and Robert Ochsenfeld. 1933. Ein neuer Effekt bei Eintritt der Supraleitfähigkeit. *Naturw.* 21: 787 f

Meitner, Lise, and Otto Robert Frisch. 1939. Disintegration of uranium by neutrons: A new type of nuclear reaction. *Nature* 143, no. 3615: 239–240

Mendelssohn, Kurt. 1973. *Walther Nernst und seine Zeit*. Weinheim: Physik-Verlag; in English: *The World of Walther Nernst. The Rise and Fall of German Science*. London: MacMillan

Mises, Richard von. 1934. Über Heisenbergs Ungenauigkeitsbeziehungen und ihre erkenntnistheoretische Bedeutung. *Naturw.* 22: 822

Möllenstedt, Gottfried, ed. 1988. Zum 100. Geburtstag von Walther Kossel. Unpaginated offprint from *„Aus Kossels Laboratorium, Danzig 1932–38" zum 100. Geburtstag am 4. Januar 1988*. Tübingen: University, Faculty of Physics

Müller-Enberga, Helmut, Jan Wielgohs, and Dieter Hoffmann, eds. 2000. *Wer war wer in der DDR? Ein biographisches Lexikon*. Berlin: Links

Pauli, Wolfgang. 1979. *Scientific Correspondence with Bohr, Einstein, Heisenberg et al. 1919–1929*, Vol. 1. Berlin, Heidelberg, New York: Springer

Pauli, Wolfgang. 1985. *Scientific Correspondence with Bohr, Einstein, Heisenberg et al. 1930–1939*, Vol. 2. Berlin, Heidelberg, New York: Springer

Planck, Max. 1932. Der Kausalbegriff in der Physik. *Sb. Preuß. Akad. Wiss.* Phys.-Math. Kl. 1932: 13

Planck, Max. 1937. Zur 25-jährigen Jubiläum der Entdeckung von W. Friedrich, P. Knipping und
    M. v. Laue. *Verh. Dt. Phys. Ges.* 18th ser., 3: 77
Planck, Max. 1944. Die Kausalität in der Natur. In M. Planck: *Wege zur physikalischen Erkenntnis.
    Reden und Vorträge*, 4th ed.: 223 f. Leipzig: S. Hirzel
Planck, Max. 1947. Mein Besuch bei Adolf Hitler. *Phys. Bl.* 5: 143; trans. in Hentschel 1996:
    359–361
Poliakov, Leon, and Joseph Wulf. 1959. *Das Dritte Reich und seine Denker*. Wiesbaden: Arani-
    Verlag
Pufendorf, Astrid von. 2006. *Die Plancks. Eine Familie zwischen Pazifismus und Widerstand*. Berlin:
    Propyläen-Verlag
Schmidt-Böcking, Horst, and Karin Reich. 2011. *Otto Stern. Physiker, Querdenker, Nobel-
    preisträger*. Frankfurt am Main: Societäts-Verlag
Schmitt, Carl. 1940. *Zum 30. Juni 1934. Positionen und Begriffe*. Hamburg: Hanseatische
    Verlagsanstalt
Schönbeck, Charlotte. 2006. Physik. In *Die Universität Heidelberg im Nationalsozialismus*. Edited
    by Wolfgang U. Eckart and Volker Sellin, pp. 1087–1149. Heidelberg: Springer
Sime, Ruth Lewin. 1996. *Lise Meitner. A Life in Physics*. Berkeley: Univ. of California Press
Sommerfeld, Arnold. 2004. *Wissenschaftlicher Briefwechsel 1919–1951*. Vol 2. Edited by Michael
    Eckert and Karl Märker. Berlin, Diepholz: Verlag für Geschichte der Naturwissenschaften und
    Technik
Stark, Johannes. 1937. Die 'Wissenschaft' versagte politisch. *Das Schwarze Korps*, 15 July 1937,
    no. 28: 6; trans. Stark, J. 1937. Science is politically bankrupt. In Hentschel 1996: 157–160
Steinhauser, Thomas, Jeremiah James, Dieter Hoffmann, and Bretislav Friedrich. 2011. *Hundert
    Jahre an der Schnittstelle von Chemie und Physik. Das Fritz-Haber-Institut der Max-Planck-
    Gesellschaft zwischen 1911 und 2011*. Berlin: de Gruyter
Stoltzenberg, Dietrich J. 1998. *Fritz Haber. Chemiker, Nobelpreisträger, Deutscher Jude. Eine
    Biographie*. Weinheim: Wiley-VCH
Szöllösi-Janze, Margit. 1998. *Fritz Haber 1886–1934. Eine Biographie*. München: C.H. Beck
Wolff, Stefan L. 2012. Marginalization and Expulsion of Physicists under National Socialism. What
    Was the German Physical Society's Role? In Hoffmann and Walker 2012: 50–95
Zeitz, Katharina. 2006. *Max von Laue (1879–1960). Seine Bedeutung für den Wiederaufbau der
    deutschen Wissenschaft nach dem Zweiten Weltkrieg*. Stuttgart: Steiner

Chapter 10
Physics and Politics During World War II

The first issue of *Die Naturwissenschaften* in 1939 appeared on January 6th. On the front page was an advertisement for the book 'The Structure of the Atomic Nucleus—Natural and Artificial Nuclear Transformations,' by Lise Meitner and Max Delbrück (1935). For inexplicable reasons, there was no rebuke from the state for displaying a volume authored by a Jewess and an "emigré" to the U.S.! On January 14th, Max von Laue wrote a letter to Lise Meitner. The postscript was: "On Mar. 14th there'll be a 60th birthday at Princeton. I can't help thinking how good the class of 1878/79 was! (Lemmerich 1998: 45. Laue also let Meitner know that names in his letters would henceforth be in code.)" In Germany, no one dared to congratulate Einstein publicly on his 60th birthday on 14 March 1939, or to mention his scientific work, let alone pay tribute to it. Max von Laue had already sent his congratulations from Berlin on February 10th (UAF, Laue papers, 9.1). Wilhelm Lenz, full professor at Hamburg, tried to salvage the theory of relativity for academic teaching by claiming that not Einstein but Henri Poincaré had been the first to propound it. He wanted to publish an article about it, since the theory was, after all, "Aryan." Laue wanted to prevent this and asked Einstein in a letter from Berlin on February 27th to take action himself if his own efforts failed, adding: "'Just that my name must not be mentioned,' or else it will be said of me, as in Mozart's Abduction from the Seraglio: 'First beheaded and then hanged, Impaled on glowing prongs […].' In any case, as you see, you are still keeping us with bated breath. Warm congratulations again on your 60th birthday (UAF, Laue papers, 9.1)."

Open anti-Semitism had been prevailing in Germany for six years. In his Reichstag speech on 30 January 1939, Hitler proclaimed: "Today, yet again, I want to be a prophet: If the international financial Jewry inside and outside of Europe should succeed once more in plunging the nations into a world war, then the result will not

© Basilisken-Presse, Natur+Text GmbH 2022
J. Lemmerich, *Max von Laue*, Springer Biographies,
https://doi.org/10.1007/978-3-030-94699-9_10

be the Bolshevization of Earth and thus the victory of Jewry, but the annihilation of the Jewish race in Europe."[1] Life in the Third Reich went on as usual, if one disregarded the unspoken breaches of law. One "had to get used to" dictatorial measures. In 1939, the National Socialist mathematician Ludwig Bieberbach succeeded Max Planck in his post, who had been the "permanent secretary" of the physico-mathematical class at the Prussian Academy for decades. Adolf Hitler's 50th birthday on April 20th was celebrated with great pomp and ceremony. Reich Minister for Science, Education, and Culture Bernhard Rust and his colleague, the head of the Office of Economics, presented a 274-page volume entitled 'German Science. Work and Mission' rendering account of their work (Rust 1939). The loss of momentum and knowledge due to the expulsion of Jewish scholars was not mentioned, of course.

At the University of Berlin, Laue's assistant, Max Kohler, held the Physics lecture course during the summer semester. He had defended his doctorate under Laue in 1932 on a topic in general relativity (Gönner and Klein 1982). The crystallographer Georg Menzer lectured on his field of expertise. Laue continued to supervise the Physics Colloquium and the Physics Seminar. In both, he endeavored to maintain the quality of the topics treated and to attract good speakers. This year Max von Laue hardly found time for publications of his own. He only wrote one paper and two reviews, because his book 'X-ray Interference' (1941a) took up all his time, as can be inferred from brief remarks in his letters to Lise Meitner. Max von Laue's first book review in 1939 was only brief and concerned the German translation of Enrico Fermi's 'Molecules and Crystals' (1938): "This book gives an excellent survey of the present knowledge on the bonding of atoms to molecules, in general, and to the giant molecules of crystals, in particular. [...] The auth[or] succeeds in bringing out in concise words the essentials of the field of phenomena just considered, without attaching importance to completeness in their derivation. [...] One has the feeling throughout that an important man is speaking here (Laue 1939a: 32)."

In the summer, Laue's review of the edition *Background to Modern Science. Ten Lectures at Cambridge arranged by the History of Science Committee 1936* (Needham and Pagel 1938) appeared. Laue's contribution shows very clearly his interest in how and why science develops. The editors of these *Lectures* opened their volume regretting the situation in the history of science in general; their volume was an attempt to repair this grievance by presenting lectures on science dated from 1895 to 1935. The reviewer admitted not being qualified to review the lectures on physiology, pathology, parasitology, tropical medicine, evolutionary theory, and genetics. However, he could very well appreciate the contributions to physics by Rutherford, Bragg, Jr., Aston, and Eddington. He found the two lectures on philosophy even more interesting. The first lecture dealt with the question of why the natural sciences, unlike mathematics, did not begin in ancient Greece. The second lecture dealt with the period from Aristotle to Galileo. "In fact, this evolution [toward today's natural sciences] waited for Copernicus and Galileo. With the latter, however, who brought

---

[1] Hitler, Adolf: Erklärung vor dem Großdeutschen Reichstag am 30. Januar 1939 in Berlin. Proceedings of the Reichstag. 4th Electoral Period, Vol. 460. Stenographic Reports 1939–1942. 1st Session, Monday, 30 January 1939: 1–21.

about the great turn, there then also arose the hitherto unknown question of how the human mind acquires the knowledge of that objective external world with its mathematically apprehensible laws, that question which has dominated epistemology as a whole ever since (Laue 1939b: 435)."

Another 60th birthday could be celebrated: This time it was Otto Hahn's, on March 8th. Max von Laue was involved in the preparations. He wrote a congratulatory piece for *Naturwissenschaften*. Certainly Laue felt the injustice that he could only insinuate Hahn's collaboration with Lise Meitner in it: "The very fact that Hahn kept *such* associates *and* maintained *such close* ties for *so long* is a merit in itself (Laue 1939c, orig. emphasis)."[2] In his letter to Lise Meitner of March 14th from Sulden he stated: "Almost all the speakers also acknowledged your 30 years' collaboration with Hahn (Lemmerich 1998: 49)."

For the Laue couple, son Theodore's sojourn in the U.S. was a great reassurance, and it was a delight to read how their son was mastering his studies. On March 3rd, he informed them that he intended to stay in America for good. His father replied with a long letter on April 25th: "It's hard for me, but I can only say: You're right; you are doing the best thing not to return home right now, at least. It supposedly has happened that young people who had been abroad for two years were first sent to a 'training camp' here. It would then hardly be conceivable to leave again. So, if you think you have the energy to take up the not easy struggle for survival over there, then do try to stay. But make sure not to block your return forever. Something could change here one day. And then nowhere would you feel more at home than in your fatherland (Lemmerich 2011: 105 f.)." Laue advised Theo to do everything he could to earn the regular degree (p. 106). In mid-March, German troops marched into what remained of Czechoslovakia. Shortly thereafter, Hitler demanded that Poland return Danzig. In June 1939 Theo passed his exams *magna cum laude*; his father congratulated him warmly from Berlin in his letter from June 28th and remarked: "That you did not receive the 'highest honors' shouldn't bother you or me. I didn't receive summa cum laude on the doctoral examination either, and I think that's entirely fine (AMPG, III rep. 50 no. 7/1. Reprinted ibid., p. 115)." The effect of Germany's increasingly critical political situation upon Theo's residence in the U.S. were discussed by letter, although any mention of "desertion" or its consequences were avoided. Laue traveled to see Niels Bohr in Copenhagen again in early June 1939, where he could write to his son without fear of censorship. It was a great relief when the permit for a third year of academic study in the U.S.A. was granted.

At the end of June, Max Planck's 60th doctoral anniversary was commemorated by an "address" on behalf of the Prussian Academy of Sciences. Max von Laue played a decisive role in formulating it.[3] It quotes some passages from Planck's speeches that could also be read in light of the current situation; for example, from

---

[2] In a letter dated 26 Feb. 1939 to his son Theodore he remarked that he was prohibited from mentioning Lise Meitner in his tribute to Hahn.

[3] Vahlen 1939. The attribution on this address is: "(gez.) Vahlen." Theodor Vahlen was a convinced National Socialist. The ministry wanted to appoint him president of the academy. However, according to the statutes he had to be elected. He lost the election and could only administer the post provisionally.

his first address in January 1913: "This is a serious time, no one can say whether it is not the harbinger of a far more serious one (Planck 1913: 65, quoted in Vahlen 1939)." And the speech of 3 July 1919, warns: "For, the more direly the need of the hour looms, the more inevitable the apparently impending doom, the heavier does the obligation weigh upon each individual member of the present generation to render account some day to posterity, about whether he has really done everything within his powers to avert the onset of total ruin (ibid.)."

The Laue family drove away on holiday in their new Steyr in August 1939. It was a long haul all the way to the Canton of Valais, as he wrote to Lise Meitner on August 19th, "in the fully occupied and packed car. Because in addition to my wife, Hilde, and me, our maid also came along, who traveled back home in this way to spend her vacation there (Lemmerich 1998: 61)."

Worried that they might not be able to buy petrol en route, which was quite scarce in Germany, they had an extra 30-L canister in the back as well. Father and daughter took turns driving on this long journey. Arrived in Switzerland, they met Otto Stern, who had emigrated, and Mrs. Schrödinger. From Champéry, Max von Laue wrote his son on August 23rd that the papers were full of war cries. "Hitler is inwardly much closer to the Communists than to any bourgeois constitutional state. His only real guideline is: power. Everything else is just hypocrisy (Lemmerich 2011: 120)." The next day he wrote to Theo from Neuchâtel: "We saw the Alps again just before Neuchâtel. *Farewells.* To you, too, we must say farewell for a long time, perhaps forever. It's unlikely that the likes of us will survive the 'total war.' (p. 121, orig. emphasis)" Georg Menzer's family had been living in the Laues' home in Berlin during their absence, because even during the Third Reich break-ins into empty houses happened. Returned home again in the final days of August, the Laues decided to let the Menzers stay on.

Political events accelerated. On August 23rd Germany concluded a pact with the USSR, the essence of which was not to intervene in the event of war. It already provided for a partitioning of Poland. This gave Hitler the chance to invade Poland without officially declaring war, on 1 September 1939, on ostensible grounds of exaggerated "provocations" by Poland. Germany deployed dive bombers, so-called *Stukas*, for the first time. The mutual assistance pacts led England and France to declare war on Germany, but they did not intervene. On September 17th, Soviet troops marched into eastern Poland and occupied it. By October 6th the hostilities had ended. Poland remained occupied.

Max von Laue was commuting a lot by bicycle during this period (Fig. 10.1). When an inner tube got damaged, there were none to be found in the shops—an indicator of how efficient the German economy in fact was. Although Laue applied to the district office of Zehlendorf via the general administration of the KWG, he could not get a new tube.

The semester started in September and Laue took on a four-hour lecture course on Theoretical Optics, because there was no one else at the university able to hold it. As a lecturer he was no better than before from the linguistic point of view. He was popular among his students and generally as well nonetheless, as the publications honoring his 60th birthday on October 9th showed. The *Annalen der Physik* presented

a commemorative publication with a photo of Max von Laue, taken by Lise Meitner's
sister-in-law Lotte Meitner-Graf in London, who had emigrated in 1933 (Festschrift
1939). Nineteen authors contributed their research results to this congratulatory issue,
among them Walther Meissner with his sister Gertrud, Max Kohler, Georg Menzer,
Walther Kossel, and Otto Hahn. Arnold Sommerfeld supplied a contribution "On the
dimensions of electromagnetic quantities" with the concluding statement: "Although
these unambitious lines do not offer any deepening of our electrodynamic knowledge,
their systematic and didactic character may still attract the interest of our jubilarian
(Sommerfeld 1939: 339)." The appreciation in *Naturwissenschaften* came from Max
Planck, also with a portrait photo of Laue. Acknowledgments of his intense activities
and his many publications were followed by a lengthy passage about his interest in
historical contexts, such as for Galileo and Newton. Planck then discussed Laue's
attitude on the live issue of causality. He praised his commitment to the seminar
and the colloquium as well as his solicitude for the younger generation in his field

(Planck 1939). Theodore wrote from Princeton to his father on October 7th: "I warmly congratulate you, as much as anyone else, on all that you have accomplished in science and its related organizations. It is true that I can only measure it indirectly by the esteem in which you are held among your colleagues everywhere; but it's not difficult for me to see. Your nonscientific life, as I have witnessed it (I believe you would distribute the weights differently—or am I mistaken?—) earns my direct respect for you, even though our values about life are different. It would be unfair to apply the values of another environment as a gauge. Certainly, I often do so, and I do have that right."[4] Two days after his birthday, Laue informed the university trustee that he had reported to the responsible police station, as a registered member of Landwehr II, in compliance with a decree. His enlistment into the military was thus prevented for the duration of the war (AMPG, III rep. 50). On October 15th, Laue thanked Lise Meitner for the flowers and two letters, which had pleased him very much: "It was a somewhat bittersweet joy though, and melancholy reigned throughout my birthday, because so many people I would have liked to see and who would have liked to come could not be there (Lemmerich 1998: 71)."

He had given his lecture that day, afterwards visited Arnold Berliner, and then gone home. Celebrations were held first at the KWI and then at home, where the Plancks, Hahns, Hertzs, Justis, Lamlas, and many others had also come (ibid.). Laue sent out a printed thank-you note after all those festivities: "For the countless congratulations on my birthday and the appreciative words written, spoken, and even printed on the occasion, I thank you most sincerely. It is not for me to judge how much my activities deserve such praise (HUA, Laue staff file)." He had always abided by his doctoral oath, he added, which he thereupon quoted in the original Latin.

Soon after his birthday, Laue's situation at the institute changed. On October 17th, a meeting was held between Senior Government Councillor Walter Basche and Kurt Diebner of the Army Ordnance Office (*Heereswaffenamt*, HWA) to negotiate a change in orientation at the KWI of Physics for the purposes of the HWA by commission of Ministerial Director Erich Schumann. "The director of the institute, Prof. Debye, would be granted leave of absence by the KWG at the request of the Ministry of Education and the Army High Command (Army Ordnance Office) (AMPG, I rep. 1A, no. 1652/1–3 (329). Debye was a Dutch citizen.)."[5] On December 12th, the Advisory Board of the KWI of Physics met and resolved: "During the period of war the main portion of the Institute will be taken over by the Army Ordnance Office (ibid.)." On the penultimate day of the year, Peter Debye wrote a long letter to Arnold Sommerfeld from Dahlem describing all the details of the negotiations, including the fact that the low-temperature laboratory would still be available to him for his purposes after his departure (Sommerfeld 2004: 469).

---

[4] Lemmerich 2011: 126. Laue heartily thanked his son for these congratulations in a letter from Berlin on 17 Nov. 1939. In Lemmerich 2011: 127.

[5] When the war broke out, Debye was asked to assume German citizenship, which he declined. As a result he was "furloughed" from his post in the KWG. For details about the conduct of the ministry and Rudolf Mentzel, see Sommerfeld 2004.

In December Laue had sent to the editors of the *Annalen* the manuscript 'Fitting the Kossel-Möllenstedt electron interference into the theory of space lattices.' The paper appeared in 1940 and was more a discussion of the figures and their interpretation by the two authors Kossel and Möllenstedt. Laue's introduction was an elaboration on the geometrical theory. He agreed with the authors' reasoning of assigning the visible minor maxima to the nearest major maxima. "This would probably put all the essential features of the new fringe systems, insofar as initially necessary, into the space lattice theory. Convergence of the incident radiation is accordingly not necessary for its formation, but admittedly a particularly advantageous approach for it (Laue 1940a: 172)."

At the turn of the year 1939/40 Max von Laue was at the health spa Hohenpeißenberg in Upper Bavaria on vacation. Whether he worked on his book *Röntgenstrahl-Interferenzen* there is not clear from the letters. In general, only occasional brief mentions of this major work can be found in his preserved papers, such as in a letter from 14 January 1939 to Lise Meitner: "I am up to my ears working on my book (Lemmerich 1998: 45)." He probably submitted the manuscript or parts of it to the publisher, *Akademische Verlagsanstalt Becker & Erler Kom.-Ges.* in Leipzig, in early 1940. The proofreading dragged on until late autumn 1940.

In the new year, on 3 January 1940, Max von Laue wrote to Theo from the Hohenpeißenberg resort, thanking him for his letter of December 15th. He was very pleased "that you don't want to simply 'resign yourself to the facts.' That's the mark of ethical people: that they counter 'That's the way it is' with 'But it should be different.' Unfortunately, there aren't many such people (Lemmerich 2011: 133)." He then described how he was feeling by reminding his son about a verse by Friedrich Rückert about a pharaoh which he had quoted earlier and which Theo had found outrageous. "It might become very modern now." He recommended that he look at the volume in the university library. It is unclear why the censors let this letter pass without complaint. If they had been familiar with this myth, Max von Laue could have gotten into serious trouble.

To the evil spirit Pharaoh spake:
    Help me and do thy part,
    That in the eyes of my people
    I seem godlike.
    The demon said: Not yet
    Has the time come for that;
    First thou must needs do much evil,
    Before it come to that.
    And much evil he still did,
    To earn that dignity;
    And at last the false demon spake:

Now the time has come.

Why now? Spake Pharaoh.

The demon said: No way

For the people to see and obey

Was left, save with the title of god.

For, so hast thou now oppressed

The young and old,

That they must needs endure more,

Save by seeing thee as god.

The myth Laue mentioned from the collection *Sieben Bücher Morgenländischer Sagen und Geschichten* by Friedrich Rückert, 1837: 36 f.

At the beginning of 1940 all privately owned automobiles had to be relinquished to the authorities. Laue handed in his Steyr on March 1st. They had driven 24,000 km with it in the last 1¼ years, as he wrote to Meitner on March 2nd (Lemmerich 1998: 82).

Laue still could not completely shake off the suspicion that Sommerfeld was judging him unfairly, which also influenced Laue's evaluation of research done by Sommerfeld's students. He complained to Walther Meissner about this in a long letter on March 1st, who in his reply from Pasing on March 20th tried to assuage his fears (DMA, NL 045/006).

The scientific publications obviously said nothing about the current political and military events; they left the impression that these were peaceful times. Max Planck undertook in this period the 'Attempt at a synthesis between wave mechanics and corpuscular mechanics': "But one may by no means simply say that for h $\rightarrow$ 0, wave mechanics passes over into corpuscular mechanics. No matter how small the quantum of action is assumed to be, a wave packet never becomes a corpuscle, at least not for any duration [...] (Planck 1940: 261)." In 1940 Laue could only boast two minor publications under the rubric "short original notices" in *Naturwissenschaften* and one longer one in *Zeitschrift für Kristallographie*, which he was now coediting with Menzer. For a long time the Swiss mineralogist Paul Niggli had assumed this task, but after taking over the presidency of the University of Zurich and due to the difficulties caused by the war, he had given it up. The journal's staff included eminent physicists and mineralogists from various countries, including states with which Germany was now at war. Laue definitely wanted to maintain the international character of the journal. Some of his colleagues did not agree and claimed the journal for themselves. In his letters on November 17th and December 4th Laue applied to Sommerfeld for his support, spelling out his reasons in detail, including that physicists also had a right to be involved with this journal (Sommerfeld 2004: 529, 533).

Laue's first paper actually written in 1940(b), entitled "The electrostatic mean potential in crystals," was related to the manuscript for the comprehensive paper 'On the electrostatics of space lattices' (Laue 1940/41). His paper became a very critical review of the previous methods of theoretical treatment, which assumed that the potential of a space lattice with triple periodicity has a mean value, adding: "theory today always assumes that this mean is: (1) a characteristic constant of the body and (2) can be readily calculated from the charge distribution inside the space lattice. A critical consideration now shows, however, that neither of these two assumptions is certain (Laue 1940b: 515)." Then he went into the feasibility of doing the calculation, which was not possible in the usual way, because the charge distribution in the lattice cell was different at the edge than inside the cell. The problem of potential would not even be solved if the displacements were ignored. This was followed by considerations about NaCl, which was not suitable for the calculation, and then about a kind of ideal crystal. Max von Laue had presented a talk 'On the electrostatics of space lattices' at the Prussian Academy of Sciences already in 1930. He now returned to the problem of the course of the potential in a space lattice: "But whereas hitherto it has been believed that this latter [part of the potential] must be constant in the interior, our considerations are intended to show that a slowly varying position function is generally superimposed on that threefold periodic portion; the nature and position of the boundaries are decisive in the determination of this function (Laue 1940/41: 54 f.; he completed this paper in July 1940.)." Although Laue cited publications by Hans Bethe, he thought that they were not pertinent to his considerations. The arrangement of the lattice building blocks and the geometry of the boundary layers were described in detail with "slight generalization." The third section dealt with the potential of the half-space lattice with the superpositions of potentials, and the fourth introduced the macroscopic consideration. The different classes of crystals were briefly mentioned in the fifth section. In section 6 Laue discussed the influence of boundary perturbations, which play a role especially in electrically conducting crystals. The boundary field of a space lattice was treated mathematically in section 7.

Only a few lines in Laue's self-portrait, 'My physical career' (1961), mention his efforts to help "sufferers" (*Betroffenen*) under the Hitler dictatorship. An assistant to Professor Gustav Hertz in the Physics Institute of the Charlottenburg polytechnic was one of them. Young Dr. Fritz Houtermans was a convinced communist and had to flee to England in 1933. In 1935 he moved to the Ukrainian Physics Institute in Kharkov but was arrested there very soon afterwards. In the course of the Germano-Russian negotiations in August 1939, he was deported to Germany in May 1940 where he was immediately arrested again and taken to the Gestapo prison at Alexanderplatz. Laue visited him there and with the support of other colleagues managed to get him released.[6]

The experimental work on superconductivity made no progress, partly because the KWI of Physics was having difficulties with the production of liquid helium, partly for lack of personnel. The situation was similar at the bureau of standards (*Physikalisch-Technische Reichsanstalt*, PTR), where there were complications, as Laue wrote to Meissner in Munich on May 4th (DMA, NL 045/036).

But still there were plenty of other things to do. On August 23rd while on vacation in Dierhagen by the Baltic Sea, Laue informed Lise Meitner about the progress he

---

[6] Frenkel 2011: 72. Houtermans is also mentioned in Lemmerich 2011: 153, 158, 162, 165, 258, 345.

was making with his book on X-ray interference (Laue 1941a): "Here, on account of the miserable weather, I am chiefly engaged in proofreading. The first corrections are thus pretty much finished here. Then they go out to Dr. Lamla, then to Professor Menzer, and in between I still take a look at what these helpers have found wanting (Lemmerich 1998: 95)." Soon afterwards, Laue's communication 'Interference birefringence of X-rays in crystal prisms' appeared. It was a brief description of the experiments by Werner Ehrenberg, Paul Peter Ewald, Hermann Mark, and others, on how two wave fields form in a plane-parallel plate from an incident X-ray beam and cause interference birefringence and how the beams are deflected when the plate is tilted. "It would be all the more interesting, since we have no other knowledge of the oscillation process in the interior. Only the Kossel effect, as the inverse phenomenon to irradiation according to the reciprocity theorem, gives an indirect notion of it, the details of which we are not yet able to decipher, however (Laue 1940c: 646)."

In a letter on 27 October 1940, Laue reported to Lise Meitner a little about wartime life: "This afternoon we had guests over for coffee. We wanted to celebrate the completion of the book proofs with Dr. Lamla and Prof. Menzer, both of whom had helped out (Lemmerich 1998: 102)." He mentioned the air raids and told about improvements to their bomb shelter. "My wife and Hilde also use it whenever there's an alarm—another one's just starting now—whereas I sleep in my study and stay put even when there's shooting going on (ibid.)." Domestic life during the war had become trying. Food shortages made shopping time-consuming; ration cards issued for the various occupational classes defined precise quantities that the holder was permitted to purchase. Since their maid Frida had not returned, Mrs. von Laue had to do the housework herself. Laue and his daughter did the shopping, which involved waiting in line in the stores.

The controversy about the theory of relativity was by no means settled. But at least a meeting could be arranged between advocates of the two views in Munich to discuss it. Max von Laue largely kept out of those debates. He wrote to Meissner on 30 October 1940 what he thought of them: "The time for public disputes about science is over—for good reasons. And I cannot imagine how anything at all is to be gained by *such* a dispute. If, as is to be expected, one party gets into a scientific quandary, it will shift the issue to politics, probably with a racist slant, and thereby gag its opponents. And when it then loudly proclaims that it has won a brilliant victory, who will stand in the way? (DMA, NL 045/36, orig. emphasis)" Nothing in Laue's scholarly writings suggests that the war had broadened its fronts, that the Nazi regime was extending its brutal rule and was intensifying its persecution of Jews. Laue was constantly worried about whether Fritz Reiche and his wife would still manage to leave the country. Laue visited Arnold Berliner often, who very rarely ventured out of his apartment. None of these worries could even be hinted at in the letters. There was no hope that the system would soon be collapsing. In April, German troops occupied Denmark. When the persecution of Jews began there, Niels Bohr was able to flee to Stockholm with his family and a few coworkers. Norway was conquered . In May, the German attack on the "Western Front" began with an invasion of the Netherlands and Belgium and fighting in France. On June 14th, Paris was occupied and eight days later the French government had to sign the armistice in Compiègne.

Peter Debye's decision to go to the U.S. turned Laue into the sole director of the KWI of Physics. He had to take on Debye's duties, which required a lot of diplomatic skill in this politically difficult situation and robbed him of his time for independent research. In his letter to Lise Meitner on 17 January 1941 discussing why he was writing less often, he commented sarcastically: "On the contrary, I am not at all seldom amused. How nice it was, for ex[ample], when on new year's eve, soon after midnight, the Führer's salute march by General Erich Schumann was broadcast on radio! (Lemmerich 1998: 110 f.)" He then drew Ms. Meitner's attention to an emigré in Stockholm, Prof. Dr. Harald Perlitz, asking whether she knew him. He mentioned Justi's experiments at the PTR with superconductors connected in parallel. At the end, he wrote: "By the way, did you see the diffraction of matter waves at an edge in the Naturwissenschaften issue of 1 Nov. 40 (H. Boersch)? That really is fine; because a skeptic could still consider all previous diffraction experiments with matter waves as a devilry of those ever mischievous atoms, you know. Now this is no longer possible (Boersch 1940; Fresnel's diffraction had been observed.)."

In the first letter in the new year to Walther Meissner on 1 February 1941, Max von Laue struck a light note, yet his earnest concern about the future was clear: "I would first like to write you about skiing. I think it would be better if I did not stray so far away from Berlin as would be necessary. Because, if peace doesn't break out by March—and it doesn't look like it will at present—we must anticipate new and more violent air raids on Berlin. But if the English then do operate carelessly with their incendiary bombs, I should like to be in a position to protect the house personally against fire (DMA, NL 045/036)." Laue mentioned the experiments by Eduard Justi and Gustav Zickner on the distribution of current in superconductivity. In the next passage, Laue-the-theoretician had very precise conceptions for an experiment: "I am also planning another experiment on the Meissner-Ochsenfeld effect (I'm afraid that name is going to stick, even though Ochsenfeld only collaborated by chance). Imagine a long thin metal cylinder insulated inside an equally long hollow cylinder, connected to it only at the ends by a lid of the same metal. We apply current to it at high temperature [i.e., room temperature], which distributes itself in the inner and outer conductor according to the resistances (ibid.)."[7] This was followed by a description of the cooling processes involved. After Walther Meissner's appointment to Munich, a certain rivalry arose between Eduard Justi in Berlin and Meissner, which intensified when Meissner was initially unable to conduct experiments at Munich. Laue could actually have arranged for others to carry out superconductivity experiments at the KWI of Physics, the "Max Planck Institute." But the personal dispute between those responsible for the production of liquid helium still caused great delays (Laue to Meitner, 3 Apr. 1941, Lemmerich 1998: 119).

---

[7] It was occasionally difficult for Walther Meissner to follow Laue's ideas. Laue's attempts to persuade him also occasionally led to misunderstandings. See Meissner to Laue, 27 Jan. 1942. DMA, NL 045/036.

It was important to Laue that his research results and findings not be lost if catastrophe struck in Germany. On 25 April 1941 he sent a copy of his book 'X-ray Interference' to Niels Bohr in Copenhagen. This topic was not exactly within Bohr's field of research, but Laue wrote: "You may gather from this book how much quantum theory now plays into the theory of X-ray interference, even though it had initially developed independently (NBA)."

Although Arnold Berliner was no longer allowed to direct *Naturwissenschaften* as chief editor, the journal maintained its quality under Fritz Süfferts backed by its proven staff of contributors for the separate disciplines, including Max von Laue for physics and Otto Hahn for chemistry. At the beginning of 1941, Laue reviewed a book by Erich Wintergerst (1940): 'The Technical Physics of Motor Vehicles.' First he detailed the manifold demands on a car, which require solutions to a plethora of problems in technical physics. This was followed by details about engine performance, vibration phenomena, the gearbox, tires and road surface, the generation of the electricity needed, the ignition, and "knocking": "This book provides the answer to how this happens, which is therefore warmly recommended to anyone who not only wants drive a car but also to understand it (Laue 1941b)." Max von Laue received mail from Lise Meitner in Stockholm at the beginning of April 1941 dated March 27th: "In Naturwissenschaften, I recently read a review of a book about automobiles. I hadn't looked at the reviewer's name yet when at a particular sentence it suddenly occurred to me: that sounds like Mr. v. Laue. And it was fun to then find the name (Lemmerich 1998: 118)."

Max von Laue, ever ready to cultivate the history of physics, reminded his colleagues "somewhat belatedly" of the discovery of the law of radiation. He did so with the facsimile of a letter dated 12 October 1900 from Ernst Pringsheim and Otto Lummer to Max Planck about the latter's formula (Laue 1941c).

At the end of March, Carl Friedrich von Weizsäcker published an essay in *Naturwissenschaften* entitled 'Contemporary physics and the physical world view,' which he commenced with the declaration that the closed view of the world in physics as had existed a few decades before no longer held. He claimed: "The present condition, then, is richer in positive findings than the former one. But it appears to be poorer in inner unity (Weizsäcker 1941: 185)." Von Weizsäcker later published a book on the subject (1943), which Max von Laue critically reviewed.

Laue's book 'X-ray Interference' appeared in the series "Physics and chemistry and their applications as monographs" in April 1941. It was not a textbook, but a critical summary of the results of very many and very different experiments and theories. The preface mades this clear:

> There are a number of books on X-ray interference; crystallographers, chemists, and experimental physicists have written some; there are also handbook articles about it from the pens of theoretical physicists. Nonetheless, a gulf threatens to gape in the minds of specialists between these interference phenomena and present-day theory; almost as if the two fields had nothing to do with each other. This book is intended to bridge this gap and connect X-ray interference to theory where it belongs, namely, to atomic physics. The necessary thinking has already been done, but it is scattered in numerous papers. The task is to weld together

the multitude of valuable individual publications into one theory; the less the reader notices the trouble this has caused, the better the book will serve its purpose (Laue 1941a: V).[8]

Max von Laue then pointed out that it could not be such a well-rounded theory as those for mechanics or electrodynamics. This was due to atomic theory and the currently insurmountable mathematical obstacles: "The bridge to be built has two piers, one on the side of crystallography, the other on the side of wave mechanics. The book must present quite much from both fields but attaches no importance either to completeness or to the presentation of proofs. The actual bridge-structure, however, should be as gapless as possible (ibid.)." And criticizing some current publications very clearly, he then continued: "For nothing would have a more devastating effect on our science than if the physicist became accustomed to believing he knew where proofs were necessary and possible. This book is simply intended to convince, not as if it contained but incontrovertible truths, certainly rather that physical knowledge today, carried out in a reasonably consistent manner, leads to the results presented here. And some of it may ultimately have the power to outlast the changes in our times (p. VI)." In the acknowledgments for the included illustrations he mentioned, among others, his American colleagues Samuel King Allison and Arthur Holly Compton. He thanked his colleague Ludwig Bieberbach for perfecting an integration method, Peter Debye for adding the interference theory of mixed crystals. Those to whom he expressed his gratitude at the KWI were his coworkers Ilse Fränz and Georg Menzer, the latter also for reading the proofs, as well as Ernst Lamla.

The content is divided into five chapters, each of which filling about 50 pages. The first chapter, an overview, was subdivided into § 1 "X-rays as electromagnetic waves" and § 2 "Quantum processes in X-rays," with critical remarks strewn throughout the chapter about the history of science. Einstein was quoted and many colleagues abroad. The second chapter, "Scattering by single atoms and molecules and in liquids," covered scattering according to Maxwell's theory, wave mechanics, the atomic form factor, interference by polyatomic molecules, and the effect of temperature. The third chapter, "X-ray interference by crystals, the geometric theory," covered, again with a thorough historical retrospective, atomic theory, space lattices, factors of structure, and Lorentz factors. His own papers and those by Friedrich, Knipping, W. H. Bragg and W. L. Bragg, and by P. P. Ewald were of course mentioned with extensive citations of the literature. In his description of the principal methods of observation, those by the Braggs took first place, and Manne Siegbahn's with reference to precision methods. In the fourth chapter, "Lattice defects and their influence on interference," first gave an overview. This was followed by interference in mixed crystals and the influence of temperature, the Debye factor as a function of temperature, and the mosaic crystal. Many papers were mentioned, including by Ivar Waller. On lattice vibrations, Laue cited publications by Max Born and Theodor von Kármán. In the fifth chapter "The dynamical theory of space lattice interference" the necessity for the dynamical theory was discussed, incorporating papers by Charles Darwin,

---

[8] Laue's correspondence with the many other scientists while at work on his book *Röntgenstrahl-Interferenzen* and his numerous requests for original illustrations are unfortunately not available in the various partial estates.

P. P. Ewald, and Ivar Waller. This chapter treated the "bridge piers" crystallography and wave mechanics; brightening lines and detour excitation, basic equations of the dynamic theory, formation of the interference with only one wave field in the crystal and with two, generalization, and the Kossel effect were other sections. Laue emphasized the importance of the value of the elementary charge. Finally, the appendix was a detailed historical table of crystallography and physics on the subject.

Laue had copies of the book sent to his friends at home and abroad. Fifty copies were sent to USA via Siberia. The publisher also sent 70 copies to Japan. Only a few weeks after the publication of his book, Laue already mentioned working on another book, on electron interference, in a letter to Lise Meitner dated July 27th (Lemmerich 1998: 130). Even so, Laue's interest in superconductivity was unabated. It is evident in his correspondence with Walther Meissner, which also shows how complex written communications could be. On May 15th, Laue replied to a letter from Meissner dated May 4th, which, among other things, touched on his attitude toward physics: "I get the impression you were a little annoyed at me when you wrote that letter. And I put it down to the fact that my previous letters were not exactly stylistic masterpieces. But I beg you please to distinguish between such letters between good friends and scientific treatises. In the latter, one should express opinions that are as clearly stated as possible. In the former, however, I like to speak to you as if I were sitting with you at dinner in Sulden (DMA, NL 045/037)." He then broached Meissner's relationship with Justi and asked if he was upset: "The whole world agrees that it would be better if *you* were still sitting here in the low-temperature laboratory. But since this is not the case, it should be all the more acknowledged that Justi has taken the low-temperature experiments energetically in hand and with growing understanding [...] (ibid., orig. emphasis)." Laue suggested to Meissner that the Meissner-Ochsenfeld effect be reexamined in a modified arrangement. In explanation he pointed out that "in the 1920s, when I had long since been firmly convinced of the correctness of the special theory of relativity, I nevertheless always advocated repetitions of the Michelson and Trouton-Noble experiments (in the Emergency Committee), because in science it is not merely important to be convinced oneself or among a small circle. Rather, that such a conviction become *common terrain* among physicists (ibid., orig. emphasis)." Max von Laue mentioned doubts about the Meissner effect, saying that the conviction that it was correct was not yet commonly held. But Meissner did not agree with Laue's remarks in his reply on May 23rd (DMA, NL 045/037).

On July 31st, the senate of the KWG met at Harnack House. An excerpt from the minutes about the "Kaiser Wilhelm Institute of Physics" reads: "The Institute has been largely taken over for the purposes of the Army Ordnance Office. Dr. *Diebner* of the Army Ordnance Office is in charge, whereas the main scientific direction is in the hands of Prof. *Hahn* (Kaiser Wilhelm Institute of Chemistry) and Prof. *Heisenberg*. The Army Ordnance Office bears a considerable part of the budget. The refrigeration laboratory operates under the direction of Dr. *Bewilogua*, assistant to Prof. *Debye*, on aeronautical research (AMPG, I rep. 1A, no. 1652/1–3, orig. emphasis)." Further details referred to Debye, who had a six year contract at the university in Ithaca, New York (ibid.).

In his correspondence with Lise Meitner, Laue occasionally mentioned his varied historical and contemporary reading. In the letter of August 18th he wrote: "Then I have before me the August issue of the 'Deutsche Rundschau' with a masterfully written article by Rudolf Pechel: 'Dr. Leete.' It will interest you very much, too, and you should try to get hold of that number. I don't want to tell you what's in the article in excerpt, because any rendering would ruin this work of art (Lemmerich 1998: 136; Pechel 1941)."[9]

The editors of the *Zeitschrift für Physik* received Laue's manuscript 'On the theoretical significance of the Justi-Zickner experiments on current branchings in superconductors' on September 12th. With reference to his publication of 1932 in *Physikalische Zeitschrift*, Max von Laue established that the current distribution given there had been confirmed by these experiments "with quite good accuracy." Without naming authors, he discussed the idea that a "higher state of order" existed in superconductivity compared to normal conduction. However, the results of Justi's and Zickner's experiments did not indicate such a state. Then Laue discussed the idea that by observing the permanent currents it ought to be possible to establish a residual resistance in the superconducting state. But this had not worked. He analyzed theoretically whether this resistance could be demonstrated in the case of two super-conductors connected in parallel. Finally, he presented the experimental arrangement proposed in his letter to Meissner of February 1st of that year. His considerations led to the assumption that the current is displaced onto the superconductor surfaces.

In September it became compulsory for the Jewish population to wear star-shaped yellow badges. It was terribly hard for those concerned to have to present themselves like that in public. Marie Rubens, the widow of the deceased professor, Heinrich Rubens, could not bare it and took her own life on September 20th.

Occasionally, Laue managed to write a letter to Meissner not containing any physics, as on 29 September 1941: "It's beautiful autumn weather here. The sky is clear even at night, and Hilde and I used the last air-raid alarm to look at Mars, Saturn, and Jupiter through the telescope. I have now taken many weeks' vacation from physics, and have occupied myself instead with the postmortem publication by C. Stumpf, which I must review for 'Naturwissenschaften.' It is called 'Epistemology' and comprises some 800 pages in 2 volumes (DMA, NL 045/037; Stumpf 1939/40)." This book review appeared in early 1942 and exemplifies Laue's own thoughts about intellectual insight and philosophy (Laue 1942a). Max von Laue quoted from the preface written by the author's son, Felix Stumpf, that his father's intention had been

---

[9] The editor of the magazine *Deutsche Rundschau*, Rudolf Pechel, used the novel Bellamy 1914 by Edward Bellamy (1850–1898) for his contribution. The novel tells the story of a man who awakens in the year 2000 after a deep sleep of 113 years. Pechel asks what the man would experience if he awoke in 1941. He describes a misanthropic dictatorship. "For, every nation remains responsible for the government it tolerates, and must pay its toll for all the misdeeds of its despots, who should have made stones do the talking where fear paralyzes people's tongues." To avoid being arrested for this outspoken criticism of the Third Reich, he had the man awaken in Soviet Russia in 1941. Pechel was arrested nonetheless and sent to a concentration camp.

"to give a general account of philosophy in a popular form."[10] But in the course of writing it, Carl Stumpf had become increasingly engrossed in the deeper problems. "For, now he [Stumpf] lays out a scientific epistemology which favorably matches all the other works of this sort, with the advantage of taking into extensive and usually cogent account *recent* mathematics and natural sciences, just as making recurrent reference to all the more important epistemological currents of the nineteenth and twentieth centuries (Laue 1942a: 123, orig. emphasis)." First Laue explained the structure of the work, divided into "basic concepts" including "thing or substance," "causation," etc., on 112 pages; and "The path to knowledge" comprising 450 pages. Laue found remarkable the definition of the term "truth" as "the conformance of quality to the matter of judgment." Laue briefly mentioned that the influence of the psychologist Stumpf was occasionally perceptible in the work. "As for every reasonable theory, outright rejection of that skepticism which casts doubt on the general possibility of knowledge serves as the starting point of this theory of knowledge. In view of the culture-changing successes of the natural sciences, which can leave no one unimpressed, because no human existence is left untouched, the question is not *whether*, but *how* science is possible (ibid., orig. emphasis)." Stumpf agreed with Kant about experience-independent knowledge. But Laue did not agree with his explanation of geometric probabilities. He did, however, grant his full approval to the remarks about the theory of relativity. He concluded, the book: "seeks to place philosophy at the center of all scientific endeavor. That it regain this position, which is clearly not generally granted it at the present time, does appear to the rev[iewer] to be a cultural question of the highest order (p. 124)."

The nightly howling of the air-raid sirens was not the only thing making life in wartime so trying for the population. It was the mostly unacknowledged uncertainty about what might happen next. Who else would be persecuted? What would happen after the "final victory"? Magda von Laue was totally exhausted by the time she was finally able to leave on a health cure in Hohenpeißenberg in October 1941. She weighed just 57.5 kg. To Lise Meitner she wrote on October 4th: "I lie in the open air six hours a day, and spend another three hours tramping about through the woods and meadows. And what a fine forest it is!—Out of sheer devotion to so much beauty and stillness I walk barefoot. Among the 11 other guests I am considered haughty and boring, because I always want to go out walking on my own. One always does want to just have another look (Lemmerich 1998: 142 f.)." When she had returned, recuperated, her husband spent his vacation in Saal, Western Pomerania, and could describe in his letter on the 21st to Lise Meitner the landscape and flocks of migrating wild geese (pp. 145 f.)

In April 1940, the president of the KWG, Carl Bosch, had died unexpectedly. The society managed to retain its influence over the election of its president. Albert Vögler, a long-time member of its senate, had been appointed president on June 31st (Hachtmann 2007: 840). Back in Berlin, Laue attended the inauguration of the new

---

[10] Carl Stumpf (1848–1936) was professor of philosophy and psychology at the Friedrich-Wilhelms-Universität Berlin from 1894. He was a friend of Max Planck. After his death, his son Felix published the work and asked Planck to arrange that it be reviewed.

president and his address on November 4th at Harnack House, which he wrote to Lise Meitner about. On 18 November 1941 Walther Nernst died in Zibelle (Upper Lusatia), and the funeral service six days later took place in Berlin at the crematorium with a large attendance. Max von Laue wrote an obituary paying tribute to Nernst's scientific achievements (Laue 1942b).

A fundamental turning point in World War II occurred in December 1941. On the 7th, Japanese bombers destroyed eight battleships and eleven other warships at the American military base of Pearl Harbor, without first declaring war. This led to the United States and England declaring war on Japan the following day. Germany declared war on USA on December 11th. This meant the end to the correspondence between father and son. Only occasional brief signs of life arrived through the mediation of the Red Cross. But there was one important piece of news: Theo had met a woman about the same age as he by the name of Hilli Hunt. She was acquainted with Germany and even spoke German. Lise Meitner now served as a relay station for letters between Laue and his son. She had to translate letters she wrote to Theo into English.

Shortly before Christmas, on December 21st, Laue wrote to Lise Meitner about the imminent funeral of Ellen Hertz, Gustav Hertz's wife. She had died unexpectedly at the beginning of December after a minor operation (Lemmerich 1998: 153). That couple were part of the circle of close friends of Ms. Meitner and the Laues. Their two sons were in boarding school, and the discussion that ensued between Meitner and Laue by letter about family life in general led Laue to revealing what he privately thought about it on 21 January 1942.

> But I still maintain that children should grow up in the parental home, unless there are quite compelling reasons to the contrary. Reasons such as those you gave here or, to be more precise, such as you have written me, do not fall into consideration at all. I think very little of any principles of upbringing, with one exception; it's called: love. And that, only the mother can give fully, possibly also the father. If children find it difficult to be with other children, well, what harm is that? Must one really start in childhood with that mental levelling which later sets in of its own accord and becomes such a great danger to any mental independence, and therefore to our entire culture? [...]
>
> I, too, was accused during my youth of not being sociable. Today I consider this to be one of my few advantages over others (pp. 161 f.).

In his report to Stockholm on 21 December 1941, Laue also mentioned the poor health of Arnold Berliner and suggested that Ms. Meitner write to him on his birthday on December 26th. He also pointed out the upcoming anniversary of the dynamo in engineering, which was being honored in *Naturwissenschaften*, adding: "This engine is just 75 years old, and we can hardly imagine life without it already! The internal combustion engine will disappear off the face of the Earth again as soon as the oil deposits are all exhausted; and the present overexploitation of it is hastening that moment. The dynamo generator, however, is likely to stay around as long as humanity exists on this Earth (p. 153)."

In acknowledgment of the 300th anniversary of Galileo's death, Laue wrote an article for the *Frankfurter Zeitung* sketching his life. It underscored his importance to astronomy for having turned the newly invented telescope into a scientific tool and realizing that sunspots are processes on the Sun, as well as for discovering four moons of Jupiter. It also emphasized how he helped physics make the transition from Aristotelian views to laws governing processes. Laue's treatment of the controversy with Pope Urban VIII and the Inquisition was very restrained, without quoting that famous line (Laue 1942c; Wohlwill 1909/26 was his main biographical source).

It is questionable whether rumors brought Max von Laue news of the top secret "Wannsee conference" on 20 January 1942. But he certainly experienced the consequences with its "final solution" to the Jewish question" when the Jewish grandmother of the Ruge family in his immediate neighborhood was taken away. She had some poison ready to take her own life, if worst came to worst. But the people who picked her up prevented that.[11]

Laue included in his compilation of his scientific oeuvre (1961) the paper: 'Statistics on X-ray irradiation through many layers of the same kind' (Laue 1942d). It presented the experimental results of exposing layers of woven material—hence, "lattices"—to X-rays, which he examined by a method indicated by Lord Rayleigh and applying Fourier series to determine the filament spacing (ibid.; Franke 1941: 288). Did Laue make the remark about amplitude and phase, just in order to be able to cite Einstein's paper in the *Annalen der Physik* of 1915?

Sir William Henry Bragg died on 10 March 1942, and Laue's attempt to send a letter of condolence from Berlin failed. But Theo received news of this indirectly from Walther Nernst's daughter in England, and wrote from the U.S. in this spirit to Bragg's son Lawrence on behalf of his friends in Berlin (Lemmerich 2011: 211; RSA, MDA 7, A 9/94).

An important meeting of the KWG senate on May 24th at Harnack House brought change for Laue. The minutes recorded: "*KWI of Physics. The President* announces that, on the basis of discussions between General *Leeb* (HWA) and himself, the OKW [Army High Command] will return the Institute, which had been at the disposal of the Army Ordnance Office since the beginning of the war, to the Kaiser Wilhelm Society."[12] But the institute retained its priority of pursuing research of importance to the war effort.

Max von Laue continued to visit Arnold Berliner regularly, whose health was deteriorating. He was depressed because there were plans to evict him from his apartment. On March 21st, Berliner committed suicide by taking potassium cyanide. Laue informed Lise Meitner on the following day on a post card from Saal in Western Pomerania. He had been using the code name "Kielgan" for Arnold Berliner for some time now. Upon returning to Berlin, he wrote to her again at the end of March: "Solitude heals, and it certainly was solitary enough there by and on the lagoon. 'On' means that the lagoon was still frozen solid and consequently easy to walk on (Lemmerich 1998: 177)." Laue drafted an obituary for Berliner, as he had promised Dr. Ludwig Ruge in Berlin-Charlottenburg. This lawyer was part of the round table

---

[11] Laue to Theodore, Castle Facqueval (on the way to Farm Hall), 22 Jun. 1945. In Lemmerich 2011: 238.

[12] AMPG, I rep. 1A, no. 1652/1–3, orig. emphasis. On the election of Heisenberg as the new director of the KWI of Physics, see AMPG, I rep. 1A, no. 1653.

that met regularly in Deutscher Club. Its members were Max von Laue, two high officials at the Ministry of Culture, Otto von Rottenburg and Wolfgang Windelband, the theologian Erich Seeberg, and the owner of the industrial bank Berliner Handels-gesellschaft, Otto Jeidels. Various guests had been invited to join them, including Arnold Berliner. Obviously, Laue's obituary could not appear at that time; it had to wait for 1946 to be published in *Naturwissenschaften*. It described his life's course in great detail, his difficult parental home, his studies of physics with the help of the Neisser family, Berliner's industrial activities at the electricity combine AEG under Rathenau, and then from 1913 as editor of *Naturwissenschaften* at Springer publishers. The obituary ended with Berliner's interest in the fine arts (Laue 1946a).

When writing a text like this one, Laue presumably consulted the wife of the deceased, reading it out to her or letting her take a look at the manuscript herself.[13]

In September of the foregoing year, Laue had published his ideas on the distribution of current in superconductivity (1941d). Now, in March, his 'Remarks about superconductivity' came as a sequel (Laue 1942e). It was a rigorous critique of the available findings by other physicists and demonstrated his intense interest in their results and his ambition to develop a complete theory of superconductivity. The problem areas from optics and X-ray interference which he had treated hitherto could be clearly settled by experiment. Now, however, neither the theoretical approach with Maxwell equations nor some of the measurement data from experiments had straightforward answers. Laue subdivided his extensive paper as follows:

I. The ordering energy.

II. The question of the displaceability of the lines of current.

III. Superconductivity and diamagnetism.

IV. On the thermodynamics of superconductivity.

V. Some recent experimental results.

VI. The intermediate state.

VII. Mathematical appendix.

Under Section I, Laue came to the conclusion that the notion of a current-dependent "ordering energy" was superfluous. Under II he mentioned the experiments by Heike Kamerlingh Onnes with spheres and by other authors with hollow spheres. He posed the following questions to investigate this theoretically: (1) Do the current paths shift when a magnetic field is applied? (2) Do the original paths of current reappear when the field is reapplied? The third question related to the

---

[13] Arnold Berliner left a letter of farewell to Ludwig Ruge: "You mustn't condemn me summarily; consider that I have been living for years as if in prison, excluded from all the cultural things I was attached to all my life […]." In Lemmerich 1998: 176.

Laue to Meissner, Berlin, 13 Apr. 1942. DMA, NL 045/037. The danger of being deported to a concentration camp had already become real more than a year before then. Laue had written to Arnold Sommerfeld on 11 Dec. 1941, asking for his help. DMA, NL 89.

assumptions about the force. Laue answered the first two questions in the affirmative, and Fritz London's theory was used to tackle the third. Under III he referred to Heinrich Welker's considerations and criticized the views held by Wander Johannes de Haas and Hendrik Casimir. Section IV covered Laue's longstanding field of work, and thus treated it at length with reference to older research by Casimir and C. Groth as well as to David Shoenberg's book from 1938. Laue concluded that London's "electrodynamics" contained an indeterminacy. Sections V and VI discussed questions about the penetration depth of a magnetic field in spheres and cylinders (Laue 1942e: 274).[14]

The next publication: 'The electric elementary quantum and X-ray interference' showed another side of von Laue's personality, that of a well-read historian of science willing to explain the significance of the correlations between the natural constants to fellow physicists and laymen alike. Laue started from the observations of electrolysis and Faraday's experiments and from Avogadro's number, to determine the magnitude of the elementary charge. H. A. Lorentz, who had reformulated Maxwell's theory for electrons, was also mentioned. Then he described the gas discharge experiments, the discovery of the angular momentum of electrons and of their magnetic moment. The main part was taken up by the exact determination of the elementary charge, including Millikan's experiments, thereafter describing the determination of the wavelength of X-rays, naming P. P. Ewald, A. H. Compton, and R. L. Doan. He noted that in 1940, F. Kirchner's space lattice calculations found a larger value for "e" than Millikan had indicated. A redetermination by this method yielded agreement in 1937 at an accuracy of 1 per mille. "Now, why is this story about the determination of e so interesting? It may appear to some as learned pedantry, or even as a kind of sport, that physicists should toil so much over one more digit behind the decimal point. But let us consider the situation that would have arisen if the discrepancy between Millikan's value of 1930 and the roentgenographic value had not been resolved. Then one would have had to conclude that there is a gap in our knowledge about the processes of interference in the space lattice of crystals (Laue 1942f: 19)."

No reference whatsoever is made to current military and political events in scientific publications, of course. Nor about the beginning physico-chemical researches into nuclear fission, with consequences that are still being felt well over a century later. In April 1942, Werner Heisenberg was appointed director at the "Max Planck Institute"—that is, as "deputy" to Peter Debye—but he did not arrive in Berlin until the autumn. Heisenberg was an extraordinarily successful researcher and had been a full professor on the chair for theoretical physics at Leipzig since 1927 (Fig. 10.2). By early 1933 he had several doctoral students from abroad. His current field of research was, broadly conceived, elementary particle physics and cosmic rays. Arrived in Berlin, he offered lecture courses on these topics at Friedrich Wilhelm University.

---

[14] Laue commented on this work in a letter to Ms. Meitner on 20 May 1942: "I at least have made some progress in the last few weeks on the electron diffraction book and also on the theory of superconductivity. I believe I have found a generalization of London's theory there, which in some respects achieves more than it and perhaps opens the way to the wave mechanics of this phenomenon." In Lemmerich 1998: 190.

Fig. 10.2 Werner
Heisenberg (1901–1976),
head of the KWI of Physics
between 1942 and 1945 and
professor at the University of
Berlin, where he became
involved in the uranium
project of the Army
Ordnance Office. Photo ca.
1939. *Source* Archive of the
Max Planck Society,
Berlin-Dahlem

The discovery of nuclear fission of uranium by neutron bombardment by Otto
Hahn and Fritz Strassmann in Berlin and the interpretation of the nuclear physics by
Lise Meitner and her nephew Otto Robert Frisch at the turn of 1938 into the beginning
of 1939 in Sweden, indicating an enormous output of energy in the process, soon
found confirmation by several physicists, who also began to speculate about using it
to construct a bomb. With the outbreak of war in 1939, the Army Ordnance Office
(HWA) had entrusted the then managing director of the "Max Planck Institute," Kurt
Diebner, with research tasks in this area. Heisenberg also participated in this with
related work at Leipzig, along with the couple Robert and Klara Döpel. One goal was
to build a uranium reactor for the generation of energy. Toward this end, Heisenberg
wrote a report as early as December (1939), on 'The potentials for technically generating energy by uranium fission'; and in December (1940) a 'Report on the first tests
of the apparatus built at the KWI of Physics.' This was followed in March (1941) by
a detailed 'Report about the experiments with layer arrangements of preparation 38
(uranium) and paraffin [as moderator] at the Kaiser Wilhelm Institute of Physics in
Berlin-Dahlem.'[15]

---

[15] Papers on the German Uranium Project (1939–1945). In Heisenberg 1989: 378, 397, 432. See
also Hermann 1976: 71.

Nothing in Laue's correspondence from this period indicates that research on the exploitation of energy in the nuclear fission of uranium was underway at his "Max Planck Institute," the KWI of Physics. Such research was top secret and Laue was not involved. Even so the construction work necessary for such experiments could not have escaped his notice. Whether he occasionally did gather something about the research and its results is not documented. It may be assumed that Otto Hahn, who was participating in the chemical analysis, informed Laue openly or by insinuation, because Hahn and his colleague Strassmann published all their findings about the chemistry of nuclear fission.

On 26 February 1942, Heisenberg presented a long talk on 'The theoretical basis for the generation of energy from uranium fission' at *Haus der Deutschen Forschung* in Berlin, as part of the 2nd Scientific Conference of the Nuclear Physics Working Group (Reich Research Council—Army Ordnance Office), referring in it to a military application of such a "machine" as a propulsion system for submarines. Then he mentioned: "As soon as such a machine is in operation—to borrow a thought by v. Weizsäcker—the question of producing an explosive also takes a new turn. Since, during the transmutation of the uranium inside the machine, a new substance is formed (element of atomic number 94 [plutonium]), which is most probably an explosive of the same unimaginable effect as pure $^{235}_{92}U$. But this substance can be obtained from uranium much more easily than $^{235}_{92}U$, because it can be chemically separated from uranium (ibid.)." Lectures on the same topic were held under strict secrecy on 2 June 1942, in Helmholtz Hall of Harnack House, about which there is no direct documentation. In mid-1942, the leadership of military research had been assigned to Albert Speer and he wanted to be informed about the status of the work done as part of this uranium project. The most important speaker was Werner Heisenberg. Whether he used the same or a modified text of his lecture of February 26th is not known. There were statements by Ernst Telschow, the secretary general of the KWG, about this meeting, but they are not verifiable. Generalfeldmarschall Erhard Milch had purportedly asked how big such a bomb would have to be in order to destroy London. Heisenberg had replied: "About as big as a pineapple." Laue was not among those invited to participate (ibid.).[16]

---

[16] A reprint of the sequence of talks is provided in Hahn, ed. 1979: 171. First Prof. Schumann spoke on 'Nuclear physics as a weapon.' In Hachtmann 2007: 902–910.

A historical note about this period: In the Federal Archives in Berlin (BAB), under call number R21/366, there is a message dated 19 Jun. 1942 from a German warship: "In the House of Commons on 19 Jun. [1942] a question was put to the government by the Conservative wing concerning the forced labor imposed by the Germans on all Poles between the ages of 18 and 60 years. Further, the Foreign Minister was asked whether he knew how many Polish university professors and scientists had been murdered or had died in concentration camps, requesting this information be provided by the German government, containing the demand that these measures be stopped immediately. It was proposed to convey that if the reply was in the negative, forced labor should be imposed in the same manner on all members of the National Socialist Party after the war, and that all German professors who had advocated National Socialist philosophies, geopolitical ideas, or racial questions should be imprisoned for a long period after the war. Eden responded that the government always remembered that retribution for German crimes was henceforth among the main war aims."

On 10 June 1942, Laue wrote the usual weekly message to Lise Meitner in Stock-holm, mentioning his newspaper article 'Energy principle and modern physics' (Laue 1942g) in *Frankfurter Zeitung* of June 3rd, which she would be receiving directly from the publisher. He asked what she as a nuclear physicist had to say about it (Lemmerich 1998: 194). He would use the text at a public Academy lecture. It began with a historical remark: "In May 1942 it has been 100 years since that famous work by Julius Robert Mayer appeared which regards physics as the historical basis of the theory of energy. Its basic law, the theorem of conservation of energy, will be discussed here, especially its role in modern and recent physics (Laue 1943a)." He mentioned the various conservation laws, momentum, charge, and heat, before acknowledging Hermann von Helmholtz and Rudolf Clausius in 1852 for the second law, and Walther Nernst for the third law in 1906. Then followed an account of the experiments in atomic physics on the main laws, such as, the collision observations in the cloud chamber and the Compton effect as well as Rutherford's discovery of nuclear transmutation (Laue 1943b).

Laue's various publications on historical events in physics encouraged Professor Erich Rothacker to ask him to write a book on the 'History of Physics.'[17]

Laue's paper on 'An elaboration of London's theory of superconductivity' began with a presentation of Fritz London's reasoning. His theory "proves its merit so well, as I have sought to show elsewhere, not only in interpreting the Meissner effect and the permanent currents in superconducting rings, but also in recent experiments, that one must draw it seriously into consideration as a basis for further research. But it is not perfect (Laue 1942h: 65)." Laue supplemented Fritz London's theories by his own ideas to apply Maxwell's extended theory and notions about the penetration depth of the magnetic field in the "Meissner effect." In summary, he postulated: "Our explanations seek to implement ideas consistently that F. London has already introduced but never completely reconciled. They assume two independent flow mechanisms in the superconductor, the ordinary one satisfying Ohm's law, and a new one to which London's equations [...] apply. These two formulas are summarized relativistically and uniformly by the newly established eq[uation]. [...] The aim of the theory of superconductivity is, of course, a wave-mechanical explanation for this phenomenon (p. 83)." Laue justified this reference to a "Meissner effect" in a letter to Walther Meissner from Berlin on September 4th: "After all, there is something new about it, especially about the intermediate state. You were by no means wrong to have complained in the past that I was rather late in recognizing the significance of the Meissner effect. The reason was that, at first, the existence of such an intermediate state was not clear to me and I was not quite able to understand some of your experimental results—namely, everything you observed about the residual fields in hollow cylinders after the external magnetic field is switched off (DMA, NL 045/038)."

In his contribution 'Once more on current distribution in superconductors,' Laue cited London's brochure only with the remark "as F. London suggests" (Laue

---

[17] Erich Rothacker (1888–1965) taught philosophy and psychology in Bonn and apparently asked Laue in 1944 to publish such a history of physics. UAF, Laue papers, 3.2. Cf. Laue 1946b.

1942/43a). Laue described an imaginary specimen consisting of an outer hollow cylinder and an inner hollow cylinder held together by flat lids. Current is conducted through this arrangement and then cooled down to superconductivity. The resulting current distribution was analysed mathematically. In the first part, Laue set the basic conditions: "It should be noted in advance that the distribution theorem does not presuppose anything about the shape of the superconducting current branches [...]. Once the distribution of current across each of the branch lines is given (it is of course confined to thin surface layers), the magnetic energy is a homogeneous quadratic function of the partial current strengths. Thus the conditions for deriving the induction equations from the principle of least action are given (p. 578)." The correspondence with Walther Meissner about superconductivity issues continued. On 21 July 1942 Laue reported about his progress:

> I once wrote you that I was working obsessively. The three superconductivity papers now in press will show what the outcome is. I am somewhat proud of them, because I believe that, firstly, I have clarified our theoretical knowledge of this strange phenomenon insofar as it is possible at the present time, and secondly, I have eliminated a vast amount of rubbish among the available theoretical considerations on it. The book by Shoenberg, about which Gerlach wrote such an appreciative paper in *Physikalische Zeitschrift*, is a particularly deplorable instance of this.
>
> A long time ago Dr. Jacobi once invited me to write a book about superconductivity. I declined on the grounds that the time for it had not yet arrived. Well, these three papers taken together roughly supply the theoretical part that such a book would have to contain (DMA, NL 045/038).

In the summer of 1942, the Laue couple took a vacation, first in Saal, then in the Baltic Sea resort of Deep in the Greifenberg district of Pomerania. Max von Laue took along the published letters of the cultural historian Jacob Burckhardt as his leisure reading. In September, he surreptitiously informed Lise Meitner about the fates of two of their Jewish acquaintances, the widow of Eugen Goldstein and Professor Wilhelm Traube, by simply stating that he did not know where they lived.[18]

Work on superconductivity continued. In October 1942, the editors of the Proceedings of the German Chemical Society received the manuscript 'Our present knowledge about superconductivity.' The character of the text suggests that it was a manuscript for a lecture tailored to a more generally interested audience, because it was headed by a historical introduction. Beginning with Kamerlingh Onnes's discovery and the attempt to apply Maxwell's theory of electrodynamics to explain these processes, it also mentioned its extension by Fritz London to include the "Meissner effect." The penetration depth of a magnetic field in superconductivity was of the order of 10^{-5} cm, and the current did not penetrate "much further under any condition" either. Laue then discussed the cancellation of superconductivity by a magnetic field, citing his own previous papers: "A truly totally superconducting sphere, as a simple homogeneous superconductor of the above description, never can sustain a permanent current; magnetic lines of force would have to permeate it in the

---

[18] Laue to Meitner, Berlin, 13 Sep. 1942. In Lemmerich 1998: 215. It was unclear whether they were in a concentration camp or dead.

process, which are incapable of penetrating so deeply inside a superconductor (see the illustration) (Laue 1942i: 1431 f.)."

Laue had someone to discuss the scientific problems of superconductivity with personally. Hartmut Kallmann had lost his position at the KWI of Physical Chemistry and Electrochemistry on the basis of the Nuremberg Laws. In his letter to Lise Meitner from September 21st, Laue confided to her that he "always got on excellently" with him (Lemmerich 1998: 219). Another of his weekly letters to her dated November 1st reads:

> Now, what shall I tell you today? I have finished a brief paper dealing with the form taken by minute crystals when they grow out of the mother liquor or vapor. These questions are actually rather remote for me, but as editor of the Zeitschrift für Kristallographie I had to study them somewhat, in order to cut short some polemics, and then found such a heap of obscurities and perversities in the literature that I could not help but set things straight. As early as 1878, Gibbs proved thermodynamically the very plausible theorem that the equilibrium form is determined by a minimum of the surface free energy (for a given volume), and in 1901 the crystallographer G. Wulff showed by certain examples that this occurs when the distances of the crystal faces from a certain point in the interior act like the free energies of these faces per unit area. The proof of this theorem was what was involved. The Munich mathematician Liebmann had already given a compelling mathematical proof in 1914, but I was now able to simplify it further and thus also draw a closer connection to thermodynamic considerations. The representation of this entire complex of questions is now extremely simplified; [...] I am proud of my proof, above all, because it yields so much about things that don't appear in it; in contrast to other types of proof, everything that was superfluous is avoided (Lemmerich 1998: 229–230, Laue 1942/43b).

In late autumn 1942 the first hesitant remarks arrived from Manne Siegbahn in Stockholm about extending an invitation to Laue to perhaps deliver some talks. According to Laue's letter to Meitner from Zehlendorf from November 17th, the date was already definitely fixed for March of the coming year (Lemmerich 1998: 232). The German Foreign Office tried to transform such lecturing occasions on cultural and scientific topics into propaganda events for Germany in neutral countries as well as in the occupied territories. National Socialists as well as such scientists as Werner Heisenberg and Max Planck were sent out to fulfill this mission.

Carl Friedrich von Weizsäcker's lecture, held on November 18th on 'The question of the infinity of the universe as an example of symbolic thinking in science,' Laue called "the event of the week" in his next letter to Lise Meitner on November 22nd (p. 235). It had been a historical retrospective on infinity, as Laue enthusiastically reported. But just three days later he wrote to the speaker with second thoughts about some issues, such as, "heat extinction" and the "indeterminacy relation" (UAF, Laue papers, 3.9).

Despite his worries about censorship, Laue still repeatedly mentioned the fates of persecuted persons in his letters. In his reply on December 11th to Sommerfeld's letter of December 8th, he first told him about Ellen Hertz's sudden death, then informed him of Arnold Berliner's fate, as having "almost been deported to the East" and having only just been saved by the intervention of his consul. Security concerns prevented him from sharing any further details (DMA, NL 089).

The war had brought almost all of Europe under Nazi control. German troops were stationed in the East deep inside the Soviet Union. Along this front line, Soviet troops succeeded in encircling German forces, the 6th Army, at Stalingrad. Was this the pivotal event in the war? For Laue a year full of worries and almost hectic scientific activity came to an end. Was his scientific work perhaps crucial to his psychological survival in enduring so much stress from the Nazi terror and incessant bombing raids on the city? Laue's sense of insecurity did not stop at physics, though, as his letter from Zehlendorf on 9 December 1942 to Walther Meissner clearly showed: "As I publicly stated in Naturwissenschaften in 1932, I firmly believe that the whole controversy about the indeterminacy relation is on a wrong track. Concepts are being sought by which to make definite statements, since one doesn't have any. I don't know whether this is a task for decades, centuries, or longer spans of time, and it's none of my business. But a noncausal physics, I mean one that basically renounces causality, isn't science at all. That's my sacred conviction, no matter how many times I may be shouted down as a heretic (DMA, NL 045/006/038)."

At the end of January 1943, the 6th Army was forced to surrender at Stalingrad, and the German troops gave up the Caucasus.

The already announced trip to Stockholm became reality. Lise Meitner wrote to Laue on February 7th that she had reserved a hotel room for him after talking with Siegbahn (Lemmerich 1998: 254). Laue flew to Stockholm on March 3rd around noon. His speaking agenda was extensive. At the University of Stockholm he lectured on Superconductivity; at Uppsala the subject was The Dynamical Theory of X-ray Interference; and at the university colloquium he spoke about How X-ray Interference Came About. He was obligated to speak in addition at the *Deutsche Gesellschaft*, the branch office of the Academic Exchange Service, as well as at the local chapter of the NSDAP. To Meissner he noted, on March 16th, the ongoing cultural competition between England and Germany; England was being smarter about it. "The Swedes very ably maintain their neutrality in this. To counterbalance my visit, they have invited W. L. Bragg for April, and I am sure they will celebrate him no less than they did me (DMA, NL 045/038)."

He purchased all sorts of things in short supply in Germany, of course, and wrote many letters to his son Theo and friends containing things that he wouldn't have been allowed to express inside Germany. He flew back home on March 14th and wrote to Meissner: "The flights were both most enjoyable (ibid.)." Lise Meitner reported to her friend Elisabeth Schiemann in Berlin from Stockholm on 21 March 1943: "Laue gave a series of talks here, one being an account of the discovery of X-rays that was simply delightful. He won the hearts of all his Swedish colleagues, also of the younger physicists and mineralogists, who came in large numbers to his presentations (CAC, MTNR 5/32; Lemmerich 2010: 285–286)." Laue wrote to Percy Quensel on March 18th to thank him "for a hospitality that goes beyond belief, for your readiness to help with the shopping, and not least for your practical assistance, which was such a favor to me (KVA, Quensel papers)."

The Reich Ministry of Education—namely, Rudolf Mentzel—nevertheless had to reprimand Laue in his circumlocutory style:

In the appreciative official reports on your lecture tour to Stockholm, my attention was drawn to the fact that you mentioned the term relativity theory in your lectures, without expressing in this connection that German research explicitly distances itself from Einstein's theory. This could have created an ambiguity about your position among the attending Swedes. Even if this in itself is insignificant compared to the cultural-political gain by your lectures, considering that Swedish physical research does not by any means generally take the standpoint of Einstein's theory and, as experience has shown, in scientific discussions in Sweden the scientific points of view weigh more heavily than the political ones, I would nevertheless not like to refrain from drawing your attention to this point for future lectures abroad.[19]

British intelligence was also watching German cultural activities. One report stated: "Two physicists from the Third Reich—von Laue and Hahn—have recently spoken against the National Socialist régime while on tour in Sweden (in private). Since von Laue is banned from speaking on the German home radio, it is clear that the Nazis were aware what risks they were running. Presumably they were prepared to connive at reasonable utterance abroad, hoping that the legend of a 'Better Germany' may come in useful, should the facade of National Socialism collapse (TNA, FO, 371/43529, p. 70, 1944, Science, original English)."

In April 1943, a retrospective article appeared in *Naturwissenschaften* by Max Planck (1943) on 'Finding the physical quantum of action,' and Laue wrote to Lise Meitner on April 10th: "It was a joy beyond description for me to have before me once more those dear old reflections by which I learned my trade as a physicist. It reminded me of such a very different world. … And such joy is truly needed right now (Lemmerich 1998: 262)."

The second issue in 1943 of *Zeitschrift für Kristallographie* was prefaced by a commemorative article by Max von Laue on the 100th birthday of the founder of that journal Paul von Groth. It was brief, since a detailed biography already existed from 1938. Laue contented himself with offering a personal recollection, von Groth having once asked him "'whether X-ray diagrams of crystals fully secured the reality of atoms […].' At that time I could only answer him that I, in any case, had only gathered the courage to do these experiments from the firm conviction that atoms were real […], without which the discovery of X-ray interference would have been a matter of sheer chance, and their interpretation quite impossible (Laue 1943c)."

The holidays could again be spent by the Baltic Sea, this time in Ahrenshoop, and Hilde's future parents-in-law in Schwerin were visited. Back home again, Laue learned that a heavy air raid on Berlin had damaged Max Planck's house so badly that it was uninhabitable. The Planck couple found refuge on a farm in Rogätz near Magdeburg owned by their friend, the entrepreneur Dr. Carl Still. The Laues took into their own home the mechanic Koch and the laboratory assistant Margarete Schrödger from Hahn's institute. Lise Meitner and the Laues were good friends with the physicist Heinrich Rausch von Traubenberg and his wife. Because she was

---

[19] Mentzel to Laue, 22 May 1943. BBAW, IV, 1–4 Laue and AMPG, III rep. 1A, Laue staff file. A few weeks before this reprimand, Rudolf Mentzel had refused to give Laue a full-time position in the academy as compensation for his early retirement (see the correspondence in April 1943 below). Laue to the Vice-President of the Prussian Academy of Sciences, Professor Dr. Grapow, 14 May 1943. BBAW, II–III, 75/2.

Jewish he had lost his university post. By inference Laue informed his pen friend in Stockholm of their fates. On 10 April 1943, he also wrote Meitner about doing the finishing touches to a volume on 'Electron Interference.' "Otherwise, I am now reading the correction proofs for the concluding chapter of my book. Although the account is such that I do not really need to change anything, nor do I find much to add or delete, nonetheless the dynamic theory of electron interference is a difficult and, one might almost say, a disagreeable method of approximation, even more so than that of X-ray interference."[20]

On the occasion of Max Planck's 85th birthday, a contribution by Laue was broadcast worldwide on radio. Planck's importance was also commemorated on television with Laue's participation, as he wrote Meitner on April 23rd, showing excerpts of a film with Planck from 1939 (Lemmerich 1998: 265–266).

Rudolf Mentzel never missed an opportunity, though, to show Laue "who was in charge during the 'Third Reich.'" The ministry ordered Laue to offer lecture courses at the university, which he had not been doing since 1925, or else go into early retirement. Max von Laue informed the general administration of the KWG of this with the quiry: "Is my employment at the Kaiser Wilhelm Institute of Physics sufficient to preclude intervention by the Labor Office? Or would I have to relinquish this office upon retiring, i.e., vacating my university post? (AMPG, II rep. 1A, Laue staff file, no. 6 fols. 19 f.)" That could be averted. Max von Laue assumed the status of emeritus and the KWG increased his salary to ensure that his income would not be reduced (Laue to Meissner, Berlin, 21 Apr. 1943. DMA, NL 045/038).

In the first part of his autobiographical contribution, which was originally intended to appear on his anniversary in 1944, Max von Laue was not free to allude to his assistance to Jewish fellow citizens. The second part—written after 1945—does mention such help, but without naming names (Laue 1944/51). One striking instance was his intervention for Professor Richard Gans (1880–1954), who as a Jew had lost his chair for theoretical physics at Königsberg. His name appears several times in Laue's letters to Lise Meitner but nothing, of course, about Laue's efforts to save him. The physicist and manufacturer Dr. Heinz Schmellenmeier became interested in engaging Gans as an important and particularly knowledgeable expert for the development of the "Rheotron" electron accelerator. For that, permission first needed to be obtained for his dispensation from debris removal duties and preventative measures taken against his being sent to a concentration camp. At the beginning of May, Laue was invited to meet with Schmellenmeier about this project and Gans was asked to prepare an expert opinion which was sent to Himmler. But the responsible authority, the Reich Aviation Ministry, did not issue a contract. For an alternative application in medicine, Laue and Walter Friedrich wrote on his behalf to the influential surgeon Ferdinand

[20] Laue to Meitner, Berlin, 17 Mar., 4 and 10 Apr. 1943. In Lemmerich 1998: 258–262. The fate of the two Traubenberg daughters is mentioned here. Their parents had managed to send them off to England in time.

Sauerbruch. This was successful. On October 4th this research was declared secret and its indispensable scientist Richard Gans survived.[21]

Laue could not jab at his opponent Mentzel outright. So he used other means. That summer he submitted to the editors of *Annalen der Physik* a short paper on 'A relativistic proof of Wien's displacement law.' Wolfgang Pauli and Laue had already derived this proof in the 1920s, but Laue found a simpler form that assumed a monochromatic polarized beam. The invariance of the three quantities used had already been shown computationally for $E/h\nu$ by Einstein in 1905; Planck had proved the invariance of the entropy S; and the third quantity Z, an approximation for the number of degrees of freedom of the beam and an integer, is therefore also invariant (Laue 1943d). Once again Laue had boldly cited Einstein and mentioned the theory of relativity, but there was no response either from the ministry or from Mentzel. To Lise Meitner, Laue cryptically explained on June 6th: "There was a reason for having the thing printed right now. You can imagine which (Lemmerich 1998: 275)." He was more specific about explaining this action in a letter to Walther Meissner dated June 13th from Pomerania: "That had I reacted to Mentzel's letter in the form of submitting a small relativistic paper to Annalen d[er] Phys.[...] Well—truly without any involvement on my part—my portrait, an ancient photograph, appeared in 'Stürmer' with a caption that is directed blatantly enough against Einstein. I am juxtaposed to him there as an 'Aryan scholar.'"[22]

In the same month Laue submitted a short mathematical addendum (1943e) to his paper 'An elaboration of London's theory of superconductivity' (1942h) to the editors of *Annalen der Physik*.

Laue's "dismissal" as a reading lecturer occurred on 16 July 1943 "due to unfitness for duty."[23] (Fig. 10.3) Lise Meitner learned of this later on rather off hand, in Laue's letter dated October 31st: "When I wrote you that letter 42/43, I completely forgot to mention my retirement. You can gather from this that it did not leave much of an impression on me; but that impression is pleasant at least. It alters almost nothing for me. I am still holding my colloquium and my seminar [...] (Lemmerich 1998: 318)." Pascual Jordan was appointed as Laue's successor on 11 March 1944 (Hoffmann 2012: 131).

In August the Laue family spent their holiday in Ahrenshoop. Comfortably settled in a canopied wicker beach chair, Laue read Carl Friedrich von Weizsäcker's volume 'On the World View of Physics.' He was so inspired by these philosophical expositions that he intended to write a review of it for *Naturwissenschaften*, as he wrote Lise Meitner from Ahrenshoop on August 5th (Lemmerich 1998: 292). He also mentioned a government appeal made to everyone not professionally tied to Berlin to leave the city, and he pointed out the consequences for the KWG. He himself, he continued,

---

[21] Swinne 1992: 115. On Gans's 70th birthday, when he had already returned to Argentina, a special issue was published in 1950 with a contribution authored by Laue.

[22] DMA, NL 045/038. The caption published in this propagandistic tabloid newspaper read: "In order to instigate disunity and discord among nations, the Jew professes that all men were equal. Who would claim that this German physicist (Max von Laue), whom science has so much to thank for, should be equated with some such vile good-for-nothing?" *Der Stürmer*, 3 Jun. 1943: 2.

[23] HUA, Laue staff file, fol. 141; on the dismissal: BAB, R 26-III/9 and R 4901/13270.

Fig. 10.3 Max von Laue, early 1940s. *Source* Photograph in the author's estate

would have to be back at the institute on September 1st to stand in for Heisenberg. The end of the letter cites a verse by Schiller:

> A torch I see aflaming
>> But not in Hymen's hand,
>> To the clouds I see it trailing,
>> But a burnt offering it is not! (p. 293)

Wartime life got even harder. Shortages worsened, air raids multiplied and were more destructive. Closer detonations shattered window panes or roof tiles of Laue's home in greater or lesser numbers, and a door frame collapsed. The Max Planck Institute was largely spared. The expansionary war tactics needed more and more soldiers. This meant that even important branches of industry could only produce anything if people were recruited from regions that had been occupied by German troops, who were then referred to as "foreign workers" (*Fremdarbeiter*). The Kaiser Wilhelm Society also employed such people. From time to time the Laues invited mainly younger persons home for a meal, and some friendships developed as a result. Classmates of their son Theo who were on leave from the frontline also numbered among their guests. Mrs. von Laue was the driving force behind this hospitality. At the end of July 1943, a young geologist in the army, Dr. Kurt Lemcke, was one of these dinner guests and would soon become their son-in-law. "Hilde is really very happy," Laue wrote to Lise Meitner from Ahrenshoop on August 10th (p. 293). Their son already had a fiancée in the US. News from America still came almost exclusively through Lise Meitner. Alternative channels were a relative of the Laues in Switzerland and the Red Cross.

Most people avoided writing candidly about the political and military situation by then. The risk of prosecution and condemnation was great. Nevertheless, Laue must have been frank in a letter to Planck dated August 24th, because the latter replied from St. Jakob on August 31st: "I certainly do share your grave concerns about the political future. It is difficult to say who bears the main blame for this disastrous development. It might very well have been avoided. The basic evil, I believe, is the rise of the rule of the masses. I already consider universal direct suffrage (from the age of 20!) a great mistake. But the wheel of history cannot be turned back, and we must take the consequences upon ourselves (AMPG, V rep. 13, Planck, no. 1300)." In spite of all these difficulties, great and small, the social circle consisting of the Planck, Laue, Hahn, and Hertz couples, as well as Clara von Simson, and Paul Rosbaud, who was involved in the publication of *Naturwissenschaften*, still tried to convene occasionally in the "Bristol" as in former days. Almost every one of Laue's letters to Lise Meitner now reported deaths on the battlefront and in air raids. In October, the Laues received a notice informing them that Klaus Grüneisen had been killed in action at the age of 22. Max von Laue informed Lise Meitner on October 4th, because she had often looked after the Grüneisen children in the evenings and was a close friend of the family (Lemmerich 1998: 311). So the news hit her hard. On the evening of his birthday, on October 9th, after three siren alarms, Laue wrote a letter to Lise Meitner by hand, because he had stowed the typewriter away in the cellar for fear of it getting damaged. "The most beautiful congratulatory letter I received came from Planck. Only that he sounds a little melancholy, as if he feared he would not be able to write often. The war affects him deeply, and I know from his son that his health is not what it used to be either. […] Our Hilde is so touchingly in love. Every day a letter goes off to her Kurt, and he replies as often as he ever can. The day before yesterday Hilde proudly presented herself in the wedding dress her mother had worn 33 years ago […] (p. 313)."

To secure the manuscript of his forthcoming book, Laue sent fair copies and the correction proofs to Lise Meitner for safekeeping. Soon afterwards, the Akademische Verlagsanstalt informed him that the book could not be published until the following year. Like others, Planck, too, was concerned about the safety of his correspondence, and despite the many obstacles imposed on travel, he set off on a lecture tour with his wife in October 1943 carrying valuable letters and writings with him in his heavy luggage. But already in Frankfurt am Main, his first stop, heavy destruction prevented him from delivering a lecture. The lecture in Koblenz was also cancelled because of a sounding siren. On October 22nd his lecture in Kassel was delivered. But shortly afterwards there was very heavy bombing and the home of relatives with whom the Plancks were staying burnt down. It took them hours to scramble out of the cellar again, but the correspondence that the Planck's had taken with them was destroyed.[24] On November 28th, Laue wrote to Lise Meitner about multiple damages to his home in Berlin during the nighttime attack on November 26th. Subsequent letters informed her about destructions of one home after the next from among their relatives and colleagues (pp. 328, 338 on 31 Dec. 1943). Towards the end of the year, bad news arrived from Leipzig from the Akademische Verlagsanstalt: the entire remaining print run of his book on X-ray interference had gone up in flames, and galley sheets of the planned book on matter waves had also been destroyed; the set type no longer existed.

Max von Laue was still working on the second edition of his book on 'X-ray Interference' when the printers managed to finish his book 'Matter Waves and Their Interference' (1944a), after all. Laue opened it with an orientational preface:

> The elder generation of physicists today has experienced more of the great, revolutionary advances in its science than ever before, and it should consider this incredible good fortune as recompense for everything else that world events have subjected it to, especially since later historical circumspection will probably discern an underlying link between the two. If I am to say what has left the deepest impression on me personally from more recent times, then from the last decade I must mention pair generation, the formation of a positive and a negative electron from gamma radiation and their annihilation into such again, and from the penultimate decade, the discovery of matter waves. For their existence spectroscopy undoubtedly furnishes the most precise proof, but the interference phenomena of these waves supports it most graphically. It is the latter that this book treats (p. V).

The book is similarly structured to the one on X-ray interference but, as the author assured, one does not need to be familiar with that book to understand it, even with respect to dynamic theory. "Research on matter waves is even less complete than on X-ray interference. Thus the present account is dated to an even greater extent than my book about the latter topic (p. VI)."

---

[24] Laue to Meitner, 31 Oct. 1943. In Lemmerich 1998: 318 f. Upon returning from Kassel, Planck told Laue about the devastating air raid on the city and the loss of the documents and letters he had brought along. Laue also wrote Lise Meitner about this, enclosing a longer excerpt from Planck's report with his letter.

Table of contents of Max von Laue's *Materiewellen und ihre Interferenzen*, appeared in Leipzig 1944.

Many illustrations, diagrams, and photographs illustrated the text. The table of contents was again rigorously systematic. Ignoring any possible admonition by the ministry, Laue's first chapter recalls Louis de Broglie, employs relativity theory for calculations, and cites a letter by Walter M.Elsasser (1925) to the editors of *Naturwissenschaften*, which began with the statement: "A while ago Einstein arrived at a physically very strange result by the byway of statistics (Being of Jewish descent, Elsasser had been forced to emigrate to the U.S. in 1936.)." A paper by the émigré Otto Stern (1939) was also cited.

About the few experiments with natural neutron sources, also about the first experiments using a cyclotron to generate a neutron beam, Laue remarked: "But since experience and some atomic and molecular beams prove the correctness of the de Broglie theory, we shall probably have to admit the interference interpretation of these neutron experiments (Laue 1944a: 222)." Max von Laue used the opportunity in Appendix I with its discussion about causality to present his opinion on this problem again inside a book, although he did not broach the philosophical issue. He began with Newton's findings about the orbit of a body irrespective of its size. They had stood the test of time for over three hundred years, he asserted. But Ernst Mach had questioned this. "And then happened what sooner or later just had to happen. Facts were found that refute the hypotheses made in mechanics. This is a process that physics has often experienced; one need only think of all the transformations in optics, thermodynamics, and electrodynamics. Did any of these instances give cause for thought about any change to the epistemological foundations? (p. 366)" By the discovery of spin by Samuel A. Goudsmit and George E. Uhlenbeck, a foreign element had entered into the notion of a point mass. Schrödinger's ideas from 1926, based on de Broglie's conception, caused the point mass to "dissolve completely." But, Laue stated: "We have, moreover, learned—and this is probably the greatest experience of the present generation of physicists—directly by interference experiments that wave processes are attached to the motion of every body; and the term is entirely unsuitable to describe this motion (ibid.)." Laue then discussed the effort to rescue the notion of the point mass, which inevitably led to the uncertainty relation. Returning again to epistemology, causality: "We mean the principle of proximate action and Newton's law of attraction, which, as is well known, argued for action at a distance. Newton himself pointed out the inadequacy of it; notwithstanding this deficiency, theoretical astronomy grew out of this law. Today, the general theory of relativity has finally set that principle right [...] (p. 367)." On the quantum puzzle he thought "that the time for its solution is not yet ripe, that rather many an undreamt-of, seemingly far-off observation must first succeed. Nor do we wish by any means to assert that the term 'point mass' alone is in need of reform (ibid.)." Laue again cited Ernst Mach and concepts in other areas of physics. "The quantum question is simply the fundamental question of physics (ibid. Quantum electrodynamics had recently made some progress, mainly in the USA.)."

At the beginning of 1944, Laue's detailed and critical review of Carl Friedrich von Weizsäcker's book 'On the Worldview of Physics' appeared in *Naturwissenschaften*. After a general introduction about the structure of the volume containing four lectures and a laudatory essay on Weizsäcker, Laue discussed the philosophical conception of quantum mechanics with its conflicting notions of the particle and the wave. Borrowing the author's arguments Laue reiterated: "So I am not allowed to say: 'The atom is a particle' or 'it is a wave,' but 'it is either a particle or a wave,' and I decide by my experimental arrangement what it manifests itself as. [...] Every single causal chain created by an experiment is completely governed by classical physics; only because the experimental arrangement is classically understandable can we comprehend the resulting measurement at all (Laue 1944b: 86)." Weizsäcker set this critically against the epistemology of Immanuel Kant. About this part Laue remarked:

"The auth[or] becomes uncertain and also lapses into visible misunderstandings [as, e.g., about 'things in themselves,' ...]. But a comparison between the portrayed view of quantum physics and Kant cannot, in my opinion, come out positively at all; both are irreconcilable opposites (ibid.)." In response to Weizsäcker's question about what atoms would look like if they were visible, Laue rephrased the question. "Physically it is: What electromagnetic radiation emanates from an atom when it is illuminated by radiation of a certain wavelength? And about this the theories of light scattering and quantum jumps provide information and, where these should fail, also direct experiment (ibid.)." The reviewer criticized the passage under the heading 'Complementarity between chemistry and mechanics,' faulting his conceptions about the properties of matter, its structure, and its analysis, in the process of which atoms would be destroyed. Laue asked: "How does this match the acknowledged fact that the atoms of a crystal can be excited to emit X-ray spectra? This emission is deeply seated inside the atoms, one learns from it much about the atomic structure. And yet the radiating atoms do not cease to be the crystal's building blocks (ibid.)." He added further critical remarks before reaching a conciliatory conclusion: "It seems to me that the value of the present book is that it shows where quantum mechanics is still lacking; this may perhaps affect future research in physics favorably. And in this sense we wish it very many and very informed readers (p. 87)."

The Allied bombardments by air continued. On 15 February 1944, Max Planck's home was completely destroyed. Laue had previously arranged for parts of his extensive library to be removed into storage as a precautionary measure (Planck to Laue, Rogätz, 18 Feb. 1944. AMPG, V rep. 13, Planck, no. 1337).

In a commemoration of the 100th birthday of Ludwig Boltzmann on February 20th, Laue had much praise for the Austrian physicist. Starting with his early work on thermodynamics in 1884, he then discussed his publications on physical statistics, the hypothesis of molecular disorder, and the theorem of the equipartition of energy. This was followed by a study of irreversibility and Boltzmann's insight into the connection between entropy and thermodynamic probability and the additivity of entropy. He concluded, with reference to Boltzmann's admiration for Shakespeare, Goethe, and Schiller, with a quote: "Through Schiller I became what I am, without him there could have been a man with the same shape of beard and nose as mine, but never me (Laue 1944c, Boltzmann 1905: V)."

The relocation of the Kaiser Wilhelm Institutes of Physics and Chemistry to southern Germany had already been planned the previous year. The move began in early 1944, but too late for the KWI of Chemistry, as it was badly damaged in a bombing raid on the night of February 14/15th (Fig. 10.4). The wing housing Otto Hahn's office and the laboratories where nuclear fission had been discovered was gutted. The directors' villa, where Lise Meitner had lived until her escape, was also badly damaged.

A few days later, Mrs. von Laue travelled to Hechingen, where the KWI of Physics was to be relocated. The furniture destined for Hechingen was picked up in early March. Four days later there was a daylight attack on Berlin, which Laue experienced in an air-raid shelter in the city center together with Walther Bothe. They had been

Fig. 10.4 The Kaiser Wilhelm Institute of Chemistry after the air raid in the night of 14/15 February, 1944. *Source* Archive of the Max Planck Society, Berlin-Dahlem

on their way to Bothe's lecture at the academy.[25] In a daytime bombing raid on the nearby radio station on 9 March 1944, the Laue family home was also very badly damaged. The windows and doors were shattered, walls collapsed, and a garage was devastated.

Their worries could be forgotten for a short while when Hilde and Kurt Lemcke were married. The wedding was held in Pomerania on March 17th as festively as if it had been peace time. The speeches were to everyone's satisfaction and the in-laws had provided plenty of food and drinks.[26]

The farewells from Berlin were over and Max von Laue arrived in Hechingen on April 13th; his wife and daughter followed a day later, the furniture a few days after that. Their daughter continued on to Tailfingen along with Otto Hahn's institute. In an undated letter to Lise Meitner from May 22nd, Magda von Laue reported about daily life in Hechingen: "But I must set right that, as far as being lonely is concerned, it does not apply to me. In the first place, nature fulfills my needs entirely; I have

---

[25] Laue to Meitner, 22 and 24 Feb., 4 and 8 Mar. 1944. In Lemmerich 1998: 353 ff.

[26] Laue to Meitner, Saal, Pomerania, 20 Mar. 1944. In Lemmerich 1998: 358 f. The correspondence with Lise Meitner during this period contains allusions to Max Planck's hernia during his stay in Amorbach and his operation performed by Prof. Sauerbruch, as well as a brief reference to the suicide attempt by Planck's granddaughter, Emma Fehling. Personal communication by Hildegard Lemcke, née von Laue, to the author.

always preferred to go on solitary walks, and it is so beautiful here and has been my dearest wish ever since I can remember to be able to live in this region, that I accordingly feel completely happy. Of course, I am more interested in the young physicists' wives than my husband is; their children and common questions about housekeeping always provide subject matter [...] (Lemmerich 1998: 368 f.)." She had a very positive assessment of the availability of fruit and vegetables. Max von Laue described to his friend in Stockholm their rather cramped lodgings in two rooms with a makeshift kitchen. But there was enough room on the walls to hang up their pictures. The space provided for the KWI of Physics was also very cramped, and the poor soundproofing disturbed Laue very much. Heisenberg lived in temporary quarters without his family. Georg Menzer and Gerhard Borrmann had also come along to Hechingen, as we learn from Laue's letters to Meitner from 24 Jun. and 2 Jul. 1944 (pp. 377, 380).

It had been a long time since any victory reports could be broadcast. There is no documentation about what Laue thought of the official announcements and news reports from the various battle fronts, which he was now also receiving from his son-in-law. What conditions for a surrender could be expected? What "peace conditions"? How much harsher would they be compared to the defeat of 1918? False propaganda about a "wonder weapon" promised a victorious end to the war. But the German air force could not overcome the Anglo-American superiority in the skies. The rocket attacks, initially deploying the V1, could not prevent further air raids on Germany. The more lethal V2 rockets did not bring about that fabulous turnaround to the war either.

In Haigerloch, the first larger uranium nuclear fission reactor using heavy water as a moderator was set up in a rock cellar; and measurements were commenced on the multiplication rate of neutrons produced by uranium nuclear fission. That experimental setup never became critical, however. The turning point of the war came on 6 June 1944 when American, British, and Canadian troops landed in Normandy. After fierce fighting, they were able to extend their bridgehead and spread out in several directions. American troops had already landed south of Rome, at Anzio and Nettuno. Soviet troops advanced ever further and had already reached the banks of the Vistula river.

In keeping with tradition, the academy in Berlin opened its "Leibniz session" on 29 June 1944, on which occasion the fiftieth anniversary of Max Planck's membership was also honored. Werner Heisenberg then spoke about the elementary quantum of action. Max von Laue flew to Berlin for this meeting, as he informed Lise Meitner (ibid.).

Heisenberg had to travel to Berlin from time to time. He had recently become a member of the "Wednesday Society." This small circle met with opponents of the regime to hear talks from the various fields of expertise of its members. Heisenberg spoke about 'Astrophysics' on July 12th and, as he recalled: "the atomic energy in the stars and its technical exploitation on Earth, insofar as I was allowed to talk about it according to the secrecy regulations. [Ludwig] Beck and [Eduard] Spranger mainly participated in the discussion. Beck saw at once that henceforth all previous military ideas would have to change fundamentally."[27]

On July 20th, Colonel Claus Schenk, Count of Stauffenberg, smuggled a bomb into the East Prussian headquarters to assassinate Hitler during a meeting. Hitler came away from the explosion with only slight injuries. Stauffenberg was already on his return trip to Berlin. He and his complotters were summarily shot on the same day. Planck's eldest son, Erwin, was also arrested. The People's Court sentenced him to death, even though he had not been an active member of Stauffenberg's circle (Hoffmann 1979, Roon 1994: 178, Pufendorf 2006: 447). In his letter of August 8th from Rogätz via Wolmirstedt, Max Planck thanked Laue for his letter of July 4th, but had to inform him that he had not received his letter of the 29th. It is very likely that Laue had mentioned something about the events of July 20th in it and the letter had been confiscated. After some news about his family and friends, Planck continued:

> We ourselves also had to experience all kinds of negative things. Since Jul. 20th, a considerably harsher tone prevails among the state leadership. Even my son Erwin—but I ask that this be treated with complete discretion—has become the object of surveillance, the only grounds for it being that some of the assassins were acquainted with him.
>
> My consolation is that quite a large number of other personalities have been affected by the same fate, such as the Prussian Minister of Finance, Prof. Popitz. One cannot possibly build a reasonable verdict on such a basis. But this matter still is very unpleasant—namely, because it will probably take weeks before a decision is reached (AMPG, V rep. 13, Planck, no. 1375 K).

Laue's cautious letters to Lise Meitner only hinted at these events and their consequences. News reports about the assassination attempt appeared in the Swedish papers, mentioning the names of those arrested and the executions.

For the fiftieth anniversary of Hermann von Helmholtz's death on 8 September 1944, Laue wrote a short biographical sketch. His admiration of Helmholtz had its roots in his school days in Strasbourg, when his mathematics and physics teacher had recommended that he obtain the two-volume edition of Helmholtz's lectures and speeches (1865/76). After an appreciation of some of Helmholtz's achievements, his ophthalmoscope, sound analysis and its basis in physiology and psychology, his hydrodynamics, the law of conservation of energy, and also his considerations about the field as conceived by Faraday and Maxwell, there followed a section on the two volumes that had enthused Laue so much in his youth (Laue 1944d).

Magda von Laue wrote to Lise Meitner on September 21st from southern Germany:

> Planck now has a great-grandchild, a little girl. The uncle [Erwin Planck] of this child is very badly off, there isn't much hope for his condition, but great-grandfather and great-grandchild are well.
>
> In spite of the war with its suffering and its sorrows, I have lived the happiest summer I am aware of. It sounds strange, because every day there was rough, hard work—carrying

---

[27] Heisenberg 1972: 258. General Ludwig Beck headed the General Staff from 1935. He resigned in 1938 because he did not approve of Hitler's expansionist plans. On 20 July 1944 Beck was arrested and, after a failed suicide attempt, shot.

30 hundredweight of coke alone from the yard to the cellar, washing all the laundry until your knuckles bleed—even so, the closeness to nature you have here helps you get over everything. Especially since it was a very fine, sunny summer. I walk to my view once a day –the castle view. It crowns the summit, steep, high, inaccessible, sometimes appearing above the fog, like Montsalvat, sometimes as if within reach in the evening sun. Like shaking off a depressing nightmare, that for 25 years I had been stuck in my close, skyless garden, whereas my husband would still like to be there now, if the bombs had not hit our property. I wouldn't mind dying thus, if the winter should bring us trouble.

My (in truth) better half would also like to say a word though, ... (Lemmerich 1998: 405)

Three days later, Max von Laue informed their friend in Stockholm of the sudden death of the baron Heinrich Rausch von Traubenberg. His Jewish wife Marie lost her protection as a result. She was sent to a concentration camp. Otto Hahn tried to improve her situation by claiming that she had to organize her husband's important scientific papers, and that she was the only one able to do it. That is how she was able to survive her stay in a concentration camp until the end of the war and the liberation.[28] Another worrying letter arrived from Max Planck in Rogätz on October 11th: "The fate of my son Erwin, who was arrested after July 20th merely because some of the assassins were among his circle of acquaintances, still has not been decided. He is still in remand prison and I am deeply worried about him. Certain signs as a result of steps I have taken [do?], however, allow me to regard as justified the hope that at least the worst will pass him by (AMPG, V rep. 13, Planck, no. 1383 K)."

A letter dated November 26th informed Lise Meitner about two manuscripts that Laue had just completed: about his "physical career," which was supposed to be published by Keiper in Berlin (cf. Laue 1944/52), and a 'History of Physics,' which was scheduled to appear at the publishing house Junker & Dünnhaupt. The war conditions prevented the latter from being printed at first. It was not a historical account of physics generally from its beginnings to modern times. Laue only dealt with the history of individual fields and their fundamental discoveries, with the emphasis placed on theory: time measurement, mechanics, gravitation and action at a distance, optics, electricity and magnetism, the reference system of physics, the fundamentals of thermodynamics, the theorem of the conservation of energy, basic thermodynamics, atomistics, nuclear physics, crystal physics, thermal radiation and quantum physics. Some historical context prefaced each area. The book also appeared in foreign language editions after the war (Laue 1946b).

In early 1945, Laue corresponded with Friedrich Möglich about superconductivity. It always involved Fritz London's theory. On January 9th he wrote him from the KWI of Physics branch office in Hechingen: "I can add, however, that I have now succeeded in confirming London's theory, which I had reworked, and which was truly unexpected (AMPG, III rep. 50 no. 1378a)." To Lise Meitner he also reported on his ideas and successes in the theory of superconductivity, still taking London's

---

[28] Laue to Meitner, 24 Sep. 1944. In Lemmerich 1998: 406. Mrs. von Traubenberg had received an official summons to register herself while in Hirschberg, where they had moved after the destruction of their Berlin apartment. Her husband suffered a fatal heart attack while accompanying her there.

theory (1937) as its basis. On the same day he wrote her about his talk at the collo-quium in Hechingen: "McLennan's curve proves not only the existence of ohmic resistance in the superconductor, but also the decrease in conductivity below the jumping point. This shares a common cause with the decrease in London's constant λ: as the conduction electrons cool down below the jump, they *gradually* transition from the ohmic conduction mechanism to the superconduction mechanism."[29] Copies of the manuscript were sent by Laue to friends for safekeeping; his anxiety about the future was becoming increasingly apparent. A few weeks later, he informed Lise Meitner from Hechingen of the death of the biologist and director of the KWI of Biology, Friedrich von Wettstein, who had died of protracted pneumonia. They had been counting on his diplomatic skills and scientific reputation for the future. Laue also insinuated Erwin Planck's execution by refering to "another, more terrible death, which causes deepest grief above all to our old teacher in Rogätz [...] (13 Feb. 1945. In Lemmerich 1998: 436)."

There was occasional respite from this constant confrontation with the loss of friends and acquaintances and the rising uncertainty about the future, as one description of a lay concert in his letter to her on February 23rd shows:

> Today you may gape at what the KWI is capable of. Hear this: Last night there was a public concert here. The performers were Miss Troll from the biological KWI (soprano), Dr. M. Pahl from the KWI of Physics (baritone), and Heisenberg on the grand piano. In addition, a local violinist and a cellist. The brunt of the evening was borne by Heisenberg, who sat at the piano for the full two hours without a break. Program: Trio in B major (Mozart), aria for soprano and violin from 'Apollo and Hyacinthus' (Mozart), duet from 'The Marriage of Figaro' (Mozart), two arias of Cherubino from the same opera (Mozart) (I have never heard 'Sagt holde Frauen, die ihr sie kennt' so nicely sung as yesterday, with the text even intelligible), the spring sonata by Beethoven, 4 songs by Schubert for baritone, two duets by Schubert, an andante from a cello sonata (Klingel), and finally two songs by Wolf and the 'Zuneignung' by Richard Strauss (p. 437).

Laue was probably continuing to work intensely on superconductor theory and sending out parts of the manuscript to Lise Meitner for safekeeping. In the past, Laue had often immersed himself in scientific work as a way to cope with diffi-cult situations. He did so in Hechingen, too. The finished paper (Laue 1948) was originally submitted to the publisher in February 1945. In 1932 John C. McLennan, in collaboration with A. C. Burton, A. Pitt, and J. O. Wilhelm had conducted very complicated experiments with superconducting coils at high frequencies within the range of 2×10^6 to 3×10^7 Hz. Compared to experiments with direct current, the high-frequency resistance dropped steeply only at considerably lower temperatures. About this Laue remarked: "The conclusion that the jump depends on the vibration number of the measured current is, taken literally, in contradiction to the present view that the normal conductor and the superconductor represent two phases in the

---

[29] Laue to Meitner, South Germany, 9 Jan. 1945, orig. emphasis. In Lemmerich 1998: 431. Here Laue writes for the first time about a transition of the conduction electrons. The publication on this appeared in the following year: Laue 1946c. The constant λ given by Laue had been introduced by Fritz London in 1937 $\lambda \times J = e$. J is the density of the electron current. Elsewhere, $\lambda = m/n\, e^2$. denotes the number of superconducting electrons.

sense of thermodynamics, whose transitions into each other cannot depend on this vibration number (Laue 1948: 136)." Applying the extended "telegraph equations" and taking the skin effect into account as well as assuming two conduction mechanisms, Laue succeeded in proving that the course of the curve given by the coauthors was also describable by theory.

At the beginning of March 1945, Max von Laue sent letters to Lise Meitner with some family news without knowing whether they would be delivered. He also wrote a long one to his son Theo on April 7th in which his nervousness about a possible evacuation by the German authorities is perceptible. It closes with remarks about his scientific legacy and about the future:

My dear son!

We are now on the verge of the long-anticipated great change and await with tense anxiety the shape it will take over here. Nothing can be predicted. We don't know whether Hechingen will be fought over, whether it will succumb to destruction, whether the residents will stay here or be transported away and how, and a whole series of other worries. We scholars in particular seem to be especially at risk of deportation and we are intently considering every possibility to prevent it. [...]

Whatever may happen now, on the whole we have in any case had a beautiful year. Mama, in particular, feels happier here in the countryside than anywhere else, and she occasionally says she never wants to return to Berlin. Well, I don't think the last word has been spoken yet. But I, too, have had it better here than many a year before. No superfluous and time-consuming meetings [...] (Lemmerich 2011: 216).

He listed the many people they had been seeing. In the first place he named the Hahn couple, then Alfred Kühn, Georg Menzer, and Werner Heisenberg with their wives. He was pleased that Max Kohler had also come to visit with his wife. "But all of this pales against what Mama has done for me here (p. 217)." He described her many strenuous chores, shopping, and fetching wood. He told Theo about their closer relatives, Laue's sister Elly and their stepmother, as well as his cousin Martha. "Provided I get through this catastrophe, my main task for the rest of my life will probably be to do my part in the spiritual renewal of Germany (p. 218)." He hoped for goodwill among his friends abroad. "I know what a mountain of hatred will have to be overcome. In the face of that, it will have to be emphasized again and again that hatred destroys humanity and its culture—we can truly see all too clearly where that has already led—and that Christian love of one's neighbors, which does not stop at the barriers of nationality, apart from all the others, is the best political principle. (Ibid.)

References

Bellamy, Edward. 1914. *Ein Rückblick aus dem Jahre 2000 auf das Jahr 1887*. 2nd ed. Stuttgart: Dietz
Boersch, Hans. 1940. Fresnel'sche Elektronenbeugung. *Naturw*. 28: 710 f
Boltzmann, Ludwig. 1905. *Populäre Schriften*. Leipzig: Barth

Einstein, Albert. 1915. Antwort auf eine Abhandlung M. v. Laues ‚Ein Satz der Wahrscheinlichkeit-srechnung und seine Anwendung auf die Strahlungstheorie'. *Ann. Phys.* 352: 879–885

Elsasser, Walter. 1925. Bemerkungen zur Quantenmechanik freier Elektronen. *Naturw.* 13: 711

Fermi, Enrico. 1938. *Moleküle und Kristalle*. Leipzig: Barth

Festschrift. 1939. Festschrift zum 60. Geburtstage Max von Laues. *Ann. Phys.* 428 (3/4): 189–380

Franke, H. 1941. *Forschung aus dem Gebiet der Röntgenstrahlen* 64: 288

Frenkel, Viktor J. 2011. Professor Friedrich Houtermans – Arbeit, Leben, Schicksal. Biographie eines Physikers des zwanzigsten Jahrhunderts. Edited and expanded by Dieter Hoffmann and Mary Beer. Max Planck Institute for History of Science, Berlin. Preprint no. 414. www.mpiwg-berlin.mpg.de/Preprints/P414.PDF [last accessed: 8 Oct. 2021]

Gönner, Hubert, and R. Klein. 1982. Nachruf auf Max Kohler. *Phys. Bl.* 38: 298 f

Hachtmann, Rüdiger. 2007. *Wissenschaftsmanagement im „Dritten Reich". Geschichte der Generalverwaltung der Kaiser-Wilhelm-Gesellschaft*. Vol. 2. Göttingen: Wallstein

Hahn, Dietrich, ed. 1979. *Otto Hahn. Begründer des Atomzeitalters. Eine Biographie in Bildern und Dokumenten*. Munich: List

Heisenberg, Werner. 1939. Die Möglichkeiten der technischen Energiegewinnung aus der Uranspaltung. Kernphysikalische Forschungsberichte. German Reports on Atomic Energy no. G-39. Dec. 1939. In Heisenberg 1989: 378

Heisenberg, Werner. 1940. Bericht über die ersten Versuche an der im KWI für Physik aufgebauten Apparatur. Kernphysikalische Forschungsberichte. German Reports on Atomic Energy, 21 Dec. 1940. In Heisenberg 1989: 397

Heisenberg, Werner. 1941. Bericht über die Versuche mit Schichtenanordnungen von Präparat 38 (Uran) und Paraffin am Kaiser-Wilhelm-Institut für Physik in Berlin-Dahlem. Kernphysikalische Forschungsberichte. German Reports on Atomic Energy, Mar. 1941. In Heisenberg 1989: 432

Heisenberg, Werner. 1972. *Der Teil und das Ganze. Gespräche im Umkreis der Atomphysik*. 4th ed. Munich: Europäische Bildungsgemeinschaft

Heisenberg, Werner. 1989. *Collected Works*. Ser. A: *Original Scientific Papers*. Vol. II. Ed. by W. Blum, H.-P. Dürr, and H. Rechenberg. Berlin, Heidelberg: Springer

Helmholtz, Hermann von. 1865/76. *Populäre wissenschaftliche Vorträge*. Braunschweig: Vieweg

Hermann, Armin. 1976. *Werner Heisenberg in Selbstzeugnissen und Bilddokumenten*. Reinbek: Rowohlt

Hoffmann, Dieter. 2012. Pascual Jordan (1902–1980). Der gute Nazi. In *Die Universität Rostock in den Jahren 1933–1945*. Edited by Gisela Boeck and Hans-Uwe Lammel. Rostocker Studien zur Universitätsgeschichte no. 21, pp. 131–162. Rostock: Universität Rostock

Hoffmann, Peter. 1979. *Widerstand, Staatsstreich, Attentat. Der Kampf der Opposition gegen Hitler*. 3rd ed. Munich: Piper

Laue, Max von. 1932. Zur Deutung einiger Versuche über Supraleitung. Lecture given at the award of the Planck Medal. *Phys. Z.* 33: 793 ff

Laue, Max von. 1939a. Fermi, Enrico. Moleküle und Kristalle. *Naturw.* 27: 32

Laue, Max von. 1939b. (Needham and Pagel.) Background to Modern Science. *Naturw.* 27: 434 f

Laue, Max von. 1940a. Einordnung der Kossel-Möllenstedtschen Elektroneninterferenzen in die Raumgittertheorie. *Ann. Phys.* 429: 169–172

Laue, Max von. 1940b. Das elektrostatische mittlere Potential in Kristallen. *Naturw.* 28: 515 f

Laue, Max von. 1940c. Interferenz-Doppelbrechung von Röntgenstrahlen in Kristallprismen. *Naturw.* 28: 645 f

Laue, Max von. 1940/41. Zur Elektrostatik der Raumgitter. *Z. Kristall.* 103 (1941), 1st issue, 1940: 54–70

Laue, Max von. 1941a. *Röntgenstrahl-Interferenzen*. Physik und Chemie und ihre Anwendungen in Einzeldarstellungen, no. 6. 1st ed., 2nd ed. Leipzig: Akademische Verlagsanstalt

Laue, Max von. 1941b. Die technische Physik des Kraftwagens. *Naturw.* 29: 184

Laue, Max von. 1941c. Zum 40-jährigen Jubiläum des Wärmestrahlungsgesetzes. *Naturw.* 29: 137

Laue, Max von. 1941d. Über die theoretische Bedeutung der Justi-Zicknerschen Versuche über Stromzweigungen in Supraleitern. *Z. Phys.* 118: 455–460

Laue, Max von. 1942a. Erkenntnislehre. *Naturw.* 30: 123 f

Laue, Max von. 1942b. Walther Nernst zum Gedächtnis. *Med. Klinik* 1942: 7 f

Laue, Max von. 1942c. Galileo Galilei. Zur dreihundertsten Wiederkehr seines Todestages. *Frankfurter Zeitung*, 8 Jan. 1942

Laue, Max von. 1942d. Statistisches über Röntgendurchstrahlung vieler gleichartiger Schichten. *Naturw.* 30: 205–207

Laue, Max von. 1942e. Bemerkungen zur Supraleitung. *Phys. Z.* 43: 274–284

Laue, Max von. 1942f. Das elektrische Elementarquantum und die Röntgenstrahl-Interferenzen. *Scientia* 72: 14–19

Laue, Max von. 1942g. Energiesatz und neuere Physik. *Frankfurter Zeitung*, 3 June 1942

Laue, Max von. 1942h. Eine Ausgestaltung der Londonschen Theorie der Supraleitung. *Ann. Phys.* 2/3: 65–83; abridged version in *Verh. Dt. Phys. Ges.* 2 (1942): 62

Laue, Max von. 1942i. Unsere heutige Kenntnis der Supraleitung. *Ber. Dt. Chem. Ges.* 75: 1427–1432

Laue, Max von. 1942/43a. Nochmals über Stromverteilung in Supraleitern. *Z. Phys.* 120 (1943): 578–587 (received 12 July 1942)

Laue, Max von. 1942/43b. Der Wulffsche Satz für die Gleichgewichtsform von Kristallen. *Z. Cryst.* (A) 105 (1943): 124–133 (received 23 Dec. 1942)

Laue, Max von. 1943a. *Energiesatz und neuere Physik. Vorträge und Schriften. Preußische Akademie der Wissenschaften.* Berlin: de Gruyter

Laue, Max von. 1943b. Energiesatz und neuere Physik. *Research and Progress. Nachrichtenblatt der Deutschen Wissenschaft* 19: 983

Laue, Max von. 1943c. Zu P. v. Groths 100. Geburtstag. *Z. Kristall.* 105: 81

Laue, Max von. 1943d. Ein relativistischer Beweis für das Wiensche Verschiebungsgesetz. *Ann. Phys.* 435: 220 ff

Laue, Max von. 1943e. Nachtrag zu meiner Arbeit ‚Eine Ausgestaltung der Londonschen Theorie der Supraleitung.' *Ann. Phys.* 435: 223 f

Laue, Max von. 1944a. *Materiewellen und ihre Interferenzen*. Physik und Chemie und ihre Anwendung in Einzeldarstellungen no. 7. Leipzig: Akademische Verlagsanstalt

Laue, Max von. 1944b. Besprechung. Zum Weltbild der Physik von Carl Friedrich von Weizsäcker. Leipzig 1943. *Naturw.* 32: 85–87

Laue, Max von. 1944c. Zu Ludwig Boltzmanns 100. Geburtstag. *Forschung und Fortschritte* 20, 4/5: 46 f

Laue, Max von. 1944d. Zum 50. Todestage von Hermann von Helmholtz. *Naturw.* 32: 206 f

Laue, Max von. 1944/52. Mein physikalischer Werdegang. (November 1944) In Hans Hartmann: *Schöpfer des neuen Weltbildes. Große Physiker unserer Zeit.* 1952: 37–70. Bonn: Athenäum-Verlag; revised in Laue 1961: V–XXXIV

Laue, Max von. 1946a. Arnold Berliner. *Naturw.* 33: 257 f

Laue, Max von. 1946b. *Geschichte der Physik.* Geschichte der Wissenschaften, no. 2. Naturwissenschaften. Bonn: Athenäum-Verlag, 2nd ed. 1947

Laue, Max von. 1946c. McLennans Hochfrequenz-Beobachtungen an Supraleitern. *Naturw.* 33: 92

Laue, Max von. 1948. Supraleitung und Hertzsche Schwingungen. *Z. Phys.* 124: 135–143

Laue, Max von. 1950. Aspectos históricos de la supraconductovidad. *Revista de la Union Mathematica Argentina* 24: 109

Laue, Max von. 1961. *Gesammelte Schriften und Vorträge.* Vol. III. Braunschweig: Vieweg

Lemmerich, Jost, ed. 1998. *Lise Meitner – Max von Laue. Briefwechsel 1938–1948.* Berliner Beiträge zur Geschichte der Naturwissenschaften und der Technik no. 22. Berlin: ERS-Verlag

Lemmerich, Jost, ed. 2010. *Bande der Freundschaft: Lise Meitner – Elisabeth Schiemann. Kommentierter Briefwechsel 1911–1947.* Österreichische Akademie der Wissenschaften, Mathematisch-Naturwissenschaftliche Klasse, no. 61. Vienna: Verlag der Österreichischen Akademie der Wissenschaften

Lemmerich, Jost, ed. 2011. *Mein lieber Sohn! Die Briefe Max von Laues an seinen Sohn Theodor in den Vereinigten Staaten von Amerika 1937–1946.* With two contributions by Christian Matthaei.

Berliner Beiträge zur Geschichte der Naturwissenschaften und der Technik no. 33. Berlin, Liebenwalde: ERS-Verlag

London, Fritz. 1937. A new conception of superconductivity. *Nature* 140: 793–797

Meitner, Lise, and Max Delbrück. 1935. *Der Aufbau der Atomkerne. Natürliche und künstliche Kernumwandlungen*. Berlin: Julius Springer

Needham, J., and W. Pagel. 1938. *Background to Modern Science. Ten Lectures at Cambridge Arranged by the History of Science Committee 1936*. Cambridge: Univ. Press

Pechel, Rudolf. 1941. Bei Dr. Leete. *Deutsche Rundschau*, August 1941

Planck, Max. 1913. Neue Bahnen der physikalischen Erkenntnis, p. 65. In M. Planck: *Vorträge und Reden*. Leipzig: Hirzel

Planck, Max. 1939. Max von Laue zum 9. Oktober 1939. *Naturw.* 27: 665 f

Planck, Max. 1940. Versuch einer Synthese zwischen Wellenmechanik und Korpuskularmechanik. *Ann. Phys.* 429: 261–277

Planck, Max. 1943. Auffindung des physikalischen Wirkungsquantums. *Naturw.* 31, nos. 14/15: 153–176

Pufendorf, Astrid von. 2006. *Die Plancks. Eine Familie zwischen Pazifismus und Widerstand*. Berlin: Propyläen-Verlag

Roon, Ger van. 1994. *Widerstand im Dritten Reich*. 6th ed. Munich: Beck

Rückert, Friedrich. 1837. *Sieben Bücher Morgenländischer Sagen und Geschichten*. Stuttgart: Liesching

Rust, Bernhard. 1939. *Deutsche Wissenschaft. Arbeit und Aufgabe*. Leipzig: Hirzel

Shoenberg, David. 1938. *Superconductivity*. Cambridge: Univ. Press

Sommerfeld, Arnold. 1939. Über die Dimensionen der elektromagnetischen Größen. *Ann. Phys.* 428: 335–339

Sommerfeld, Arnold. 2004. *Wissenschaftlicher Briefwechsel 1919–1951*. Vol. 2. Edited by Michael Eckert and Karl Märker. Berlin, Diepholz: Verlag für Geschichte der Naturwissenschaften und Technik

Stern, Otto. 1929. Beugung von Molekularstrahlen am Gitter einer Krystallspaltfläche. *Naturw.* 17: 391

Stumpf, Carl. 1939/40. *Erkenntnislehre*. 2 vols. Leipzig: Barth

Vahlen, Theodor. 1939. Adresse der Preußischen Akademie der Wissenschaften zu Max Plancks 60. Doktorjubiläum am 28. Juni 1939. *Naturw.* 27: 441 f

Weizsäcker, Carl Friedrich von. 1941. Die Physik der Gegenwart und das physikalische Weltbild. *Naturw.* 29: 185–193

Weizsäcker, Carl Friedrich von. 1943. *Zum Weltbild der Physik*. Leipzig: Hirzel

Wintergerst, Erich. 1940. *Die technische Physik des Kraftwagens*. Berlin: Springer

Wohlwill, Emil. 1909/26. *Galilei und sein Kampf für die Copernicanische Lehre*. 2 vols. Hamburg, Leipzig: Voss

Chapter 11
War's End and Farm Hall

Max von Laue wrote another letter to his son on 21 April 1945, again not knowing whether he would ever receive it. He reported that the French were now in Tübingen, Horb, Balingen, and Ebingen and thus Tailfingen was surrounded. Werner Heisenberg had left by bicycle to join his family in Urfeld. Four days later he wrote about the peaceful occupation of Hechingen on April 23rd and described the details:

> House searches followed at the homes of most of the members of these [K.W.] institutes; at my apartment, too, an officer and another man spent three hours looking for papers (on that occasion he also took away that letter to you I mentioned). The Americans treated us curtly and very matter-of-factually, but without rudeness. The officer working here, when Mama addressed him in French, turned out to be a native Swiss from near Neuchâtel. [...]
>
> We hope they [the Americans] will stay here for a longer time or at least leave an institute contingent behind so that we can continue to work under their protection. [...]
>
> The state of affairs is by no means ideal, but the end of the 3rd Reich is inevitable. The main thing is and remains: We're rid of Hitler.[1]

The Americans had already found documents about the German uranium project in the physics institute of the University of Strasbourg when they had occupied that city and knew that no project for the production of a bomb existed yet. Nevertheless, the scientists involved were taken into a kind of "protective custody" and driven away from Hechingen in cars without giving any indication of where they were going. This inspired Laue to draw a comparison to Martin Luther's secret stay at Wartburg Castle in his letter to Theo dated April 28th. He was worried about the precarious situation of the families left behind. He described the trip from Hechingen via ruined Stuttgart and Bruchsal to an unknown destination. Laue mentioned with praise the treatment by the British officers who constantly accompanied them. They also received a brief

---

[1] Laue to Theodore, Hechingen, 21 and 25 Apr. 1945. In Lemmerich (2011: 218, 220 f.), here p. 221. On Hechingen see also Laue (1944/52): "Even the efforts to defend Hechingen, childish as they were, would have meant great danger if it really had come to conflict. Fortunately, insight into the frivolity of such hopeless resistance won out, and on 23 April 1945, despite all the military measures associated with such a 'conquest,' the inhabitants watched the French and Red Spanish troops entering without a struggle with a sense of relief. All of this was no surprise to any one."

© Basilisken-Presse, Natur+Text GmbH 2022
J. Lemmerich, *Max von Laue*, Springer Biographies,
https://doi.org/10.1007/978-3-030-94699-9_11

Fig. 11.1 Farm Hall, the country estate near the city of Cambridge, where the ten physicists were held from 3 July 1945 to 3 January 1946. *Source* Archive of the Max Planck Society, Berlin-Dahlem

visit from Samuel Abraham Goudsmit. Laue was confined together with Otto Hahn, Carl Friedrich von Weizsäcker, Karl Wirtz, Horst Kosching, and Erich Bagge. They were joined at Versailles by Werner Heisenberg, Kurt Diebner, and Paul Harteck, and later by Walther Gerlach. They were being "detained at His Majesty's pleasure."

Among these scientists Otto Hahn and Max von Laue assumed a special role. They were allowed, under guard, of course, to visit the cathedral in Reims, the palace in Versailles, and even briefly visit the unscathed city of Paris. Laue wrote: "I can't say how many beautiful things, and above all the interior of Notre Dame—That was impressive!!"[2] In his letter to Theo dated May 26th from Le Vésinet near Paris, Laue thought: "Everyone but me collaborated on the 'uranium machine,' and that's probably why we were detained. I seem to have been taken along by mistake. Or do they suspect me of some particularly deep secret, since they know nothing at all about my relevant activities? If so, the present state of affairs could drag on for quite a long time; as no one is able to expose a secret that doesn't exist (Lemmerich 2011: 225)." The final destination of their journey was a remote country estate called Farm Hall north of Cambridge (Fig. 11.1). During the sojourn that followed, they were allowed four times to write strictly censored letters to their relatives. Therefore, their content cannot be counted as a reliable portrayal of Max von Laue's real circumstances and

---

[2] Laue to Theodore, Le Vésinet near Paris (on his way to internment in England), 26 May 1945. In Lemmerich (2011: 226). Laue wrote this message "in stock" when he had already been picked up by the British from Hechingen.

mental state. Laue wrote to his son at length about the air raids and their destruction, the fates of his relatives and friends, the events in Berlin with the persecution of his Jewish friends. These letters were several pages long and document the pent-up need to render an account of what he had experienced.[3]

The letter indicating the location Huntingdon dated August 7th was important: "Dear Theodore, This letter refers to a globally historic event of incalculable significance. Although I was far away from Hiroshima when the uranium bomb exploded there, I experienced the news of it in the midst of colleagues who have worked on the uranium problem for years and who, like me, all were living in quite exceptional circumstances, namely 'detained at His Majesty's pleasure' (Lemmerich 2011: 241)." Laue described their transfer by airplane to England and the good accommodations, the excellent board, and ample access to newspapers, books, and English radio broadcasts. There was also a playing field behind the house for sports.

> At 19:45 hours yesterday began, as usual, our dinner, which is regularly attended by the two English officers guarding us, Major [Thomas Hardwick] Rittner and Captain [Patrick Lennox] Brodie. [...]
>
> Earlier the Major had already insinuated something to my colleague Hahn about a radio report that the Americans had used an atomic bomb. At table he added to this announcement and a lively discussion immediately arose, of course. We weren't quite willing to believe it. Some voices said that if there was any truth to it at all, the name 'atomic bomb' meant something different from what we understand by it; in any case, this thing could have nothing to do with uranium fission.
>
> But then we listened to the news on English radio at 21 hours. And there it was stated in no uncertain terms that the English and the Americans had worked out uranium fission for the construction of a bomb in years of joint, laborious, and extremely expensive developmental research (p. 242).

It might even have been a good thing that Otto Hahn's name had not been mentioned in this context, he thought.

> According to one such [newspaper] commentary, headlined 'A Jewess found the clue,' Lise Meitner was accordingly the discoverer of uranium fission.
>
> This announcement moved the physicists assembled here very deeply, of course. Although I—if I may start with myself—was relatively uninvolved; since I had only ever played the role of an observer in the entire research on uranium fission, whom the participants sometimes, but by no means always, kept somewhat up to date. Even Otto Hahn, about whose mood Major Rittner was seriously worried, remained quite calm, saying only that he was pleased to be uninvolved in the construction of such a murderous weapon. But Walther Gerlach was very agitated, who as a former 'Plenipotentiary of the Reich Marshal for Nuclear Physics,' felt something like a defeated commander and, moreover, was embarrassingly affected by a few careless remarks made by one of the younger ones among us (ibid.).

This was followed by a lengthy passage about why a bomb had not been developed in Germany. Reading it, one gets the impression that this is a version that they had agreed upon amongst themselves. After their return to Germany, the discussion

---

[3] Laue to Theodore, Le Vésinet near Paris, 29 May 1945, and Château de Facqueval, 21, 22 and 27 Jun. 1945. In Lemmerich (2011: 219–240).

among those who had been involved was apparently continued. "All our uranium research was directed toward the creation of a uranium machine as a source of energy, firstly because nobody believed in the feasibility of a bomb in the foreseeable future, secondly because basically nobody among us wanted to place such a weapon in Hitler's hands (Walker and Rechenberg 1992; Hoffmann 1993)."[4]

On August 9th, the second atomic bomb—a plutonium bomb—was dropped on Nagasaki. The reaction to this event found no mention in Max von Laue's subsequent letters. The plutonium bomb (comparable to the explosive power of the uranium-235 bomb) mentioned by Heisenberg on 26 February 1942 and then very likely again during his lecture in Harnack House on 2 June 1942, was evidently deliberately not discussed by the internees, or else the topic was intentionally omitted from the published transcripts of their conversations.

Based on the incomplete information from newspaper reports and from a visit by Charles G. Darwin, the director of the British metrological institution, the National Physical Laboratory, they had learned that Franz Simon and Rudolf Peierls had originally instigated the construction of the bomb, Laue wrote to his son from Huntingdon on August 19: "This is how the history of the atomic bomb appears, based on everything that we have been able to gather about it. It was thus the impassioned hatred of Hitler by those two emigrés that set everything in motion. In this Hitler also proved his worth as an essential element of that ever evil-seeking force—and thereby ever eliciting good (Lemmerich 2011: 245)." Theodore was probably more skeptical about this; he probably couldn't quite share his father's sense of vindication. It had not been Franz Simon but Otto Robert Frisch, however, who together with Rudolf Peierls had assessed the feasibility of building an atomic bomb in a memorandum of a sort, had recognized it could be done, and presumed that similar ideas were surely being considered in the "Third Reich" as well. Both saw the danger that Hitler, in possession of such a bomb, would win the war. They also already pointed out that contamination from the radioactive substances would prevent any safe reentry into the impact area for decades to come. It had not been hatred but fear of a German atomic bomb that had led to the development of the atomic bomb by British and American scientists. The fact that it was an extremely interesting scientific problem was an additional factor.

---

[4] On 8 Aug. 1945 these German scientists wrote a kind of declaration, setting forth in five very brief points the research conducted in Germany on uranium nuclear fission: "By comparison, the preconditions for the production of a bomb within the context of the technical means available in Germany do not appear to have been given at that time." (DMA, 1976-20/8). The original audio recordings of the conversation were not published. It therefore cannot be established whether the subsequent translation into English follows the literal sense exactly. Laue is quoted as making only unsubstantial, brief contributions, see Hinsley (1996: 122) and TNA, TATC Papers.

It is clear from meeting minutes that a small group of British politicians and scientists became aware that Germans were actively working toward developing an atomic bomb in the spring of 1941. German interest in heavy water from Norsk Hydro, which was producing 120 kg a month, indicated a reactor designed to breed plutonium. A report dated 23 Apr. 1942 stated: "Since recent experiments have confirmed that element 93 would be as good as U_{235} for military purposes, and since this element is best prepared in systems involving the use of heavy water, …". See also Hinsley 1996 and CAC, IV 12/2, Chadwick Papers.

There is no record left of the debates between the main participants, Heisenberg and Weizsäcker, and the other physicists in the days that followed. But they all continued to be strictly isolated from the rest of the world. Years later, Laue wrote to Paul Rosbaud in a letter dated 14 April 1959 from Berlin: "The reading then gradually evolved out of table conversations that German nuclear physicists had not wanted to have the atomic bomb at all, be it because they deemed it impossible within the anticipated duration of the war, or because they did not want it generally. These discussions were headed by Weizsäcker (UAF, Laue papers, 3.5)."

Writing to his son was apparently a form of relaxation for Laue. He probably had no choice but may also not even have intended to send the letters out from Farm Hall. In any event, they reported about all kinds of circumstances involving his family and circle of friends. At last, on 28 August 1945, the first mail arrived from Germany with news from their families. Magda had enclosed a kind of diary with her letter, 18 double-sided sheets of paper. Max von Laue learned that Hilde's husband, his son-in-law, had returned without having suffered confinement. Soon a palpable uneasiness spread among "His Majesty's guests." The oath they had taken not to flee was, of course, a barrier. The English head of the atomic bomb project, Wallace A. Akers, asked Patrick Blackett, an English physicist, for help on September 5th: "The reason I am writing to you now, at a time when you are on holiday, is because these German 'visitors' have suddenly become extremely restive and today Heisenberg said that he thought he would withdraw his parole. As they were not very strictly guarded, it is probable that Heisenberg could get away from the house, if he set his mind in this."[5] First Charles Darwin had been sent to Farm Hall, but that reassurance did not last long. Consideration was also given to offering some Germans scientific positions in England. Then Patrick Blackett visited the internees and first had a long conversation with Heisenberg about the future of physics in Germany upon the return of the group. Heisenberg never mentioned Laue on that occasion. Afterwards Heisenberg drew up two outlines exclusively dealing with research in the field of nuclear physics. Blackett also spoke with Otto Hahn, Max von Laue, and Carl Friedrich von Weizsäcker.

Blackett: "You are very keen to get back to Germany?"

"Yes," was Laue's reply, and Blackett retorted: "It ought to happen soon. It is all very unfortunate. Great efforts have been made over here by the people concerned to get everything cleared up, but, of course, we have had a very difficult position. We have not, in fact, been able to do what we want and I think, they tried to make it as comfortable as you can be under conditions which we did not invent."

To Laue's statement that some younger physicists wished to go to America or England, he answered evasively, that first they should return home. Laue

---

[5] NARA, CAB 126/333. Also in Lemmerich (2011: 250). Patrick Blackett wrote down his recollections of this time. They reveal that he had been briefed at the Intelligence Dept. in London prior to his visit at Farm Hall (RSA, P.M.S., Blackett Papers, A 10).

then asked: "Do you think it is possible for me and my wife to go to our son at Princeton?"

Stalling was again the only response. To Blackett's question, "If you went back to Germany, what would you do?" Laue replied: "I would go to the seat of the 'Kaiser Wilhelm Institut.'" Blackett replied: "This is now moved to Hechingen."

To which Laue replied, "Only a part of it has moved to Hechingen and I don't think it is possible to go to Berlin yet."

Blackett then added: "I feel that you are one of the people who should, in my own personal view, be back there helping to re-build academic and scientific life. Your reputation is very high as a very wise man who has taken a very good line and you are respected enormously. I remember talking about your attitude to the Nazi-movement—do you remember—right back in '38, I feel you ought to be one of the people back there, trying to get the right views and re-building things. That you would like to do?"

Laue answered: "Yes."

Excerpt of a recorded conversation between the English physicist Patrick Blackett and Max von Laue (Dialogue in the original English translation, NARA, CAB 126/333: 16, also cited in Lemmerich 2011: 251 f.).

Weizsäcker: "There is just one snag: if you don't get to Hechingen while Laue does, the institute will have the most incompetent leadership imaginable."

Heisenberg: "In fact, Wirtz would have the actual control. We would have to think over carefully how to work that. It will be difficult officially to take away Laue's authority. I might well do it in a friendly manner, saying: 'You are the official head, but I recommend that, for the time being, you let Wirtz do everything.'".

Weizsäcker: "I would even undertake a little intrigue to see that Laue doesn't get back either until you get back, but I should say that is rather difficult to bring about, as one should not do too much in that direction."

Heisenberg: "As Laue is the only one who would gladly go to America, it might be possible to do it in that way."

Excerpt from the transcript of a conversation between Heisenberg and Weizsäcker about Max von Laue and the eventual return to Hechingen (Original English translation, ATA, CAB 126/333, also cited in Lemmerich 2011: 253).

The original German transcript of the conversation between Heisenberg and Weizsäcker about Laue is not available. Thus it is not possible to judge whether Weizsäcker's opinion of Laue's ability to lead the Institute of Physics was in fact as harsh as it sounds. A few years later, Laue would very successfully head the much larger Fritz Haber Institute of the Max Planck Society in Berlin. Both parties in this conversation evidently were unwilling to acknowledge the importance of the stance that Laue had taken during the Nazi regime to readmission of German physicists into the international community.

Laue attentively followed the news about the British government's promotion of physics, including theoretical physics. To his son he remarked in a letter from September 22nd: "When I began my studies, physics, especially theoretical physics, was considered by many to be a somewhat better form of indolence: I had to suffer all my life for not having become a lieutenant, as my father had wanted. And now here comes this importance for physics! Of course, this also has its risks. In former times, when physics was still a 'bread-and-butter art,' only those young people entered it who were dedicated to it heart and soul, who more or less 'couldn't help it.' Thus the level of physicists was very high (Lemmerich 2011: 255)."

A few British scientists who were privy to the internments of Hahn, Heisenberg, and Laue received the three in strict secrecy for a discussion, not at the Royal Society, which might have been noticed by the press, but at the Royal Institution of Great Britain on Albemarle Street. The subjects were the Kaiser Wilhelm Society and German research. The war was over and life in the countries of the victorious powers was gradually returning to normal. So they also wanted to celebrate the 300th birthday of Isaac Newton and the 50th anniversary of the discovery of X-rays with guests, naturally also including Laue. But that is where difficulties began! Nobody was supposed to know that Laue was in England. Simply releasing him for the celebration would very likely lead to the press finding out about Laue's sojourn at Farm Hall. A more adventurous scheme was to bring him to Hamburg first and then back to England for the festivities. But one agency was unwilling to approve it. The English then reached the pragmatic decision to postpone the celebration for a year.[6]

A return home was still out of the question, indeed, it was said that it could still take years. Nevertheless, in his letters to Theo, Laue was careful not to write much about the oppressive situation—perhaps he was not permitted to do so, as the letters still were being "censored." In the first half of November 1945, Max von Laue wrote his son a very detailed description of the life he was living and—with critical commentary—about the politics of the time in Strasbourg and Alsace-Lorraine. Just two days later he wrote him how he and his wife had experienced the beginning of the war in 1914.[7]

On November 16th, the internees celebrated the news that Otto Hahn had been conferred the Nobel prize in chemistry for 1944. Max von Laue, holding the address at table, recalled having once asked Hahn for a portrait photo. Laue had then written on the back of it not only the date of his birth but also Theodor Fontane's praise for Adolf von Menzel: "Giftedness, who does not have it? Talents, toys for children! Only earnestness makes the man, only diligence the genius." Laue added: "When I

[6] NARA, 70, 800/566, TOP SECRET. There Laue mentions the invitation to the Röntgen celebration, which was later postponed, cf. Lemmerich (2011: 267); see also TNA, (War Office) WO 22/22/17 on the erroneous internment of Max von Laue in England.

[7] Laue to Theodore, Huntingdon, 11, 12, 14 and 18 Nov. 1945. In Lemmerich (2011: 267–275).

survey the scholars whose careers I believe I know somewhat, I find none whom it would suit as well as you (Lemmerich 2011: 273)." Afterwards he proposed a toast to Otto and Edith Hahn. He told Theo about this along with some other background information about the award in his letter dated November 18th. (pp. 273 f.) Strangely enough, he did not mention the award of the Nobel prize in physics to Wolfgang Pauli for having formulated the exclusion principle in quantum theory.

From Max von Laue's correspondence one cannot conclude whether there had been any discussions at Farm Hall about the past years under the National Socialist regime of terror or about the concentration camps. Neither do we know about any discussion of the questions of guilt or moral complicity in the crimes. The published English tape recording transcripts contain no such passages either.

On October 18th, the war crimes tribunals at the International Court of Justice began in Nuremberg, first against the 22 main defendants. Further trials followed in 1946 and later. Carl Friedrich von Weizsäcker's father Ernst was also among the defendants in one of the subsequent trials. Max von Laue's letters made no mention of this, but perhaps it was forbidden for him to do so. Only in the letter of November 25th did he record for his son a conversation with Major Rittner on the "question of guilt":

> He pointed out, albeit with his singularly typical great tact, that the non-Nazis among the Germans had also been complicit in Hitler's war. I replied that the worst thing one could reproach them for was for following the motto '*Right or wrong, my country*,' and that this hardly originated in Germany. I think he took it rather badly; but you have to risk that when speaking the truth. And it actually is true that the Nazi spirit is everywhere to be found in the world today. That motto reveals in classic brevity the deep moral depravity of such exaggerated nationalism (I pass over here the question of the extent to which the Germans' participation in the war was in fact voluntary). […]
>
> '*Right or wrong, my country*.' That is the spirit I had to struggle against from my youth, first with my father, then to an even greater extent with my stepmother. Whereby, I do concede, instead of *country* for homeland, it would be better to put: military standing. According to the genuine military conception, there is no crime at all that cannot be legitimized by the assessment of the officer caste. For this we have the concept of an officer's honor, that honor which supercedes all else and which no civilian can ever share.—When I had really become acquainted with it, I flung it back at the feet of those high and mighty gentlemen (pp. 275 f., emphasis added to indicate original English).

Max von Laue received a letter from Theo dated November 4th, which he answered in delight from Huntington on December 9th: "I would like to tell you again that I was pleased beyond words about your letter of 4 Nov. 1945 and Lise Meitner's additions to it of 13 Nov. 1945; especially about the news that you have earned your doctorate. And thus you have completed your studies in a regular manner. I was always silently worried that you might not be able to make much progress professionally over there, in a foreign country (pp. 280 f.)." As a father, he was pleased that his son had already held a lecture. Looking back on All Souls' Day, he told his son about people he had seen leave this world, including his mother and Arnold Berliner. In the last section he mentioned the execution of Max Planck's son Erwin (pp. 280–285).

Shortly before Christmas 1945, the internees were informed that they would be returning to Germany at the beginning of the year.

References

Hinsley, Francis H, ed. 1996. *British Intelligence in the Second World War*, Vol. II. London: Her Majesty's Stationery Office

Hoffmann, Dieter, ed. 1993. *Operation Epsilon. Die Farm-Hall-Protokolle oder Die Angst der Alliierten vor der Deutschen Atombombe.* Berlin: Rowohlt

Laue, Max von. 1944/52. Mein physikalischer Werdegang. (November 1944) In Hans Hartmann: *Schöpfer des neuen Weltbildes. Große Physiker unserer Zeit.* 1952: 37–70. Bonn: Athenäum-Verlag

Lemmerich, Jost, ed. 2011. *Mein lieber Sohn! Die Briefe Max von Laues an seinen Sohn Theodor in den Vereinigten Staaten von Amerika 1937–1946.* With two contributions by Christian Matthaei. Berliner Beiträge zur Geschichte der Naturwissenschaften und der Technik, no. 33. Berlin, Liebenwalde: ERS-Verlag

Walker, Mark, and Hellmut Rechenberg. 1992. Farm Hall tapes: On the uranium bomb. Werner Heisenberg's intercepted lecture at Farm Hall on August 14, 1945. *Phys. Bl.* 12: 994–1001.

Chapter 12
Back in Germany

The returnees from Farm Hall were not taken to Berlin, but to Westphalia in the British zone of Germany. The Germany they had been forced to leave and to which they were returning had changed. It had surrendered unconditionally. The National Socialist regime of terror had collapsed. What many knew, some out of fear only suspected, others did not want to know—the horrific murder of millions in the concentration camps—had indeed taken place. Collective guilt befell the Germans.

The four Allies determined the four zones into which Germany was partitioned: in the north, the British zone with the industrial Ruhr area and Hamburg, Göttingen, and Hanover; in central Germany, the American zone; and in the south (thus also Hechingen and Tailfingen), the French zone. The east became the Soviet zone. Soviet troops had conquered Berlin in April 1945. The Soviets largely dismantled the industrial plants and research laboratories located in the later Western parts of the city. In July, Berlin was then divided into four sectors. The Soviets got the largely destroyed central area with the Friedrich Wilhelm University and the eastern sector. The less devastated southern area including the Dahlem suburb, where the Kaiser Wilhelm Institutes of Physics, Chemistry, and Physical Chemistry among others were located, as well as Zehlendorf went to the Americans. The western parts of the city with the polytechnic (*Technische Hochschule*) and the bureau of standards (*Physikalisch-Technische Reichsanstalt*), to the British, and the French sector comprised the northwestern regions. The borders between these sectors initially existed only administratively. The German agencies only had subordinate decision-making powers.

The many millions of party comrades were—by necessity—to be "denazified." Guilt had to be distinguished from mere obedience. At first, this was done by the occupying powers; then this task was handed over to German courts, which reached very different verdicts. But if a specialist was important to the Allies, his political past was suddenly insignificant and he was free to accept an "invitation" to cooperate.

The Allies redistributed "Greater Germany." France got the Saarland back; Poland, as compensation for losing eastern Poland to the Soviet Union, received parts of East Prussia, Pomerania, and Silesia. The German inhabitants of these areas—including

© Basilisken-Presse, Natur+Text GmbH 2022
J. Lemmerich, *Max von Laue*, Springer Biographies,
https://doi.org/10.1007/978-3-030-94699-9_12

Germans who had been living in Poland for a very long time—were expelled and fled westwards; there were over six million of them. It took a long time for them to feel at home again elsewhere in Germany. Their way of life, their religion, and their dialect often stood in the way. But this and other major waves of immigration caused an increase in the population in the three western zones, because people were also moving westwards away from the Soviet-occupied zone, and it was soon to become an essential factor in the country's reconstruction (Jäckel 1996: 242).

The Berlin Declaration by the victorious powers of 5 June 1945 included the following passage: "There is no central Government or authority in Germany capable of accepting responsibility for the maintenance of order, the administration of the country and compliance with the requirements of the victorious Powers."[1] And the Allied Control Council in Berlin issued Law No. 25 on 29 April 1946:

Regulation and Control of Scientific Research

In order to prevent scientific research for military purposes and its practical application for such purposes, and in order to control it in other fields where it might create a potential for war, and to direct it into peaceful channels, the Control Council has passed the following law:

Article I

All technical military organizations are hereby disbanded and prohibited. Buildings and equipment of a purely military character shall be destroyed or removed. Buildings and equipment whose peacetime use is possible may be made useful for such purposes by permission of the Military Government.

Article II

1.   Applied scientific research shall be prohibited in fields which: (a) are purely or essentially military in nature; (b) are specifically listed in Schedule "A" attached hereto.

2.   Applied scientific research in any of the fields specifically listed in Schedule "B" attached hereto shall be permitted only with the prior written approval of the Zone Commander in whose Zone the research institute is located."[2]

Applied nuclear physics was also listed under "A." This meant that throughout Germany many industrial enterprises and sectors of the chemical industry were prohibited from either manufacturing or conducting research in such fields. Compliance was monitored in unannounced inspections. Most of the armament factories were dismantled.

"His Majesty's guests" arrived in the British zone by plane on 3 January 1946 (Fig. 12.1). They were taken to Alswede, an undamaged village in Westphalia. They found accommodations in a requisitioned department store. There were no more restrictions on corresponding with their relatives and friends. Max von Laue's first letter to Theo from Alswede on January 13th described his new surroundings. Laue also mentioned two church services. Only gradually could the events of the

---

[1] Declaration Regarding the Defeat of Germany and the Assumption of Supreme Authority with Respect to Germany", 5 June 1945. Online: https://usa.usembassy.de/etexts/decla3945.htm [last accessed: 21 Oct. 2021].

[2] Allied Control Council, Law No. 25. Online: http://www.verfassungen.de/de45-49/kr-gesetz25. htm [last accessed 21 Oct. 2021].

Fig. 12.1 Werner Heisenberg, Max von Laue, and Otto Hahn (*from left to right*) shortly after their return to Germany from internment in January 1946. *Source* Archive of the Max Planck Society, Berlin-Dahlem

war be viewed in retrospect. True liberation from the experiences and events of the years 1933 to 1945 only came much later though and probably never quite completely. There were no comments about the internment and his relationship with his colleagues at Farm Hall. The events of the day were the main focus (Lemmerich 2011: 286).

In his letter dated 20 January 1946, Laue reported to his son: "Here one doesn't see much of the misery of the war, although some of the refugees from East Germany who have been housed here look pitiful; but one often encounters the native residents, who never suffered any need. But from the accounts given us, for instance, by one of our orderlies, a now released former prisoner of war, of his journey home, it must look dreadful at railway stations and other heavily frequented junctions (p. 288)."

Three days later Otto Hahn, Paul Harteck, and Max von Laue were able to travel to a celebration in Hamburg, with English escort. It was held in commemoration of Röntgen's lecture in Würzburg on the discovery of the rays named after him, and Laue delivered a talk on his discovery. On the way there they saw the many ruins in Hanover and Minden and then the remains of heavily bombarded Hamburg.[3] Laue also met the Hamburg historian of science Hans Schimank there, who told him that only one room in their family apartment could be heated. The hardship of the times, the catastrophe they had just survived, frequently guided the conversation

---

[3] Laue to Theodore, Alswede, undated. In Lemmerich (2011: 291). The temperature in the lecture hall was 14 °C.

to the question: How could this have happened? This subject also arose with Laue and Schimank, as Laue's letter to his son from April 6th, not long after his visit to Hamburg, shows. "The misfortune of humanity seems to me (we talked about this in January) ultimately to be based on the Christian churches (both of them!) not having banished the belief in miracles from their professed [doctrines] about the spiritual attitude, etc., of modern man reshaped by scientific knowledge. Thus the majority of people look upon Christianity as grounded in a belief in miracles, and is inclined to throw that overboard. And this belief in miracles has indeed been thrown overboard by every modern person, even if he doesn't admit it to himself (Göttingen, handwritten, GNT, Schimank papers)." And Laue set the successes of natural science, Newton's laws of planetary motion, against this belief in miracles.

> If Christianity is not neatly sundered from it, however, the masses will lose their religion, their reverence for something higher, and will become those 'masses' in the sense of Gustave Le Bon, those 'masses' who are feared with good reason. I have no hope that such an ailing humankind can be cured now by political organizations, however well-intentioned and shrewdly devised. At best, one can thus postpone the fatal crisis. The cure succeeds only by clearing the way to Christianity again for people, for the great majority, by abolition, by abolition of any 'miracle' by dictum. [...] *That* would be the task of the churches at this time; so far I don't see that they have understood it as such (ibid., emphasis added).

Among the returnees from Farm Hall, Paul Harteck was the first to be allowed to accept a professorship, in Hamburg, and Kurt Diebner a position in industry. Both left on 28 January 1946. Walther Gerlach, who had been a professor in Munich, had to stay within the British zone. He accepted a "guest professorship" in Bonn, which was part of the British zone. It became clear to German nuclear physicists very soon that permission would not be granted to operate research equipment such as high-voltage generators and neutron sources. How should things continue? In which areas of physics could research be carried out in future?

His wartime experiences and its terrible end with so many deaths and suicides during the period of collapse were the subject of Laue's letter to Theo on February 3rd. Among their friends and acquaintances, he counted Paul Kirchberger, Wolfgang Windelband and his wife, Peter Paul Koch in Hamburg, Otto Hönigschmid in Munich, and in Berlin the chemist Hermann Leuchs and seven of his assistants (Lemmerich 2011: 290).

At the beginning of that month Otto Hahn and Werner Heisenberg were granted permission to travel to Göttingen. They wanted to make the preparations for the transferral of the Kaiser Wilhelm Institutes from Hechingen and Tailfingen to Göttingen and the relocation of their families there, as Laue wrote to Theo from Alswede on February 17th. About his future in Göttingen, reunited with his wife, he remarked: "It'll be especially hard for Mama. She won't be able to take Hilde and her baby along. And if it should come to Kurt having to separate from Hilde again for a longer period of time for the sake of a job, it's beyond me how that should work out (p. 293)."

On March 13th, Max von Laue also moved to Göttingen, where a certain order had already established itself, as the university had reopened since September 1945. The British occupation forces still largely determined developments, however. Laue, like Otto Hahn and Werner Heisenberg, was assigned a room of his own with a bedstead and a sack of straw in the now empty Aerodynamic Design Testing Station (AVA). The others moved into double rooms. There was also a guest room where Magda could stay if she came from Hechingen. As he informed his son on March 18th, he had the Physics Institute with its library diagonally across the street from him (p. 295). For decades he, just as Hahn and Heisenberg, had lived amongst his family, had regularly exchanged invitations with their friends. Almost an entire year had passed since then and the future was still very uncertain. In that same letter Laue informed Theo about having visited Max Planck and his wife that day, who were living with Planck's niece Hildegard Seidel and her husband in Göttingen.

"Planck himself has aged a lot. His wife thinks also mentally. I couldn't see that. His demeanor is still quite quick. But of course his head of hair is thinned out more than before. He suffers, I am told here, from protein deficiency. So I brought him three eggs from Alswede, which he was delighted about (p. 296)." This was followed by a description for Theo of Planck's experiences during the last days of the war in Rogätz and his transferral to Göttingen by the Americans at the initiative of Robert Pohl (ibid., cf. Hermann 1973: 114). On March 21st, Mrs. Laue took the long train ride to Göttingen to stay for a fortnight, and the couple was able to discuss the move.

Laue had received the correction proofs to his small volume 'History of Physics' from his publisher. The complicated postal conditions and controls delayed the return shipment with his corrections, however, and so the first edition still contained some errors. An important statement about the discovery of nuclear fission is made in it: "Physically, it was the greatest experiment ever conducted by man. It was the brilliant confirmation of a bold scientific prediction. How profoundly the consequences will reshape mankind externally and internally cannot yet be estimated. It is possible that later historians will regard the above-mentioned dates [the discovery of nuclear fission and the atomic bombings] as the most important of a whole epoch (Laue 1946a)."

Multifarious activities in science were still essential for Max von Laue. On March 28th, he wrote from Göttingen to Springer Publishers, mentioning a commitment he had made with Paul Rosbaud in early 1945 about a book on the theory of superconductivity. He now had completed a manuscript of eighty pages with twenty illustrations. By the beginning of July the first galley proofs were ready. After approval by the British authorities, 1572 copies were printed by the end of the year. Laue requested plenty of free copies in exchange for an author's honorarium (UAS; Laue 1947a).

Ernst Telschow, the general secretary of the Kaiser Wilhelm Society (KWG), who had been a member of the Nazi party since 1933, had evidently known at an early stage that after the end of the war the western territories of Germany would be occupied by the Western allies, because he had moved the society's general administration to Göttingen before the end of the war. The last president of the society, Albert Vögler, had taken his own life at the end of the war. The KWG was thus headless. At 87 years of age Max Planck agreed to assume the office of presiding secretary again. He soon wrote to Otto Hahn—still unaware that Hahn was at Farm Hall at the time—that, with the approval of some directors, he had proposed Hahn for the position of president. Hesitantly Hahn had accepted the office, while still in England.[4]

According to the Allied Control Council law, the KWG fell under law no. 25, but the four Allies treated such decisions differently. In Berlin, after some back and forth, the city's magistrate appointed the physico-chemist Robert Havemann, an opponent of National Socialism, as head of the Kaiser Wilhelm Society in July 1945 (Hoffmann 1991). At the same time, all powers of the former president and secretary general were revoked. Although this met with protest from some former Berlin KWI members, it did not change the order (Steinhauser et al. 2011: 140).

Some of the English officers who were responsible for monitoring and reorganizing the universities and other scientific institutions, such as the KWG, in the British zone had studied various fields in the natural sciences in Germany prior to 1933. Colonel Bertie Blount and Ronald Fraser were among these.[5] They were very committed to preserving the Kaiser Wilhelm Society. But this was of little help, as at a meeting of the Allied Control Council in the spring of 1946, the Americans demanded its dissolution which the Soviets and French thereupon approved. According to the rules the British then ought to have agreed to it as well. A solution arose unexpectedly, though, because during a visit to England by Colonel Blount, Sir Henry Dale suggested that the KWG be renamed 'Max Planck Society.' That was the way out! But some members of the KWG, for example, Werner Heisenberg, did not like the idea. Otto Hahn then turned to Sir Henry Dale in July to ask him to support the retention of the name Kaiser-Wilhelm-Gesellschaft. But that was to no avail. Finally, in order to avert the dissolution of the society, the members reluctantly agreed, and the negotiations about formulating the new statutes began.[6]

Max von Laue had studied in Göttingen for a few semesters more than four decades ago, but none of his academic teachers were still alive. Professor David Hilbert had also died in 1943. Richard Becker had been "punitively transferred" from the Berlin polytechnic to Göttingen in 1936 to replace Max Born. A second chair for the subject of theoretical physics could not be created for Laue. Göttingen was a small town back

---

[4] On Ernst Telschow: AMPG, II rep. 1A, staff file no. 12, Telschow. On Ernst Telschow and the KWG, see also Przyrembel (2004); Planck to Hahn, 25 Jul. 1945. AMPG, II rep. 14 no. 3318 fol. 43; Meitner to Hahn, Stockholm, 20 Oct. 1946. AMPG, III rep. 14A no. 4897.

[5] The physicist Ronald Fraser was not unknown to Laue. In 1930, Fraser had asked Arnold Sommerfeld for an "opinion" about his scientific achievements. Sommerfeld informed Laue about this in a letter from Munich, on 2 Jul. 1930: "The canal-ray experiment by Mr. Fraser particularly interested me, since it affords direct confirmation of the new wave-mechanical views." AMPG, III rep. 50 no. 2393.

[6] Laue to Meitner, Göttingen, 21 Sep. 1946. In Lemmerich (1998: 461 f.); Hahn to Sir Henry Dale, Göttingen, 11 Jul. 1946, draft. AMPG, III rep. 53, Otto Hahn papers.

Laue mentioned the renaming issue in a letter to Meissner, Göttingen, 30 Jul. 1946. DMA, NL 045 006/1: "Incidentally, I also saw Niels Bohr and Vilhelm Bjerknes from Oslo in the second week (after the 15th). Both of them heftily urged that the Kaiser Wilhelm Society should change its name, and hinted that something could then be done for it abroad. Lise Meitner […] also sounded the same bugel. Here in Göttingen, the suggestion is not exactly warmly received. However, I am of the opinion that the change of name is an absolute necessity and have proposed the name 'Harnack Society.'"

then. You could almost say, "Everybody knew each other." There were only a few jobs in industry, but many city and state administrations. The university had long been world-renowned. This tradition had been continued in the 1920s to 1933 by Max Born and the Nobel laureate James Franck in the subject of theoretical physics, by Robert Pohl in experimental physics, and by Richard Courant in mathematics. The university with its professors, assistants, and international student body, as well as many visiting scholars, was the intellectual center of the town. Despite the high intellectual level there, however, the ideas of National Socialism had soon found a growing following in the 1920s, not only among the teaching assistants but also among students and the general population. In the final years before 1933, the Nazi party received more than fifty percent of the vote in elections in several instances. Rudolf Mentzel had also studied chemistry at Göttingen and had joined the NSDAP there (Rasch 1994: 96 ff.). Thus, the Nazi's "grasp for power" had been cheered on by many Göttingen residents. A "book burning" event was also staged in their town in 1933. The "nationalistic purging" of the university, the emigrations of Professors Max Born and James Franck as well as Richard Courant wiped out its international reputation. As in the "Third Reich" generally, many civil servants in Göttingen only applied for membership in the NSDAP in order to retain their positions. In 1945 these so-called 'fellow travelers' (*Mitläufer*) were dismissed at first; they needed to be denazified. That was an arduous process. Witnesses were summoned to the hearings, which led to tensions among the population (Becker et al. 1998: 229, 552, 709).

On 3 May 1946, Laue wrote to his son from Göttingen about having visited a relative near Wolfenbüttel and told him about the family's hard fate: "And it's unthinkable what they have all been through, including the children! When I once took 7-year-old Elke for a walk across the fields, she said: 'This is just like being on the run.' So this child already experienced fleeing over hill and dale! Her one-year elder brother Jochen crossed the border in a different way—namely, on the train, hidden under the coats of the other passengers (Lemmerich 2011: 298)." In a postscript, he added: "Right now I am reading a book called 'The German Catastrophe' by the 83-year-old Friedrich Meinecke [...]. This text corresponds so completely to my views about the background to what I have witnessed in grand politics since my youth that I recommend it to you most warmly (p. 299)."[7]

---

[7] Meineke (1946); Laue had chapter VI: 'Militarism and Hitlerism' photocopied, as print runs were low at the time due to the scarcity of paper, and there was much interest in the book. Meinecke, born in 1862, had studied history and German language and literature. From 1887 to 1901 he was archivist in the Prussian Secret State Archives in Berlin, then professor of history in Strasbourg, Freiburg im Breisgau, and from 1914 in Berlin. Chapter VI began with a historical retrospective: "As far as we can see, no attention has ever been paid to the fact that the modern technically utilitarian spirit, we have just been thinking about in connection with the Hitler view of humanity, already had a 1½ to 2 centuries older precursor in Prussian militarism, as was created by Friedrich Wilhelm I [1688–1740]. And this precursor, about which we have already said a few things, imprinted itself extraordinarily firmly and deeply in people's minds. The Prussian officer type came into being, which stood out sharply against both the officer type in other countries and other professional types in his own country. The decisive thing was that a certain rational idea gained absolute dominion over the essence of being human—the idea of unconditional devotion to the profession, looking

Max von Laue was still deputy director of the KWI of Physics as well as professor emeritus at the University of Berlin. He now began to receive a pension from the state of Lower Saxony. He had plenty of time, free from professional duties or family obligations, to summarize his research on superconductivity in book form for publication the following year by Springer. Otherwise, he was scheduled to present a talk on Newton in Hamburg in July. The university administration in Hamburg had invited him, as he wrote his son, and: "Among the good friends from the militaristic clique who were still surrounding me in Alswede, there was gossip because you had changed your nationality during the war. I pointed out to them that there was no obligation of loyalty to a criminal state. You don't have to worry about that."[8]

Lise Meitner had been employed as "visiting professor" at the Catholic University in Washington since January at the invitation of Karl Ferdinand Herzfeld. On June 28th she found time to write a long letter to Laue. She had met many old friends:

> Your ears must have been ringing often, because so many people wanted to hear from you. At Princeton, besides Ladenburg, I met Einstein and the Weyls. Einstein I found depressingly aged and tired, but as fascinating as ever, even where I disagreed with him. [...]
>
> A really unalloyed joy was the reunion with Franck. He has really grown as a personality through those bitter times. He reminded me that in our youth I would often tell him, for heaven's sake don't become a physics machine, and for all his love of his research, he does indeed have an open-mindedness to the grave problems of the day and a sense of responsibility that I greatly admire. [...]
>
> It all looks like great general confusion. Some seem to have forgotten what they promised, others to have forgotten what has happened and what has been allowed to happen (undated, Lemmerich 1998: 451 f.).

Meitner also told Laue what she had told understanding Americans: a statement that Planck had made in Stockholm in 1943: "Terrible things must happen to us, we've done terrible things (p. 452)."[9]

Planck held his own admirably. In spite of his old age and many disabilities, he felt obliged to give scientific lectures that were comprehensible to a general audience. On

---

neither left nor right, and devotion to the one who gave him this profession, the supreme warlord, thus producing a maximum of professional achievement as an absolute value—whether on the parade grounds or on the battlefield, and on the parade grounds with specially selected calculation and technique. For it was here that man was trained to perfection, that is, remodeled according to a rational scheme into a being that has learned to blindly sacrifice its life for a purpose not set by itself." In this chapter Meinecke discussed the importance of the Army to Hitler's success in seizing power as well as in subsuming it and forming the paramilitary SS. "The spirit—rather the evil spirit—of the Waffen-SS henceforth determined the nature and will of those in power and led us into the abyss." Chapter XIII was given the heading 'Does Hitlerism have a future?' which he denied, proposing instead a path to renewal in chapter XV. At that time, all publications were subject to Allied approval.

[8] Laue to Theodore, Göttingen, 19 May, 21 Jun. 1946. In Lemmerich (2011: 306); Laue (1946b); he appreciated Newton's optical experiments and the three editions of the *Philosophiae Naturalis Principia Mathematica*.

[9] Lise Meitner herself was rather late in starting to reflect about what she experienced in the "Third Reich" until her escape, or about why she had decided to remain in Germany for so long despite all that brutality by the regime.

June 17th, for instance, he spoke in Göttingen about 'Apparent problems in science' following a few introductory words by Laue. British scientists honored him with an invitation to London for the Newton festivities. Planck accepted. He wanted to say something about Newton on that occasion so Laue lent him his manuscript for his coming Hamburg speech.

Daily life was still far from normal, though. There could be no "coming to terms" with the past; denazification could only have a limited effect. In a letter to Theodore in June, Laue criticized the denazification methods used by the Allies.[10] Laue received a reply to his letter from Erwin Schrödinger in Dublin dated June 9th. The latter had left Berlin and Germany in 1933 for England, moving on to Austria in 1936. But after the *Anschluss* to the "Third Reich" he had emigrated to Ireland where he had been offered a professorship in the capital city. "First of all, though, I want to congratulate you personally—and us—that you are still alive. When they were talking about you during the period before the end, I had often said (where it seemed permissible without endangering you even more): that was the bravest fellow of all; if they let him live at the end, that'll be a miracle. Thank God that miracle has happened." Schrödinger wrote about an upcoming meeting for the Newton jubilee in London. Perhaps they could see each other in Cambridge, he suggested. "I'd give a lot to be able to talk to you. The whole situation is so bleak. [...] Most ardent wish to see the hated régime destroyed—a wish pending for years—finally fulfilled—and now what? A good friend wrote from Vienna that the attitude of the intellectual supporters of the former régime was now a cool and condescending: Well—is it better now?—And—*is* it better now? (orig. emphasis, Sommerfeld 2004: 592)."

In the company of a British escort, Laue attended the Leibniz celebration in Hamburg, where he gave his lecture on Newton. Walking through the destroyed city, he was seized by old memories and despair. He wrote to his son from there on July 3rd: "Here, on my way to city hall, I keep passing by the spot where I saw you with Mama for the last time. Who would have thought then of a goodbye for 9 years! And who knows whether it won't be a goodbye for life. Of course you are right, a thousand times right, that you don't want to return to Germany. When I suggested this to you from Stockholm in 1943, I knew that the United Nations would win, but I had no idea that the end of the war would look like this. Just since seeing the people on the train, I am convinced that all is lost here. Make sure to take in Kurt and Hilde in U.S.A. They'd only perish here (Lemmerich 2011: 310)."

But his despair passed. On July 8th Max von Laue was in London, following a two-hour flight, to attend the International Congress of Crystallographers at the Royal Society, as had already been planned in 1945. He stayed with a colleague in a suburb of London. To his son he wrote on July 10th from there about the scientific topic: "The subject of all the talks is the study of crystal structures with X-rays. I look upon this my brainchild with feelings similar to upon you, my son. I am glad that it has developed by itself, and that it has become much more than I ever suspected

---

[10] Laue to Theodore, Göttingen, 22 Jun. 1946. In Lemmerich (2011: 308). Such blunders were understandable, because the Allies had very little idea, if at all, about life under the "Third Reich." It took some time for them to understand somewhat better what it takes to survive such a dictatorship.

(p. 311)." A formal dinner was held that evening, and in the "after-dinner speech" Sir William Lawrence Bragg mentioned Laue's "personal integrity." In his letter to Theo from London on July 12th, Laue added: "My rejoinder was: 'That is high praise; but I must affirm, there are thousands of men and women in Germany who deserve it at least as much' (p. 313)." Two days later he found the time to report in greater detail about whom among the emigrés he had met, and that he had been allowed to stay two days longer in order to speak with Lise Meitner, who was returning from the U.S. (p. 314). He had already written Theo in anticipation, "I'm looking forward to seeing Lise again, I'm enormously pleased. In general, everything is turning out better here than I thought (p. 313, cf. Laue 1945/46)." He met Paul Rosbaud and Marie Rausch von Traubenberg, who had survived the concentration camp, accompanied by her daughter Helen. Her parents had brought her and her sister to safety in England in time. In the evening Laue was sitting together with his friends and a French woman. They were talking about Sophocles' Antigone and her answer to Creon. Laue quoted the Greek text, which reads: "To league with love not hatred was I born" (p. 313).

The festivities honoring the tricentennial of Newton's birth commenced on July 15th. There was a reception at the Royal Society, and Planck was very warmly welcomed. The scientific part began with a lecture by Sir Edward Andrade. Laue saw friends again, such as Niels Bohr and Vilhelm Bjerknes. Afterwards he visited Lise Meitner's sister-in-law, Lotte Meitner-Graf, in her studio to have his photograph taken. That was where he also met Lise Meitner, as he recalled to Theo in his letter from Purley on July 16th. "That reunion was something! Almost as if I had encountered you there. But one doesn't speak much about that. I'd only like to touch on some politics (p. 314)."

This marked the beginning of the psychologically difficult discussion about guilt and complicity with his son, and also with Lise Meitner, without the words "collective guilt" coming up. Max von Laue did not need to reproach himself. He had helped Jewish colleagues in several instances. But he now tried to find a solution to the question of whether Germans should confess their guilt and denazify their mentality. He did not succeed in finding one. It was a decades-long process and never could come to complete closure; each individual fate was too unique. His own actions showed that his main concern was the reconstruction of his beloved science. "Only now do I understand a remark you wrote me," he added to his son, "that in the U.S. it is not understood why no one in our country is publicly repudiating the Hitler era and its atrocities. How do you know that this isn't happening? (p. 315)" He tried to explain by pointing out that at the official opening of the semester in Göttingen the university rector had delivered a speech to that effect and that newspaper articles on the subject had also appeared. But—after twelve years of propaganda—people thought that these articles were just some more propaganda. Lise Meitner had taken offence, he continued, at an interview that Otto Hahn and Werner Heisenberg had apparently given about the construction of the atomic bomb. Since at this time Laue was not aware of how fearful the English and Americans had been about a German atomic bomb being deployed as a weapon of assault, he drew a false comparison. He set the wartime research by German scientists against the researches by British, American, and French scientists.

At best they can say: 'You worked *for* Hitler, we *against* him.' But although in this regard I myself kept my distance from all war efforts during Hitler's war (as opposed to the 1st World War), this point of view is not so commonly accepted as to be able to derive from it great censure for actions to the contrary. *I am not sure that all those who are making this rebuke would have acted differently if they had happened to have come into this world in Germany.*—By the way, this statement also applies to many other activities during the 'Third Reich' (ibid., orig. emphasis).

Then, returning to denazification, he wrote that the major criminals, of course, "will be eliminated one way or another." He argued for large-scale amnesty for all the rest. "Because, a person's mentality—and that ultimately is what it comes down to—cannot be established by administrative measures (ibid.)." Laue also protested to Theo about the many obligatory questionnaires, not only those related to the denazification procedure. His opinion was the same as regarded public confessions of guilt. One did not spell out such self-evidential things specifically, least of all in a solemn declaration (ibid.).[11]

On the following day, the 17th, he happily told Theo about all the people he had met in London, including Max Planck and his wife. With Lord Cherwell—Frederick Lindemann, Nernst's former assistant—he had discussed politics (p. 316). His flight was scheduled to leave London the next day July 18th. But before leaving Purley in Surrey, he wrote a brief letter to Einstein (UAF, Laue papers, 9.1).

Only a few days later, the Laue couple was able to move into an apartment of their own on the second floor of a rental building at Bunsenstraße no. 16 in Göttingen. There were two rooms, a pantry and kitchen with an electrical stove, and a WC, but they had to go down to the basement to bathe! Some windows faced north, as Laue had wanted. The furniture from Hechingen could be set up. About the apartment Laue wrote to Theo on July 28th: "It's beginning to look habitable thanks to Mama's skill and diligence, and I am beginning to reconcile myself to it a little. But I can't quite get over it that the nightmare I had for many years of possibly losing our Zehlendorf house has come true. Well: 'He who possesses, let him learn loss, and he who is happy, let him learn pain.'" (p. 318) After a while, a very large part of the library arrived in Göttingen, which the mathematician Alexander Dinghas had managed to salvage in Berlin. Carl Friedrich von Weizsäcker and his wife lived in the same building as the Laues in Göttingen. However, the relations between the two couples did not become particularly close. Laue's home in Zehlendorf had been looted when the Soviet troops had occupied the area; but the Dutch couple who had been living there had managed to flee in time. From then on, the house was administered by the senate of the city of Berlin.

The KWI of Physics was reestablished in Göttingen. Werner Heisenberg took over as director, his deputy was again Max von Laue. Carl Friedrich von Weizsäcker

---

[11] On Göttingen see Thadden and Trittel (1999), esp. Tollmien (1999: 275–290), Dahms (1999: 395–443), Gidion (1999: 539–586), Manthey and Tollmien (1999: 675–760). See also Hentschel and Rammer (2000). The Protestant church (*Deutsche Evangelische Kirche*) had published a confession of guilt already in October 1945. On the postwar mentality of German physicists, cf. Hentschel (2007).

assumed the Theory Department and Karl Wirtz the Experimental Physics Department (Henning and Kazemi 2011: 281). Scientific life continued; the meetings of the Göttingen Academy of Sciences took place again with lectures by its members. On June 7th, Laue spoke about 'The distinguishability theorems in the theory of superconductivity'—a topic that had been preoccupying him during his time at Farm Hall. With reference to considerations by Fritz London from 1937, he studied a thick superconducting ring. He asked which determinants were needed to calculate the field and current distribution unambiguously. This could not be achieved by Maxwell's electrodynamics. The coupling of magnetic field and superconductivity prevented any decomposition into subproblems. "The state in the superconductor is accordingly uniquely determined by the tangential components of either J^1 [the current density of the supercurrent] or H [the magnetic field strength] (Laue 1946c: 88. The calculation was based on London's equation and a stationary current.)."

Actually more pressing than these scientific issues were three major political and administrative problems awaiting some solution. All of them were urgent. Laue was also involved in these tasks. They were: the reestablishment of the Kaiser Wilhelm Society, the founding of the German Physical Society, and the rebuilding of a new bureau of standards, the former *Physikalisch-Technische Reichsanstalt*, which the British wanted. With regard to the reopening of the *Physikalisch-Technische Anstalt* (PTA), as it was now called, Laue was set before the problem of assessing which scientist would be scientifically and administratively suitable for the post of president. In addition, there was the equally important question of what the latter's conduct had been during the "Third Reich" and how it should be judged now. Psychologically, these were very complex questions. According to which criteria should the decision be reached? In addition, younger academics who had been trained during the National Socialist period ought to be judged differently, because the younger lecturers had been obligated to attend a political lecturers' camp and had usually been members of the party. The henceforth commonly used word "reeducation" only proved how little was understood about a solid change of heart, about insight, mentality. If Laue already knew his colleagues well for a longer while, his judgment was naturally more reliable[12] (Fig. 12.2).

In addition to sketching family affairs, Laue often described to his son events that had impressed him. On the tenth Sunday after Holy Trinity he attended a service in the university chapel. A formerly East Prussian pastor preached about the destruction of Jerusalem by Titus: "A passage from Elijah was also read which predicts the destruction of the Jewish state as punishment for impiety. Now all this was the presentation, for the sermon to apply to the present state of Germany. [...] The collection was being made for the mission for Jews. The pastor recalled the terrible

---

[12] The foundings of the three institutions mentioned has been described by Dr. Zeitz (2006). The great burden placed on Laue by these necessary, various, and often simultaneous activities is not as clear there as it is in what follows in the present book. The simultaneity of the decisions required of him becomes clearer here, but also perhaps demands somewhat closer attention by the reader.

Fig. 12.2 Max von Laue, around 1950. *Source* photograph in the author's estate

injustice done by Germans to the Jews in recent years, and called for a kind of atonement offering to be made at this collection."[13]

The professional organizations of the Third Reich had been banned by the Allies and were forced to dissolve. This also included the German Physical Society (DPG) with its headquarters in Berlin. In 1938 its Jewish members had been asked to leave, and in 1939 it had been compelled to install the *"Führerprinzip."* The DPG also had numerous members who were more or less convinced supporters of the Nazi party, often even card-holding members. The society had to denazify itself if it wanted to gain permission to refound itself. In Berlin, Carl Ramsauer was trying to solve this problem with Robert Pohl in Göttingen. Werner Heisenberg wrote to Colonel Blount requesting a permit to found a German Physical Society in the British-Occupied Zone, which was approved. This required that new statutes be formulated. Laue vehemently lobbied for a slightly altered version of the statutes from the 1920s. Its foundation brought back oral debate within the framework of a conference.[14] During the first Physicists' Convention after the war, from 4 to 6 October 1946, the society was founded in an open session on October 5th. Max von Laue was elected chairman,

---

[13] Laue to Theodore, Göttingen, 25 Aug. 1946. In Lemmerich (2011: 323). In the same letter Laue mentions a village of originally 600 inhabitants, where now up to 800 refugees from the East were living.

[14] Detailed in Rammer (2012: 367, 379–380). Laue to Theodore, Göttingen, 5 Oct. 1946. In Lemmerich (2011: 330).

Clemens Schaefer from Cologne as deputy. At the same time, *Neue Physikalische Blätter* appeared as a small-format journal with Ernst Brüche as editor (Fig. 12.3).

Max Planck attended the inaugural meeting of the Physicists' Convention, and Laue held an address with memories about the meeting in Würzburg in 1933. The presence of guests from England—Nevill Mott, who had studied at Göttingen in the 1930s, and Wolfgang Berg, who knew Laue from Nernst's institute in Berlin— as well as from the Netherlands, contributed to the success of the meeting with over eighty participants, despite the difficult postwar conditions. On October 4th, the foreigners and two other stationed Englishmen as well as some German friends were guests of the Laues for an evening beer. Food rationing prevented them from hosting a formal dinner, but Mrs. von Laue unexpectedly produced a snack of spread roles. It was very difficult to find accommodations for the German physicists who had arrived from elsewhere, as no hotel rooms were available. They were lodged in frugal student accommodations with bring-your-own bed linen and towels. On the third day Max von Laue presented a talk about Intuitive Aspects on Superconductivity Theory (*Anschauliches über die Supraleitungs-Theorie*), Eduard Justi spoke about Superconductivity and the Periodic System, and Werner Heisenberg, On the Electron Theory of Superconductivity.[15]

Laue wasn't exempted from denazification either, as all former civil servants were affected. He wrote to his son from Göttingen on September 5th: "They are continuing to 'denazify' here. My colleagues and I are now supposed to fill out the fourth questionnaire, a monstrosity of 12 pages and containing 133 questions! We have declared our refusal to fill it out. This matter is gradually becoming unedifying (Lemmerich 2011: 325, Hentschel 2007: 102–106, 116 f.)." All trips abroad and all publications from 1924 onwards had to be listed. Theo had replied on August 4th to his father's letter of July 16th from London containing remarks about the not-always-inevitable cooperation in the war by German scientists. But Laue was of a different opinion and remarked skeptically in his next letter on September 6th: "I can already think of a good method for this ['reeducation'], which, however, is quite different from the one being applied now. Namely: the broadest possible access to a humanistic education, or, since only few can be guided directly to Homer and Sophocles, at least the closest possible return to the German classics, whereby the 'Romanticists' should by no means be excluded. It is no coincidence that militarism spread as classical education declined. [...] The main aim, in my op[inion], must be to raise men to know what is good and what is evil in any situation—and who act accordingly (Lemmerich 2011: 325)."

---

[15] Laue to Meitner, Göttingen, 7 Oct. 1946. In Lemmerich (1998: 464). Cf. *Neue Physikalische Blätter* 2 (1946): 178 f., therein the list of speakers and their topics, followed by Marga Planck's account of Max Planck's attendance at the Newton festivities in London (pp. 179 f.). Laue's abstract in the following issue (no. 8, p. 220) reads: "In superconductivity the forces exerted on the material cannot produce volume effects but can only act on the boundary layers. It is discussed how these relations affect volumes diminishing relative to the surface." AMPG, III rep. 50 nos. 2388–90; DPGA, 40,037, Tagung and DPGA, 40,038, Protokoll, 5 Oct. 1946, Bekanntmachung 14 Oct. 1946; see also Zeitz (2006: 89).

Laue's letter to Einstein had evidently arrived, since he received a response dated August 9th: "Dear Laue, It is very nice of you to have such an positive recollection of me, although I have been neglecting all my faraway friends out of laziness and indolence from a curious sluggishness about writing. I am fully aware that you held your ground wonderfully during those unspeakably difficult years, that you made no compromises and remained true to your friends and convictions as very few others did. It's only right and fair that people abroad appreciate and even admire this. As, everyone knows that it wasn't easy." As an afterthought he added: "I actually rarely see your son. But I can tell you that he's become quite some fellow, too (UAF, Laue papers, 9.1)."

Their debate about the Jewish persecution did not end in unanimity though. Einstein could only remark in response to Laue's answer: "Now just imagine the following: Imagine that the Jews had massacred 3.5 million Germans, merely out of some impulse. Imagine, furthermore, that the remainder were robbed and abandoned to misery, and imagine you were now being asked to give money for a Jew."[16] Max von Laue sent Einstein a lengthy quote from Lessing's play *Nathan the Wise* with the passage about reconciliation in Act 4, Scene 7. But Einstein replied: "Now, as for Lessing's model, Moses Mendelsohn, he was indeed a Good model for wise Nathan, who thought he could handle hard reality with ingratiating words. Self-deception simply arises in situations that are otherwise intolerable."[17] On September 9th, Max von Laue proudly informed his son of Einstein's opinion of him. He then told Theo that Paul A. Schilpp, a professor at Northwestern University in Evanstone, Illinois, and editor of the series Library of Living Philosophers, had asked him to write a contribution on "Inertia and Energy" for the volume *Albert Einstein, Philosopher–Scientist*. Max von Laue had accepted and that the honorarium was supposed to go out to Theo (Lemmerich 2011: 327, cf. Schilpp 1949/55).

The lack of mention of the Nuremberg tribunals is striking in Laue's letters. He probably rightly assumed that the American and Swedish press was publishing extensive reports about them. The letter to Theo dated 15 October 1946 does briefly mention the executions of those sentenced to death by hanging.

There were repeated revisits of the past for Laue. In October 1946, he received an letter from Paul Rosbaud in London, who during the "Third Reich" had smuggled news about armament projects abroad: "Remember how often we used to talk about Europe's future during the dark years, when we were sitting anxiously by the radio every day listening to the English broadcasts; it was a very small circle of friends, but what a fine and unforgettable circle (undated, AMPG, III rep. 50 no. 1666)."

Joyful news arrived from America: Theo and Hilli informed his parents of the birth of a son whom they were going to call Christopher.

The question of guilt for the crimes of National Socialism was still a topic of very heated debate. The Heidelberg professor of philosophy, Karl Jaspers, offered a lecture

---

[16] Einstein to Laue, 15 Aug. 1946 (the month is probably September) Laue's note: "postmark 10.10.46". UAF, Laue papers, 9.1.

[17] Laue to Einstein, Göttingen, 15 Oct. 1946, and Einstein to Laue, [Princeton], 9 Jan. 1947. UAF, Laue papers, 9.1.

course on the subject, which was published as a book entitled *The Question of German Guilt*.[18] Max von Laue read it and wanted to send a copy to his son from Göttingen, as he wrote him on November 8th. "For today, just one quote from p. 22: 'The difference' (i.e., among people) 'upon losing faith is tremendous. The ground has been pulled out from under all of us in one way or another. Only a transcendentally grounded or philosophical faith can persist through all these catastrophes.'" (Lemmerich 2011: 335) In response to an earlier letter by his son in which he had referred to faith as capable of "evolving," Laue countered, "We remain what we are through all disasters. That doesn't mean we don't learn anything from them; but it doesn't touch the depth of our being, even if things got considerably worse (ibid.)." Only four days later, in his letter from the 12th, Laue addressed the reproach by the Allies about lacking resistance against the regime and defended the view that any implementation of those ideas by the Allies would have been tantamount to suicide. That would, of course, be contrary to Christian convictions. Edith Hahn had once given him a motto: "Learn to button up without busting" ("Lerne schweigen ohne zu platzen"), which hung framed above his desk. Laue thought he had sometimes been unnecessarily uncautious (p. 337).

In the letters to his son there is no mention of the activities involving the refounding of the Kaiser Wilhelm Society, which had not yet been dissolved. Perhaps that was in order to avoid drawing any censor's attention to how much the British occupying force was tolerating by granting its permission. A few changes had to be made to the old statutes, but on 11 September 1946 the "Max Planck Society for the Advancement of Science in the British Occupied Zone" officially came into being (Figs. 12.4 and 12.5). Otto Hahn became its president. The Clementinum, the theological seminary building in Bad Driburg, was chosen for its official founding. The document also bore Max von Laue's signature, as he had collaborated on the formulation of its statutes. A total of 13 institutes of the former Kaiser Wilhelm Society were located in the British-occupied zone. Nothing had yet been declared about the specific scope of their research, however.[19]

In the Soviet-occupied zone, a German Office of Weights and Measures (*Deutsches Amt für Maß und Gewicht*) had already been established in Weida as a

---

[18] Jaspers (1946). Karl Jaspers divided the question of guilt into four concepts: criminal guilt, political guilt, moral guilt, and metaphysical guilt. Jaspers wrote: "Law can only apply to guilt in the sense of crime and in the sense of political liability, not moral and metaphysical guilt. [...] Morally, guilt can only be self-assigned." As a psychologist, he noted, "Our human makeup—at least in Europe—is such that we are as sensitive to reproach as we are ready to reproach others." His statement was quite contemporary: "Some allow themselves to be led astray by momentary disinterest. It then seems advantageous to confess guilt. The world's indignation at morally depraved Germany is proportionate to its will to concede guilt (Our trans.)."

The guilt question could not be resolved even in the years that followed. It was still a psychologically complex issue. The *Deutsche Rundschau*, an important literary magazine edited by Rudolf Pechel, published essays commenting on the rise of National Socialism and its consequences.

[19] AMPG, II rep. 1A, Gründung Nr. 5 (2–16), Gründung Nr. 1, 3. It is not clear whether the naming is solely due to Sir Henry Dale. Lise Meitner had supposedly originally made the suggestion. She was a great admirer of Max Planck. See also Lemmerich (1981) (without sources); Heinemann (1990: 407), Zeitz (2006: 79).

Tagungen und Veranstaltungen

PHYSIKERTAGUNG IN GÖTTINGEN

Zum erstenmal nach Kriegsende trafen sich am 4. Oktober 1946 die deutschen Physiker der britischen Zone zu einer dreitägigen Tagung in Göttingen. Die Teilnehmerzahl mußte im Hinblick auf Unterbringungs- und Verpflegungsschwierigkeiten beschränkt bleiben. Trotzdem nahmen etwa ·80 Physiker aus der britischen Zone teil. Einige wenige Teilnehmer waren aus der russischen Zone gekommen, unter ihnen der letzte Vorsitzende der DPG vor dem Zusammenbruch, Professor *Ramsauer*-Berlin. Am 4. Oktober wurde die Tagung um 10 Uhr durch Professor *v. Laue* eröffnet. Im Anschluß an die Begrüßung der Teilnehmer und Gäste, unter ·denen sich neben dem Vertreter der britischen Militär-Regierung, Dr. *Frazer*, u. a. auch Prof. *Mott*-Bristol sowie Dr. *Michels*-Eindhoven befanden, war der erste Tag bereits eine Reihe von Vorträgen gewidmet. Am Sonnabendvormittag fand dann eine Sitzung der Mitglieder der früheren Deutschen Physikalischen Gesellschaft statt, im Anschluß daran wiederum Vorträge, ebenso am Sonntagvormittag. Der Sonnabendnachmittag stand für die Besichtigung der verschiedenen Institute sowie für die persönliche Aussprache zur Verfügung. Das Tagesprogramm zeigt, daß es der Leitung (Professor *v. Laue* und Professor *Heisenberg*) gelungen war, eine größere Zahl interessanter Vorträge zusammenzustellen.

Vortragsverzeichnis. *Meixner* (Aachen): Theorie der Beugung elektromagnetischer Wellen an der Kreisscheibe. *Lauterjung* (Köln): Empfindlichkeitsänderungen an Lichtzählrohren. *Meyer* (Göttingen): Über eine Methode zur Bestimmung der Energie und der Energieverteilung ionisierender Teilchen. *Houtermans* (Göttingen): Über das Alter der Welt. *Kopfermann* u. *Meyer* (Göttingen): Ueber den Isotopie-Verschiebungseffekt an W.I.-Spektren. *Bartels* (Hannover): Stoßentladungen in Hg-Höchstdrucklampen. *Mannkopf* und *Peetz* (Göttingen): Die Akkommodationszeit der Elektronen-Temperatur in einer stationären elektrischen Entladung. *Maecker* (Kiel): Experimenteller Nachweis der von Schmidt'schen Kopfwelle in der Optik. *Mott* (Bristol): Oxydschichten auf Metallen. ·*Mollwo* (Göttingen): Zur Dichte aufgedampfter Salzschichten. *König* (Göttingen): Ändert sich die Gitterkonstante bei sehr kleinen Teilchen? *Jensen* (Hannover) und *Houtermans* (Göttingen): Zur thermischen Dissoziation des Strahlungshohlraums (vorgetragen von *Jensen*). *v. Laue* (Göttingen): Anschauliches über die Supraleitungs-Theorie. *Michels* und *De Groot* (Eindhoven): Einfluß

Fig. 12.3 Article about the first Physicists' Convention after World War II. *Source Neue Physikalische Blätter*, vol. 2, no. 7, 1946, p. 178

branch of the former *Physikalisch-Technische Reichsanstalt* (PTR). Wilhelm Steinhaus, who had been employed at the PTR, was acting "president." With the support of the British occupying power, Laue intensively lobbied for the refounding of the PTR branch within the British-occupied zone. The "Acoustics" department had been transferred from Berlin to Bad Warmbrunn in Silesia and finally to Göttingen before war's end. The local premises of the Aerodynamics Design Testing Station (AVA) could not come into consideration for this new *Physikalisch-Technische Anstalt* (PTA— also divested of its imperial designation *Reich*). Ronald Fraser drew Laue's attention to the former *Luftwaffe* grounds with some vacant buildings in Völkenrode near Braunschweig as a potentially suitable location for this institution. A joint inspection yielded a positive result. The German Air Force had erected several buildings there for various research activities, which were now supposed to be blown up. The

Fig. 12.4 Founding meeting of the Max Planck Society in the Clementinum in Bad Driburg in the British zone, 11 September 1946. Max von Laue and Otto Hahn are standing in the second row. *Source* Archive of the Max Planck Society, Berlin-Dahlem

English demolition squad was scheduled to arrive there just while Laue would be inspecting the site. He lept into action, and with the help of Sir Charles Darwin arranged for a telegram to be dispatched from the British government cancelling the demolition order. The English lieutenant in charge remarked, not without a hint of sarcasm, that the Germans ought to attach a plaque to the oak tree under which he had been shown the telegram: "*von Laue vicit*" (conquered by von Laue). That actually even came to pass![20] (Fig. 12.6).

The search for a president for the PTA turned out to be very difficult. Walther Gerlach, whom Laue asked, declined, saying he was not suitable for political negotiations. In spite of these various obstacles, the scientific work in Völkenrode got underway. It was later even possible to publish an official report for 1947 about the

---

[20] Report on Physikalisch-Technische Reichsanstalt, June 1945, R. H. Ranger, Combines Intelligence Objectives Sub-Committee (CIOS), Item No. 1. On page 3 it states: "It is believed that this group is an important aid to rehabilitate the German equivalent of the Bureau of Standards, and should be tied in to whatever Government Institution is established to look towards economic technical rehabilitation of Germany." Quoted from Kern (1994: 271); therein also the site plan of the area with marked buildings in Völkenrode (p. 276). On the oak tree, see PTB (1961: 4). Gerlach to Laue, Bonn, 2 Dec. 1946. AMPG, III rep. 50 nos. 2390–91; Laue to Gerlach, Göttingen, 13 Feb. 1946. DMA, NL 80/282/1.

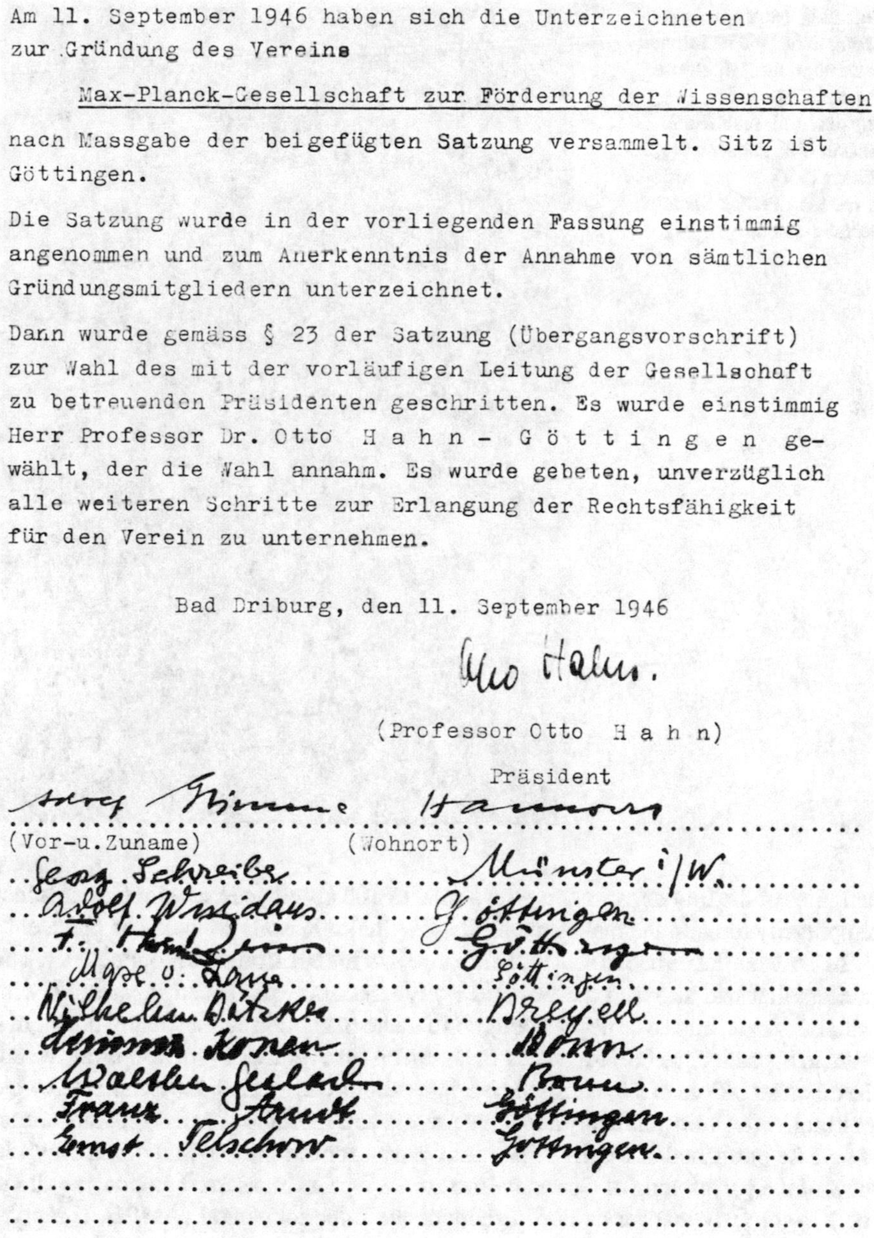

Am 11. September 1946 haben sich die Unterzeichneten
zur .Gründung des Vereins

    Max-Planck-Gesellschaft zur Förderung der Wissenschaften

nach Massgabe der beigefügten Satzung versammelt. Sitz ist
Göttingen.

Die Satzung wurde in der vorliegenden Fassung einstimmig
angenommen und zum Anerkenntnis der Annahme von sämtlichen
Gründungsmitgliedern unterzeichnet.

Dann wurde gemäss § 23 der Satzung (Übergangsvorschrift)
zur Wahl des mit der vorläufigen Leitung der Gesellschaft
zu betreuenden Präsidenten geschritten. Es wurde einstimmig
Herr Professor Dr. Otto H a h n - G ö t t i n g e n ge-
wählt, der die Wahl annahm. Es wurde gebeten, unverzüglich
alle weiteren Schritte zur Erlangung der Rechtsfähigkeit
für den Verein zu unternehmen.

            Bad Driburg, den 11. September 1946

                        (signature: Otto Hahn)

                    (Professor Otto H a h n)

                             Präsident
(signatures)
(Vor-u.Zuname)   (Wohnort)
Georg Schreiber   _Münster i/W._
Adolf Windaus   _Göttingen_
F. Kienle   _Göttingen_
Max v. Laue   _Göttingen_
Wilhelm Dittler   _Dreyell_
Hermann Konen   _Bonn_
Walther Gerlach   _Bonn_
Franz Arndt   _Göttingen_
Ernst Telschow   _Göttingen_

Fig. 12.5 A page from the founding minutes of the Max Planck Society for the Advancement of
Science with headquarters in Göttingen, Bad Driburg, 11 September 1946. *Source* Archive of the
Max Planck Society, Berlin-Dahlem

Fig. 12.6 Max von Laue sitting under the mentioned oak tree in front of House "M 5" of the Physikalisch-Technische Anstalt in Braunschweig, 20 March 1950. *Source* Archive of the Max Planck Society, Berlin-Dahlem

activities of the four departments with a total of 108 employees. Martin Grützmacher temporarily took on the management, then Wilhelm Kösters followed as president.

In between, Laue had to deal with the proofs to his book on superconductivity. The printers that had survived the war had no typesetters experienced in working with scientific texts, and so Laue was obliged to make many corrections many times. In a letter to his colleague Carl Ramsauer in Berlin on 4 November 1946, discussing where the editorial offices of *Physikalische Berichte* should be located, Laue indicated what was motivating him politically about the present and the future. He thought the journal should be published in Berlin: "The main reason for this we see in the following: In view of the partitioning of Germany into zones, we Germans must emphasize all the more vigorously that for us Berlin remains the imperial capital (AMPG, II rep. 50 nos. 2390–91 fol. 12)."

Three days before Christmas Eve, Laue was able to impart some pleasant news to Theo: "From the English, I have been informed that I might meet Otto Stern in Switzerland in one of the coming months, and that an exit permit there is promised. By arrangement, Stern will be inviting me to Stockholm for the Nobel award ceremony." Theo should write to Wolfgang Pauli in Zurich, he continued, if he wanted to tell his father things that weren't suitable for other letters. (Lemmerich 2011: 345; the physicist Otto Stern had been compelled to emigrate to the U.S. in 1933.)

A week before, on December 14th, Laue had replied to a letter from Otto Meyerhof, who had managed to escape to the U.S. at the last minute. As in his letters to other emigrés, Laue pointed out the lack of protests from abroad either in 1934 after the Röhm putsch or for the 1936 Olympic Games. He believed that biologists such as Eugen Fischer, the recent director of the KWI of Anthropology, Human Heredity, and Eugenics, had played a considerable part in developing the Nazi ideology with their racist theories. "The human being was regarded there merely as a strange two-legged perambulatory mammal. There was no question of human dignity and other ethical concepts (AMPG, II rep. 50)." To Lise Meitner, Laue wrote on December 25th about what Otto Hahn had told him about the Nobel prize awards and the fabulous festivities in Stockholm. The Laues' little Christmas tree had only two candles, because they needed the others for the frequent power cuts (Lemmerich 1998: 476).

'On beginning the year 1947' was the heading of Max von Laue's welcome into the new year in the first issue of *Physikalische Blätter*. In connection with the German Physical Society (DPG), he mentioned the necessity for intellectual intercourse. "Now the exchange of ideas is more necessary than ever—namely, as encouragement for us along the difficult path ahead of us in sober and illusionless research and teaching, to be performed with an open eye and a cool mind, despite a thousand hardships and tribulations. This path is illuminated again, of course, by a precious gift that the times now lying behind us had most severely jeopardized: intellectual freedom (Laue and Regener 1947)."

The trip to Switzerland became reality. On January 24th, Laue was able to write to Bohr from Zurich that Otto Stern had informed him about the restoration of the Nobel medals, deposited in Copenhagen during the war.—Georg von Hevesy had dissolved them in aqua regia to keep them safe during the German occupation. Since Germans were forbidden from possessing gold following the surrender in 1945, Laue asked that his restored medal be sent to his son Theodore. Then he asked about his Nobel certificate. He assumed it, too, had been lost and wondered whether it could be replaced as well.[21] In conversing with Stern about politics, the deep distrust towards Germans arose without their being able to find common ground, though.

Now able to correspond with his son in the United States again, Max von Laue also resumed writing to James Franck in Chicago, who had sent him a political circular. Laue remarked from Göttingen: "In my opinion, all political discussion is superfluous at the moment. The current crisis in the world—in the *whole* world—is a religious one. And religious crises last a long time. 'Good luck' and 'prosperity,' and whatever other such buzzwords there are, won't come out of it, but certainly the elevation of humanity onto a higher plane—after much hard suffering (orig. emphasis, JRL, Franck papers)." He also wrote Franck about his efforts on behalf of the PTA and a possible trip to the U.S. in August 1948. This interested Mrs. Ilse Rosenthal-Schneider, an acquaintance since Berlin prior to her emigration to Sydney.

---

[21] NBA, N. Bohr papers; see also: Schwarz (2015: 8 f). The Nobel certificate had not been destroyed. It was transferred to the archive of the Max Planck Society much later.

She wrote Laue that she would be delighted if she could have an opportunity to see him in Australia.[22]

The gradual stabilization of conditions led to gathering general interest in things in print, and so Laue worked on an improved second edition of his 'History of Physics.' His letter to Hans Schimank on March 5th settled a few questions in this matter (GNT, Schimank papers).

Johannes Stark now started to try to present himself as a victim of National Socialism. He claimed that he had fought for the freedom of research (Stark 1947). Laue's published commentary (1947b) on his article quoted verbatim his speech of 14 December 1933 before the Prussian Academy of Sciences. First he discussed Stark's teaching at Göttingen, where he had personally attended some of his lectures. Then Laue mentioned with regret various tenureship opportunities that had passed over Stark's candidacy. But now, Laue stressed, the future of science was at stake. He referred to Stark's counterproductive activities in the PTR, such as obstructing experimental investigations that had not been to his liking. His handling of publications in physics had been equally damaging. In conclusion, Laue remarked: "In the name of the freedom of research, I therefore move that the Academy members not elect him."[23] Nevertheless, when Stark's denazification proceedings ended in an unusually severe conviction, Sommerfeld and Laue lobbied for a milder sentence.

People often spoke and wrote about "coming to grips with the past," but to a contemplative person those were just empty words. In the new year Laue received mail from his colleague Gerhard Borrmann, who had joined him at the KWI in Berlin from his position with Kossel in Gdansk and conducted research on X-ray interference in ideal crystals. Later he had accompanied Laue to Hechingen. In his letter from February 28th, Borrmann apologized for his long silence, saying that it was not due to any lack of respect: "The situation is somewhat as follows: When a scientist looks about in the 'world' today (he has reason enough to do so), he automatically applies his usual standards to his observations, i.e., he looks for law & order. In so doing, however, he runs up against a sea of questions, and if he doesn't have an innate compass, it may well turn into an odyssey (UAF, Laue papers, 3.1)."

At the start of the year 1947, the American and British governments agreed to unite their zones to form a bizone. This was advantageous for founding the new bureau of standards, the PTA, because it could be erected in Völkenrode near Braunschweig for both the American and British zones. At the beginning of March, Laue wrote to Walther Gerlach informing him of a meeting in Minden about the future PTA, at which the necessary financing was set in motion. A provisional presiding panel consisting of Werner Heisenberg, Hans Kopfermann, Robert W. Pohl, and Laue as chairman, was appointed. He still hoped that Gerlach would assume the presidency, but these hopes were dashed. Undaunted, Laue continued to search for a president and tried to bring about a meeting about it. A list of three, which Gerlach later

---

[22] Ilse Rosenthal-Schneider had studied physics in Berlin and later published a book about her conversations with Einstein, Laue, and Planck. See Rosenthal-Schneider (1980).

[23] Laue to Meissner, Göttingen, 13 Jun. 1947. DMA, NL 045; Sommerfeld to Laue, Munich, 24 Jul. 1947. AMPG, III rep. 50 no. 2393. See also Kleinert (1983).

approved in his letter of May 28th, named Karl Willy Wagner, followed by Paul Harteck and Klaus Clusius (DMA, NL 80/282). The meeting was scheduled to be held on May 21st, but Laue received cancellations, and in his letter to Gerlach on March 4th admonished: "There is no doubt, of course, that difficulties are attached to attendance at this meeting as have never before been demanded of executive board members of a scientific association and are unlikely to be demanded again in future. However, the unique importance of the meeting does justify some sacrifice. An opportunity is being offered to our society here, by a cooperation with the military government and German ministries, to secure for itself an importance far higher than it has ever possessed before in a vital question of German physics and technology (DMA, NL 80/282/2)." By summer a president had still not been found. A meeting of the preliminary presiding committee, now with twelve members, two of them from industry, was set for September 4th. As chairman of this presiding committee of the PTA, Laue extended an invitation to this meeting in Göttingen on August 9th from Tailfingen (APTB, G30/2-1, part 2). But his actions again led nowhere, because shortly before Christmas Professor Wagner also declined the appointment as president (Zeitz 2006: 142).

The second annual meeting of the "German Physical Society in the British Zone" was to take place from the 5th to the 7th of September. Max von Laue informed Lise Meitner in good time on June 30th from Göttingen, writing that he would be very pleased if she came and even more so if she were to hold a talk (Lemmerich 1998: 491). This prompted Lise Meitner to make a statement about the then still acute and unfinished discussion about the Germans' behavior under National Socialism. Responding to Laue's invitation on August 4th from Hjortnäs, she explained her difficulties as a quasi stateless person and added: "I need not tell you what it would mean to me to be able to see my old friends again and speak with them at length sometime. And I hope just as much that you know how much Germany's fate is close to my heart and how affected I am about the great problems that Germany must contend with (p. 499)." She mentioned that she had sent shipments of clothing to Germany. "But still I cannot look at the complicated global problems *only* from the German point of view. And who among my old friends is not as much under the pressure of daily need to be able to understand my attitude properly? (Ibid., orig. emphasis)." Her meetings with her close friend Elisabeth Schiemann in London and with Mr. and Mrs. Hahn in Stockholm had made her very anxious about this. "Please do understand me correctly: I don't want to lose my friends just because I look at some problems differently from the way they do. The risk is not on my part. I am willing to consider conscientiously any honest view put forward, and if I cannot share it, that has nothing to do with my friendship. But how is it on the other part? (pp. 499 f.)." Laue replied by hand on August 19th in a letter from Tailfingen: "But then you really should come, precisely because you think we here may have a point of view that differs from yours. Because then, given the importance of the issue, you have a duty not to duck under with your opinion. And even if an argument does remain unresolved, is that so bad? Over here differences of opinion on politics have been existing in almost every family, in almost every group of persons since 1933 (pp. 500 f.)." In these arguments with Lise Meitner, Max von Laue avoided flaunting his own

Fig. 12.7 Max Planck and
Max von Laue in Göttingen,
1947. *Source* Archive of the
Max Planck Society,
Berlin-Dahlem

opposition to National Socialism and the regime of terror or the help he had given
to persecuted persons. During this time Laue also corresponded with Otto Stern in
the United States. After Stern sent the Laues some CARE packages, Laue responded
from Göttingen with reports about his family and academic life on August 15th and
September 23rd (UAF) (Fig. 12.7).

When the physicists in the British-occupied zone convened in Göttingen, Laue
first welcomed the foreign guests in attendance, mentioning Amsterdam, Cambridge,
Copenhagen, London, Manchester, and Stockholm. He then reminded the audience that on July 23rd one hundred years ago Hermann von Helmholtz had
published his work on the conservation of energy. He called to mind those members
who had recently passed away: Heinrich Rausch von Traubenberg, Hans Geiger,
Friedrich Harms, Werner Kolhörster, Walter Kaufmann, Friedrich Paschen, Friedrich
Rogowski, Bernhard Bavink, and also Philipp Lenard, who had died in May. "We
cannot and do not want to skip or excuse the transgressions of Lenard the pseudo-
politician, but as a physicist he is among the greatest (DPG 1947)." Ronald Fraser
spoke about how impressive the conference program was with its numerous new
and important topics, showing "that—in physics, at least—stone is again being laid
upon stone here in Germany." He quoted Niels Bohr: "Physics is the treatment of

open questions about nature—everything else is technology." The true physicist, he said, must be consciously supranational (orig. emphasis, Fraser 1947). The topics treated at this conference were cosmic rays, nuclear physics, superconductivity, and acoustics. With some three hundred participants it was well attended. Afterwards Max von Laue sent David Shoenberg in Cambridge his manuscript: 'Once again on the thermodynamics of superconductivity,' in which he quoted Shoenberg's talk. But their two theories had little in common.[24]

Even at this conference, about 800 days since the National Socialist regime of terror had ended, some participants, whether they admitted it or not, could not speak unreservedly with just anyone among their professional colleagues. Some of them had become convinced Nazis—perhaps now already former ones—, many had only been fellow travelers in order to keep or acquire their positions. But was membership in the NSDAP or one of its affiliations really the decisive criterion? Weren't there also "mind-mates" of the principles of Nazism? The DPG could not bring about a so-called "coming to grips with the past" (*Bewältigung*) either. But one participant at the conference, Ursula Maria Martius, gave food for thought. She had been a coworker of Hartmut Kallmann in Berlin. The *Deutsche Rundschau* published her impressions of the conference in an article entitled '*Videant consules*' (the formulation granting Roman consuls special supervisory powers). It began with a recollection:

> It is not yet three years since I stood before the barbed wire of the camp where my father was being held. At that time I promised myself two things in case I should survive: first, that I would work with all the energy I could muster, in order to relieve my parents of the burdens of daily living as quickly as possible, and second, that I would do everything I could to spare other young people a similar physical and mental fate. For, this evening before the camp was, after all, but one station on a long road, the decisive point of which for me had been my expulsion from university. However, I would not have believed then that I would be reminded so quickly of the hard lessons learned along this road.
>
> In September of this year, I took part in the *Annual Meeting of the Physical Society in Göttingen*, which was both stimulating and upsetting for me. In the first place, though, it was upsetting because of the constant encounters with the past. People who still appear to me in my nightmares were sitting there alive and unchanged in the front rows. Unchanged, if you don't consider the simple blue suit instead of the uniform and the missing party badge a 'change.' (Martius 1947, cf. Hentschel 2007: 108, Rammer 2012: 393 f.)

Martius listed several physicists who had held their lectures in party uniform during the Third Reich, and who had advocated Nazi ideology in their writings and speeches. She impressed upon professors to be fully aware of their responsibility as teachers and role models. "The objection that one cannot get along at universities without these people [meaning National Socialist professors] isn't valid at all. There are enough others, as many young academics did not embark on a university career during those past twelve years because they didn't want to be part of that 'rabble-rousing.' [...] The German intelligentsia has resolutely rejected the idea of collective

---

[24] DPG (1947); see also Rammer (2012: 367 f., 386 f.) listing the participants. The following documents are later in time, but deal with issues related to the meeting: Laue to Shoenberg, Göttingen, 2 Oct. 1947. AMPG, III rep. 50 no. 1863; Laue's manuscript: Nochmals zur Thermodynamik der Supraleitung (1947). There was subsequently an exchange of letters on very specific questions, such as the work by Edgar Thomas Snowden Appleyard.

guilt for the misdeeds of National Socialism. One of the strongest arguments was that at the time nothing could be done. Today, however, something can be done [...] (Martius 1947)."

Max von Laue felt obliged to respond. He wrote her a letter from Göttingen on December 26th:

> Dear Miss Martius,
>
> Prof. Hahn gave me your letter of 27 Nov. '47 to read. From it and the enclosures I learned many things that are new to me and quite unpleasant, especially the reprehensible nonsense that Jordan produced in his book on the Mystery of Life. The G. Phys. Soc. in the British Zone will not reply to your article in Deutsche Rundschau. But in this private letter I would like to comment on it and on other points in your letter.
>
> First of all, there is the case of Schumann. That this charlatan was denazified, he owes to a letter of recommendation from a scholar who is surely esteemed by you as much as by us, who in his old age had fallen victim to his charming ways just as formerly Schumann's teacher Carl Stumpf had (AMPG, III rep. 50 no. 23595; cf. Jordan 1941).

Laue then described the appearance Schumann had made in the cafeteria during the convention and his attempt to run for office, which Richard Becker had thwarted. He reassured her that the authorities in the Western zone were anxious "that those so heavily incriminated not regain a footing in the academic establishment (ibid.)." The second part of his letter dealt with legal principles in an orderly state:

> About the other cases you mention, I would like to specify my personal attitude thus: In any orderly state, two principles of justice must be strictly observed. Neither may anyone interfere with a pending court case in the press, nor may anyone reopen the matter after the verdict has been given in accordance with the law; criticism is, of course, permissible against the legal provisions upon which the case is based. No reasonable and fair-minded person denies that the denazification proceedings offer ample occasion for such criticism. But one doesn't make anything better; on the contrary, one merely undermines confidence in a judiciary, if individual cases are repeatedly brought forward again. *The defendants have a right not to be molested after their case has been officially disposed of. And this right is violated by articles such as the one you have written.* This consideration is also the reason, by the way, why we don't peal the bells about our proceedings in the Schumann case, but settle it quietly (ibid., orig. emphasis).

Laue argued for the reinstatement of legal principles and concluded: "Besides, by no means do I mistake the good intentions that motivated you to write your article (ibid.)." Max von Laue's remarks were an expression of his longing for the rule of law he had so sorely missed during the Third Reich. It is evident in his letters to his son written from abroad. The one he wrote while in Tisvilde, Denmark, on 1 August 1939 reads: "Germany needs people of your mentality and talent to reestablish a constitutional state, no matter what that state may look like in detail (Lemmerich 2011: 117)." And from Neuchâtel, Switzerland, on 24 August 1939, he downright adjures Theo to "strive to contribute toward the installation of a constitutional state, a state in which freedom is based on the law, in which the law comes *first*. How the younger generation obliged to do this sets it up in detail is its own concern. But law must prevail, *law, the law!*" (p. 121, orig. emphasis) What Laue's wishes for his own personal future were he confided to Hartmut Kallmann in a letter dated 5 May 1947:

"I cling with every fiber of my soul to the memories of Berlin and Zehlendorf and Dahlem, even though I know that I must not and cannot succumb to these feelings (UAF, Laue papers, 3.4)."

In the winter term of 1947/48, Laue began offering lecture courses again at Göttingen University. But this semester instead of lecturing about X-ray interference, he chose to present the history of physics in the broader sense of the word. It also included discoverers, such as Columbus and Vasco da Gama. The handwritten notes for this lecture mainly contain biographical dates and only brief passages of text (AMPG, III rep. 50 no. 14).

Max von Laue invited Lise Meitner, Max Born, Hendrik A. Kramers, and Nevill Mott by letter on September 26th to contribute to a special issue of the *Annalen* celebrating Max Planck's 90th birthday. "We realize that in all likelihood it will become a memorial issue, because his current condition is unlikely to last until then (Lemmerich 1998: 502)." Max Planck died in Göttingen on 4 October 1947. Max von Laue paid tribute to him in an obituary (Laue 1947c). Lise Meitner was unable to attend the funeral so Laue described the ceremony to her, which had taken place on the morning of October 7th in St. Alban's Church in Göttingen. In this letter of October 9th he also wrote about the speaker, Professor Friedrich Gogarten, "who was very sensitive and skillful about citing passages from Planck's addresses. I have seldom heard such a good funeral oration (p. 503)." On the musical accompaniment to the coffin bearing, he noted that it was "the second paraphrase of the chorale 'Now thank we all our God.' This probably followed the wishes of his family. And in between the Klingler Quartet played once, the other time deeply-moved Klingler on his own. Other addresses held were only by Hahn and by me. I intend to enclose mine for you, as I have tried to reconstruct it today. I did not manage to finish it beforehand (pp. 503 f.)." Max von Laue later had his address published (Laue 1947/58: 417 f., cf. Laue 1947g). His speech sketched the stages in Planck's life and ended by identifying the wreaths laid by the various organizations. "And then there is another simpler wreath without a bow. I have laid it for all his pupils, among whom I count myself, of course, as an ephemeral token of our undying love and gratitude (ibid.)."

Work on the theory of superconductivity continued. Until then, the theory had been limited to cubic crystals. Laue extended it by applying λ as a symmetric tensor. However, this led to an asymmetric stress tensor. He succeeded in finding an explanation for this from hydrodynamics and dynamics in relativity theory (Laue 1947d, extended further in 1948a).

Full "reckoning" with the dozen-year past was by no means over. Attempts to "clarify" things repeatedly led to new accusations, not just in the political arena. Even so, it was not comparable to the academic "War of the Intellects" during World War I. The unfinished research on the German atomic bomb and on other projects by physicists for the war effort were among these open questions. Samuel A. Goudsmit published a detailed report about the American bomb project under the codeword "Alsos" (1947a). On October 20th an article by him appeared in the widely read American magazine *Life* under the subheading 'The chief of a top-secret U.S. wartime mission tells how and why German science failed in the international race to produce the bomb' (1947b). Goudsmit's book was reviewed by Philip Morrison for the *Bulletin of the Atomic Scientists* (1947: 354 f.). Upon comparing the collaboration by Germans in their atomic-bomb project with that of their counterparts in America, his general indictment was: "But the difference, which it will never be possible to forgive, is that they worked for the cause of Himmler and Auschwitz, for the burners of books and the takers of hostages." (p. 365) Laue felt obliged to comment on this passage in addition to the other text. Again, a deep misunderstanding about living and *surviving* under the National Socialist dictatorship emerged. Laue started by mentioning the tragic fates of Goudsmit's parents and relatives who had perished in concentration camps. "We realize fully what unutterable pain the mere word Auschwitz must always evoke in him (orig. English, Laue 1948b: 103)." He then mentioned the "barbaric total war" to which the nation as a whole was subjected. "If in the last war one or other of the German scientists succeeded in keeping his work out of the war maelstrom, we must not conclude that this was possible for all of them (ibid.)."[25]

Laue mentioned the options then available of deploying politically and racially persecuted persons in research "of importance to the war," in order to shield them from internment in a concentration camp; and he cited larger institutes at which this was practiced. He had personally observed it at the KWI. At the end of this short contribution, he questioned the *Bulletin*'s purpose of creating lasting peace. This purpose, he contended, had been ill served by Morrison's article, and he quoted Antigone's statement again: "To league with love not hatred was I born." Laue wrote about this to Lise Meitner on 15 June 1948:

> Now about my article in the 'Bulletin.' I simply had to get it off my chest; I couldn't help it. I know that I shall encounter some opposition over there. But I also know that the 'Alsos' book has not won overly much influence and appeal. It is not being widely distributed.
>
> It is so easy to misjudge the situation in the Third Reich because, of course, every utterance in favor of Hitler was widely disseminated, whereas everything in opposition was suppressed. After all, the power of a modern-day tyrant (such even exist in very 'democratic' states) is largely based on control of all means of communication (Lemmerich 1998: 517).

Laue's invitation on December 20th to the board meeting of the German Physical Society was also sent to Walther Gerlach in Bonn. As Laue had written him from Göttingen on December 12th, his colleague Karl Willy Wagner had declined the appointment to the presidential post of the PTA. So the search had to go on. Richard Vieweg had been named in a private preliminary meeting (DMA, NL 080/282/2).

The establishment of a second headquarters for the Berlin-based publishing house Springer in Göttingen was another sign of the political and economic uncertainty in Germany. The publisher did manage to produce Laue's 'Theory of Superconductivity' in 1947, though. On June 18th of the foregoing year he had written to Max Born that this book on superconductivity was his "life-work" (*Lebensaufgabe*, SBPK, Born papers, 429). Permission to print it still had to be obtained from the

---

[25] The article appeared in English in the *Bulletin of the Atomic Scientists* in April 1948. Also in AMPG, III rep. 50 no. 125. Comp. Laue (1947e). Philip Morrison replied to Laue that he had not attacked him, of course. There was also a comment by the editor, Morrison (1948); see also Henning (1992).

British Control Commission, of course. It was a slim volume of 123 pages. Laue emphasized the purpose of the work in the introduction: "This text strives to clarify our picture of superconductivity by extending Maxwell's electrodynamics to super-conductors according to ideas proposed by Fritz London in 1935 and later. It will go just as far as this extension of the facts leads. The reader will see how large the area it embraces is (Laue 1947a)." He presented the various difficulties with other theories and then continued:

> This theory seeks to approach reality by means of successive approximation. Moreover, many inexplicable results give us the suspicion that the experimental sample material is somewhat imperfect. [...]

> Thus all atomic, i.e., quantum-theoretical approaches toward explaining superconduc-tivity stay out of consideration. The author knows well enough that only such an explanation will fully satisfy our demands. Only, the time does not seem to be ripe for it yet. [...]

> One should not forget, though, that quantum mechanical considerations are what had led Fritz London not only to the mathematical form of the theory presented here, but also to the characteristic constant of this theory, the superconduction constant λ, which, according to present knowledge, yielded the correct order of magnitude. It is difficult to account for this agreement as purely accidental (ibid.).

The table of contents divides the text into 18 sections, most of which are only a few pages long. Many of his own papers were incorporated along with the findings of others. The various geometric forms of superconductors had chapters of their own.

Max von Laue sent copies of his books on superconductivity and on the history of physics to many friends. The book on superconductivity went to Erwin Schrödinger, among others, who called it "extraordinarily valuable." The book on the history of physics had gone out to Einstein, who thanked him from Princeton on 15 May 1947: "It is truly commendable that some one who has such an informed survey of the entire course should take the historical account of human memory out of the hands of philologists and wordsmiths, and present the grand drama, brushed free of the dust of trivial detail (UAF, Laue papers, 9.1)."[26] The book on super-conductivity was reviewed by a fellow physicist of Laue's, Erwin Fues. He had the following praise: "Numerous important examples are calculated through to the point that direct comparison with the experimental results is possible, thereby also providing preparation and stimulation for a wealth of new experiments. [...] The author deliberately limits himself to the theoretically transparent conditions of pure elements [...]. Atomistic theories of superconductivity are not addressed, but the book will be a good guide for all specialists working on the atomistic or phenomeno-logical aspects of this interesting phenomenon (Fues 1947)." The review mentioned its application of the theories by Maxwell and London as well as of thermodynamics in the final section (ibid.). Walther Meissner discussed Laue's book for *Naturwis-senschaften* also mentioning its point of departure, the phenomenological theory of

---

[26] In the preface to Laue (1946a) the author recalled Erich Rothacker's invitation to write the book, as well as Arnold Berliner's, who had also encouraged him to do so. (The anti-Jewish persecution cornered Berliner into committing suicide in 1942.) There was also a separate chapter on crystal physics with a historical introduction. Neither in the second nor in the following editions did Laue thank Hans Schimank for his advice.

Fig. 12.8 Diagram of a
"Barlow's wheel"
demonstrating the friction of
electrons in a metallic
conductor. *Source* Pohl
(1912)

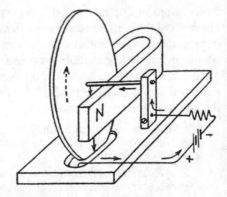

Fritz London, and Laue's extension of it on the basis of Maxwell's theory: "Thus conceived, Laue's text presents an admirably clear and uniformly executed theory that incorporates and explains all the relevant processes in a surprisingly impeccable manner. Moreover, it suggests many ideas for new experiments, which generally is a very special quality of good theories" (Meissner 1947). There followed an extensive synopsis and notes on interesting details. "The decomposition of the current into a normal conducting portion and a superconducting portion explains the experiments on the optically normal behavior of superconductors and the dependence of their high-frequency resistance on temperature (ibid.)."

To an outside observer, scientific life at Göttingen operated as if it were still in the golden age of the Twenties. There was a wide range of lectures on offer on political and scientific topics, not only at the academy. Heisenberg shared Laue's interest in superconductivity and had also presented talks on the subject. He published a paper jointly with Laue on the question: 'Can one build a superconducting Barlow wheel?' (Heisenberg and Laue 1948, received on 1 Sep. 1947). The complex question of the "current collectors" was discussed in detail, because the torque of the wheel depended on them. London's theory provided no information about this. When the transition to superconducting current occurred on the wheel, there would be no angular momentum. When a current occurred otherwise—the authors left open the question of the material involved—the wheel would turn (ibid.)[27] (Fig. 12.8).

A more generally accessible article reveals Laue's thoughts about the development of his science and its presentation. It was the publication 'What is matter? (Laue 1947f)." In the preface to his book 'Matter Waves and Their Interference' (1944) he had confessed that pair production and pair annihilation had left "the deepest personal impression" on him. He now embarked on an analysis of the subject: "To any question: 'What is …', natural science can only answer by enumerating as many properties as possible of the object in question, and that way draw a picture of its behavior under as many different conditions as possible (Laue 1947f: 1217)." He

---

[27] The idea and realization of a conductor rotating in a magnetic field originates from Michael Faraday in 1821. Peter Barlow invented this "wheel" in 1822. The torque is determined by the strength of the current flow, the magnetic field, and the radius of the wheel.

followed these questions back to their historical roots in 19th-century chemistry and physics. The successes of spectroscopy and Bohr's atomic model, the discovery of radioactivity and Rutherford's researches followed, before he moved on to elementary particles. The wave–particle issue and the problem of pair formation and pair anni-hilation were treated. All-pervasive cosmic radiation was briefly mentioned before concluding: "And yet, as if lit up in a flash of lightning, we suddenly see before us here the full gravity of the question: *What is matter?*" (orig. emphasis, p. 1224) Towards the end of the year, on November 27th, Laue wrote to Lise Meitner from Göttingen: "I am now offering the history of physics in one-hour lectures and am working much on my little book about it at the same time. The second edition obviously cannot profit from it anymore, though, which is already in print […] (Lemmerich 1998: 509)." There was still room for Mrs. von Laue to add a postscript: "Dear Miss Meitner, How sweet of you to write me such a warm-hearted letter! Since everything here revolves around my husband, I feel so much like a quantité négligeable and household cook. My husband is busier than ever here and never has time for himself, let alone for his wife. Since I have been suffering for months from an ailing foot and an ailing right arm—the best orthopaedists in the area can't help—that doesn't exactly boost my spirits either, which is why I unfortunately cannot ever go out on a walk anymore because of it, which could have brought back my equanimity. […] Since her illness, Mrs. Hahn sees no farther than her duster, so one can't do anything with her (p. 510)."

A commemorative event in honor of Max Planck was scheduled for 23 April 1948. That required timely preparation. Addresses were to be held by representatives of the Berlin Academy, the Royal Society of London, and the University of Berlin, as well as speeches by Otto Hahn, Max von Laue, Werner Heisenberg, and Richard Becker. The Swedish Academy was invited as well as Lise Meitner.

At the end of November, Laue's investigation entitled 'London's theory for noncubic superconductors' went out to the *Annalen*. By way of introduction he remarked that thus far only a scalar mathematical treatment had been available for the superconduction constant λ for elements forming cubic crystals. Now he showed that λ (the constant which Fritz London had introduced in his superconductivity theory) can be applied as an asymmetric tensor of classical electrodynamics for elements forming noncubic crystals. Following this introduction, he divided the paper into five sections: (I) The energy principle, (II) continuous currents, (III) Meissner effect, (IV) law of momentum and London's potentials, and (V) thermodynamics of the transi-tion from normal conductivity to superconductivity. About the Meissner effect he noted "profound changes"; about IV he remarked: "The forces exerted by a stationary magnetic field on the superconductor must act exclusively at the boundary. Theory must therefore find a mechanism which cancels out the volume forces of the field and transfers them to the boundary surface. […] Thus, the asymmetry of the tensor […] ensures that the supercurrent can flow in a stationary manner (Laue 1948a, IV: 36, 38)."

The rising generation was called upon to shape the future of his beloved physics, and Laue had helped promote promising talent during his Berlin period. In November 1940 he had mentioned to Lise Meitner his appreciation of experiments by Hans Boersch with electrons diffracted off an edge. He had continued to follow his scientific

Fig. 12.9 Founding meeting of the Max Planck Society in Göttingen, 26 February 1948. *From left to right*: Theo Goldschmidt, Wilhelm Bötzkes, Erich Regener, Adolf Grimme, Otto Hahn, Max von Laue, Richard Kuhn, and Adolf Windhaus. *Source* Archive of the Max Planck Society, Berlin-Dahlem

investigations since and, in order to secure for Boersch a position, he now offered him on December 9th a post in the new PTA as senior government councillor, with one collaborator, a mechanic, two assistants, and laboratory space with access to the workshop (AMPG, III rep. 50 no. 300 fol. 9). A few years later—when Laue had returned again to Berlin—Boersch would receive a call from there to the Technical University (TU) in Berlin as full professor.

The political tensions between the three Western Allies and the Soviet power rose significantly. Their alliance had practically fallen apart. The American and English zones had already united to form the bizone at the beginning of 1947. Now the French zone was added as well. The administration of these three Western zones was gradually transferred over to German authorities. The Allies had already dissolved the State of Prussia in February 1947.

On 16 January 1948, Otto Hahn received word that the establishment of a Max Planck Society for the Advancement of Science in the Bizone (MPG) had been approved (Fig. 12.9). The military governments appointed a committee consisting of Otto Hahn, Ernst Telschow, Max von Laue, Richard Kuhn, and Erich Regener to constitute the society, as Hahn informed Laue on February 12th (AMPG, III rep. 50, no. 771). Thus once again Laue was distracted away from his primary focus on physics. However, his publications henceforth received the adscript, "*Göttingen*, Max Planck Institute of Physics."

Two papers on superconductivity were completed in early 1948. In March, Laue delivered a detailed mathematical investigation on 'Superconductivity and electrodynamic potential' to the editors of Zeitschrift für Physik (1949a). It again departed from Maxwell's equations and posed the question (in the case of a superconductor) of the work that must be performed for a quasi-stationary magnetic field to undergo a change in its geometry. Two months later a tabular survey of superconductivity and its crystal classes appeared (1948c), following on the heels of his paper on 'London's theory for noncubic superconductors' (1948a).

No sooner had a Max Planck Institute of Physical Chemistry been officially established, Laue submitted a proposal to the president on March 8th—very probably after consulting with Otto Hahn—to found a Max Planck Institute of Structural Research. "It is intended primarily for the investigation of matter using X-ray and electron interference, but by no means excluding other methods for the same purpose, such as, e.g., electron spectroscopy. Later, perhaps, when the restrictions now imposed on German research are lifted, the extremely important investigations with neutron interference can be added (AMPG, 102.046, BiophCh 01)." Laue would later be able to adopt a large part of this program for a total of five departments four years later at the Institute in Berlin (ibid.).

On April 23rd, the planned memorial event in honor of Max Planck took place in Göttingen with musical accompaniment. Addresses were held by Otto Hahn as president of the MPG, and by the representative of the German Academy of Sciences in Berlin (in the Eastern sector of the city). Sir Charles Darwin spoke on behalf of the Royal Society, followed by the rector of the University of Berlin. The commemorative speeches were delivered by Max von Laue, Werner Heisenberg, and Richard Becker. Lise Meitner was also able to attend in Göttingen and, back home in Stockholm, she wrote to her friend Elisabeth Schiemann on June 5th: "But on the whole I was sincerely glad to have been back in Germany once again, and to have spoken with Hahn and Laue and some others (Lemmerich 2010: 354)." It was more difficult for her to talk about the past with the rest.[28]

There was also the invitation to speak about Max Planck in Tübingen which Laue had received the previous year. On 8 April 1948 Laue obliged and gave a detailed account of the life of his revered teacher. It turns out that Planck had told him many interesting details about his life, such as, Adolf von Baeyer's critical attitude towards theoretical physics which Planck had observed in Munich and about Kirchhoff's opinion of Planck's dissertation in Berlin. Laue especially praised his teacher's endeavors to establish entropy in thermodynamics. For those familiar with Laue's early work, the passage about Mosengeil's dissertation under Planck is interesting because Laue left out his own involvement in revising it for publication after Mosengeil's untimely death in 1906. The talk was mainly devoted to Planck's research on radiation, his insight in recognizing the correlation between energy and entropy. This was followed by an appreciation of Planck's activities at the Prussian Academy of

---

[28] AMPG, II rep. 1A, Planck staff file, no. 42 fol. 1; Laue to Meitner, Göttingen, 1 Jan. 1948. CAC, MTNR 5/32.

Sciences. Laue briefly touched on the period after 1933 before mentioning the many public lectures his teacher had held (Laue 1948d).[29]

Despite—or perhaps because of—her recent visit to Göttingen, Lise Meitner resumed the debate with Laue about the past under the Third Reich in her letter from Stockholm on June 24th (Lemmerich 1998: 517 f.). After a few remarks about Planck's decision not to return home from Italy in January 1933 and Meitner's criticism of it, her friend replied on June 30th:

> You and wide circles abroad reproach the Germans and especially German scholars for not having shown enough resistance to Hitler. Did those circles abroad ever attempt any form of resistance themselves?—which wouldn't have been dangerous for them at all. I don't remember any foreign government making any form of official protest about the murders after June 30th 1934; and yet it would certainly have been worth the trouble to see whether by breaking off diplomatic relations with him Hitler could have been toppled. And in 1936 those supposedly so indignant circles didn't want to forego that little pleasure of coming to the Olympics, traveling rather in hoards to Berlin. It is undoubtedly right that we here have sinned, but even more sins were committed abroad, more because opposition there would have been entirely unrisky (p. 519).[30]

Laue then addressed the possibility of Lise Meitner returning to the MPI of Chemistry. It had been reestablished in Mainz on the left bank of the Rhine, and Fritz Strassmann, with Otto Hahn's approval, had asked Lise Meitner if she would return to her post as head of her department. But she had declined. Laue wrote about this: "You mention some foolish remarks you had heard after the first World War. I don't doubt for a moment that you would hear something similar now and then today, too; but what does that matter to you? Where would we all be if we heeded everything that is spoken or written here? I don't want to influence your decisions in any way, but this one argument against Mainz does not impress me (ibid.)."

Physicists wanted to resume the DPG's awards of the "Planck Medal." But gold was still far too scarce and costly. Max Born had been selected for this year's distinction. Sommerfeld also proposed Lise Meitner, but Laue wrote him from Göttingen on May 25th that he thought it better to award medals to both Otto Hahn and Lise Meitner (DMA, NL 89/010).

Laue had been occupied with the theory of relativity for over four decades. When exactly he had come across the physicist Ludwig Lange (1863–1936) and his publication on the problem of relativity theory, he did not state in his next brief paper on the topic. "Ludwig Lange furthered the problem of the physical frame of reference, incompletely solved by Copernicus, Kepler, and Newton, to the point that only Einstein's theory of relativity could add anything new (Laue 1948e: 193; cf. Lange 1885a, b, 1886, Laue 1982)." Laue analysed the contributions by those aforementioned three scientists as well as Lange's texts:

---

[29] Laue presented far more personal details from Planck's life than in the latter's own 'Scientific selfportrait,' see Scriba 1990: 9–20.

[30] Also in AMPG, III rep. 50 no. 1330. In this correspondence Laue never reproached Lise Meitner for having remained in Germany after 1934 and for not having resigned her position in protest and emigrating, as their mutual friend James Franck had done in 1933 despite initially being permitted to stay as a war veteran.

Lange's solution to the problem sets in place of that 'somewhat spooky' absolute space and absolute time the clear concepts of an inertial system and inertial time. Everything is contained within his two definitions and two theorems. [...]

Theorem I: With respect to an inertial system, the path of every fourth isolated point is also rectilinear. [...]

Theorem II: With respect to an inertial time scale, any other isolated point is also in uniform motion along its inertial path. [...]

These definitions make sense irrespective of experience; obviously they would have little value if experience did not confirm the correctness of the two theorems (Laue 1948e: 193).

Laue then briefly discussed results from long-term astronomical observations: "It is Lange's definitions which lend meaning to such investigations (p. 194)." A few details about Lange's life as well as his relationship to the psychologist Wilhelm Wundt and a synopsis of Wundt's life and illness history complete Laue's report (pp. 194 ff.).

This year Laue had to finish his contribution on 'Inertia and energy' for the volume on Albert Einstein to appear the following year in the series Library of Living Philosophers, edited by Paul Arthur Schilpp. The text was translated into English (Laue 1949b: 503). Laue also made sure the article was published in German, which the publishing house Kohlhammer in Stuttgart was able to realize soon afterwards. In Section I, Laue introduced the topic in outline: "In physics today, the laws of conservation play a fundamental role. There are essentially three: The principle of inertia, which asserts the conservation of momentum; the principle of energy, which asserts the conservation of energy; and the law of the conservation of the quantity of electricity. It is true that there are others, too, e.g., that of the constancy of inertial mass. But, insofar as we still acknowledge it as correct, that has become identical with the theorem of the conservation of energy. Finally, for the modern, relativistic interpretation, the momentum and energy theorems also merge into one. These unifications form the subject of the following arguments (Laue 1949/55: 364, our trans.)." In this introduction Laue discussed the invariability of electric charge, which Michael Faraday had compellingly demonstrated in 1843 with the "ice bucket experiment." This and the energy theorem with its eventful history and entanglements were treated by Laue in the subsequent parts. Part II, on 'The principle of momentum in Newtonian mechanics,' includes a historical survey of chemical reactions for the conservation of mass, with an appreciation of Galileo Galilei, René Descartes, Christiaan Huygens, Leonhard Euler, John Wallis, Louis Poinsot, and Hans Landolt. "As a *physical* problem, mechanics was settled by Newton. Mathematicians were still kept busy with it for another century and a half. But in the process they erected an edifice of such architectonic beauty as Newton himself would never have suspected (p. 368, orig. emphasis). Section III, entitled 'The energy theorem,' again contained a historical review naming the main participants: Gottfried Wilhelm Leibniz, Johannes Bernoulli, Leonhard Euler, Jean-Victor Poncelet, Thomas Young, Sadi Carnot, Émile Clapeyron, Robert Mayer, Ludwig August Colding, James Prescott Joule, Hermann von Helmholtz, William Thomson (Lord Kelvin), Wilhelm Ostwald, Hans Geiger, and Walther Bothe. Under IV, 'The doctrine of the flow of energy,' we read: "The times changed, physical knowledge delved deeper. The epoch that ended with the break-through of the energy principle was not sharply sundered from the following period, characterized by the displacement of the theories of action-at-a-distance by the notion of proximate action, which suited the principle of causality better (p. 375)." Laue named: Heinrich Hertz, Gustav Mie, Hermann von Helmholtz, Michael Faraday, James Clerk Maxwell, John Henry Poynting, Pyotr N. Lebedev, Ernest Fox Nichols and Gordon Ferrie Hull, Walther Gerlach and Alice Golsen, Henri Poincaré, Max Abraham, Max Planck, Kurd von Mosengeil, Frederick Thomas Trouton and Henry R. Noble, Rudolf Tomaschek, and Albert Einstein. Section V, 'The inertia of energy' states: "The insight into the inertia of energy per se was published by Einstein in 1905 (*Annalen der Physik 18, p. 639*), hence in the same year as the foundation of the theory of relativity. He derived it relativistically. And a rigorous derivation of this assumption is indeed required (Laue 1949/55: 380, orig. emphasis; Einstein 1905)." Laue's explanation was a simple thought experiment, starting by supposing an empty cylindrical cavity with a plate at each end. Hertzian oscillations or light is sent from one end A to the other end B, which is completely absorbed at B. Motion from A to B then returns the energy stored at B.

At the end, the energy is distributed as at the beginning, but there remains a center-of-mass displacement in the amount $L/M(\Delta E/c^2 + m_2 - m_1)$ in the direction $B \rightarrow A$.

Now, is it credible that the cylinder could shift its center of gravity, that is, in effect itself, without any external influence and without any change in its interior? This contradicts not only mechanics, but also our entire view of physics, which does, after all, contain venerable and valuable, if not always conscious, empiricism (Laue 1949/55: 381 f.)."

Laue then solved this problem: "Therefore, energy determines inertial mass. Can *all* inertial mass be traced back to the energy content of a body? Does the more general relation $m = E/c^2$ hold? The theory of relativity has answered this question in the affirmative from the beginning as a hypothesis (p. 384, orig. emphasis)." There followed praise for Hermann Minkowski's mathematical formulation as the energy–momentum tensor. In Section VI, Laue briefly discussed the inertia of energy in nuclear physics. In conclusion he stated:

Moreover: Can the concepts of momentum and energy be adopted in any future physics as well? W. Heisenberg's indeterminacy relation, according to which it is impossible to determine with precision the location and momentum of a particle simultaneously—a law of nature prevents it—, can only mean for any causally-minded physicist that at least one of the two concepts occurring in it, locality and momentum, is an insufficient description of the facts. But present-day physics does not know of any substitute. We perceive here, probably particularly vividly, that physics never is finished, but approaches step-by-step eternally evolving Truth (p. 388).

On June 21st the *Deutsche Mark* was introduced in the Western zones. Bank accounts in *Reichsmark* currency were devalued 10:1. Not so "surprisingly," the supply of wares shot up. In reaction to this order, the Soviet administration imposed a blockade on the three Western sectors of Berlin and denied entry. West Berlin was thereupon supplied as best as possible with food, coal for the electricity and gas plants, and other essential goods by airlift. Electricity was intermittently available twice a day, in two-hour intervals at alternating times for households, shops, and many public institutions, such as, schools, which led to considerable restrictions, especially in the autumn and winter.

After the capitulation in 1945, the Kaiser Wilhelm Society still possessed the former Institute of Physical Chemistry and Electrochemistry in Berlin, which under the direction of Peter Adolf Thiessen had been pursuing research important to the war effort for the Nazis. After the war ended, Thiessen accepted an offer to work in the Soviet Union. The KWI and various other research establishments were merged in 1948 to form a research institute called *Deutsche Forschungshochschule*. In 1947, the physicochemist Robert Havemann was appointed director of the Haber Institute. During the Third Reich he had belonged to the political resistance. Sentenced to death, Havemann had been kept busy with war-related work in his cell until his liberation in 1945 (Hecht et al. 1991; Müller-Enberga et al. 2000: 230).

Max von Laue was probably aware of the problems in Berlin. Whether it was discussed among his circle in Göttingen including Otto Hahn, Werner Heisenberg, Ernst Telschow, and Carl Friedrich von Weizsäcker has unfortunately not come down to us, although many of them had private property in Berlin. The Laues' home, for example, had been requisitioned by the authorities and rented out to four families.

Normally, a scientist pursues *one* scientific problem, especially in old age. Laue, however, continued to make important contributions to superconductivity, X-ray interference, and expositions of relativity theory. Prior to his trip to the U.S.A., he completed an important paper which gave rise to a number of new publications throughout the 1950s, on 'The absorption of X-rays in crystals in the case of inter-ference' (Laue 1949c). Gerhard Borrmann had been able to continue his investiga-tions on perfect crystals under the difficult temporary conditions at Hechingen. The abstract of Laue's paper appeared in English: "Experiments by Borrmann show that the absorption of the primary X-ray in a perfect crystal is affected, and sometimes reduced, by the presence of diffracted rays. The paper discusses the case where both the primary and the diffracted rays leave the crystal slab by the same face; the change of absorption of the primary ray is calculated according to the dynamical theory of X-ray diffraction and is found to be in general agreement with Borrmann's observations (Laue 1949c: 106; Borrmann 1941)." In the section on 'Posing the problem,' Laue pointed out: "The *wave fields* which thereby take the place of the simple plane wave, undergo a fundamentally different kind of absorption. Since, a plane wave, when it impinges on a crystal, produces two wave fields in it, which also differ in absorption; and this causes a complication in the relations that calls for exact calculation [...] (Laue 1949c: 106 f., orig. emphasis)." He briefly described Borrmann's experiments "on the passage of X-rays through thin, strictly single-crystal layers characterizing interference effects also in the intensity of the beam passing directly through (p. 107)." Then Laue discussed the various blackenings on the photographic plate. He said that these phenomena were the grounds for the following theoretical exposition. For the solution he referred to the fifth chapter of his book on 'X-ray Interference' (1941). In Chap. 3, 'The absorption of individual wave fields,' the formation of two wave fields, which he designated + and −, was analyzed mathematically. "The two wave fields, excited simultaneously by the incident wave, hence undergo absorption differently (p. 109). The fourth section treated the interaction between the two wave fields mathematically. The intensities of the directly traversing beam through the sample Laue had chosen as his example (strangely enough NaCl) and the diffracted beam were depicted graphically (p. 112).

In the meantime, the preparatory activities for the Laue couple's trip to America had taken on somewhat hectic proportions and Ronald Fraser was fully occupied with settling everything. Laue had been politically screened in advance by the Military Government Department of Public Safety (Special Branch), (AMPG, II rep. 1A, Laue staff file, fol. 21). Just before their departure, an important distinction arrived from the States. On May 12th the American Academy of Arts and Sciences nominated Max von Laue as "Foreign Honorary Member." The great voyage began on July 9th, leading first to the Netherlands, where the couple spent four days visiting friends and museums, as well as seeing Paul Peter Ewald and Franz Simon. As Laue reminisced in a letter to Robert Ruge on board the Niew Amsterdam on July 18th, "A particularly nice reunion was with a young Dutchman who had been abducted to Berlin in 1942–1943 and had often been our guest at our home on Sundays (SUB, R. Ruge papers, 4" Cod. Philos., 221)."

On July 14th they left Rotterdam by steamer for New York. They were picked up there and soon met Theo. Together they continued on to East Sandwich, Cape Cod, where their daughter-in-law was already waiting with their grandson. Max von Laue was delighted with 22-month-old Christopher with whom he could play well on the beach, as he reported to Lise Meitner in a letter written together with his wife Magda on August 27th from the community of West Dover. "But it is Theo who occupies himself most with him. He was always very fond of children (Lemmerich 1998: 520–525, AMPG, II rep. 1A, Laue staff file, fol. 37)." There they visited James Franck, who was quite ill, and Laue took part in the first congress of the International Union of Crystallographers in Cambridge, Massachusetts. Max von Laue was elected its honorary president (as a German he could not be elected onto its board). He had to share with Lise Meitner his enthusiasm about the congress: "There was such an abundance of good presentations as I have seldom experienced at a congress. You probably know that crystal interference is now being applied to neutron beams. Wonderful new results were announced and the corresponding exposures were shown around. What a pity that uranium piles are needed for this! So we poor Europeans have little prospect of participating in such investigations. It gives me special pleasure to find that the elementary theory of these interferences offers even better approximation for neutrons than for X-rays (p. 521)." Laue mentioned he was working on this for the second edition of his book on 'Matter Waves.' The results about crystal growth, especially large quartz crystals, also excited him. A few years later, that became important for Borrmann's research in Berlin at the MPI headed by Laue.

From Cambridge, James Franck's son-in-law, Arthur von Hippel, drove them to the log cabin in the New Hampshire mountains that had been a gift to Theo's wife from her aunts. Max von Laue had to go to Chicago for two days, where he was awarded an honorary doctorate from the university. He did not tell Lise Meitner that he had been titled "champion of freedom." He did write her that he could read the papers a lot. "There is much in them about Berlin and Moscow; it isn't pleasant. The Americans are sure that they would win a possible war against Russia, albeit with great casualties, which they'd prefer to avoid. The aspect that the Western European nations would then have to suffer horrendously [...] is rarely expressed (p. 523)." The Max Planck Society was completely unknown in America, he said.

In a separate message to Meitner of the same date Mrs. von Laue reported somewhat appalled about daily life, especially about the completely different division of labor between husband and wife in the household, but also about alienation due to different ways of thinking: "This goes so far that Theo is influenced by his wife and her family. And that he is ashamed of us, and we absolutely don't understand each other—he hardly listens when we talk to him about Europe, and then we find it really hard to say anything when the other party doesn't want to listen. So we haven't become closer, rather more estranged because of different attitudes (CAC, MTNR 5/32)." She then added that her husband would never have gotten through his workload if she hadn't assumed more than half of the housework (ibid.).

Almost a month later, Lise Meitner received the next report from the USA, this time from Princeton at the Institute for Advanced Study, with Laue's cordial wishes on her birthday. He began with a retrospective on their first meeting in 1908, if it hadn't rather been 1919, in Berlin. "And since then we have experienced many things together in loyal friendship, from advising our despairing colleague Haber, to the ascent of the Caputschin. It is beyond estimation how often we visited our highly revered teacher and friend Max Planck together [...] (Lemmerich 1998: 525)." He paid tribute to Lise Meitner's achievements for science. "In any case, the history of science will one day speak of two great women physicists of this period and, if it is sensible, will not pose the question which of them had been greater but will be pleased that two of this sort existed. The German Physical Society in the British Zone, at any rate, has conferred upon you honorary membership in recognition of your high merits (ibid.)." And he congratulated her on this distinction. All told, Laue visited twelve universities in the USA. Laue did not write either to Lise Meitner or to Walther Meissner about his encounters with Einstein, which very probably occurred.

At the beginning of the new year, on 8 January 1949, Laue wrote from Pasadena to Dr. Ilse Rosenthal-Schneider in Australia. She had formerly audited his courses, as well as those of Planck and Einstein during the twenties. He thanked her for her Christmas letter and in response to her question about whether he wouldn't like to stay in the U.S. his answer was a decided "no." First citing reasons of health, he then came to the main reasons, "that I can never be anything else here than the spare wheel on the cart, and that I am needed in Germany. The core of the German problem seems to me to be the reestablishment of a culture-bearing stratum, after the former one, the so badly denigrated bourgeoisie, has collapsed under the hostilities from the right and the left, as well as from abroad, admittedly not without bearing some fault of its own. The value of any nation is determined by its cultural achievements, and this applies doubly and triply to us now, as we are barred from every other activity. But there are not many left in Germany who can work at this task with some prospect of success (DMA, NL 264)."[31] Laue considered it his duty to persevere in Germany, despite the hardships and risks involved. Talking about German philosophy of history, he mentioned to her a talk that his son had held about Leopold von Ranke and historicism. Theo had given him the manuscript to read and he much appreciated this enlightening instruction (ibid.).

Writing "in good time" on February 20th from Berkeley, Laue sent Einstein his congratulations on his 70th birthday, as he would be on board ship on the Atlantic Ocean on the day itself. He recalled their first meeting in August 1906 at the Swiss Patent Office, where Einstein had welcomed him in the hallway in his shirtsleeves. He hoped he would still be able to see him at Princeton on March 3rd (UAF, Laue papers, 9.1). Somewhat belatedly, Laue's public tribute to him in honor of that occasion appeared in German in *Reviews of Modern Physics* with a compressed list of his scientific achievements and their reception (Laue 1949d). Laue asked: What do people admire most about Einstein? After mentioning his mathematical aptitude, the versatility of his talents, and his impartiality towards existing theories, he concluded: "Rather above this is the sheer genius, the immediate and, having once been expressed, so simple insight into that which is of essence in nature. It has proved equally effective in both its branches. But, in order to be effective, another character trait has to be added—namely, an absolute honesty and uncommon courage of conviction to fight against age-old, ingrained, powerful notions—initially entirely on his own (ibid.)."

While the von Laue couple were still away from home, an American relief program known as the "Marshall Plan," named after its initiator Secretary of State George Catlett Marshall, became reality. After the Eastern Bloc rejected participation in it, only Western European states, including West Germany and West Berlin, received financial and material support as well as access to research equipment. Max von Laue first learned about it confidentially from the Ford Foundation, which was planning another aid program. He immediately let some of his colleagues know about these new opportunities, which also included the exchange of researchers and scientific journals. Letters addressed to Bothe, Carl Wilhelm Correns, Hahn, and Meissner left Pasadena on 10 February 1949 (DMA, NL 045/016).

The von Laue couple returned to Germany in March. Much had fundamentally changed. In the interim, the three Western powers had given the provincial governments the mandate to form a constituent assembly. This was seen by some German politicians as a dangerous step toward the division of Germany, but the Western Allies did not recognize this. So the Parliamentary Council was formed and the constitution of the Federal Republic of Germany—its Basic Law—was drawn up. Bonn was to become the federal capital. The tense political situation, the East–West conflict, very soon led to the formation of military units and the production of war equipment on both sides of the inner German border. West Berlin recovered only gradually after the blockade was lifted on May 12th. In the American sector there was the "Free

---

[31] I am grateful to Dr. Wilhelm Füßl for having pointed out to me the existence of the Rosenthal-Schneider collection.

University of Berlin" (FU)—newly founded in protest against the ongoing politicization of the University of Berlin located in the Eastern sector of the city. The polytechnic *Technische Hochschule* was renamed *Technische Universität* (TU) from 1946. Initially, nuclear physics was not taught at the universities. The Berlin blockade and the uncertainty about its political future made West Berlin lose many scientists, scholars, and other experts. Appointments to West Berlin became complicated due to its insular political situation.

Soon after his return to Göttingen from the U.S., Laue informed Lise Meitner by telegram on April 24th that the West German Physical Societies had approved the award of that year's "Planck Medal" to Lise Meitner and Otto Hahn (CAC, MTNR 5/32). She replied to Laue's message from Stockholm on the following day: "If any award can please me, then it certainly is this one. I am sure you, who—I believe—also owns the medal, will understand this fully and, who—as I know for sure—shares my great love and veneration of Planck. [...] Any tie that binds me to the old Germany I love so much, the Germany which I cannot thank enough for the decisive years of my scientific development, for the deep pleasure in doing scientific research, and for a very dear circle of friends, is a very precious gift to me (Lemmerich 1998: 528)." She wrote that she intended to come to Bonn for the conference—which was scheduled from September 5th to 16th—and also asked for news about his stay in America. Laue wrote back from Göttingen on April 30th with a detailed account, including about meetings he had had in the U.S. with Ladenburg and Franck, Debye, Max Delbrück, and Einstein: "He is still compulsively pondering about unified field theory, but without hope of overcoming the mathematical difficulties. Nevertheless, his loyal secretary, Miss Dukas, told me he had just finished a very fine new publication. Otherwise he is delightfully relaxed and serene (CAC, MTNR 5/32)." Laue told Meitner also that Hartmut Kallmann, who had not been in the U.S. for long, scolded no end about the political circumstances in Germany. That the Nazis were gaining influence everywhere, etc. And the only thing Laue could say was: "Unfortunately, he is not entirely wrong (ibid.)."

On April 24th Einstein replied, apparently rather belatedly, to a question that Laue had posed about the general theory of relativity: "I feel guilty about not having written to you, either. But in the end it does work. If you do treat general relativity, I think you should approach it from the view that the law of gravitation in empty space implies the law of motion (UAF, Laue papers, 9.1)." Considerations about abandoning symmetry followed (ibid.).

Immediately upon returning from a short trip to West Berlin, still at Bückeburg airfield, Laue wrote on July 2nd to Carl Ramsauer, who held the chair for experimental physics at the Technical University in West Berlin and had been the last chairman of the German Physical Society until 1945. He informed him that the five regional associations in the Federal Republic would now unite and invited the "Berliners" to the meeting in Munich.

In a letter from Göttingen dated July 3rd Mrs. Rosenthal-Schneider received an account from Laue about his experiences in Berlin. They show what life was like at that time in this area of Berlin:

From the 29th of Jun. to the 2nd of Jul. I was in Berlin for the first time since the war. I wanted to stay a day longer; but the gods did not want to bear the responsibility for my safety there any longer than necessary. Besides, I never came anywhere near the Russian zone, of course, or even the Russian sector. A 'shadow' accompanied me from the moment I landed at Gatow until I boarded the omnibus by which the Royal Air Force transports its protégés from the travel agency at Fehrbelliner Platz to Gatow airport. I constantly drove along in a car belonging to the American military government and lived with a member of it, which naturally had some very pleasant aspects to it. I just was not allowed to send out any telegrams from this apartment into Russian-occupied territory; that could have endangered the recipient. Once I went for a short walk along the Wannsee shore with a juvenile relative; my shadow always remained within eyeshot. If I then add that in almost undamaged Dahlem, where I was living, one suffers from the constant buzzing of planes ascending and flying low over Berlin from Tempelhof, then you have a rough picture of the situation in Berlin; the permanent residents hardly hear it anymore, though (DMA, NL 264).

Laue also described to her the three rival universities in Berlin and reported that he had presented a talk on superconductivity and the wavelength measurement of γ-rays at the former Haber Institute. He mentioned that the academy was getting increasingly swept up in the Bolshevistic wake (ibid.).

Around this time, a staff member of the *Darmstädter Echo* published an interview on July 27th quoting Max von Laue as criticizing the situation of German scientists in USA. This was noticed in the United States and caused a stir. Paul Peter Ewald wrote Laue about it from Brooklyn on November 23rd and 19 January 1950 in protest (AMPG, II rep. 50 nos. 562 f.). However, such an "interview" had never taken place. While on a trip in England, Laue read the first letter by his friend and answered on December 5th from Kensington: "Dear Ewald, I got to read your letter about the alleged interview on the train between Cambridge and London today. I am outraged. The 'Darmstädter Echo' is as unfamiliar to me as is 'Dr. W.' Nor do I remember having received a journalist in July, that is 4 months after my return, and talked to him about U.S.A. Several appeared before then, but I certainly turned them *all away* (orig. emphasis, UAF, Laue papers, 3.2. Laue wrote again from Göttingen on 28 Dec. 1949.)."

In mid-June, Max von Laue had sent a lengthy manuscript from the Göttingen Max Planck Institute of Physics to the editors of the *Annalen*: 'A nonlinear phenomenological theory of superconductivity' (Laue 1949e). His theory set out from what had already been achieved, "but replaces the hitherto linear relation between superconducting momentum and superconducting current density by a nonlinear one, which still leaves them undefined to a large extent. […] This new form is suitable for incorporating the idea of a maximum current density, thus permitting a quantum theory of superconductivity (p. 197)." He simultaneously extended the theory to noncubic crystals. He had been able to consult a related consideration by Heinz Koppe prior to publication and critically remarked that its author had not touched the basic question of any theory of superconductivity, "namely, whether there is a system of potentials dependent on the vector J^1 which, in the stationary case, absorbs the forces of the magnetic field on the carriers of the supercurrent and transmits them to the surface of the conductor […] (p. 198)." The further mathematical treatment by Laue led to a type of solution that describes the Meissner effect, i.e., the displacement of the field from the interior to the layer below the surface. Heisenberg's theory of a maximum, unsurpassable current density could be incorporated.

The Laue couple spent their summer vacation in Großholzleute near the town Isny im Allgäu in southern Germany to be close to their daughter. They spent several hours together on long hikes. Max von Laue read his son's first book about Leopold von Ranke with interest, as he told Theo in a letter dated Aug. 4th (AMPG, III rep. 50 suppl. 7/10).

In September, the Physicists' Convention was held in the federal capital Bonn (Figs. 12.10 and 12.11). The highpoint was on September 23rd. In the presence of the newly elected first federal president, Theodor Heuss, the Golden Planck Medal was awarded to Lise Meitner and Otto Hahn. The laudatio was held by Max von Laue as chairman of the DPG. He sketched the history of the medal, which had initially only been intended to honor research in theoretical physics. According to Laue, Planck himself had abolished this prerogative after a while. Laue's address on the two laureates offered an overview of their biographies and their research, including the discovery of nuclear fission and its consequences. He concluded with the optimistic prospect of applying nuclear fission as a source of energy. The first to congratulate the medalists was President Theodor Heuss, who briefly said something about the importance of research (see the proceedings of the convention in Bonn, DPG 1949: 511).

One festive occasion followed the next. Max von Laue celebrated his 70th birthday with his wife and their daughter and grandchildren in Tailfingen. Back in Göttingen, he found the stack of birthday correspondence waiting, but some letters had arrived early. He had already thanked Einstein for his congratulations on October 4th: "You write me some rather flattering words of praise in your birthday letter. But the very best things I have accomplished, I did not perceive myself to have been the driving force of, rather, having been led there; sometimes quite differently from what I had imagined. I suppose that happens to many people. I have never been able to discard my attachment to Germany, and even now I hope for a spiritual rebirth, even in the event that we do become more closely acquainted with Bolshevism. That case, incidentally, does not yet seem to me to be inevitably imminent (UAF, Laue papers, 9.1)." Upon his return to Göttingen, Laue also found a congratulatory letter from Mrs. Rosenthal-Scheider, to whom he replied on October 19th: "Incidentally, Thersites was not absent among the chorus of well-wishers. Johannes Stark had a diatribe printed against Sommerfeld, W. Voigt, Heisenberg, and me (DMA, NL 264)." Lise Meitner wrote a long letter on October 5th from Stockholm recalling their first meeting, the motorcycle tours, their last excursion by car together before her escape, before mentioning their very personal friendship.

[We had] many happy hours together, we discussed many a factual or personal sorrow with one another in confidence and thereby were able to resolve or at least alleviate it. That I was able to endure the years 33–38 in Germany, I really owe to a considerable part to you and your wife.

Not just that I was warmly welcomed whenever I wanted to come and could count on your understanding for my principal attitudes. It meant almost more to me that both of you saw so clearly the unfortunate effects of Nazi ideology and did not allow yourselves to be

Fig. 12.10 Lise Meitner in the Physics Institute in Bonn giving a lecture after the award of the Planck Medal, 23 September 1949. *Source* Archive of the Max Planck Society, Berlin-Dahlem

Fig. 12.11 *First row from the right*: Otto Hahn, Theodor Heuss, Lise Meitner, Nelly Planck, *second row from the right*: Clemens Schaefer, M. Pahl, and Richard Becker during the address by Max von Laue in Bonn, 23 September 1949. *Source* Archives of the Max Planck Society, Berlin-Dahlem

Fig. 12.12 Max von Laue, about 1950. *Source* Laue (1961, vol. I, frontispiece)

distracted throughout the war years and tried to help wherever it was possible to help. This greatly bolstered my confidence in people and helped me to maintain the moral ground upon which we must stand if life is to retain any deeper meaning (CAC, MTNR 5/32).

James Franck also wished him all the best on his 70th birthday from Chicago on September 28th: "It meant much to me and continues to do so unchanged that friends like you exist in Germany who constantly use their great gifts for the enrichment of our science, and in doing so don't forget their primary duty of being a human being. It cannot and will not be forgotten that you, undeterred by death and the devil, have always stood up for genuine and noble humanity with your whole person (AMPG, III rep. 50 no. 626)." Laue received a Saba radio set, the costs for the banquet in the Ratskeller in Göttingen, as well as the travel expenses to Tailfingen as a gift from the Max Planck Society. Arnold Sommerfeld's birthday wishes were published in the tenth issue of *Physikalische Blätter*. Sommerfeld emphasized Laue's close relationship with Planck. "The essence and oeuvre of Max von Laue can probably best be characterized by the label *Planck's most outstanding and loyal pupil*. Beginning with his Berlin doctoral thesis, he deepened the concept of entropy by introducing it into the optics of interference phenomena. From Planck he has also inherited that conceptual purity and absolute conscientiousness in all his writings (Sommerfeld 1949, orig. emphasis)." Sommerfeld's presentation focused on the multilayered significance of Laue's discovery of X-ray interference (Fig. 12.12).

William L. Bragg also paid tribute to the discovery, which became public just as he was embarking on his own career and which had then formed the basis of his research. "He has also won for himself a place in our hearts. To those who have the privilege of counting him as a personal friend, he is warmly loved for his kindness, his courtesy, his courage and his staunch adherence to his principles. If I may so express it, he has gained international affection and esteem as well as international fame (Bragg 1949: 444, orig. English)." In October at the "Laue colloquium" in Göttingen, Walther Meissner characterized his friend by the words: "If I personally wanted to laud you as a scientist, I couldn't do it better than by saying: Max Planck entirely rightly and deeply justifiably called you his favorite pupil. You have not only adopted Planck's crystal-clear, penetrating scientific way of thinking, which I too was privileged to acquaint myself with as his student, but also—I am surely not saying too much by this—you have extended that easy mastery of mathematics along a certain line of very strict self-critique (DMA, NL 045, Meissner)."

Notwithstanding the paper shortages, a second edition of *Röntgenstrahl-Interferenzen* had been able to appear at the Akademische Verlagsanstalt in 1948. Laue's article, 'Historical notes on superconductivity' in *Forschungen und Fortschritte* (Laue 1949f), now offered instruction to readers without the mathematics. After looking back on the many people involved and acknowledging their contributions towards its resolution, it discussed still open questions. On the new phenomenological theory Laue remarked: "Admittedly, this theory does lose a feature which it had hitherto had in common with Maxwell's theory—namely, that its equations and differential equations be linear. It is now no longer the case that by superposition we could form a third field from two fields possible according to it. For the mathematical elaboration of the theory, this means a great complication (p. 280)." This intense occupation with superconductivity by no means suppressed Laue's interest in X-ray diffraction, however, as the paper published in *Acta Crystallographica* on 'The absorption of X-rays in crystals in the case of interference,' for instance, demonstrates (Laue 1949g, cf. Laue 1952). Inspired by the results of Gerhard Borrmann's investigations on nearly ideal crystals, Laue showed that the refractive index and normal absorption lose their significance.

In the autumn the eldest granddaughter moved in with her grandparents in Göttingen, and after Christmas the second also, because the Laues' daughter Hildegard Lemcke was expecting her third child. The Laues engaged a kindergarten teacher to look after them, and their grandfather read them fairy tales at bedtime. Laue's Christmas and New Year's greetings to Sommerfeld left Göttingen as late as 23 December 1949 (DMA, NL 089). Besides writing him about family events, he also mentioned his participation as a German delegate in the conference of the "European Movement" in Lausanne. Much sincere goodwill about coming to an understanding among European nations had been expressed, he noted. He urged that the German language maintain its importance in communications around the world. Then he called Sommerfeld's attention to Borrmann's success in measuring the anomalous absorption of X-rays in a crystal, which had been diminished by a factor of $10^{-22.7}$, in the interference case (ibid., Borrmann 1950: 299).

At the beginning of the new year Laue received a copy of the "Einstein volume" in the series Library of Living Philosophers (Schilpp 1949) and wrote to Einstein on 4 February 1950 that up to that point he had mainly read his Autobiographical Notes (Einstein 1949a): "Let me tell you that I have seldom read anything so interesting and full of substance. By the way, this book comes just at the right moment for me, as I am about to follow the request by Vieweg publishers and revise my book on relativity: in particular, I intend to alter much in Volume II on the general theory of relativity (UAF, Laue papers, 9.1)." He took the liberty to criticize Einstein's remark about concepts "close to experience." At first he had been afraid of reading Philipp Frank's book on Einstein, but now he wanted to mention his account of the 1933 academy session in Einstein's absence, with the vacant chairs (ibid., Frank 1947). On February 27th Laue also sent his congratulations to Richard Gans on his 70th birthday, who had been living in Argentina with his sons since 1947 (AMPG, III rep. 50 no. 669). The well-wisher also wrote him about his impressions of the Einstein volume edited by Paul Arthur Schilpp and Einstein's epilogue about Kant's epistemology with the citation: "The world of physics is not given to man, but assigned." And Laue continued: "The way physicists have gone about this task is something I felt very strongly as I was writing my History of Physics. But in no epoch has it been more apparent than in the last 50 years, that we are all working together on solving it—whether we know it or not (ibid., Einstein 1949b)." A historical retrospective by Laue on the development of superconductivity also appeared in a scientific journal in Argentina in honor of Gans (Laue 1950a).

Gerhard Borrmann had accompanied Laue to Hechingen in 1944 and had been conducting experimental research there since then on the "anomalous absorption of X-rays in crystals in the interference case." Now the position there in Hechingen was supposed to be discontinued. Laue applied to the Emergency Association of German Science (*Notgemeinschaft*) in a letter dated January 10th, attaching an expert opinion, stating that competent English experts had found Borrmann's findings highly interesting (AMPG, III rep. 50 no. 324).

The invitations to speak kept coming in. The Society of German Natural Scientists and Physicians intended to honor its member, Professor Paul Pfeiffer, at a conference of chemistry lecturers in Bonn on April 22nd. Laue agreed to give the honorary address. Since crystal structures had played a part in Pfeiffer's scientific activities, Laue chose this topic for his speech. Johannes Kepler had presented a short text about the hexagonally shaped snowflake to his protector at the court of Emperor Rudolf on New Year's Day: "In his youth he had attempted to correlate the distances between the planets and the Sun with the dimensions of certain regular polyhedra, and one of his later major works bore the significant title 'Harmonices mundi' (AMPG, III rep. 50 no. 1514)." Kepler's pursuit of geometric order led into a discussion of symmetrical atomic arrangements in inorganic crystals. The closing statement was: "And thus we see in the work of our jubilarian one of the most impressive pieces of evidence of truth in science (ibid.)."

Laue's letter on March 5th was timely in sending his friend Einstein his birthday greetings. Having finished reading the Einstein volume of the Library of Living Philosophers series, he remarked to him that he had gathered from it "what you actually had been doing during those years after your disappearance to the U.S. I, and probably all German physicists, had no idea that you were publishing many a philosophical paper and giving many a speech on the same topic." He continued, "Nothing pleased me more than your discussion of Kant, and that you make the a[lpha] and o[mega] of his epistemology your own—namely, that the physical world is not given, but assigned. This I do indeed regard as the lasting gain of the Critique of Pure Reason, which otherwise proceeds as the Bible." Continuing this train of thought further down he recollected, "You last spoke to me at Princeton about a closed universal field theory not being allowed to assume any physical constants from experience, but must be able to define theoretically the velocity of light, the elementary charge, the quantum of action, or the gravitational constant (UAF, Laue papers, 9.1)." He mentioned that Mrs.Rosenthal-Schneider (1949) had presented this in the Schilpp volume as well and, asking for his response, Laue wondered: "How should this work? Purely mathematical operations with dimensionless quantities never get you out of the realm of dimensionlessness! (Ibid.)."

Laue was just as punctilious after his 70th birthday as before. If in his opinion someone within his fields of expertise had not expressed himself precisely enough, he had to count on Laue raising objection. When Eduard Rüchardt published (1950) an interpretation of light fluxes and distinguished them from interference phenomena, Laue proved that there was no essential difference between those two forms at all. "Whether we speak about one or the other is merely a matter of the frame of reference used. And nothing changes when we consider the unavoidable spectral width of any optical radiation; the two waves only have to be coherent (Laue 1950b)."

Science notwithstanding, daily life continued to take its course. The Laue grandparents enjoyed a visit by their eldest granddaughter Grete. Another visitor, a physicist friend of theirs from Vienna, Hans Thirring, gave occasion for an excursion by car to Kassel. In his letter from Göttingen on May 15th, Max von Laue enthusiastically told his son Theo about it: "I have seldom felt so comfortable in my life as I do now (AMPG, III rep. 50 suppl. 7/10)."

The next publication that summer was a purely mathematical consideration 'On Minkowski's electrodynamics of moving bodies' (Laue 1950c). It was the fruit of Laue's revision work for a new edition of his 'Theory of Relativity' (1921). The summary of that paper presents the question at issue: "In this electrodynamics, no decision has yet been reached between the Minkowskian world tensor and other approaches to it. The former is asymmetrical and therefore contradicts Planck's version of the law of the inertia of energy: momentum density equals energy flux divided by the square of the velocity of light. The other approaches are intended to save precisely this symmetry of the tensor, that is, this version of the law of inertia. This paper aims to prove that the Minkowskian approach is the correct one (Laue 1950c: 387)."

The former Prussian Academy of Sciences in Berlin, renamed the German Academy of Sciences, intended to celebrate its 250th anniversary in the Eastern sector of the city and invited former members, such as Max von Laue. Apparently Laue had informed Hartmut Kallmann in New York about this. Kallmann wrote back very agitatedly on June 6th and implored Laue not to attend under any circumstances.

He reminded him of how foreigners had flocked to Berlin in 1936 and bolstered Hitler's prestige and power. "And today! Instead of the Nazis, the Communists are committing the same misdeeds, perhaps even greater ones. Thus, any participation in an institution promoted by the Communists only helps the Communists and their crimes." He added: "I know from the Nazi era how terrible it is to be oppressed and to see other people who are free voluntarily come and sing along (AMPG, III rep. 50 no. 991)."[32] Max von Laue had already refused to participate from the outset, however, but for reasons of personal safety.

Before leaving on vacation, Laue wrote to Einstein on July 12th to inform him about the progress he was making on the new edition of his book on relativity theory. He was sure, he told Einstein, that Minkowski was right about introducing an asymmetric tensor, "because his approach leads to the addition theorem of velocities holding for the beam velocity (energy flux/energy density), which is absolutely necessary (UAF, Laue papers, 9.1)." From mid-July to mid-August, the couple recuperated again in the Allgäu region near Isny, close enough to be able to visit their daughter and her family in Biberach. However, Laue had probably been continuing to think about the reedition of the volume on the general theory of relativity during that vacation, because on September 8th he wrote again to Einstein from Göttingen. This time it was about the mathematical treatment of co-rotating measuring rods, and Laue told him how he conceived it: "I'd like to think that one may only measure with free gauges, i.e., ones whose parts all have geodesic world lines, therefore fall freely. Any force exerted on a gauge, e.g., the one which forces it onto the circular orbit prescribed by you, falsifies it. At each precision measurement of distances, where the gauges cannot be allowed to fall freely, of course, the experimenter considers very carefully the error that can arise by resting the gauge on a support, and chooses the support in such a way that the resulting, basically unavoidable falsifications remain below his accuracy limit. That's why your proof seems to me erroneous (ibid.; cf. Einstein 1950: 59 f.)."

Mrs. Rosenthal-Schneider received news from Laue from Bad Nauheim on October 9th about the busy year. He had to attend meetings by the *Physikalisch-Technische Bundesanstalt* (PTB), as the federal bureau of standards was now called, in Bonn. He had driven the institute car from Göttingen to Bonn and then onwards to Bad Nauheim. "It is a fine season for motoring; pleasant temperatures, and the foliage more gloriously colorful than at any other time of the year (DMA, NL 264)." The reason for this stop was that the Physicists' Convention was taking place in Bad Nauheim from October 11th to the 15th, but, as he told her, he hoped to be able to leave early, as he had to be in Berlin on the 16th. "For, I have been asked whether I would like to take over the former Haber Institute as its director, and I must study the conditions on site; as, the decision is difficult. Bear in mind that if I moved to West Berlin, I could only get out again by air. I can't coast through the Mark Brandenburg on Sunday excursions, as I used to love to do, but would be locked up inside a vast field of rubble. That isn't easy. On the other hand, so many attach importance to my going to Dahlem, and for valid reasons [...] (ibid.)." After his stay in Berlin, he would

---

[32] See also Laue to Meissner, Göttingen, 7 Mar. 1950. DMA, NL 045/006/1.

be attending the conference of German Natural Scientists and Physicians in Munich, but before that, he had to take care of some business in Stuttgart. On December 10th, he and his wife were planning to go to Stockholm for the Nobel awards (ibid.).

At the Physicists' Convention, Laue managed to spearhead the foundation of the Federation of German Physical Societies (*Verband Deutscher Physikalischer Gesellschaften*). This prevented the coexistence of several regional associations. On October 16th, Laue flew to Berlin—for safety reasons he was not allowed to travel by car or train through the Soviet-occupied zone—to spend four days informing himself about the work and equipment at the Kaiser Wilhelm Institute of Physical Chemistry. Hans-Dietrich Schmidt-Ott gave him a tour of the institute. As they were viewing the empty library, which had been cleared out at Otto Hahn's orders, Schmidt-Ott commented, "We call this the *Hahnebücherei*"—a pun on a word meaning absurdly impractical—which made Laue burst out laughing. Besides his own future research, Laue was greatly interested in having Borrmann's investigations continue. Back in Göttingen, he impatiently informed Borrmann of his impressions of the research facilities in Berlin. He offered Borrmann an assistant position with the salary of senior assistant. He described "a large laboratory now set up for chemical testing, but which could be remodeled. It would have enough space for you and one or two coworkers. [...] Whether apparatus is available for you, I could not find out exactly given the short time. It could be that there is a high-voltage X-ray apparatus."[33] Laue suggested that he go to Berlin and see for himself (ibid.).

In the interval between the side trip to Berlin and the letter to Borrmann, Laue was in Munich to join the Assembly of German Natural Scientists and Physicians and present a talk on 'Matter waves' with a strong historical bent on October 23rd. He introduced his remarks by showing several images observed in the microscope with objects in bright or dark field illumination. Then came images from an electron microscope, and one could see the diffraction patterns at the boundaries as published by Hans Boersch in electron diffraction off an edge. There followed a detailed historical survey of theories and interference experiments, first with electrons assuming de Broglie waves, Walter Elsasser's ideas about them, and Erwin Schrödinger's theoretical contribution. Then Laue treated neutron diffraction, conducted in the USA with neutron sources in nuclear reactors, and showed the corresponding pictures. He speculated about a probability distribution because, he argued, there is no particle independent of its environment. He doubted the notion of complementarity—of wave/particle—as being the solution and also the current interpretation of abandoning determinacy (Laue 1951).

On 6 November 1950, Karl Friedrich Bonhoeffer informed the chairman of the Foundation Council of the German Research University (*Deutsche Forschungshochschule*) of his intention to relinquish the direction of the Kaiser Wilhelm Institute of Physical Chemistry. He had already communicated this decision to the department heads and asked the Foundation Council to offer Professor Max von Laue the

---

[33] Laue to Borrmann, no loc., 28 Oct. and 7 Nov. 1950. AMPG, III rep. 50 no. 324. Personal communication to the author. Otto Hahn had ordered that the entire collection be transferred to Göttingen.

Fig. 12.13 Lise Meitner, Magda and Max von Laue on a visit in Stockholm for the Nobel awards. *Source* Archive of the Max Planck Society, Berlin-Dahlem

directorship of the institute, which had been "most enthusiastically" welcomed by all those gentlemen. Laue had already visited the Institute, he added, and "we have the impression that it is possible to win him over for this position."[34] On December 4th, the magistrate of West Berlin decided to dissolve the "Dahlem Research Group," which included the Kaiser Wilhelm Institute of Physical Chemistry and Electro-chemistry, still named as such. It was subsumed by the German Research University (AMPG, II rep. 1A, Az. 1 A 9/1 subfile 1, decision no. 2104).

Major festivities were organized in Stockholm from 5 to 13 December 1950 to celebrate 50 years of the Nobel prize. Max von Laue was invited as one of the seniors and traveled there together with his wife (Fig. 12.13). Lise Meitner helped them settle the question of what to wear. The report that Magda von Laue wrote about these events for her children provides many details (UAF, Laue papers, 1.17). On December 6th Laue spoke before the Swedish Academy of Sciences. He reported about the dissolution of the Kaiser Wilhelm Society at the decision of the Control Council in 1946, the reorganization and resuscitation of research in Germany, the troubling political pasts of researchers, and the rush of students on universities. About the Eastern sector of Berlin, Laue mentioned the hard political stance held by the Academy of Sciences and the drain of its membership into the Federal Republic of Germany. He also acknowledged the federal bureau of standards, the *Physikalisch Technische Bundesanstalt*, and Fraser's active engagement. Other topics included the

---

[34] Bonhoeffer to Minister Stein, no loc., 6 Nov. 1950. AMPG, I rep. 36 no. 14/1. Bonhoeffer had already informed the President on 20 Sep. that he had written to Laue asking whether he would be willing to go to Berlin: AMPG, 102 046, BiophCh 01.

importance of the Emergency Association of German Science, the *Notgemeinschaft*, primarily in student aid. He closed by expressing the wish that true, lasting peace become reality. His honorarium sufficed to cover the incidental expenses of the trip.

A reception hosted by the president of the Swedish Academy was held on December 9th and the grand banquet on December 10th welcomed 900 guests at the concert hall, 250 of whom were invitees, the others had to pay entry. Although Max von Laue had no female companion at table, he cut a fine figure in his borrowed tailcoat. The Nobel laureates were asked to give a short speech. Laue spoke last and, in his wife's opinion, "the best, with the most dignity, substance, and clarity" as she wrote in a letter to their children. Laue had said: "I have to begin by expressing our deeply felt gratitude. Everyone has done so at the Nobel lectures, but it is quite appropriate to reiterate these thanks today. For, in the meantime each of us has repeatedly experienced what the prize signifies—namely, unequalled support in the pursuit of scientific inquiry even during periods that were sometimes very discouraging (ibid.)." He thanked the Swedes for their hospitality. Then Laue mentioned Röntgen as the first laureate and asked whether the Swedish Academy had in fact been able to make such a happy choice every year. And the answer he could give was "yes." He noted the failure of politicians to materialize lasting peace around the world. He ended with an appreciation of the great merits of the Swedish Academy. This trip was naturally a good opportunity to meet Lise Meitner again, but the Laues also went shopping for clothing and other wares not available at home (ibid.).

The new year had hardly begun when Laue, back home in Göttingen, wrote on 8 January 1951 a rather substantial letter to Albert Einstein. He forewarned in preparation: "Today I have to write to you about two very different matters. The first question you'll perhaps write off as quite stupid, and the second matter will, I fear, upset you very much. Nevertheless, I must bring up both of them." Starting with his scientific question, he continued: "In general relativity the g-brackets are supposed to be something like the field strength of the gravitational field. Now we introduce cylindrical space coordinates or some other curvilinear coordinates into the normal pseudo-Euclidean scale of the special theory of relativity without changing the time variable. Consequence: g-brackets appear in the equations of the geodesic world lines. But surely one cannot say in any physically meaningful way that such a purely mathematical operation produces a gravitational field (UAF, Laue papers, 9.1, cf. there his letter of 11 Mar. 1951.)." Laue presented another example.

Then he discussed his and Fritz London's phenomenological theory of superconductivity, citing troublingly regular instances of incomplete citation and priority issues by Fritz London and American physicists. He went so far as to suspect a kind of conspiracy by emigrés to create a scientific "Israel" to which he could have no access. Einstein replied very sympathetically and promptly from Princeton on January 16th. "Dear Laue, The falsification of the history of science on national grounds is an ancient art in which laudable nations outrank each other (similar to political history). Now that we Jews also have a state, it would actually be high time for us to engage in this art as well. If researchers living in Israel did so on their own, the effect would apparently be quite minimal, and the actions by the other Jews are entirely unrelated! (UAF, Laue papers, 9.1.)." Laue's hypothesis seemed implausible to him and he advised against it. Then he addressed his question about relativity. "Now about the gravitational field. One must keep different terms properly apart there. In Newton's theory, the gravitational field would be all expressions obtainable from the potential. The field strength in particular would be understood as the first derivative of the potential. In the relativistic theory of gravitation, the gravitational field is everything that is formed from the symmetrical g_{ik}'s." And about the existence of a real gravitational field, he wrote. "In relativity, the dimensionality of the field is all that remains of the former physically independent (absolute) space." Most people hadn't understood that, he added (ibid.).

On January 14th Laue wrote to Meissner to inform him about the status of his appointment at the former KWI of Physical Chemistry and Electrochemistry, pointing out that "whether I come to Berlin is up to the founder's association of the 'German Research University,' and it will meet in Wiesbaden at the end of January. Now all the German provinces seem to be involved in it. Because among their ministers of culture there are some who dislike me, [...] the matter is quite doubtful, although my Berlin colleagues claim the contrary (no loc., AMPG, III rep. 50 no. 2803)."

References

Becker, Heinrich, Hans-Joachim Dahms, and Cornelia Wegeler, eds. 1998. *Die Universität Göttingen unter dem Nationalsozialismus.* 2nd ed. Munich: De Gruyter

Borrmann, Gerhard. 1941. Über Extinktionsdiagramme von Quarz. *Phys. Z.* 42: 157–162

Borrmann, Gerhard. 1950. Die Absorption von Röntgenstrahlen im Fall der Interferenz. *Z. Phys.* 127: 297–323

Bragg, William L. 1949. Glückwünsche aus England. *Phys. Bl.* 10: 444 (in English)

DPG. 1947. Physiker-Tagung in Göttingen, *Phys. Bl.* 3: 317–325

DPG. 1949. Überreichung der Planck-Medaille an Otto Hahn und Lise Meitner durch Max von Laue. *Phys. Bl.* 10: 471 ff.

Dahms, Hans Joachim. 1999. Die Universität Göttingen 1918 bis 1989. In Thadden and Trittel, eds. 1999, vol. 3: 395–443

Einstein, Albert. 1905. Ist die Trägheit eines Körpers von seinem Energieinhalt abhängig? *Ann. Phys.* 18: 639–641. Reprinted in *Collected Papers of Albert Einstein*, vol. 2, 1989: 311–315, Princeton: Univ. Press. English version: Einstein, A. Does the inertia of a body depend upon its energy content? Trans. by Anna Beck, idem, vol. 2: 172–174, doc. 24

Einstein, Albert. 1949. Autobiographical Notes. In Schilpp 1949: 1–95

Einstein, Albert. 1949. Remarks to the Essays Appearing in the Collective Volume. In Schilpp 1949: 663–688

Einstein, Albert. 1950. *The Meaning of Relativity.* 3rd ed., Princeton: Univ. Press

Frank, Philipp. 1947. *Einstein. His Life and Times.* New York: Knopf

Fraser, Ronald. 1947. Begrüßungswort zur Physikertagung in Göttingen, September 1947. *Phys. Bl.* 3: 289

Fucs, Erwin. 1947. Theorie und Supraleitung. M. v. Laue. Book review. *Phys. Bl.* 3: 365 f.

Gidion, Jürgen. 1999. Kulturelles Leben in Göttingen. In Thadden and Trittel, eds. 1999, vol. 3: 539–586

Goudsmit, Samuel A. 1947a. *Alsos.* New York: Schuman

Goudsmit, Samuel A. 1947b. Nazis' atomic secrets. The chief of a top-secret U.S. wartime mission tells how and why German science failed in the international race to produce the bomb. 20 Oct. *Life* 23: 123–134; annotated in Hentschel 1996: 379–392

Hecht, Hartmut, Dieter Hoffmann, and Klaus Richter. 1991. *Robert Havemann. Dokumente eines Lebens*. Compiled and introduced by Dirk Draheim. Berlin: Links

Heinemann, Manfred. 1990. Der Wiederaufbau der Kaiser-Wilhelm-Gesellschaft und die Neugründung der Max-Planck-Gesellschaft (1945–1949). In *Forschung im Spannungsfeld von Politik und Gesellschaft. Geschichte und Struktur der Kaiser-Wilhelm/Max-Planck-Gesellschaft. Aus Anlaß ihres 75jährigen Bestehens*, ed. Rudolf Vierhaus and Bernhard vom Brocke, 407–472. Stuttgart: Deutsche Verlagsanstalt

Heisenberg, Werner, and Max von Laue. 1948. Das Barlowsche Rad aus supraleitendem Material. *Z. Phys.* 124: 514–518 (received 1 Sep. 1947)

Henning, Eckart. 1992. Der Nachlaß Max von Laues. Neue Quellen im Archiv zur Geschichte der Max-Planck-Gesellschaft (Berlin). *Phys. Bl.* 11: 938 ff.

Henning, Eckart, and Marion Kazemi, eds. 2011. 100 Jahre Kaiser-Wilhelm-/Max-Planck-Gesellschaft zur Förderung der Wissenschaften. Part 1: Chronik der Kaiser-Wilhelm / Max-Planck-Gesellschaft zur Förderung der Wissenschaften 1911–2011. Daten und Quellen. Berlin: Duncker & Humblot

Hentschel, Klaus. 2007. *The Mental Aftermath. The Mentality of German Physicists 1945–1949*. Oxford: Univ. Press

Hentschel, Klaus and Ann M. Hentschel, eds. 1996. *Physics and National Socialism. An Anthology of Primary Sources*. Revised reprint 2010. Basel, Boston, Berlin: Birkhäuser

Hentschel, Klaus, and Gerhard Rammer. 2000. Kein Neuanfang. Physiker an der Universität Göttingen 1945–1955. *Zeitschrift für Geschichtswissenschaft* 8: 718–741. English version: K. Hentschel and G. Rammer. 2001. Physicists at the University of Göttingen, 1945–1955 (trans. Hentschel, A.M.), *Physics in Perspective*, **3**: 189–209

Hermann, Armin. 1973. *Max Planck in Selbstzeugnissen und Bilddokumenten*. Reinbek: Rowohlt

Hoffmann, Dieter, ed. 1991. *Robert Havemann: Dokumente eines Lebens*. Berlin: Links

Jäckel, Eberhard. 1996. *Das deutsche Jahrhundert*. Stuttgart: Deutsche Verlagsanstalt

Jaspers, Karl. 1946. *Die Schuldfrage. Von der politischen Hoffnung Deutschlands*. Heidelberg: Schneider. English version: K. Jaspers. 1947. *The Question of German Guilt* (trans. Ashton, E.B.) New York: Dial Press

Jordan, Pasqual. 1941. *Die Physik und das Geheimnis des organischen Lebens*. Stuttgart: Vieweg

Kern, Ulrich. 1994. *Forschung und Präzisionsmessung. Die Physikalisch-Technische Reichsanstalt zwischen 1918 und 1948*. Weinheim: VCH

Kleinert, Andreas. 1983. Das Spruchkammerverfahren gegen Johannes Stark. *Sudhoffs Archiv* 67: 13

Lange, Ludwig. 1885a. Über die wissenschaftliche Fassung des Galileischen Beharrungsgesetzes. *Philosophische Studien* 2: 266–279

Lange, Ludwig. 1885b. Nochmals über das Beharrungsgesetz. *Philosophische Studien* 2: 539–549

Lange, Ludwig. 1886. *Die geschichtliche Entwickelung des Bewegungsbegriffes und ihr voraussichtliches Endergebnis. Ein Beitrag zur historischen Kritik der mechanischen Principien*. Leipzig: Engelmann

Laue, Max. 1921. *Die Relativitätstheorie*. Vol. 1: *Das Relativitätsprinzip der Lorentztransformation*. 1st ed. 1921, 2nd ed. 1952; Vol. 2: *Die Allgemeine Relativitätstheorie und Einsteins Lehre der Schwerkraft*. 1st, ed. 1921, 2nd rev. ed. 1923, 3rd rev. ed. 1953. Braunschweig: Vieweg

Laue, Max von. 1941. *Röntgenstrahl-Interferenzen*. Physik und Chemie und ihre Anwendungen in Einzeldarstellungen, no. 6. 1st ed., 2nd ed. 1948, 3rd ed. 1960. Leipzig: Akademische Verlagsanstalt

Laue, Max von. 1944. *Materiewellen und ihre Interferenzen*. Physik und Chemie und ihre Anwendung in Einzeldarstellungen, no. 7. 2nd ed. 1948. Leipzig: Akademische Verlagsanstalt

Laue, Max von. 1945/46. Deutsche Physiker auf britischen Tagungen. *Göttinger Universitäts-Zeitung* 18: 12

Laue, Max von. 1946a. *Geschichte der Physik.* Geschichte der Wissenschaften, no. 2. Naturwissenschaften. 2nd ed. 1947. Bonn: Athenäum-Verlag

Laue, Max von. 1946b. Isaac Newton (1642–1727). *Zeitschrift für Kultur und Technik* 1: 11 f.

Laue, Max von. 1946c. Die Eindeutigkeitssätze in der Theorie der Supraleitung. *Nachr. Akad. Wiss. Göttingen*, math.-phys. class, 1946: 86 ff.

Laue, Max von. 1947a. *Theorie der Supraleitung.* Berlin, Göttingen: Springer

Laue, Max von. 1947b. Bemerkung zu der bevorstehenden Veröffentlichung von J. Stark. *Phys. Bl.* 8: 272 f.

Laue, Max von. 1947c. Max Planck (†). *Phys. Bl.* 3: 249–252

Laue, Max von. 1947d. Eine Erweiterung der Theorie der Supraleitung. *Naturw.* 34: 186

Laue, Max von. 1947e. Die Kriegstätigkeit der deutschen Physiker, *Phys. Bl.* 3: 424 f.

Laue, Max von. 1947f. Was ist Materie? *Universitas* 2: 1217–1224

Laue, Max von. 1947/58. Funerary address in Planck, Max. 1958. *Physikalische Abhandlungen und Vorträge. Aus Anlass seines 100. Geburtstages (23. April 1958).* Published by Verband Deutscher Physikalischer Gesellschaften and Max-Planck-Gesellschaft zur Förderung der Wissenschaften, 3 vols., pp. 417 f. Braunschweig: Vieweg; reprinted in Scriba 1990: 9–10

Laue, Max von. 1948a. Londons Theorie für nicht-kubische Supraleiter. *Ann. Phys.* 1/438: 31–39

Laue, Max von. 1948b. The wartime activities of German scientists. *Bulletin of the Atomic Scientists* 4: 103; comp. Laue 1947e; annotated in Hentschel 1996: 393–385

Laue, Max von. 1948c. Supraleitung und Kristallklasse. *Ann. Phys.* 438: 40 ff.

Laue, Max von. 1948d. Max Planck. *Naturw.* 35: 1

Laue, Max von. 1948e. Dr. Ludwig Lange (1863–1936): Ein zu Unrecht Vergessener. *Naturw.* 35: 193–196

Laue, Max von. 1949a. Supraleitung und elektrodynamisches Potential. *Z. Phys.* 125: 517–530

Laue, Max von. 1949b. Inertia and Energy. In Schilpp 1949: 503, cf. Laue 1949/55

Laue, Max von. 1949c. Die Absorption der Röntgenstrahlen in Kristallen im Interferenzfall. *Acta Cryst.* 2: 106–113

Laue, Max von. 1949d. Zu Albert Einsteins 70. Geburtstag. *Reviews of Modern Physics* 21: 348 f.

Laue, Max von. 1949e. Eine nicht-lineare phänomenologische Theorie der Supraleitung. *Ann. Phys.* 440: 197–207

Laue, Max von. 1949f. Geschichtliches über Supraleitung. *Forschungen und Fortschritte* 25: 278 ff.

Laue, Max von. 1949g. Die Absorption von Röntgenstrahlen in Kristallen im Interferenzfall. *Acta Cryst.* 2: 106–113

Laue, Max von. 1949/55. Trägheit und Energie. In Schilpp 1949, 3rd ed. 1955: 364–388, cf. Laue 1949b

Laue, Max von. 1950a. Aspectos históricos de la supraconductividad. *Revista de la Unión Mathemática Argentina* 14: 109–117

Laue, Max von. 1950b. Lichtschwebungen. *Optik* 7: 125 ff. (reply to Rüchardt 1950)

Laue, Max von. 1950c. Zur Minkowskischen Elektrodynamik der bewegten Körper. *Z. Phys.* 128: 387–394

Laue, Max von. 1951. Materiewellen. *Naturw.* 38: 55–61

Laue, Max von. 1952. Die Energieströmung bei Röntgenstrahl-Interferenzen an Kristallen. *Acta Cryst.* 5: 619–625

Laue, Max von. 1961. *Gesammelte Schriften und Vorträge.* Vols. I–III, Braunschweig: Vieweg

Laue, Max von. 1982. Lange, Ludwig. *Neue Deutsche Biographie.* Berlin, 13: 551 f.

Laue, Max von, and E. Regener. 1947. Zum Jahresbeginn 1947. *Phys. Bl.* 1: 1

Lemmerich, Jost. 1981. *Dokumente zur Gründung der Kaiser-Wilhelm-Gesellschaft und der Max-Planck-Gesellschaft zur Förderung der Wissenschaften. Ausstellung in der Staatsbibliothek zu Berlin, Preußischer Kulturbesitz, Berlin, 21. Mai–19. Juni 1981,* organized by Max-Planck-Gesellschaft zur Förderung der Wissenschaften. Munich, Berlin: MPG zur Förderung der Wissenschaften

Lemmerich, Jost, ed. 1998. *Lise Meitner – Max von Laue. Briefwechsel 1938–1948*. Berliner Beiträge zur Geschichte der Naturwissenschaften und der Technik, no. 22. Berlin: ERS-Verlag

Lemmerich, Jost, ed. 2010: *Bande der Freundschaft: Lise Meitner – Elisabeth Schiemann. Kommentierter Briefwechsel 1911–1947*. Österreichische Akademie der Wissenschaften, Mathematisch-Naturwissenschaftliche Klasse, no. 61. Vienna: Verlag der Österr. Akad. der Wissenschaften

Lemmerich, Jost, ed. 2011. *Mein lieber Sohn! Die Briefe Max von Laues an seinen Sohn Theodor in den Vereinigten Staaten von Amerika 1937–1946*. With two contributions by Christian Matthaei. Berliner Beiträge zur Geschichte der Naturwissenschaften und der Technik, no. 33. Berlin, Liebenwalde: ERS-Verlag

Manthey, Matthias, and Cordula Tollmien. 1999. Juden in Göttingen. In Thadden and Trittel, eds. 1999, vol. 3: 675–760

Martius, Ursula Maria. 1947. Videant consules. *Deutsche Rundschau* 70: 99–102

Meinecke, Friedrich. 1946. *Die deutsche Katastrophe. Betrachtungen und Erinnerungen*. Wiesbaden: Brockhaus

Meissner, Walther. 1947. Theorie der Supraleitung. M. v. Laue. Book review. *Naturw.* 34: 127

Morrison, Philip. 1947. Alsos. The story of German science. *Bulletin of the Atomic Scientists* 3: 354, 365

Morrison, Philip. 1948. A reply to Dr. von Laue. *Bulletin of the Atomic Scientists* 4: 104; annotated in Hentschel 1996: 396–397

Müller-Enberga, Helmut, Jan Wielgohs, and Dieter Hoffmann, eds. 2000. *Wer war wer in der DDR? Ein biographisches Lexikon*. Berlin: Links

Pohl, Robert. 1912. *Die Physik der Röntgenstrahlen*. Braunschweig: F.Vieweg & Sohn

Przyrembel, Alexandra. 2004. *Friedrich Glum und Ernst Telschow. Die Generalsekretäre der Kaiser-Wilhelm-Gesellschaft: Handlungsfelder und Handlungsoptionen der „Verwaltenden" von Wissen während des Nationalsozialismus*. Berlin: Max Planck Institute for History of Science

PTB. 1961. Verabschiedung des Präsidenten der Physikalisch-Technischen Bundesanstalt Prof. Dr. Richard Vieweg. *Amtsblatt der Physikalisch-Technischen Bundesanstalt*, no. 1 suppl., 1961: 4

Rammer, Gerhard. 2012. "Cleanliness among Our Circle of Colleagues": The German Physical Society's Policy toward Its Past. In Hoffmann, Dieter, and Mark Walker, eds. 2012. *The German Physical Society in the Third Reich. Physicists between Autonomy and Accommodation*. Trans. Hentschel, A.M., 367–421. Cambridge: Univ. Press

Rasch, Manfred. 1994. Mentzel, Rudolf. *Neue Deutsche Biographie* vol. 17: 96 ff. Berlin: Duncker & Humblot

Rosenthal-Schneider, Ilse. 1949. Presuppositions and anticipations in Einstein's physics. In Schilpp 1949: 131–146

Rosenthal-Schneider, Ilse. 1980. *Reality and Scientific Truth. Discussions with Einstein, von Laue, and Planck*. Detroit: Wayne State Univ. Press

Rüchardt, Eduard. 1950. Lichtschwebung und Lichtmodulation *Optik* 6: 238–244

Schilpp, Paul Arthur, ed. 1949. *Albert Einstein. Philosopher – Scientist*. Library of Living Philosophers, vol. VII. 1st ed. La Salle, Illinois: Open Court. German edition: P.A. Schilpp, ed. *Albert Einstein als Philosoph und Naturforscher*. Philosophen des 20. Jahrhunderts, no. 1, 3rd ed. 1955 (1st ed. 1949) Stuttgart: Kohlhammer

Schwarz, Stephan. 2015. The case of the bottled Nobel medals. *Gamma* 150: 8–14; online at https://schwarzstephan.files.wordpress.com/2015/11/medals-gamma4.pdf [last access: 23 Oct. 2021]

Scriba, Christoph J., ed. 1990. *Max Planck Wissenschaftliche Selbstbiographie*, Acta Historica Leopoldina no. 19. Halle/Saale: Deutsche Akademie der Naturforscher Leopoldina

Sommerfeld, Arnold. 1949. Max von Laue zum 70. Geburtstag. *Phys. Bl.* 10: 443

Sommerfeld, Arnold. 2004. *Wissenschaftlicher Briefwechsel 1919–1951*, Vol. 2. Edited by Michael Eckert und Karl Märker. Berlin, Diepholz: Verlag für GNT

Stark, Johannes. 1947. Zu den Kämpfen in der Physik während der Hitler-Zeit. *Phys. Bl.* 8: 271 f.

Steinhauser, Thomas, Jeremiah James, Dieter Hoffmann, and Bretislav Friedrich. 2011. Ein Patchwork-Institut. Konsolidierung und Überleitung in die MPG. In *Hundert Jahre an der Schnittstelle von Chemie und Physik. Das Fritz-Haber-Institut der Max-Planck-Gesellschaft zwischen 1911 und 2011*, 139–157. Berlin: de Gruyter

Thadden, Rudolf von, and Günter J. Trittel, eds. 1999. *Göttingen. Geschichte einer Universitätsstadt.* Vol. 3: *Von der preußischen Mittelstadt zur südniedersächsischen Großstadt 1866–1989.* Göttingen: Vandenhoeck & Ruprecht

Tollmien, Cordula.1999. Nationalsozialismus in Göttingen (1933–1945). In Thadden and Trittel, eds., 1999, vol. 3: 127–274, 275–290

Zeitz, Katharina. 2006. *Max von Laue (1879–1960). Seine Bedeutung für den Wiederaufbau der deutschen Wissenschaft nach dem Zweiten Weltkrieg.* Stuttgart: Steiner

Chapter 13
Back in Berlin—West Berlin

Fundamental changes occurred in the Laue couple's lives in 1951. Karl Friedrich Bonhoeffer left his post as director of the KWI of Physical Chemistry and Electrochemistry on March 31st, and Max von Laue was appointed as his successor. In his birthday letter to Einstein on March 11th, Laue informed him of this from Göttingen: "I don't have so much leisure. Now, in all probability, I'll be going to the former Haber Institute in Dahlem, Berlin, and there I'm going to have to deal with all kinds of adverse conditions that have arisen there since the political unrest in 1933. And yet, I look forward to going there; because there I shall finally be coming into an environment that wants me, where I don't have to consider where every utterance of mine could be taken and what harm it could do there. What profit science will gain from this remains to be seen (UAF, Laue papers, 9.1)."

In 1951 the Max Planck Society (MPG) was given a say in filling the institute director posts at the German Research University, the *Deutsche Forschungshochschule*. The MPG's "Scientific Council" met on April 5th. At that meeting, Otto Warburg and Max von Laue were elected to the founding board of the Research University, and Laue accepted the appointment as director of the Institute of Physical Chemistry (Fig. 13.1). He proposed to name it the "Fritz Haber Institute of the Max Planck Society within the Framework of the German Research University."[1]

Magda von Laue's 60th birthday party on March 13th could still be celebrated in Göttingen. Then the packing began in earnest, because preparations had to be made for the move to the Berlin suburb of Dahlem in April. To be on the safe side, Magda von Laue packed the books and much more herself, especially since the moving company decided to shift the pickup date to three days earlier! Laue was able to inform his friends that their move to Berlin had gone well. James Franck in Chicago received a letter from Laue in Dahlem on April 30th recalling memories going back to the grand old days during the Weimar Republic when Franck joined

---

[1] AMPG, II. SP, MPG, 6 Apr. 1951, part II, p. 9; AMPG, III rep. 50, no. 2553, fol. 22, and AMPG, II rep. 1A, Laue staff file.

© Basilisken-Presse, Natur+Text GmbH 2022
J. Lemmerich, *Max von Laue*, Springer Biographies,
https://doi.org/10.1007/978-3-030-94699-9_13

Fig. 13.1 Max von Laue
with Rudolf Ladenburg
(1882–1952), around 1946.
Ladenburg was head of the
Department of Atomic
Physics at the Kaiser
Wilhelm Institute of Physical
Chemistry and
Electrochemistry in
Berlin-Dahlem from 1924
until his emigration to the
USA in 1932. *Source*
Photograph in the author's
estate

his new workplace for a brief period under Fritz Haber, joining Hartmut Kallmann
and Rudolf Ladenburg at his institute. "My wife had a hard time saying goodbye to
Göttingen but is pleasantly surprised by the apartment we got here. We live on the
first storey of Haber's villa and look out on all four sides onto beautiful gardens and
grounds (JRL, Franck papers, B 9, F 10)." (Fig. 13.2) His wife had had to accompany
the furniture to its destination. The controls at the zone border were harmless; she was
only asked if she had a copy of Hitler's *Mein Kampf* with her. Everything arrived
in good order, but they still had to buy more furniture. "Yesterday we also went
on our first excursion," he continued. "To Pfaueninsel, Moorlake, and to Glienicke
Bridge, which is now called 'Bridge of Unity.' I don't know if this irony is intended
or unintended (ibid.)."[2] Lise Meitner also got news about the move. Magda von

---

[2] The Laue couple did not move back into their house on Albertinenstraße, which had been seriously
damaged and was still in disrepair, even after it had been rebuilt with a loan. Mrs. von Laue did

Fig. 13.2 The "Fritz Haber Villa" in 1952, the director's residence of the Kaiser Wilhelm Institute of Physical Chemistry and Electrochemistry, where Fritz Haber had also once lived. The Laue family only occupied the first storey.
Source Archive of the Max Planck Society, Berlin-Dahlem

Laue, writing to her from Berlin on May 7th, remembered her first visit in the villa at the time when Haber was still alive, but about the present she remarked: "I'm still grumbling and don't want to leave the house—but I have to admit that this apartment, with its terrace overlooking the park-like garden, couldn't be nicer (CAC, MTNR 5/28)." The furnishings weren't in place yet, the glaziers and upholsterers were still busy in their six-room apartment, she continued. Her husband's letter to Lise Meitner from Dahlem two days later described the other tenants: Dr. Else Knake lived on the ground floor, and the institute glassblower and a mechanic lived upstairs. Laue himself had already taken a bicycle tour through the Grunewald woods: "The lakes do always leave a deep impression in their somewhat melancholy beauty, which suits the political situation so well (ibid.)."

Max von Laue took over the direction of a very large institute with several departments and some very broad fields of research by such independent researchers as Ivan N. Stranski, Kurt Ueberreiter, and later, Ernst Ruska. In addition, there were over a hundred employees in the various departments for the fields of X-ray and electron-beam investigations of solids and liquids, crystal structure research, X-ray diffraction, crystal growth, electron microscopy, and field electron microscopy. More independent research areas covered high-polymer structure, electrochemistry and its applications, synthesis and application of electron exchangers, redox resins, and so-called ion exchangers.

---

not want to live in a house in which she and Hilde had endured a heavy bombing raid. (Personal communication to the author by Hildegard Lemcke).

Fig. 13.3 Arnold
Sommerfeld, a long-time
companion of von Laue,
shown here in a photo from
1949. *Source* Archive of the
Max Planck Society,
Berlin-Dahlem

The administration was headed by Hans-Dietrich Schmidt-Ott, a son of Friedrich
Schmidt-Ott. It was self-evident that Laue could not focus his interest equally on all
these fields, but he very soon succeeded in promoting cohesion and also in getting
quite complex personalities involved. He was easily able to pursue his own scientific
interests. Laue now occupied Haber's former office, which strangely enough had not
been ransacked by the Soviet troops. His secretary, Miss Liselotte Fritsche, handled
the extensive correspondence (Steinhauser et al. 2011: 151 f.). In Laue's own field
of expertise, X-ray interference and crystal structure, the physicist Rolf Hosemann
joined the institute in 1951, initially as a senior assistant (Hosemann to Laue, 24 Jul.
1950. AMPG, III rep. 50 no. 924).

Laue regularly took part in the events of the Physical Society in Berlin, the
Physikalische Gesellschaft. On June 17th, he paid tribute to Arnold Sommerfeld
(Fig. 13.3) and his oeuvre, as Sommerfeld had died on April 25th after a traffic
accident (DPGA, 10318). Laue's remarks about Sommerfeld went beyond a mere
biographical outline of his colleague's activities. At several points, he incorporated
his own ideas about the interrelation between the man as a person and his physics.
They are thus quoted here somewhat more extensively.

Sommerfeld, a native of Königsberg (Kaliningrad), studied in his hometown, and his most important teachers in his major in mathematics were Ferdinand Lindemann, Adolf Hurwitz, and David Hilbert. After earning his doctorate, he went to Göttingen in 1893 as an assistant at the Institute of Mineralogy. "The obligatory study of crystallography bore fruit in 1912 when Sommerfeld was able to give valuable advice to W. Friedrich, P. Knipping, and me in performing the first experiments on X-ray interference in crystals (Laue 1951: 514; AMPG, III rep. 50 no. 90, for the ms.)." Then he became assistant to Felix Klein. Laue cited as a "highly significant" result of this period his strictly mathematical treatment of light diffraction at an edge. The old method of approximation had nevertheless produced sufficient results, he contended. "But a rigorous solution now does afford far greater satisfaction than an approximate one (ibid.)." Other stations in Sommerfeld's career then followed, at the Mining Academy in Clausthal and then, from 1900, at Aachen in the subject of engineering mechanics. "But his love of pure science could not be subdued (ibid.)." Laue then described the paths and bypaths to the theory of the electron during that period and Sommerfeld's contact with Hendrik A. Lorentz, before finally appreciating his contributions to the theory of relativity and the X-ray diffraction experiments at a slit. He went into greater detail about Sommerfeld's contributions to the quantum theory of the atom and the extension of Bohr's model in his major work, 'Atomic Structure and Spectral Lines.' Its introduction states: "The scientific activity of a theoretical physicist is by no means exhausted by the shorter or longer original papers that appear in scientific journals. An essential duty of his is the writing of larger works, which present whole chapters of his science within context. Only in such books can he communicate to his fellow man and to posterity his conceptions of the inner structure, the reliability, and the scope of a theory (Sommerfeld 1919)." In an interesting comparison between Planck, Einstein, and Sommerfeld, Laue concluded:

> Planck, 10 years older than Sommerfeld, was essentially self-taught; he had educated himself from the printed works of Rudolf Clausius, but had never had any personal contact with him. Even Helmholtz, who much appreciated his abilites and supported him, did not exert any decisive intellectual influence on him. Later, too, he never found, or even ever sought any closer scientific collaboration. What he achieved, he owed to deep pondering on his own. In his lectures, the material he presented was, so to speak, an end in itself. The auditor could learn an infinite amount, and even become enthusiastic. But for most of them theory remained a distant ideal, similar to the high peaks of an Alpine chain seen from afar. Not many of them decided to tread the long, arduous path there.
>
> The same basically applies to Einstein, who is about 10 years younger than Sommerfeld; add to that his exuberant genius, which refused to be tamed at all, indeed it occasionally jeopardized students who tried to emulate it.
>
> Sommerfeld, on the other hand, had much closer contacts with his surroundings from the outset. He had become a member of a scientific *school* at Göttingen, and to found a *school* of his own was an essential concern of his. His lectures were far more personally tuned to his auditors, and therefore more apt to attract them to pursue further study of science; with him one could view the Alps from close up, and consequently perhaps did not get an overview of the mountain ranges as a whole, but one could learn how to climb mountains with him (Laue 1951: 518, orig. emphasis).

Sommerfeld's work, Laue asserted, was being continued by all of his former students. He concluded by repeating the honorary title that a colleague had once given him: *Praeceptor Germaniae* (p. 515).

At the beginning of September, Laue had to attend the meetings of the Scientific Council in Munich. The official transferral of the German Research University into the MPG was the topic of their negotiations. At the society's session honoring the foundation of the Kaiser Wilhelm Society forty years before, Otto Hahn delivered the related speech. The Laue couple then visited their daughter in Biberach. The KWI of Cell Physiology, opened in 1931, with Otto Warburg as director, was also subsumed under the Research University. Although the "racial laws" had actually applied to Warburg during the Third Reich, he had been allowed to retain his position. He had always played a rather idiosyncratic role in the KWG as well. Now, tensions arose between him and Laue over details concerning the Research University. Laue communicated his concerns to the president Otto Hahn in a letter from Berlin on 13 October 1951. However, he did not want a quarrel with Warburg. "In spite of all this, Otto II must become a senator of the MPG. I have promised to back him on this, and have spoken to you and Telschow in Munich about it. If Telschow's assertion that there will be a vacancy for him in 52 is not true, then I simply *must* vacate my seat, however much I would have liked to have been a member of the Senate. I cannot break my promise (AMPG, II rep. 1A)."

When Max von Laue received an invitation by Francis Simon to attend a conference on superconductivity at Oxford, Laue declined because, in his opinion, current atomic theories of superconductivity contained invalid claims. Any such theory would have to take the magnetic field into account, he told Simon in a letter dated March 15th from Göttingen (AMPG, III rep. 50 no. 1877). In an English review of his book on superconductivity, Laue was accused of not having taken the English-language literature into account, as he informed the publisher in a letter to Ferdinand Springer on October 23rd. Laue protested: "The whole thing is explicable when one considers that Fritz London published a book of similar purpose [to mine] in 1950. I have had occasion to study it more closely in recent weeks. I could return to him his reproach highly amplified about not having taken certain publications into account. But worse still is the opacity in many matters of principle, and a want of care leading even to contradictions within the book and culminating in quite incomprehensible calculation errors (UAS)." Laue also mentioned that the two American translators were no longer willing to attach their names to his own book because it had been superseded by London's. Six days later, however, he was able to inform Springer that his book would be published by Academic Press after all (ibid., Laue 1952a).

At the beginning of November, Laue finished his essay 'Critical remarks about the theory of superconductivity' and sent the manuscript to Friedrich Möglich for publication in the *Annalen* (Laue 1951/52a, received 2 Nov. 1951). It contained the objections he had already mentioned to Ferdinand Springer about Fritz London's book on *The Macroscopic Theory of Superconductivity* published in America in 1950. Laue thought that two remarks should not remain standing unchallenged: the proof of the existence of a surface energy between the superconducting and normal-conducting phases, and London's arguments about an intermediate state.

Laue quoted very extensively the introduction to London's book, where alternating normal and superconducting layers are assumed as a kind of domain effect. Laue claimed that his own book had already shown that this could only exist in the case of curved superconductivity. He was of the opinion that no surface energy of any notable amount existed between the layers (phases) and cited the analyses byMeshkovski and Shalnikov (1947) of thin bismuth wire indicating superconducting "single jumpers" (*Einspringlinge*) of less than 10^{-3} cm, whereas London only assumed 10^{-1}. (Fritz London was in the USA at the time of writing.) Laue's concluding sentence about this revealed his own skepticism, however: "Besides, probably no physicist today knows the intermediate state well enough yet to reach a final verdict (Laue 1951/52a: 301)." To this Fritz London replied: "Although the reader can easily see for himself that the critical remarks in the foregoing paper are all based on misunderstandings, I shall, following a suggestion by the editor of the 'Annalen,' voice my opinion (London 1952a)."

Max von Laue was not the only one convinced that his solution was right. In an undated letter to Werner Heisenberg, he briefly explained his conception: "Isn't the atomic theory of metallic conduction as a whole on the wrong track? Superconductivity seems to me to be the simple, original thing, as a typical quantum problem whose meaning results from Schrödinger's equation and the like, irrespective of thermal motion. And Ohmic conduction, according to this view, is just a disturbance of superconduction by thermal motion, which *perhaps* in some metals is already so strong at absolute zero that it cancels superconductivity (AMPG, III rep. 50 no. 825, fol. 22, orig. emphasis, no loc.)." Heisenberg replied from Göttingen on November 12th: "With the fundamental formulation of the superconductivity problem in your letter I agree much more than you think. I am as convinced as you are that the super-conducting state is an ordered state, and in that respect simpler than the disordered state of normal conduction. Only I do not think that one should conclude from this that the present atomic theory of metallic conduction is on the wrong track (ibid., no. 826 fol. 23)."

Max von Laue could not accept Fritz London's objections, because his coworker Friedrich Beck provided a hitherto missing proof of the stable state. London's layer arrangement was critically examined and it was found that the Nth normal conducting layer disappears completely. The equilibrium conditions for the superconducting and normally conducting layers were declared wrong (Beck 1951). Shortly before Christmas, Laue wrote to Friedrich Möglich enclosing two manuscripts for publication: 'Remarks on the theory of superconductivity II' (Laue 1952b) and one by Friedrich Beck on 'A special form of the electrodynamic potential of superconductors' (1952a). Prior to that, Möglich had already received Beck's paper 'On the phase transformation between superconductor and normal conductor in the critical magnetic field' (1952b). The scientific controversy with Fritz London continued, and in his letter to Möglich dated 21 December 1951 Laue asked the editor that the publications in the *Annalen* appear in a particular order (AMPG, III rep. 50 no. 1378). The editor, in turn, then asked Fritz London to comment on these papers. Their opponent's public response was: "It seems to me that all the objections raised in v. Laue's first note can now be regarded as settled. The application of the thermodynamic potential *g*, which had originally been faulted, has now also been recognized by Mr. Beck as correct (London 1952b)." Then London noted inconsistencies in the thermodynamic equilibrium considerations, which he intended to clarify, citing a publication by his brother Heinz.

Max von Laue wrote a review of London's book, which he sent to the editors of *Naturwissenschaften* on 7 November 1951. He criticized the restriction to linear forms of electrodynamics operating with scalar superconduction constants. Laue also attacked the thermodynamic conceptions: "The old notion of the superconductor as a 'perfect' diamagnetic substance is often used, although London's own equations prove the fundamental difference between them (Laue 1952c: 288)." His proof of a surface energy between superconductor and normal conductor seemed to Laue completely erroneous. In conclusion, he remarked that the book was interesting to a reader who had some mastery of the subject matter; an unqualified reader, he feared, would be led astray (ibid.). All the participants were convinced that their views of the processes of superconductivity were correct. But a few years later, the problem of metallic superconductivity would be solved by a new theory on a completely different basis. For that a new and unaccustomed conception of the state of electrons in the superconducting state had to be adopted.[3]

In the midst of this dispute Laue was busy at work on the fifth edition of his book 'The Theory of Relativity' (1952d). His doctoral student Friedrich Beck remembered seeing how Laue made modifications to the text to incorporate his current ideas by cutting up manuscript pages and pasting them together on sheets with handwritten emendations that were difficult to decipher, oblivious of the difficulties it would present for the typesetter. About the substantive changes, Beck noted:

> More important than these superficialities were the discussions about the content of the book. About this we had differences of opinion stemming from the 'generational problem.' I thought that the book had to be modernized, adapted to the level of mathematical proficiency that physicists had reached in the meantime. But Laue, himself a mathematically extremely meticulous theorist, resisted most resolutely. And as we were arguing I realized what the aim of his 'Theory of Relativity' was. He regarded it entirely as a service to his friend Albert Einstein, whom he admired immensely, and to his ideas. They had both witnessed controversies over the theory of relativity, some of which had been conducted out of ignorance, others by political malice. Einstein did not let it affect him. But Laue wanted his book to open the way to the theory of relativity precisely for persons attached to experiment, to intuitive interpretation of physical results, by means of compelling logic and empirical consistency (Beck 1979: 143 f.).

The brief preface to the volume already referred to the second part on 'General Relativity' (1953a): "Specifically, the theorem of the inertia of energy has found unexpected confirmation." Laue continued: "More than any other physical theory, both parts of relativity theory are the work of one man. His name stands immovably above all the changes which the theory has undergone and may perhaps undergo in

---

[3] Laue did not include these two papers in his 'Collected Works and Speeches' (1961a). Perhaps they were not essential enough for Laue or else he did not want to have the controversy with Fritz London documented that way. In the USA numerous physicists were working theoretically and experimentally on superconductivity. See Blatt 1964, Cooper and Feldman 2011, and Dahl 1992.

the future. That he has permitted me to dedicate my book to him, and to adorn it with the youthful photograph from the time when he created the special theory of relativity, is one of the finest rewards I have ever received in my career as a researcher (1952d)." Hilde came to Berlin with her three daughters for Christmas. Although strenuous for the grandparents, the joy of seeing their grandchildren was great, as Laue wrote to Lise Meitner from Berlin on 22 December 1951 (CAC, MTNR 5/28).

Max von Laue's first letter to his son Theodore in 1952, written on January 20th, was an account of one of the many business trips that Laue had to take as a result of the isolated situation of Berlin and the demands of his position: "On the morning of Jan. 16th, I flew to Düsseldorf and immediately took the train to Bonn. The reason was meetings of the Max Planck Society in Düsseldorf on Jan. 17th and 18th; but in Bonn I intended to talk to the Federal Minister of Economic Affairs about the PTB and possibly visit the Federal President. The federal plenipotentiary Dr. Vockel had suggested the latter to me. Both meetings took place on the morning of January 17th. I was with Heuß from 10 o'clock to 10:30 (AMPG, II rep. 50 suppl. 7/12)." He described to his son how such visits used to be done with monarchs, never going beyond formalities. "Heuß, on the contrary, expected me to report about the purpose of my coming, and he willingly discussed each point. They weren't highly political matters, of course, but a small selection from among the ones in which I am professionally involved. All in all, the conversation was a great pleasure for me (ibid.)." He also mentioned the location of the president's home and the simple furnishings of Heuss's office.

That same month, a letter arrived from Einstein in Princeton dated January 17th: "Dear Laue, I have now received your book on special r[elativity] th[eory] and think it is very good. I especially like the exhaustive treatment of the empirical foundations (UAF, Laue papers, 9.1)." Einstein wondered about Laue's dismissive treatment of Ludwig Lange's critique of Newton's concept of absolute space and time and of Maxwell's equations. "But in 1905 I already knew for certain that they lead to spurious fluctuations in the pressure of the rays, and hence to spurious Brownian motion of a mirror in a Planck radiation field (ibid.)." Laue appreciated Einstein's objections as he corrected them in his reply on January 23rd:

Dear Albert Einstein,

Yesterday your letter of 17 Jan. 52 arrived, and I thank you most sincerely for the candid presentation of your views. If my book seems to give you a too favorable impression of the performance of Maxwell's theory, then this shows me above all how much you have outgrown your own work in the course of your life, which did set out from Maxwell's theory, you know. (See the title 'Electrodynamics of Moving Bodies' of your first publication about it.) This not-letting-yourself-be-satisfied with your own successes undoubtedly is the core to your greatness as a scientist. But nobody can deny that Maxwell's theory really has accomplished the things about which my book reports, and that the failure you speak about lies in quite other areas which remain beyond the frame set by my book. So I had no occasion, nor even an opportunity, to point out that it, like all our physics, is just one piece of the whole (UAF, Laue papers, 9.1; Einstein 1905a).

He then commented on Einstein's objection about Lange's work. In his birthday letter to him from March 7th Laue wished Einstein good health and more: "May you experience a decisive advance in the field that once was dearest to your heart, namely in the development of the General Theory of Relativity. I personally nurture the hope that you can perceive as an advance the paper by [Max] Kohler, which you recently received as a correction proof (UAF, Laue papers, 9.1)."[4] He then gave a very positive report about his visit to the federal bureau of standards, the *Physikalisch-Technische Bundesanstalt* (PTB) in Braunschweig for a meeting of the board of trustees and about its structure. He praised the work of its president Richard Vieweg (ibid.).

The past continued to revisit him from time to time. He could not let the guilt of some of his colleagues for their misdeeds during the period from 1933 to 1945 be forgotten. On 3 March 1952, he wrote to Dr. Ferdinand Springer:

> The booksellers Lange & Springer, in Berlin W35, have just sent me an announcement of the forthcoming book 'The Physics of High-Polymers, Volume I, by H. A. Stuart, ord. professor of physics, Hannover, former polytechnic in Dresden.' This designation of the author is apt to raise the false impression that he were now full professor at the Technical University of Hannover. One does have to think for a moment to realize that, strictly speaking, this is not what it says.
>
> I would like to ask you if the Springer publishing house wants to have a hand in publicizing this surely intentional misrepresentation by the author. As, the gravest objections to employing H.A. Stuart in any teaching capacity have been expressed by various colleagues during several recent appointment proceedings.[5]

Brief reminiscences of bygone years crop up repeatedly in his correspondence with his son. In response to a comment by Theo about babysitting his children, his father recalled: "You write in addition that you sometimes lose patience when you're looking after the kids. Yes, yes… You're now going through what I also experienced with you and Hilde (undated, AMPG, II rep. 50, suppl. 7/12)." But other such reflections weren't so lighthearted; he had to inform him about the death of one of his closest friends, Rudolf Ladenburg (ibid.).

In 1936, Erwin Wilhelm Müller had earned his doctorate under Gustav Hertz at the Charlottenburg Polytechnic in Berlin with a thesis on the field electron microscope, which made it possible to visualize atoms and atomic arrangements. After the war Müller was employed at the German Research University before accepting a professorship in the USA in 1952. When the Braunschweig Scientific Society awarded him the Gauss Medal in 1952, Laue attended the conferral on Müller's behalf on April 30th and delivered a talk about 'The atomic concept in physics.' After introducing the topic with a historical account including the forgotten Ludwig August Seeber as well as Dalton and Avogadro from the first half of the nineteenth century, he discussed the inquiries into crystal structure by Arthur Moritz Schoenflies and Paul

---

[4] Laue wrote to Ilse Rosenthal-Schneider on 27 Dec. 1951 that his manuscript for Vol. II, the fifth edition of 'General Theory of Relativity' (Laue 1953) would have been at the printer's long ago, "if Max Kohler in Braunschweig had not produced a fundamental reformulation of the whole general theory in the last few months, which, for example, affords a better formulation of the conservation laws of energy and momentum." Quoted in Rosenthal-Schneider 1988: 40.

[5] AMPG, III rep. 36 no. 14, about Stuart 1952. Herbert Arthur Stuart had been the directing chair of the department of experimental physics at the *Technische Hochschule* in Dresden 1939–1945, cf. Hentschel 1996 appendix.

Heinrich von Groth. He described Müller's initial idea in detail and its realization around 1935 employing a very fine tungsten tip to generate a high field strength. He showed several exposures taken by this method, including one with a barium-covered tip (Laue 1951/52b: 229 f.).

Physics was not so much the subject in the correspondence between father and son as was politics. "You speak of the necessity for massive exchanges among European nations as a necessary preparation for a united Europe," Laue wrote him from Berlin on May 5th. "Well yes, but unfortunately the French and English aren't using the favorable opportunity of their occupation of Germany for this, but rather keep their soldiers strictly separated from the German population, just as the Americans are doing. We only really get to know a few people, those involved in science, among all the rest (AMPG, III rep. 50 no. 7/12 fol. 22).[6] He also noted the fortieth anniversary on June 8th of his first presentation on X-ray interference before the Berlin Academy, followed six days later by his talk before the DPG (ibid., see here Chap. 5).

Travel was still an important part of rebuilding international contacts for Max von Laue. Thus he accepted an invitation to a conference on solid state physics in Gothenburg, Sweden, from June 9th to 13th. He chose to speak about 'The interference absorption of X-rays in crystals' (Laue 1952/53). The Laue couple traveled by car, taking Mrs. Justi along as their passenger. Since Max von Laue was not allowed to drive through the Eastern "zone" but could only fly out of Berlin, Dr. Karl Plieth, assistant to Prof. Stranski, drove the car through the "zone" for them. Laue could then return to his seat behind the wheel and drive through Denmark and onto the ferry to Sweden.

Laue's lecture was intended for a larger audience with various fields of expertise, and so he adapted the content accordingly. His introduction described the difference between the photoelectric effect with X-rays and the absorption of electrons. The former was independent of the motion of the bombarded atom, which differed from the interference case, he explained. A simple consideration showed that in interference the absorption can be considerably larger, but also considerably smaller. Borrmann could demonstrate this experimentally on a thin plate of calcareous spar. An analysis of the data obtained from Carleton C. Murdock's exposures (1934), which were displayed as an illustration, revealed further details about the absorption process for his audience. These experiments opened the way for a previously lacking theory on interference in ideal crystals (Laue 1952/53).

Publication of this lecture offered Laue the opportunity to prepare an extended version with detailed mathematical parts in which he applied the Poynting vector to the theoretical energy flux (ibid.). He examined the single wave field in the crystal without absorption, then the simultaneously forming wave fields, and finally the case of absorption and multiple interference. These considerations were discussed in connection with the experimental data provided by other researchers. He thanked in particular Gerhard Borrmann for his collaboration in the theoretical discussion.

---

[6] The designation "out of bounds" applied to many restaurants for military personnel. The marriages between German women and members of the Western occupying forces showed, however, that some contact was nevertheless possible.

Towards the end of the year Laue wrote a short supplement on this topic, in which he showed that for matter waves, too, the trajectory of the particles runs similarly to the energy flux for X-rays (Laue, 1952e, 1953b).[7]

The next trip in July led first to the general administration of the MPG in Göttingen and then onwards to the vicinity of their daughter's family near Isny for recreation. In a reply letter to Theo on June 7th from Berlin, he remarked about his son's plan to hear a lecture by Carl Friedrich von Weizsäcker in the USA: "Hopefully you did so. Hopefully so, because one always learns something from him. But be quite critical about it! I and some others cannot get over the impression that he often says what his audience expects of him. His biblical scholarship is part of this image. Nonetheless he is a very estimable man (AMPG, III rep. 50 suppl. 7/12)."

The August issue of *Naturwissenschaften* commemorated the discovery of X-ray interference with a reprint of the publications forty years ago (Lamla 1952; Friedrich, Knipping, and Laue 1952).

The Federation of German Physical Societies planned to hold this year's Physicists' Convention in Berlin from September 28th to October 3rd. The brunt of the organizational work was shared between Heinrich Gobrecht, the chair for experimental physics at the Technical University (TU), and Laue. Laue sent a draft of his opening address to Heisenberg in a letter dated September 22nd: "As representative of the Berlin physicists, I welcome all my compatriots who have come here from East and West to this German Physicists' Convention, and certainly also our foreign colleagues, some of whom have long journeys spanning up to a third of the Earth's circumference behind them, in order to show us the latest results of their research (DPGA, 10372)." Regarding the conference venue, the entire institute building, he remarked that it—just like the surrounding area—was a pile of ruins, but the lecture halls had been restored and bore witness to the ongoing reconstruction. "By reconstruction I mean not only the activities of architects, masons, and carpenters. You will, I think, soon realize, too, that the spirit of inquiry, truth, and freedom can flourish even among ruins (ibid.)." Laue then quoted Goethe's poem "Cowardly thoughts, timid shaking ...", [a call for courage in the face of hardship].[8] 113 talks were offered to the 1,543 participants, for instance, on neutron physics and astrophysics. The rector of the TU, Jean D'Ans, gave a memorial speech on Jacobus Henricus van 't Hoff. The address held by the DPG chairman, Karl August Wolf, briefly discussed the dangers posed by nuclear weapons and their development. He mentioned the inclusion of a passage to this effect in the society's statutes. A steamboat ride on the river Havel was the organized social event (ibid.).

The DPG tried to persuade the emigrés who had been excluded from among its ranks during the Third Reich to rejoin. This was a morally difficult issue—for both sides. Again Laue was called upon by virtue of the ethical stance he had held during

---

[7] The indicated address was still: Kaiser Wilhelm Institute of Physical Chemistry and Electrochemistry, Faradayweg 4–6, Berlin-Dahlem, Germany.

[8] "Feiger Gedanken/Bängliches Schwanken/Weibisches Zagen/Ängstliches Klagen/Wendet kein Elend/Macht dich nicht frei. Allen Gewalten/Zum Trutz sich erhalten/Nimmer sich beugen/Kräftig sich zeigen/Rufet die Arme/Der Götter herbei!"

the Third Reich to help formulate this appeal (Hoffmann and Walker 2012; Wolff 2012: 50 f., 88 f.). Many accepted, such as Lise Meitner and James Franck.

The occasion for travel in October brought Laue to the capital Bonn in order to accept membership in the prestigeous Order "Pour le mérite" for Science and Arts recently refounded by Federal President Theodor Heuss. This question of the order had probably been the subject of discussion between Laue and Heuss on more than one occasion already. The physicist Walther Bothe and the artist Renée Sintenis, for instance, were among those so honored in 1952 but both were unable to attend the ceremony for health reasons, as we learn from Laue's letter to Theo from October 10th (AMPG, III rep. 50 suppl. 7/12).

Later that month Laue was in London for a commemoration of the discovery of X-ray interference on October 24th and 25th. Sir Lawrence Bragg opened the meeting with a historical review, and Max von Laue was the guest of honor. The first reports about evaluating data by electronic computers were made at that meeting as well as about the feasibility of determining the structure of large organic molecules. Right after this conference, Laue ceremoniously handed the "Planck Medal" to Paul Dirac in Cambridge in the name of the German Physical Society, as had already been pronounced at the Physicists' Convention in Berlin (DPG 1952: 512 f.).

A long, pensive letter was sent to Theodore on November 17th from Dahlem about the postwar 'German Catastrophe' (Meinecke 1946): "I'm reading a short book by Fr.[iedrich] Meinicke on 'Strasbourg, Freiburg, Berlin 1901–19.' The final chapters, which deal with the events of World War I and the attempts by Berlin historians and other Berlin professors to exert some influence on them, are extraordinarily gripping. The first chapters are much more peaceful, and yet I find them so interesting because the circumstances they describe are familiar to me and I also know many of the people. But throughout I constantly feel how much easier life is for a humanist than for a natural scientist, whose work is 'nonhuman.' For the former there is not nearly as much difference between research and life, as between life and science. Indeed, it's precisely this remoteness from all things 'human' which gives the latter their special aura. But this occupation is not good preparation for dealing with contemporaries, which is simply inevitable (AMPG, III rep. 50 suppl. 7/12 fol. 52. Laue had visited Meinicke several times in Berlin.)."

A special celebration marked the end of the year. At Max von Laue's initiative a prominent sign of distinction was made for Fritz Haber as the first director of the Institute of Physical Chemistry and Electrochemistry. On his birthday a memorial plaque designed by the sculptor Richard Scheibe was unveiled in the institute's entrance area (Fig. 13.4). For the inscription Laue had chosen the passage in his obituary from 1934 comparing Haber's renown to Themistocles's lasting fame (see Chap. 9). In his speech at the unveiling Laue commemorated the man, his quick intellect, his conscientious sense of duty, and his personal attachment to the institute. He quoted Haber's letter of resignation in 1933 which expressed his appreciation of his staff. On this occasion Laue gave the audience a glimpse of his very personal attachment to Fritz Haber: "I saw with my own eyes the long weeks of struggle in which Haber arrived at the decision to submit his resignation. [...] Already in August

Fig. 13.4 Unveiling of the memorial plaque for Fritz Haber at the Kaiser Wilhelm Institute of Physical Chemistry on 9 December 1952. *Source* Archive of the Max Planck Society, Berlin-Dahlem

1933, when I saw him again in his Zurich hotel, he was but a shadow of his former self."[9] Karl-Friedrich Bonhoeffer spoke about Fritz Haber's multifaceted scientific oeuvre.

1953 was to become another year full of commitments, lectures, and travel. But good news awaited him right at the beginning. The publisher sent him the first copy of his revised third edition of 'The General Theory of Relativity' (Laue 1953a) and informed him that numerous orders had already been placed. Regarding the third, improved edition of his 'History of Physics' (Laue 1950) there was an exchange of letters with Hans Schimank with several suggested corrections, which Laue gratefully accepted on 12 January 1953 (GNT, Schimank papers). Best of all was the news in early February that Theo had been invited to lecture at the Free University (FU)

---

[9] Laue 1953c. Fritz Haber's very intense personal commitment to deploying chlorine as a "weapon" in World War I was very critically questioned only later, in the 1960s accompanied by calls to rename the institute. For Haber's letter of resignation at the end of April 1933, see Hentschel 1996: 44–45.

of Berlin and that his family was coming along. Of course, they should stay with them. "As for the travel expenses," Laue replied on February 19th, "we would contribute if need be. See if you can arrange it without an allowance from us. And then: Come very early and stay very long!" (AMPG, III rep. 50 Laue papers, suppl. 7/13).

This elated mood reemerged in Laue's letter to Erwin Schrödinger on February 20th: "I just recently was thinking again about the painful problem of the physical interpretation of wave mechanics when the book came into my hands about Louis de Broglie, in which Einstein, you, and many others once again give an opinion. When my head really started to ache from it, my friend Mephistopheles appeared to me and said: 'I recognize this all right, that's how the whole book sounds; I have wasted quite some time over it, for, a perfect contradiction remains just as mysterious to the wise as to fools.'"[10]

Fundamental questions kept arising to keep the correspondence with Einstein going. On March 27th, Laue thanked his friend for his letter of March 21st:

> To return again to the question of the definition of temperature in r[elativity] th[eory]: I happened to have occasion today to apply Le Chatelier-Braun's principle to the dynamics of the mass point. It states for the 'mass point' that the differential quotient of momentum with respect to velocity is greater if one keeps the temperature constant than if one keeps the entropy constant. This makes good sense. But the very possibility of applying that principle without modification to this example depends essentially on the digition which Planck gave for the temperature of the moving body, and which I have adopted. This again speaks for it being the most expedient among various possible definitions (UAF, Laue papers, 9.1).

After acknowledging receipt of Einstein's manuscript for the Born festschrift (1953),Laue remarked: "By the way, do you know that the Kaiser Wilhelm Institutes in Dahlem will be subsumed under the Max Planck Society on July 1st? Only mine is an exception in that it will be called the 'Fritz Haber Institute of the Max Planck Society.' Despite the general esteem in which Planck's name is held here, there are quite a few who regret this name change and ironically ask whether, perhaps, the Kaiser Wilhelm Tower in Grunewald woods should also be renamed the Max Planck Tower. That wouldn't be entirely unjustified, either; for there certainly has never been an emperor on it, whereas Planck, considering his fondness for towers, has certainly climbed it often (ibid.)."

His 'Haber Institute' featured in another one of Laue's frequent recollections about events from two decades hence, in his letter to Theo dated April 3rd: "On 1 Apr. 53, exactly 20 years after the boycott of the Jews, in which a glassblower at the KWI had barred 'the Jew Haber' entry to the Institute, his former wife Charlotte Haber was here, both at the Institute to view the memorial plaque, and at home for lunch (AMPG, III rep. 50 Laue papers, suppl. 7/13)."

Lise Meitner stopped in Berlin on her way back from Vienna—her first visit to her native city in fifteen years—to stay with the Laues from April 2nd to the 10th. It was another visit full of reminiscences. They went to the former KWI of Chemistry,

---

[10] Laue to Schrödinger, 20 Feb. 1953. In Pauli 1999: 73. Cf. George 1953; a verse from Goethe's 'Faust': "Ich kenn es wohl, so klingt das ganze Buch; Ich habe manche Zeit damit verloren, Denn ein vollkommner Widerspruch/Bleibt gleich geheimnisvoll für Kluge wie für Toren.".

Fig. 13.5 Federal President Theodor Heuss and Max von Laue after the tour through the Fritz Haber Institute. *On the right*: Ernst Reuter, the governing mayor of West Berlin, whom Laue had met during a flight in November 1952. *Source* Archive of the Max Planck Society, Berlin-Dahlem

rebuilt after its partial destruction in the war and now in use by the FU. The director's villa where she had once lived had not been rebuilt. Lise Meitner gave a talk 'On the neutrino' at the Physical Society on April 8th. The large auditorium of the Technical University was packed full, and her talk triggered a lively discussion.[11]

The next social obligation for Laue soon followed. On April 15th Federal President Theodor Heuss inaugurated the rebuilt main building of the TU and gave a speech about university issues. At this occasion Otto Warburg, the Berlin senator of finance, Friedrich Haas, and Laue were conferred an honorary doctorate, the title *Dr. rer. nat. honoris causa*. Whether Franz Bachér, who had been dismissed for political reasons in 1945 and had been teaching at the TU again since 1953, was among the professors present could not be ascertained. After that inaugural ceremony, Federal President Heuss visited the Fritz Haber Institute (Fig. 13.5).

The political conditions in the GDR caused the flow of refugees into West Berlin to increase steadily. Scientists were among them, and some found employment at the Fritz Haber Institute.[12] Just how wide-ranging Laue's activities and interests in physics still were, despite the burdens on a science administrator of the Fritz Haber Institute, is demonstrated by two papers: on 'Relativity theory, Doppler effect, and other spectral shift effects' and on 'Le Chatelier-Braun principle and relativity theory,' which he completed in June and late August, respectively (Laue 1954a, b;

---

[11] Invitation to Professor Lise Meitner to the board meeting and to present a talk on 8 Apr. 1953. DPGA, 10052; Meitner to Laue, no loc., 21 Jan. 1953. CAC, MTNR 5/32.

[12] Laue to Theodore, Berlin, 8 Mar. and 16 Apr. 1953. AMPG, III rep. 50, Laue papers, suppl. 7/13, fols. 22, 25, 28.

the second was dedicated to Otto Hahn for his 75th birthday). The first publication began with a preliminary note: "The present paper—arising from discussions with astronomer friends—is intended to compile the results of the special and general theories of relativity on spectral changes that in principle are of relevance to astronomy, whether or not they fall within the range of present-day measurement accuracy (Laue 1954a: 25)." He first briefly treated the Doppler shift in connection with special relativity, once with the spectral apparatus at rest, once with the spectral apparatus not at rest against the frame of reference. The problem was more complex for general relativity, he wrote: "As, firstly, no electromagnetic waves propagate at an invariable frequency (we shall discuss exceptions later). Secondly, unlike the special theory, it is not able to define the term 'relative velocity' for two separate bodies (p. 26)." This was elaborated further with reference to Einstein's 1911 publication. On the redshift in a three-dimensional space with radius of curvature R, he obtained the result: "$v R$ = const. If R increases with time, in accordance with Hubble's observation, then v decreases as the wave propagates. This is quite a different influence from the one that figures in Doppler's consideration (p. 27)." Laue also cited the paper by Aleksander A. Friedmann on the dependence of the radius of curvature on time (Friedmann 1924). The second publication on the Le Chatelier-Braun principle and the theory of relativity also sets the stage with a prefatory purpose:

> The Le Chatelier-Braun principle has been applied so often to mechanical, electrical, magnetic, thermodynamic, and especially thermochemical problems that one would like to think its sphere of applicability is exhausted. However, the (special) theory of relativity yields some new potential applications, because—unlike the older theory, in which the internal and kinetic energy of bodies are additive to the total energy, so that dynamics and thermodynamics could be treated independently of each other—it couples these latter two fields; the 'internal' energy depends on the velocity, and a separate kinetic energy no longer exists (Laue 1954b: 113).

Laue noted that Planck had already pointed this out in his papers on relativistic thermodynamics. After a brief calculation he arrived at the statements: "At isothermal acceleration, momentum increases faster than at adiabatic acceleration. [...] At isobaric acceleration, momentum increases more than at isochoric acceleration (ibid.)." These differences, he concluded, are based on the inertia of energy. In the limiting case, Newtonian dynamics emerges. This, he argued, does not contradict the Le Chatelier-Braun principle (p. 116).

In the middle of the year, James Franck left Chicago to participate in the 6th Congress on "Science and Freedom" in Hamburg before coming to Berlin to report about his group's research on photosynthesis, in particular, its quantum requirements. Franck recalled in a letter on July 18th:

> Then in Berlin: I was supposed to speak at Laue's colloquium and answered that Warburg would probably come and that would produce a controversy. That was right by Laue, as he was anyway steaming mad at the arrogant fellow. It turned out to be worse, though, because at the last minute the meeting was relocated to the Berlin Physical Society and that means a very large auditorium in which approx. 500–700 people came, including Warburg, Dean Burk & company. So I spoke, as I believe, quite well; that was also the opinion of my wife [Hertha Sponer] & a group of others. Very nicely about Warburg's merits in general, yet in particular he and Burk were simply wrong in that they had simply measured something

other than they had assumed, etc., etc. Then came Warburg's arguments, counterfactually arrogant; but I became quite forthright and, purely on the face of it, I had the upper hand. It got so far that the audience scuffled at his rudeness and stomped their feet at me.[13]

Laue's abhorrence of Otto Warburg occasionally became so strong that he would lock himself inside his office when Warburg was expected.[14]

In July, the third meeting of Nobel laureates took place in Lindau at the invitation of Count Lennart Bernadotte. The first meeting in 1951 had been for physicians, the second in 1952 for chemists. Now it was the physicists' turn to present talks amongst themselves:

Paul A. M. Dirac	Quantum dynamics and aether
Otto Hahn	Modern alchemy—the path from the imponderable to the ponderable
Werner Heisenberg	Progress and difficulties in the quantum theory of elementary particles
Georg von Hevesy	Biochemical effects of ionizing radiation
Max von Laue	X-ray interference
Cecil F. Powell	Free balloon flights at high altitudes
Frederick Soddy	Discovery of the natural transmutation of the radio elements
Hideki Yukawa	Attempt at a unified theory of elementary particles.

Talks held at the third meeting of Nobel laureates in Lindau, 1953 (source: Sterio 1963).

The interlude in Lindau by Lake Constance was a welcome opportunity for Laue to relax from the stress of managing so many different research areas at his institute. It was by no means easy to combine the so very different researcher personalities, some anxious to publish prolifically, such as Rolf Hosemann, and others of a much more withdrawn nature such as Gerhard Borrmann, who preferred rather to first perform that next experiment and measurement. When this invitation to the Nobel physicists for the conference in Lindau arrived, Laue gladly accepted.

The Laues' vacation in July was spent traveling in northern Germany. The Schimanks were visited in Hamburg. Laue had received a novel by Bruce Marshall (1953) as a gift, *Father Malachy's Miracle*, which he enjoyed reading during that trip, as we learn from his letter to Hans Schimank from Berlin on August 7th (GNT, Schimank papers). From the old lighthouse in Neuwerk, Laue was able to while his time away watching the shipping traffic at the mouth of the river Elbe. Then they went to Bremen to meet Theo and his family on the quay. Together they drove to Steinhorst near Celle, where the Lemcke family was now living. This reunion with all their

---

[13] Franck to Hans Gaffron, no loc., 18 Jul. 1953. Private collection. Quoted in Lemmerich 2007: 286; see also Krebs 1979: 84.

[14] Personal communication to the author by Laue's secretary Liselotte Fritsche.

Fig. 13.6 Max von Laue delivering a lecture at the Fritz Haber Institute on 9 October 1953. *Source* Archive of the Max Planck Society, Berlin-Dahlem

grandchildren was a true pleasure for the grandparents, as Laue wrote to Friedrich Möglich on August 10th (AMPG, III rep. 50 no. 1379 fol. 79).

This time the fall Physicists' Convention took place in Innsbruck. The German Society for Electron Microscopy was also meeting there beforehand and invited Max von Laue to join them for the conferral of his honorary membership, motivating Laue to plan to attend both meetings. Friedrich Beck was behind the wheel on the car journey there so Laue was able to devote himself fully to enjoying the autumnal scenery, as he told Theo in his letter from Aschach Castle on September 21st (AMPG, III rep. 50 Laue papers, suppl. 7/13 fols. 42 f.). At Innsbruck, after accepting his honorary membership, he himself awarded the Planck Medal to Walther Bothe at the plenary session of German physicists, in acknowledgment of Bothe's work on the coincidence method for determining the direction of cosmic radiation (DPG 1953).

Occasionally, in the confusion of political convictions Laue's judgment was not always entirely fair. This happened, for instance, when it was a matter of appointing Hartmut Kallmann as foreign member of the Fritz Haber Institute (Fig. 13.6). Laue had reservations about his animosities towards Germany, as he wrote to Paul Peter Ewald in Brooklyn, who, however, corrected him on November 4th: "It is thus natural for those who had emigrated to be more anxiously sensitive to the return of the former mentality than for those living in the political bustle of the day inside Germany. I think it unjustified to call his attitude, which admittedly often does differ from the present assessment of Germany, animosity. It is rather an expression of concern and perhaps even of a lack of trust (AMPG, III rep. 50 no. 56)." Cross-border contacts were generally a priority. For the *Physikalische Blätter* Laue wrote a short tribute to Walter Friedrich on his 70th birthday for his achievements related to the discovery of

X-ray interference. In the meantime, Friedrich had become president of the former Prussian Academy in East Berlin (Laue 1953d; AMPG, III rep. 50 no. 93).

The English edition of Laue's book on superconductivity (1952a) was critically received not only in England but also in America. This prompted Laue to write a long letter to James Franck in Chicago on 12 December 1953, asking about the context to his critics. He first briefly addressed the critique by David Shoenberg before explaining: "The purpose of my book, as stated point-blank on the first few pages, is to extend Maxwell's theory to superconductivity. This object has, in my opinion, been accomplished fairly well, and along a quite inevitable route, at that, without arbitrariness. All those newer theories are atomic, and have—or claim to have—the same advantage over my discussions as, for inst., the theory of […] dispersion has over the former Maxwellian theory (AMPG, III rep. 50 no. 627, text loss due to perforation)." "Superconductivity is the simple phenomenon, ohmic conduction its falsification by the thermal motion of the atomic arrangement," he continued further down. "You may gather from this my deep hesitation to expand my book further. The fact that important superconductivity experiments have become known in recent years is an entirely different matter (ibid.)."

In his letters to son Theo and also to Lise Meitner, Laue makes no mention of the arms race between the United States and the Soviet Union in the field of nuclear weapons. There was enough information about that in the newspapers. The hydrogen bomb tests in the South Pacific in 1951 by the USA had not only demonstrated the greatly increased destructive power of such bombs but also the large-scale radioactive contamination. It is very likely, though, that he did discuss these problems with his friends.

A letter from 15 January 1954 brought glad tidings in the new year. The *Société Francaise de Physique* had elected Laue as its honorary member. He informed the general administration of the MPG of this very special honor by letter from Berlin, pointing out that in 1915 Max Planck and Arnold Sommerfeld had been expelled from the Société and had not been readmitted since (AMPG). The Laue couple traveled to Göttingen for the major event in celebration of Otto Hahn's 75th birthday on March 8th. There they also met Lise Meitner. But, as they had already suspected, it was so overcrowded that they did not manage to speak with the birthday child in peace. In a letter to his son Theo from Dahlem on April 4th, Laue recalled: "On Apr. 15th it will be 3 years since I took over the Institute. Looking back on this period, I see it has undergone several important improvements; the most important one, the affiliation with the Max Planck Society is, of course, not my work alone, but I did have to make quite an effort for that, for ex., in order to overcome Otto Warburg's obstinacy, who was strongly opposed to it initially (AMPG, III rep. 50 suppl. 7/14 fol. 18)." On the subject of rearming the Federal Republic of Germany, Laue alluded to the negative influence that the officer corps had exerted on politics during the *Kaiserreich* and afterwards: "But also my own fate, that constant conflict with my father beginning in my earliest youth, was basically a struggle for freedom on my part, and an attempt to suppress it by the other part (ibid.)." Just six days later Laue wrote a letter to Walther Meissner, "because you obviously don't have a piece of news that will hit you quite hard, I think. Fritz London died in Durham on Mar. 30th. He was only 54 years old, but given the climate in Durham it's no wonder he couldn't take it any longer. I have written a letter of condolence to his wife (DMA, NL 045/006/40)." Then he told him that he intended to retire from the many offices he was currently holding (ibid.).

Postcards kept the Lemcke and Laue families informed about his many travel engagements. On April 17th, for example, the one to his son-in-law Kurt about his trip to Goslar for a physics conference: "We are returning home on Apr. 26th, but then a geological excavation of my desk under several sedimentary layers will again be necessary (Private collection)." In July, Laue replied to Einstein's letter of July 21st containing an inquiry about Planck's son Hermann from his second marriage. Then Laue reported about his trip to Sweden to observe the solar eclipse: "So, at the end of June, I together with my wife and a granddaughter [Grete Lemcke], was on Öland for the solar eclipse, where many astronomical expeditions had also set up their apparatus. E. Freundlich, together with Wempe from the Potsdam observatory, had brought along the most elaborate apparatus there. You've probably already heard that clouds prevented the observation. Even so, the journey was worth it for us, who were merely 'bystanders.' For, I had not expected to observe how the lunar umbra races over the Earth *at the Moon's orbital velocity* (orig. emphasis, UAF, Laue papers, 9.1)." Laue described the wandering of the shadow from west to east as leaving a deep "impression that one does not forget again." He let him know that they would be spending the holidays at Lake Achen in the Austrian Tyrol together with the Lemcke family (ibid.).

In October the Laues were away from home for his 75th birthday. It was a trip full of memories, since they visited his mother's grave in Magdeburg in East Germany. Back in Berlin, Laue was awarded the Grand Cross with Star of the Order of Merit, and there were personal congratulations, many written appreciations, and a great many congratulatory letters. The October issue of *Zeitschrift für Kristallographie* was dedicated to Max von Laue in celebration of his 75th birthday. It contained contributions by the department heads at the "Fritz Haber Institute" of the MPG. His granddaughter Grete sent Laue a travel report of the solar eclipse as a birthday present, which delighted him. He thanked her and her sisters for their congratulations on October 30th, writing: "But I have to work very hard now. For several days I've been doing nothing but answering congratulations (Private collection)."

After his birthday, Laue presented a talk before the Berlin chapter of the Physical Society on the topic 'Novelties on X-ray interference' (1954c). The first section covered a not-that-simple field, "Friedel's law according to the dynamical theory," the second and third topics were related to Borrmann's experiments.[15] Georg Menzer, who had obtained a professorship at the Ludwig-Maximilians-Universität in Munich, lobbied in favor of Max von Laue being awarded an honorary doctorate from there for his 75th birthday. He thought Sommerfeld had probably deemed Laue too young for it to have been conferred this honor earlier (cf. his letter to Laue on 2 Oct. 1954, AMPG, III rep. 50 no. 1345). The ceremony was set for November 23rd and Laue came to Munich in order to receive this distinction personally.

---

[15] Invitation to the meeting of the Physikalische Gesellschaft zu Berlin on 22 Oct. 1954. DPGA, 100321.

One by one, the German scientists and their collaborators who had "followed the call by the Soviet government" after the war were allowed to return from the Soviet Union to the East German Democratic Republic (GDR). Prof. Max Volmer and his wife Lotte moved back into their home near Babelsberg in 1955, and Gustav Hertz accepted a professorship in Leipzig. In divided Berlin, even apolitical commemorative events were problematic. Organizing any joint event required diplomatic skill and patience. That's why the board of the Physical Society resolved well in time in December 1954: "On 18 March 1955 a festive session is to be held on the occasion of the 50th anniversary of Einstein's papers on light quanta and relativity. The Physical Society of the GDR has proposed to arrange a parallel event in the East. The board resolves to host a festive session on the history of quantum theory on 18 Mar. 55 in the large auditorium of Physics Institute I, for which Professor von Laue wants to win Professor Max Born as speaker. The invitation to this should include a reference to a festive session of the Phys. Soc. in the GDR, at which Prof. Infeld, Warsaw will speak about the theory of relativity (Minutes of the meetings on 3 and 9 Dec. 1954. DPGA, 10053)."

In his Christmas letter to Lise Meitner on 20 December 1954 from Dahlem, Laue informed her that Otto Hahn had proposed to him to hand in his resignation on 31 March 1956. However, it was not certain whether the senate of the MPG would agree. Laue was amenable to it, he told her. "So far so good. But here comes the 'club.' Quite a number of members of the senate, whom I had spoken to informatorially outside the meeting, advocated that I not get a successor, but that the Haber Institute be split up into 5 or 6 small institutes; this proposal shows more clearly than anything else from whence the wind blows (CAC, MTNR 5/28)." Laue suspected that Otto Warburg was behind this, as his own institute was much smaller. At Christmas, Theo and his family came to Berlin again to travel to Helsinki via Copenhagen. After their departure, Laue wrote melancholically on 5 January 1955 to his son-in-law Kurt: "It's become hard for me to say goodbye, the silence in the apartment is painful for me (Private collection)."

The new year began for Max von Laue with complaints of high blood pressure and for his wife with recurring fatigue and frequent asthma attacks. But that did not stop them from attending Federal President Theodor Heuss's 71st birthday party on January 31st. Lise Meitner responded to Laue's Christmas letter about the general administration's plan to chop up the Fritz Haber Institute into several small institutes. Writing at length from Stockholm on January 18th, she was not only sympathetic about Laue's situation, but also offered him support on the difficult issue concerning Otto Hahn in defense of the current structure (CAC, Meitner 5/32). Laue had already informed the staff about the future of the institute and about his planned resignation on 1 April 1956. This worried the department heads and they wrote a joint letter to MPG-President Otto Hahn about it on January 31st: "It is our wish that you, Esteemed Mr. President, find some means of retaining Professor von Laue as director even beyond the arranged term, in some looser or firmer form, entirely at his own discretion, whether for life or until he wishes to retire for good. This would be fine acknowledgment of his decision, taken four years ago, to place himself at the disposal of a Berlin institute at his advanced age, due thanks for the work performed to his utmost by a great personality (AMPG, II rep. 1A Laue staff file, fols. 4/5)." The signatories pointed out that then the institute could continue to develop further. The letter was cosigned by Ivan Nikolov Stranski, Ernst Ruska, Kurt Überreiter, Gert Molière, and Gerhard Borrmann.

Laue actually still had too many obligations to meet. For the 28th conference on glass technology in West Berlin, the organizers had convinced him to deliver the keynote speech. He chose as his topic 'The beam path of X-rays in crystals,' which thanks to Borrmann's experiments was particularly easy to illustrate (Laue 1955a).

On January 14th, the DPG board had met again and decided the following under point no. 4 of its minutes: "Concerning the meeting in honor of Einstein on March 18th inst., Professor von Laue announces that the Physical Society in the GDR would like to extend a joint invitation to Professor Einstein. The Board resolves that Professor von Laue and Professor Hertz will write jointly to Einstein on behalf of their Boards and invite him (DPGA, 10054. The cosigned draft letter was passed.)." Their invitation to Einstein to the ceremonial speeches on March 18th and 19th cited his publications 'On a heuristic point of view concerning the production and transformation of light' (1905b) and 'On the electrodynamics of moving bodies' (1905a): "The two undersigned Physical Societies request this high honor of adding special splendor to these celebrations by your presence (ibid.)." Laue had already written to his friend on January 15th about the idea of a joint celebration and the invitation to him, arguing that this could be one way to make a contribution toward general political détente. "As this is the aim of your own political wishes too, it seemed certain to us that you could do something towards furthering it by coming to see us in March (UAF, Laue papers, 9.1)." But Einstein's response to his "Dear Colleagues" at the Berlin branch of the DPG on February 10th was: "Thank you very much for the kind invitation to the planned keynote speeches. It would have been a great pleasure for me to attend. But 50 years of intervening time have left only a dilapidated remnant of myself, so I can no longer undertake any major travels. This, however, does not obstruct my joy at this proof of your amicable disposition. In wishing all those involved and, moreover, all those attending good success throughout the course of this celebration, I am, with collegial greetings, yours, A. Einstein (UAF, Laue papers, 9.1)."

Despite the spatial division, the event was a great success (Fig. 13.7). Max Born lectured on 'Einstein and light quanta' in West Berlin and Leopold Infeld on 'The history of relativity' in East Berlin. A telegram was sent out to Einstein bearing greetings. Laue was able to attend Infeld's lecture in East Berlin without restrictions, as he wrote to his son in detail from Bonn on March 21st (AMPG, III rep. 50 suppl. 7/15; cf. Hoffmann 1995a: 157, fig. p. 163). Einstein received a full account of the celebrations from Clara von Simson, who had attended both events, as we learn from Laue's letter from Dahlem to Einstein on April 1st (UAF, Laue papers, 9.1).

After a brief hospital stay in Princeton, Einstein died of an aneurysm on 18 April 1955. Max von Laue had lost his friend. They had met each other as young men and their unusually casual way of addressing each other in the familiar "*Du*" form was an expression of their mutual trust. On the evening of his passing away, on April 18th, Laue personally delivered a funerary address at the broadcasting station Radio Free

Physikalische Gesellschaft zu Berlin e. V.

im Verbande Deutscher Physikalischer Gesellschaften
Berlin-Charlottenburg 2, Hardenbergstr. 34, Fernsprecher: 32 51 81, App. 247

Postscheckkonto: Berlin-West Nr. 465 82

Die „Physikalische Gesellschaft zu Berlin e.V." lädt ein zu einer

Festsitzung

aus Anlaß des 50. Jahrestages der Einsteinschen Arbeit

„Über einen die Erzeugung und Verwandlung des Lichtes betreffenden heuristischen Gesichtspunkt"

am Freitag, dem 18. März 1955, um 17 Uhr c.t.
im großen Hörsaal (P 270) des I. Physikalischen Instituts
der Techn. Universität Berlin-Charlottenburg, Hardenbergstr. 34.

Es spricht Herr Professor **Dr. Max Born** (Bad Pyrmont)

über das Thema:

„Einstein und die Lichtquanten."

Die Nachsitzung findet in der „Hardenberg-Hütte" Schillerstr. 5, statt.

Physikalische Gesellschaft zu Berlin e. V.
Der Vorstand

Hinweis:

Die Physikalische Gesellschaft in der Deutschen Demokratischen Republik veranstaltet aus Anlaß des 50. Jahrestages der Einsteinschen Arbeit „Zur Elektrodynamik bewegter Körper" eine Festsitzung am 19. März 1955 um 16^{30} Uhr im Vortragssaal der Deutschen Akademie der Wissenschaften zu Berlin, Unter den Linden 8, bei welcher Professor Dr. L. Infeld (Warschau) über das Thema: „Die Geschichte der Relativitätstheorie" (in deutscher Sprache) vortragen wird

Fig. 13.7 Program of the honorary session of the Physikalische Gesellschaft zu Berlin e. V., 18 March 1955. *Source* Document in the author's estate

Berlin (*Sender Freies Berlin*). His feelings and memories, his lasting admiration for his friend are expressed in a longer necrology, which he wrote in just under three days for the papers: "Albert Einstein died on April 18th at the age of 77. Physicists, who proudly count him among their ranks, indeed all natural scientists, pause from their work; as, one of their greatest has passed away. But humanists, too, especially professional philosophers, are stirred; for Einstein's significance reaches deep into the philosophy of scientific knowledge. Indeed, anyone surmising the influence of the mind on the course of history should look up (Laue 1955b)." A description of Einstein's career in life then followed before Laue returned to the impact of Einstein's publications on physicists (ibid.).

Not long afterward, on May 5th, the Federal Republic of Germany regained full sovereignty from the Western Allies; a politically momentus event. Still under the impression of a visit by Nelly Planck, the widow of Planck's son Erwin, Laue wrote to Theo from Dahlem on May 11th:

> We talked for over 2 hours, in connection with the Goerdeler book by G.[erhard] Ritter (1954), about Erwin Planck and his experiences during the Hitler era. It was deeply moving. What an unusually stoical woman is this who, with great trouble finds out about the execution of her beloved husband; who the following night lets herself be driven to Rogätz, to her parents-in-law, by the Swiss legation, brings them the news, and returns the next night to Berlin in the car of the Swiss legation; and only finds her first tears when her father-in-law, to whom she had given the terrible news early that morning, sits down at the piano and plays the favorite tunes of the murder victim!
>
> And what a proud man is he, namely his father Planck, who, after his son has been condemned, refuses to submit an appeal to Himmler for clemency, saying that the death of his son would break him completely!
>
> At last, Mrs. Nelly Pl[anck] had him give her two blank signatures, and then filled them out herself when the time came, with texts that he would hardly have approved of entirely. I am unable to reproduce the many details she reported to me (AMPG, III rep. 50 Laue papers, suppl. 7/15 1955, fols. 19, 21).

Concentrating on scientific work helped distract Laue from the painful loss of his friend Einstein. Quantum theory had made the question of determinism in physics topical again; and depending on one's preferences, it was treated either philosophically or epistemologically. Max Born published the paper 'Is classical mechanics really deterministic?' in 1955, and Laue submitted a rebuttal that substituted the word "physics" for "mechanics" in his colleague's question (1955c). The motion of a mass point was chosen as an example. The exact values are "physical realities, independent of what people know about them," Laue stated. Then he presented computations of an astronomical orbit with its determination progressively improving the accuracy by more and more measurements. "This case occurs in planetary motion when the initial determination of position and momentum does not yet allow one to make the decision between an elliptical and a hyperbolic orbit. [...] Things are different in quantum theory; it declares impossible this retrospective calculation of the initial position and initial momentum from a later observation, because each observation influences the state (Laue 1955c: 270)."

At the end of June Theo arrived with his family and stayed until July. The Lemcke and Laue families spent their holidays together in Lüneburg heath. The children were able to communicate well despite the language barrier. Theo had actually planned to go to Paris afterwards, but that was abandoned because of strikes there. They drove to Berchtesgaden instead, as we learn from Laue's letter to Otto Stern on August 23rd (UAF, Stern papers).

The manuscript of the completely revised fourth edition of his book 'General Theory of Relativity' (1953/56) was almost ready, and this provided the opportunity for him to travel to Bern for the jubilee conference of the Swiss Physical Society on "Fifty Years of Relativity." It was an illustrious gathering, including many of von Laue's friends. Wolfgang Pauli served as president. The scientific sessions were each divided into a main lecture followed by short communications. On the first day, July 11th, Walter Baade of Mount Palomar Observatory gave the main talk on the expansion of the universe. In the first short communication, Laue presented a theory of redshift using the covariant notation of Maxwell's equation. "The theory of Hubble's redshift in distant nebulae inheres in it. It differs from the Doppler effect in that the change occurs gradually over the time that the light travels, whereas the Doppler effect proper occurs all at once, at the location of the source of light (Mercier and Kervaire 1956)." After Max Born had finished presenting the final talk on July 15th, Wolfgang Pauli gave a detailed and critical closing speech, not lacking in his characteristic humor (ibid.). Laue reported about these talks to Mrs. Rosenthal-Schneider in Australia in his letter dated July 20th from Berlin: "No one commemorated Einstein's achievements on quanta; the quantum physicists are cross with him for not wanting to adopt the statistical interpretation. They did take a lot of trouble to 'convert' him. But to think that such a conversion is possible would be to underestimate Einstein's independence, not to say his delightful stubbornness. Determinism was rooted far too firmly in his limbs—as it is for me. If someone had seriously doubted it during my youth, who knows whether I would have become a physicist!" Further down he continued, "Objectively, the conference was important insofar as it gave me the certainty that all of whatever still remains standing in the general theory of relativity is also contained in my book (DMA, NL 264)." Laue also briefly alluded to the ongoing American-Soviet talks in Geneva: "At least the danger of war with atomic bombs and other such niceties has been postponed a bit (ibid.)." The Laue couple combined their participation in the festivities in Bern with a stop in Strasbourg and visits with old school friends. The next place on their agenda was Lake Thun, but bad weather and health complaints by Magda von Laue made them decide to return home early, as he wrote Theo on April 15th (AMPG, III rep. 50 suppl. 7/15 fol. 17).

Laue knew that he would be stepping down as director of the institute in the coming year, but that did not diminish his concern about improving the institute's equipment. Although he did not have the funds to buy a helium liquefier yet, he hoped to raise them next year, as he wrote Meissner on June 23rd (DMA, NL 045/40).

Traveling was important to Laue, despite the exertions and discomforts of old age, especially when there was the prospect of meeting with colleagues. He took part again in the next scheduled meeting of Nobel laureates in Lindau. Up to that point, only a few isolated scientists had warned against the devastation of a war deploying nuclear weapons. On this occasion 18 Nobel laureate scientists in Lindau decided on July

15th to issue a warning appeal, later known as the "Mainau declaration," the most important statements of which were:

> - We, the undersigned, are scientists from different countries, different races, different creeds, different political persuasions.
> - We have gladly put our lives at the service of science. It is, we believe, a path to a happier life for mankind. We see with horror that this very science is giving mankind the means to destroy itself.
> - We do not deny that today peace is perhaps maintained precisely by the fear of these deadly weapons.
> - All nations must arrive at the decision to voluntarily renounce violence as a last resort in politics.
>   The Mainau declaration by Nobel laureates (source: D. Hahn 1975: 217).

The original declaration bore 18 signatures, in the end there were 52. Laue's signature was among them.

It was a matter of course for Laue to participate at the German Physicists' Convention in Wiesbaden at the end of September, because he wanted to speak to his colleagues about Einstein. The typescript draft of his speech on 'Einstein and relativity theory' emphasized his friend's ties to Germany: "We Germans have special reason to commemorate him, because in spite of everything that happened during the Hitler era, we are proud to regard him as a German physicist. After all, almost all of his publications, and especially the most brilliant ones, appeared in German journals; he primarily used the German language throughout his life, both in publications and orally. Whether he liked it or not, he was rooted in the heritage of German thought."[16] (Fig. 13.8) Laue put forward as evidence of this Einstein's knowledge of the writings of the important physicists from the late 19th and early twentieth centuries. Einstein had not worked with models, as Bohr did, but according to principles (Laue 1956a).[17]

The usual Sunday letter to Theo on November 6th described a walk between Wannsee and Pfaueninsel in autumnal weather at 10 °C. "Now the Federal Republic and Berlin want to promote nuclear physics mightily. And it turns out that nowhere nearly enough physicists can be found for that. Money seems to be abundantly available now or in the near future. But that alone doesn't do it either. Some people also seem to be afraid of nuclear physics, because there is too much state involvement (AMPG, III rep. 50, Laue papers, suppl. 7/15 fol. 47)." Laue mentioned protests by women in Karlsruhe against the construction of a reactor. The peaceful use of nuclear

---

[16] Laue, M.: Einstein und die Relativitätstheorie. Vortrag gehalten am 23.9.1955 auf dem Deutschen Physikertag in Wiesbaden. Typescript with numerous deletions and insertions. AMPG, III rep. 50.

[17] In the autumn of 1955 proofs arrived for the fourth edition of the 'General Theory of Relativity' (published 1956), and Laue performed corrections, additions and changes, occasionally assisted by Beck. As these old proof sheets show, it was quite a major undertaking. AMPG, III rep. 50 nos. 40–56.

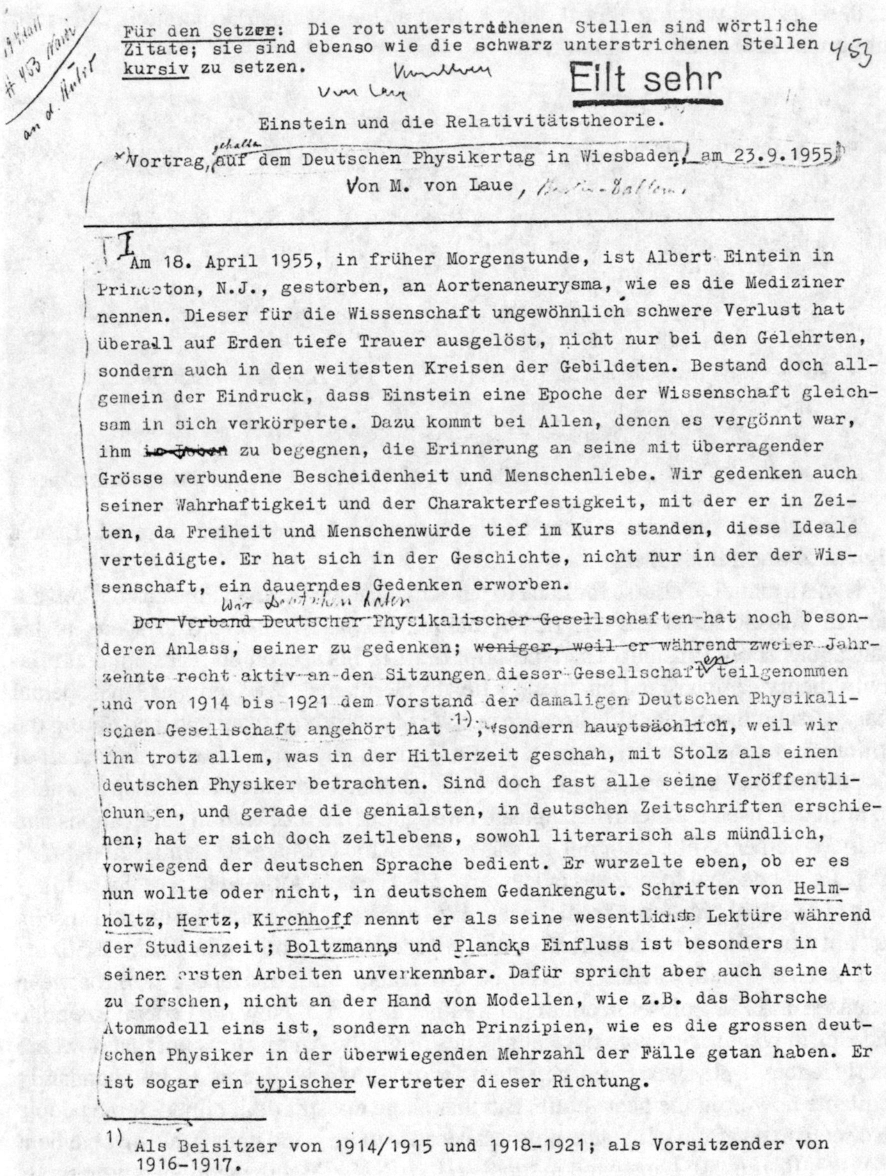

Für den Setzer: Die rot unterstrichenen Stellen sind wörtliche
Zitate; sie sind ebenso wie die schwarz unterstrichenen Stellen
kursiv zu setzen.

Eilt sehr

Einstein und die Relativitätstheorie.

Vortrag auf dem Deutschen Physikertag in Wiesbaden, am 23.9.1955.

Von M. von Laue,

Am 18. April 1955, in früher Morgenstunde, ist Albert Eintein in
Princeton, N.J., gestorben, an Aortenaneurysma, wie es die Mediziner
nennen. Dieser für die Wissenschaft ungewöhnlich schwere Verlust hat
überall auf Erden tiefe Trauer ausgelöst, nicht nur bei den Gelehrten,
sondern auch in den weitesten Kreisen der Gebildeten. Bestand doch all-
gemein der Eindruck, dass Einstein eine Epoche der Wissenschaft gleich-
sam in sich verkörperte. Dazu kommt bei Allen, denen es vergönnt war,
ihm zu begegnen, die Erinnerung an seine mit überragender
Grösse verbundene Bescheidenheit und Menschenliebe. Wir gedenken auch
seiner Wahrhaftigkeit und der Charakterfestigkeit, mit der er in Zei-
ten, da Freiheit und Menschenwürde tief im Kurs standen, diese Ideale
verteidigte. Er hat sich in der Geschichte, nicht nur in der der Wissenschaft, ein dauerndes Gedenken erworben.

Der Verband Deutscher Physikalischer Gesellschaften hat noch beson-
deren Anlass, seiner zu gedenken; weniger, weil er während zweier Jahr-
zehnte recht aktiv an den Sitzungen dieser Gesellschaft teilgenommen
und von 1914 bis 1921 dem Vorstand der damaligen Deutschen Physikalischen Gesellschaft angehört hat [1]; sondern hauptsächlich, weil wir
ihn trotz allem, was in der Hitlerzeit geschah, mit Stolz als einen
deutschen Physiker betrachten. Sind doch fast alle seine Veröffentlichungen, und gerade die genialsten, in deutschen Zeitschriften erschie-
nen; hat er sich doch zeitlebens, sowohl literarisch als mündlich,
vorwiegend der deutschen Sprache bedient. Er wurzelte eben, ob er es
nun wollte oder nicht, in deutschem Gedankengut. Schriften von Helmholtz, Hertz, Kirchhoff nennt er als seine wesentliche Lektüre während
der Studienzeit; Boltzmanns und Plancks Einfluss ist besonders in
seinen ersten Arbeiten unverkennbar. Dafür spricht aber auch seine Art
zu forschen, nicht an der Hand von Modellen, wie z.B. das Bohrsche
Atommodell eins ist, sondern nach Prinzipien, wie es die grossen deutschen Physiker in der überwiegenden Mehrzahl der Fälle getan haben. Er
ist sogar ein typischer Vertreter dieser Richtung.

[1] Als Beisitzer von 1914/1915 und 1918-1921; als Vorsitzender von
1916-1917.

Fig. 13.8 The first page of the typescript of the talk Laue delivered at the German Physicists'
Convention in Wiesbaden in 1955. *Source* Photograph in the author's estate

energy was supposed to reduce the environmental pollution caused by the exhaust gases of coal-fired power plants.

The impetus for the peaceful use of nuclear energy had come from the USA. Under the administration headed by Dwight D. Eisenhower, research on the utilization of nuclear energy continued. In a speech before the United Nations General Assembly in December 1953, President Eisenhower had initiated a program for its peaceful use under the slogan "Atoms for Peace." Professors at the Free and Technical Universities in Berlin initiated the founding of a nuclear research institute in West Berlin in December 1955. Laue was not a cosignatory, but he was involved in the search for suitable directors.[18]

The letter-writing continued in the new year. On 7 January 1956, Max von Laue informed Hermann Heimpel of the MPI for History in Göttingen: "You know that the Federal President [Heuss] has asked me to write the article on Heinrich Hertz for the work 'Great Germans.' I have used the Christmas holidays to finish the manuscript—except for one small thing, about which I would like your advice (Laue 1956b)." He asked for information about the fate of Mrs. Hertz and their two daughters after their emigration and suggested that they receive financial support. Later that month he then had to set something straight in Mr. Heimpel's reply: "[I] see with deep regret that there has been a misunderstanding concerning the article on Albert Einstein for 'Great Germans.' I did not promise Professor Hahn that I would take on this article; on the contrary, I maintain the opinion that it is not worthy of so great a deceased figure as Albert Einstein to have a large number of obituaries written about him by the same individual (ibid.)." He sent him his three obituaries.

Laue's contribution on Heinrich Hertz began with a detailed historical introduction to the notions held by naturalist scientists of the 18th and early nineteenth centuries up to Clerk Maxwell about electricity and magnetism. This was followed by his biography. Hertz had first wanted to become a civil engineer but had then joined Hermann von Helmholtz in Berlin in 1878, where he obtained his doctorate at the age of 23. The main part describes the period in Karlsruhe with the discovery and analysis of electromagnetic waves and their scientific significance. The theoretian Max von Laue did not omit to mention the problems that Hertz faced as an experimenter. In the last paragraph Laue asked: Why do we now count Hertz among the great Germans? First, he cited wireless engineering and its importance for communications; second, our understanding of cosmic rays; and third that "the discovery has reshaped the fundamental views of physics, which are of incalculable importance not only to physics itself, but also to philosophy, and deeply affect the world-view of every individual, whether or not he is aware of it (ibid.)."

Magda von Laue's health continued to deteriorate, and a longer stay in hospital became necessary in spring. After attending the spring meeting of the German Physical Society in Bad Nauheim, where Laue spoke about the radiation path in crystals,

---

[18] Atomforschung und Nutzbarmachung der Atomenergie und radioaktiver Isotope in Berlin. Denkschrift der Technischen Universität und der freien Universität Berlin. December 1955. Landesarchiv, Rep. 14, acc. 2881, Az. 318580. See also Weiss 1994.

he used the opportunity to continue driving to Göttingen to meet with the general administrators of the MPG and visit friends.

His obligations, some of which were self-imposed, did not diminish. On June 4th, Max von Laue spoke on RIAS (Radio in the American Sector) in memory of the revered Friedrich Schmidt-Ott, who had died on April 28th at the age of 95 (Fig. 13.9). He sketched his life and oeuvre. In it Laue's historical interest becomes evident again (Laue 1956c): In the late nineteenth century, *a single individual* organized the promotion of science in the Kingdom of Prussia: Ministerial Director Professor Friedrich Althoff. The 28-year old Friedrich Schmidt-Ott began working with him in 1888 and continued Althoff's work with a gentler hand and played a significant role in the founding of the Kaiser Wilhelm Society. At the end of the Empire, in 1918, he retired from government service and very soon afterwards became involved in the founding of an "Emergency Association of German Science" (*Notgemeinschaft*) for the advancement of science. He became a member of its presiding board together with Fritz Haber and Walter van Dyck. They founded twenty expert committees for the various fields and thus were able to fund many important scientific projects. Laue also mentioned Schmidt-Ott's promotion of young scientists: This funding organ "has done an extraordinary amount of good, not least because its president, while maintaining strictest conformity with the statutory order of business, has shown almost fatherly love for these young aspiring professionals (ibid.)."

Patrick M. S. Blackett – Rock magnetism and movement of continents

Max Born – Remarks on the fundamentals of the kinetic theory of gas

Sir John D. Cockcroft – Science and technical problems in the development of nuclear power

Paul A. M. Dirac – Electrons and the vacuum

Werner Heisenberg – Problems in the theory of elementary particles

Gustav Hertz – Physical methods for isotope separation

Max von Laue – From Copernicus to Einstein

Sir C. V. Raman – Physics of crystals

Hideki Yukawa – Elementary particles

Frits Zernike – Advancements in physical optics.

The speakers and their topics at the second meeting of Nobel laureates in physics at Lindau 1956.

In June 1956, the second meeting of Nobel laureates in physics took place in Lindau, at which Laue was again to speak (ms. in AMPG, III rep. 50 no. 102). (Fig. 13.10) As its section headings clearly indicate, Laue's contribution was an epistemological consideration, an account without formulas:

History of the dispute Ptolemy – Copernicus

Epistemological perspective

Fig. 13.9 Friedrich Schmidt-Ott (1860–1956), lawyer, politician, and science organizer. In 1920, together with Fritz Haber, he initiated the foundation of an Emergency Association of German Science (forerunner of today's German Research Foundation, DFG), of which he became the first president. *Source* Archive of the Max Planck Society, Berlin-Dahlem

Physical considerations

The inertial systems

The speed of light in different inertial frames

The special theory of relativity

Minkowski's "world"

The general theory of relativity

Field equations of gravitation

Conclusions from the field equations

Cosmology

Synopsis.

The brief synopsis was another eulogy of Einstein. It read: "*Visualizations* of space and time are ingrained forms of human perception, properties of our cognitive faculty, which no experience can alter. Space and time *measurement*, on the other hand, must be gathered from empirical science, that is, from physics. The long history of this process has recently reached a conclusion with the general theory of relativity, beyond which there is at present neither reason nor possibility to go. The credit for this belongs without a doubt to Albert Einstein alone (Laue 1957a, orig. emphasis)."

Laue had asked the historian of science Hans Schimank for literature on Copernicus. In his letter from 6 May 1955 thanking him for having sent what he had requested, he expressed his condolences on the death of his daughter, who had been ill for a long time: "But I am deeply saddened that you and your wife are finding it so difficult to get over the loss of your daughter. After all, a natural death is the result of a scientific process, and we scientists are used to accepting the existence of the laws of nature—especially when we see in them an expression of God's will. With very

Fig. 13.10 Max Born, Max von Laue, and Otto Hahn in Lindau on the occasion of the meeting of the Nobel laureates in physics. *Source* Photograph, in the author's estate

cordial greetings, I extend to you my hand, sincerely yours, M. v. Laue" (AMPG, III rep. 50 no. 1750).

The Fritz Haber memorial served as a model for another opportunity that Laue found to draw the public's attention to significant scientific achievements in Berlin. It concerned an immediately neighboring building to the Fritz Haber Institute, the former Kaiser Wilhelm Institute of Chemistry, where Otto Hahn and Fritz Strassmann had discovered—actually contrary to the conceptions held by nuclear physicists—the fission of a uranium atom upon bombardment with neutrons in 1938. Their highly sophisticated chemical analyses succeeded in detecting barium among the fission products. Hahn had written to Lise Meitner on 19 December 1938, after setting forth the results of their chemical analyses of uranium irradiated with neutrons: "Perhaps you can propose some fantastic explanation. We ourselves know that it can't actually burst asunder into Ba."[19] The exterior of the badly war-damaged building was restored in the old style and used by the chemistry department of the FU in Berlin. The plan was to attach a memorial plaque with a short text on the outside of the building and another memorial plaque with a longer inscription inside the building about the discovery made at that site by Otto Hahn and Fritz Strassmann. Laue toured the building together with the sculptor Richard Scheibe to select suitable locations for the two plaques. The chemists had to be involved in this tribute, and Laue approached the Society of German Chemists and the German Bunsen Society for support, which was approvingly received. The local chemists in West Berlin also

---

[19] CAC, MTNR. Reprinted in Lemmerich 1988: 170f. with a facsimile of Lise Meitner's letter to Otto Hahn of 21 Dec. 1938 expressing her doubts about the chemical results: "The assumption of such an extensive bursting apart seems to me very problematic right now [...]." The sought transuranium elements were only actually discovered later, in 1940/41 by Edwin McMillan and Glenn Theodor Seaborg.

came on board.[20] Having thus obtained the approval of the chemists, the idea now had to be materialized; Laue informed Lise Meitner about the status of the negotiations in his letters from 17 and 18 January 1956: "Hahn and Straßmann have agreed to the 'indoor' plaque, but not to the 'outdoor' one. And yet it is precisely that one which seems to us to be particularly important, in order to make clear to wide circles in Berlin by means of a compelling example the importance of the science conducted in Berlin, also in order to strengthen the self-confidence people need so much given the precarious situation of Berlin (CAC, MTNR 5/32)." But Lise Meitner also responded negatively on the 23rd from Stockholm, proposing instead that an outdoor plaque perhaps be put up on Hahn's one hundredth birthday (ibid. and AMPG, III rep. 50 no. 772). Von Laue took this difference of opinion in his stride. In his letter on March 5th he was able to inform her that Karl Erik Zimen had been appointed head of the chemistry department of the planned Institute for Nuclear Research. Otto Hahn made various proposals for the nuclear physics department, but they turned out not to be feasible. That was largely due to the political "insular situation" of West Berlin (CAC, MTNR 5/32).

As part of the lecture series at the Fritz Haber Institute, Ludwig Waldmann from the MPI of Chemistry in Mainz spoke in July 1956 about 'Irreversible processes.' This initiated an exchange of letters between Laue and the speaker on the problem of irreversibility. In his letter of the 7th, Laue argued: In classical mechanics every mechanical process is strictly reversible, and in quantum mechanics only a probability exists for Schrödinger's equation, for example, but no certainty (AMPG, III rep. 50 no. 2074). Waldmann replied in some detail taking the example of electron reflection (ibid.). This reply did not satisfy Laue though. On the 27th he wrote: "That one can make an outgoing wave into an incoming one mathematically is undoubtedly correct. But since a Schrödinger wave only yields probabilities and leaves the individual event largely open, I do not think it can be said that a thus-described inverse collision will reverse the original collision. It *could* do so, but the probability of that seems to me to be minimal (orig. emphasis, ibid.)." It was not until several months later that Waldmann was able to reconsider Laue's objection carefully and respond by letter from Mainz on December 21st, whereupon Laue replied ten days later that his argument: "has largely convinced me. I had hitherto been thinking of the reversibility of a single collision and this, in my opinion, does not exist in quantum theory. But your point of view that reversibility holds for observable averages is indeed valid (AMPG, III rev. 50 no. 2074. I thank Prof. Siegfried Hess for this reference.)."

Political conditions had little or no influence on scientific life and scientific activities, which is actually quite surprising in retrospect. The danger of a war between the major powers utilizing atomic and/or hydrogen bombs had by no means diminished. Moreover, the destructive power of such bombs wrapped in cobalt could achieve enduring lethal contamination. But the desire to have a share in the future development of nuclear energy for peaceful purposes existed in West Berlin, and Berlin parliamentarians put the question of founding a local Institute for Nuclear Research

---

[20] Laue to D'Ans, no loc., 4 Jul. 1955; Hahn to Lautsch, Göttingen, 7 Oct. 1954. AMPG, III rep. 540, nos. 776/777.

on the agenda for 1 October 1956. Otto Hahn and Fritz Strassmann were invited to speak. Hahn spoke about 'The importance of the peaceful use of nuclear energy for the future' and Strassmann's topic was 'Berlin's vocation as a center for research and training.'

It was Laue's wish to involve Lise Meitner in the celebration at the former KWI of Chemistry. She ought to be present and deliver a talk on the occasion of the unveiling of the "indoor" plaque. That is how Laue put it to her, but she could not be persuaded to attend. Laue then proposed that the Free University of Berlin confer an honorary doctorate on Lise Meitner during the dedication ceremony of the plaque. The university complied with this request, but Mrs. Meitner fell ill with an ulcerated jaw and was unable to attend. (The conferral of her doctorate took place later, on 11 May 1957.) So she sent a telegram for Laue to read out during the ceremony. The speeches were held by Siegfried Balke, Federal Minister of Atomic Questions, Otto Hahn, and Fritz Strassmann. Laue mentioned in his introductory address the idea of attaching a larger plaque to the front of the building. However, this idea did not win Otto Hahn's applause and fell dormant. Laue welcomed Otto Hahn and Fritz Strassmann with the words: "It is their attendance which lends this celebration today its true dignity. But one personality, who is unfortunately not among us today, must certainly receive at least as much acknowledgment. I refer to Lise Meitner. For 30 years she was Otto Hahn's closest colleague in a bond of friendship that is probably historically unique in science (Laue 1957b)." Laue recalled Meitner's escape and her theoretical interpretation of nuclear fission developed together with her nephew Otto Robert Frisch. Otto Hahn then spoke, and before discussing that discovery and Lise Meitner, he remarked: "Our joy today at this celebration is mixed with an oppressive sense. We do not stand alone here on the grounds of the more recent pertinent knowledge. What's the situation? We continue to build upon what came before us, find something new in the process, others pursue it further and find some other novelty. Thus the present state of modern atomic research is the sequential arrangement of many building blocks into a grand edifice (Lemmerich 1988; Zeitz 2006: 199–204; AMPG, III rep. 50 nos. 771 f. and 780 f.)."

No successor to Max von Laue had yet been appointed and Laue now had doubts about whether the choice of Rudolf Brill was the right decision. He wrote to Otto Hahn on 15 January 1957: "I have changed my mind since and would now like to propose Prof. Rolf Hosemann, who has been working radiographically here since 1951. He is very energetic, employs almost half a dozen younger coworkers, and his structural investigations, carried out together with them, are undoubtedly the best and most trenchant among those being worked on at present. Prof. R. Brill, on the other hand, has hardly come forward with any scientific investigations in recent years (AMPG, III rep. 50 no. 773.)."[21] He asked Hahn about the latest on the ongoing

---

[21] Otto Hahn apparently knew Rudolf Brill from a collaboration with the Ludwigshafen chemical plant of I.G. Farben on catalytic effects in precipitation. See D. Hahn 1975: 39.

deliberations by the MPG's Scientific Council. Since thus far Hosemann's research had received too little attention, Laue asked some colleagues for expert opinions.[22]

Another important appointment was that of staff physicist at the Institute for Nuclear Research. After Arnold Flammersfeld accepted a call to the University of Göttingen, Laue asked Léo Szilárd on January 27th whether he would be willing to accept the position (AMPG, III rep. 50). But after an exchange of letters and a visit by Szilárd in Berlin with a lecture, he decided not to accept the post after all on the reason that his current interest lay in biological problems.[23]

For all this administrative work, Max von Laue remained very much the physicist. His greetings to Max Born on his golden doctoral anniversary show how strongly he believed in causality. On January 19th after congratulating him warmly, Laue wrote: "You know that I do not entirely agree with your probability interpretation of quantum theory, insofar as I retain the goal of causally determined physics. This opposition is less physical than epistemological. But from the history of philosophy I know that the most diverse currents have contributed toward its progress. I should think that your view, too, will in any event find an honorable place in this history (AMPG, III rep. 50 no. 318)."

Another particularly high honor was bestowed on Max von Laue: the Officer's Cross of the *Légion d'Honneur* of the French Republic. It was ceremoniously presented to him at the Maison de France on Kurfürstendamm in Berlin on February 22nd. Maurice Bayen came from Paris and Adolphe Lutz from the French Embassy in Mainz. The five department heads from the Fritz Haber Institute and Laue's secretary also attended the bestowal. On the same day, the Physikalische Gesellschaft zu Berlin commemorated the centennial of the birth of Heinrich Hertz. Laue spoke at this jubilee session about 'The discovery of Hertzian waves' and Richard Honerjäger about their application in physics. After the official function, there was a gathering in Hardenberg-Hütte, where wine but no beer was served, essentially barring convinced beer drinkers from their party.[24]

From the political point of view 1957 became quite a turbulent year. The Cold War was threatening to heat up because the USA and the Soviet Union were building up their arsenals of nuclear weapons; and the Federal Republic of Germany and the German Democratic Republic were the "interface" between West and East. As a result discussions arose about arming the West-German military, the *Bundeswehr*, with tactical nuclear weapons. Was the Mainau Declaration of 1955 already forgotten? On April 8th Otto Hahn wrote to Max von Laue from Göttingen with a request. "Some months ago a number of physicists, namely the working group on 'Nuclear Physics' of the German Atomic Commission, wrote a letter to the Minister of Atomic Energy

---

[22] Laue to Menzer, Berlin, 15 Jan. 1957. AMPG, III rep. 50 no. 1346 fol. 20; Laue to Ewald, no loc., 25 Mar. 1957. AMPG, III rep. 50 no. 565.

[23] Szilárd to Laue, Chicago, 8 Feb. and 4 Apr. 1957. AMPG, III rep. 50 nos. 1970 f., see also Frank 2005: 232.

[24] Laue to Theodore, Berlin, 23 Feb. 1957. AMPG, III rep. 50 suppl. 7/17 fol. 9. For Laue's talk on Heinrich Hertz, see DPGA, 10324.

and the Minister of Defense concerning the usage of nuclear weapons. After consultations with the two ministers, this letter was not publicized. We—Heisenberg, Born, Gerlach, Weizsäcker, and I—now intend to submit to the public a new formulation of the earlier letter and would like to ask a number of well-known physicists for their support (AMPG, III rep. 50 no. 773)." Hahn listed twelve scientists. "I'd be very grateful to you, of course, if you also could give your consent, even though you are not directly engaged in nuclear physics or in nuclear research; we need your name, if you do sign (ibid.)." Hahn asked for a reply by telegram, as the statement was to be published before Easter. Laue gave the requested signature, and the Manifesto by eighteen West German nuclear scientists was published on 12 April 1957. It stated:

> The plans to equip the Bundeswehr with nuclear arms deeply trouble the undersigned nuclear researchers. Some of them have already communicated their concerns to the responsible federal ministers several months ago. Today this issue has become the topic of general debate. The undersigned therefore feel obliged to point out some facts to the public which all experts know but which the public does not seem to know well enough yet.
>
> 1. Tactical nuclear weapons have the destructive power of normal atomic bombs. They are called 'tactical' to express that they are deployable not only against human settlements but also against troops in ground combat. Each single tactical nuclear bomb or grenade is similarly powerful to the first atomic bomb that destroyed Hiroshima (D. Hahn 1975: 221 f.).

The word "small" used in the context of such weaponry was shown to be misleading.

> 2. There is no known natural limit to the developmental potential of the annihilatingly lethal power of strategic nuclear weapons. Nowadays a tactical atomic bomb can destroy a small town, but a hydrogen bomb can make a stretch of land the size of the Ruhr region uninhabitable for some time. By spreading radioactivity, a hydrogen bomb could probably wipe out the population of the Federal Republic of Germany already now. We know of no technical means to protect large populations from this danger (ibid.).

The passage that followed pointed out the magnitude of the political decisions arising from the presented facts. That was why they felt compelled not to remain silent about the political issues: "We profess our commitment to freedom as represented today by the Western world against communism. We do not deny that mutual fear of hydrogen bombs is an essential factor today in maintaining peace throughout the world and freedom in part of it (ibid.)." The penultimate passage contained a recommendation to the West German government: "We do not feel competent to make any concrete proposals for the policy between the major powers. For a small country like the Federal Republic, we do believe though that today it would be safest and would promote world peace best, if it expressly and voluntarily renounced possession of nuclear weapons of any kind (ibid.)."[25] None of the signatories were willing to participate in the development and testing of such weapons, they added.

---

[25] This "Göttingen Manifesto" of 12 April 1957 was signed by the following renowned nuclear scientists: Fritz Bopp, Max Born, Rudolf Fleischmann, Walther Gerlach, Otto Hahn, Otto Haxel, Werner Heisenberg, Hans Kopfermann, Max von Laue, Heinz Maier-Leibnitz, Josef Mattauch, Friedrich-Adolf Paneth, Wolfgang Paul, Wolfgang Riezler, Fritz Strassmann, Wilhelm Walcher, Carl Friedrich Freiherr von Weizsäcker, and Karl Wirtz. See Lorenz 2011.

However, everything should be done to promote the peaceful applications of nuclear energy. Most of the undersigned were civil servants of the Federal Republic. Their criticism of the government would have been punishable under the Kaiserreich, and fatal under Hitler. Federal Chancellor Adenauer was not pleased about the declaration and even less so Defense Secretary Strauss.

Max von Laue learned of Chancellor Adenauer's invitation to the signatories for a discussion on April 17th from the papers. Otto Hahn, Walther Gerlach, Max von Laue, and Carl Friedrich von Weizsäcker met in advance in Bonn on April 16th. They agreed not to retract anything. Wolfgang Riezler, nuclear physicist and head of the Commission for the Protection of the Civilian Population at the Federal Ministry of the Interior, joined them later. Werner Heisenberg could not be there for health reasons. The meeting took place in Palais Schaumburg. The government was represented by Konrad Adenauer, Franz Josef Strauss, Walter Hallstein, Hans Globke, press spokesman Felix von Eckardt, a chancery official, and Adenauer's personal advisor Hans Kilb. The military side was represented by Generals Adolf Heusinger and Hans Speidel and State Secretary Josef Rust. Adenauer opened the discussion on a conciliatory note. But Strauss, Heusinger, and Speidel did not follow suit. They impressed upon them that nuclear weapons were a necessity for the Federal Armed Forces and staked out the strategic situation on their maps. The counterarguments made by Hahn and especially by Weizsäcker did not lead to a convergence of positions. No minutes could be taken of the meeting by order of Adenauer. The meeting was interrupted because Adenauer had to receive a high-ranking visitor. After lunch together, Adenauer presented a draft declaration, which was adopted with a few amendments. The joint press statement was made on April 18th and clearly shows Adenauer's concern and understanding attitude: "The Chancellor expressed the wish to stay in touch with representatives of the scientific community on these matters and to keep them informed of developments in the fields mentioned and of developments in the international situation (ibid.)." The press carried extensive reports and statements on the armament problem. Max von Laue did not speak publicly on the subject in the period that followed. But many of his colleagues, above all Max Born, Walther Gerlach, Otto Hahn, and Carl Friedrich von Weizsäcker, were intensely involved. The future then showed that this initiative by the scientists had the desired effect: The *Bundeswehr* did not receive any nuclear weapons.[26]

The conferral of Lise Meitner's honorary doctorate took place at the Free University of Berlin on May 11th. In his closing address, Laue first mentioned the two usual motives of research: to earn one's daily bread and to make a name for oneself. Then he addressed the honoree: "The situation is different in Lise Meitner's case. All of you probably have the impression that her primary motive is the pure pursuit of research. It was not always easy for her to live accordingly (AMPG, III rep. 50 no. 1332)." Laue mentioned the difficulties she faced working as a woman in the former carpenter's workshop with Otto Hahn, her service during World War I, the long

---

[26] *Mitteilungen aus der Max-Planck-Gesellschaft zur Förderung der Wissenschaften* 1957: 62 f.; *Frankfurter Allgemeine Zeitung*, 18 Apr. 1957, p. 2; O.Hahn 1968: 221. See the material on the nuclear armament of German and foreign troops in DMA, NL 80 (Gerlach).

Fig. 13.11 Cornerstone ceremony of the Institute for Electron Microscopy (Ernst Ruska Building) on 5 July 1957. *From left*: Governing Mayor of West Berlin Otto Suhr, Max von Laue, Ernst Ruska. Max von Laue had Emil Orlik's "Haber" etching set in the wall. *Source* Archive of the Max Planck Society, Berlin-Dahlem

overdue habilitation degree in 1922, and the hardest blow, her flight from Germany. He finally mentioned her perpetual interest in advancements in nuclear physics:

> It is the genuine spirit of inquiry, which asserts itself against all external odds and does not expire even in old age. Such a mentality bears its value in itself, but also bears fruit in other ways. Elderly Goethe was often asked what recipe he used to preserve his mental vigor. He answered with a little verse entitled 'Panacea.'
>
> 'Speak! How do you forever yourself remake?
>
> You can too, if you ever delight in things great.
>
> Greatness stays fresh, warm, vibrates;
>
> Pettiness makes the pedant shiver and quake.'
>
> We have just been able to see for ourselves how well this panacea works for Lise Meitner (ibid.).[27]

The success by professors from the Free and Technical Universities of Berlin in initiating the establishment of a Hahn-Meitner Institute for Nuclear Research in December 1955 had become concrete. The laying of the foundation stone for the new building was celebrated on 25 May 1957 with Max von Laue in attendance (Weiss 1994. Laue was not a cosignatory). (Fig. 13.11) Shortly afterwards, preparations began for a major centennial celebration of Max Planck's birthday on 23 April 1958.

---

[27] "Sprich! Wie du dich immer und immer erneue'st? Kannst's auch, wenn du immer am Grossen dich freust. Das Grosse bleibt frisch, erwärmend, belebend; Im Kleinlichen fröstelt der Kleinliche bebend." Goethe.

From the outset, it had been the wish of Laue and some members of the Physikalische Gesellschaft that East and West celebrate it together. This was politics, but politicians were nevertheless not supposed to be part of it. The choice of speakers had to be agreed and Laue invited Lise Meitner to speak about Planck before a smaller audience in the Eastern sector.[28]

Research at the electron microscopy department under Ernst Ruska suffered from a lack of space and from the tremors in the building caused by the vacuum pumps and machine tooling. To achieve better resolving power in order to be able to see individual atoms, a separate building was therefore needed with foundations specially isolated from sources of vibration. The Siemens company and the Carl Zeiss Foundation contributed funds for the new building, as did the regional government of Berlin. Laue's authority and tenacity was the key to success in these difficult negotiations. He was unable to accept Paul Peter Ewald's invitation by telegram to a temptingly interesting congress in the USA. The work on his book, Laue explained in his letter from July 29th, prohibited it. "As of 1 Oct. 58 I may then finally resign my office. I *gladly* filled it, *very much* so. But now my time is up (orig. emphasis. AMPG, III rep. 50 no. 565)." The book that Laue had mentioned in addition to all these other commitments was the third edition of his 'X-ray Interference,' incorporating the research findings of Hosemann and Borrmann. The question of his successor also still preoccupied him, and he wrote to Otto Hahn on November 21st to point out that Professor Brill had only done solid routine research and that both Hosemann and Borrmann at the Fritz Haber Institute had done comparable work, and even Stranski, who had been excluded as a candidate, had done even better. Then he suggested his former doctoral student Léo Szilárd, whose appointment to the nuclear research institute had failed. He would be happy to discuss this with Hahn when they met in Frankfurt am Main on December 17th for the meeting of the society's senate. At this meeting, Otto Hahn reaffirmed the so-called 'Harnack principle'—"We build an institute around one man (AMPG, II rep. 1A, IB dossiers, FHI, 6 Feb. 1956)."

In his Christmas letter to his son Theo from Munich on December 21st, Laue reported about his trip to Frankfurt, where he had not only attended the senate meeting, which did not yield any clear decisions about his succession, but had also gone to the Academischer Verlag, where his volumes on the theories of relativity had also appeared. They were having difficulty again with his numerous indecipherable corrections and emendations to the second edition of his book on X-rays (AMPG, III rep. 50 suppl. 7/17). Laue had traveled onward to Munich for the celebration of Walther Meissner's 75th birthday. Mrs. von Laue, who was a close friend of Mrs. Meissner's, joined him by plane from Berlin. They could then spend Christmas with the Lemcke family from there, who were now living in the Pasing district of Munich.

The beginning of the new year was filled with correspondence and negotiations about the Max Planck birthday centennial. Lise Meitner wrote several times with questions about the celebrations in Göttingen and in East Berlin, her accommodations, and her speech: "I find it not so easy to give a reasonably adequate picture of Planck while neither omitting my personal recollections altogether nor letting them

---

[28] Laue to Meitner, Berlin-Dahlem, 7 Jul., 13 Jul., and 16 Nov. 1957. AMPG, III rep. 50 no. 1331.

Fig. 13.12 At the centennial
birthday celebration for Max
Planck in the Berlin Opera
House, 24 April 1958: Mrs.
Charlotte Hertz, Max von
Laue, and Gustav Hertz.
Source Archive of the Max
Planck Society,
Berlin-Dahlem

speak too strongly," she wrote Laue from Stockholm on 17 January 1958 (AMPG,
III rep. 50 no. 1331). In March, Laue's sister Elly died. She had been living in a
nursing home near Potsdam for some time. Although the relationship between the
siblings had apparently never been warm, Laue wanted to travel to Potsdam for the
funeral service. She was finally laid to rest in the family grave in Magdeburg.

The choice of venues for the "Planck Celebration" was important to enable as
many fellow scientists as possible to attend (Fig. 13.12). In East Berlin, the State
Opera House was designated for the main function. There was also a historic building
which had survived the war: the Magnus House. In the nineteenth century it had
belonged to the physicist Gustav Magnus, who had also taught there and introduced
the famous colloquium in 1843. The building had been used for various other purposes
since. The GDR branch of the German Physical Society put much effort into arranging
that the building be made available for the centenary and succeeded in doing so in time
for 1958. There was a lecture hall that could accommodate about a hundred persons. In
West Berlin, the new convention center on the zoo grounds was completed just in time
with enough seating to allow numerous students to also take part in the celebration
(Becker 1995a: 99 f.). Many emigrés came to this event, including James Franck from
the USA with his second wife Hertha Sponer. Max Planck had originally wanted him
to succeed Fritz Haber in 1932/33. Franck gave a report about the jubilee in a letter
addressed to his children, the Hippel and Lisco families of his two daughters:

> The festivities, in the style of eastern potentates in the wonderfully beautiful and ostenta-
> tiously renovated opera house in Berlin, in glaring contrast to the surrounding heaps of rubble
> of the shot-up neighborhood and the poverty of the Eastern population, was a sad combina-
> tion. It wasn't made better by the political speeches by some of the Eastern representatives,
> held in the East contrary to firm commitments to avoid politics [Otto Grotewohl and Walter
> Ulbricht had private box seats], but were avoided in the West, revealing that it is not easy
> to negotiate with the Russians. In the East, Hahn spoke briefly and extremely nicely, Laue
> unfortunately long, surely well, but so quietly that even people with normal hearing could
> not understand him (Lemmerich 2011: 297).[29]

---

[29] James Franck to his children, no loc., undated, partial collection of J. Franck's papers, Lisa Lisco.
The loudspeaker system was malfunctioning—was it intentional or accidental?

Fig. 13.13 Max von Laue wearing his insignia 'Pour le Mérite' and 'Order of Merit of the Federal Republic of Germany' in Berlin in the late 1950s. *Source* Archive of the Max Planck Society, Berlin-Dahlem

Franck then discussed Lise Meitner's speech before a select audience at Magnus House: Then spoke "Lise Meitner about Planck's personality, simply wonderfully. She spoke freely in front of a small group of the East[ern] Phys[ical] Society. Sadly she is quite weak and looked as if it was an effort for her to keep standing upright. (She was decorated with the Pour le Mérite, just as Hahn and Laue; they were honorary guests of Heuß at noon.) I said to her: 'Lise, you were simply loveable.' She quipped in reply: 'A little late, dear Franck.' (pp. 297–298) (Fig. 13.13)"

James Franck had to hold Max Born back from expressing in an emotional outburst his opinion about the "philistine propaganda speech" (p. 298) held by East Berlin Mayor Friedrich Ebert. About the festivities in West Berlin, Franck reported:

On the 2nd day (in the conference hall) Trendelenburg spoke briefly and properly (he sends Arthur [von Hippel] his regards), then for almost an hour long Heisenberg, in the main about his new theory. Wonderful, as far as I can understand it, far away from the goal of being a proven theory or even being completely calculated out, nevertheless magnificent as a general conception. Then Hertz spoke about the influence of Planck's elementary quantum on experimental research. After the great, but somewhat misplaced Heisenberg lecture, Hertz's natural, light, and unassuming manner was like a breath of fresh air in an oppressively odorous, overheated hall. Charming. Then Westphal spoke, simply and neatly about Planck's personality; entirely nice, but not comparable with the beautiful, profound, and short talk of the day before by Lise Meitner (pp. 298–299; cf. Hoffmann 1995b).

Max von Laue had his detailed speech appear in *Naturwissenschaften* (1958).[30] His recent editorial work for the collection of Planck's 'Physical Papers and Speeches' (1958) had prepared him thoroughly for this overview. It was a perceptive account of Planck's life, starting with a few words about his funeral, about his parental home, and the influence of his father who had been a jurist. This paternal influence was perceptible in the way Planck himself had worked at the Prussian Academy, Laue recalled. This was followed by an account of Planck's studies from 1874 onwards with his focus on thermodynamics, the concepts of entropy, and enthalpy, and his various contributions to this problem area. Laue added several very personal details, such as Planck's sensitive ear to music, which made him dislike going to concerts because every little mistake became too disagreeably noticeable. He retraced in some detail Planck's path toward an explanation for radiation with his grand derivation of the radiation formula. Laue's speech ended with Planck's attempts to bridge the gap between classical physics and quantum physics (Laue 1958). After that event Laue traveled onwards to Leipzig to listen to Heisenberg's presentation of his current ideas on quantum mechanics, staying with Gustav Hertz, who had a professorship at that university.

On 30 September 1958 Max von Laue ended his official active service as director, but continued to stay on as interim head of the institute until March 1st of the following year. The commencement of operations of the helium liquefaction facility, known as the "Meissner system," formed a fine conclusion to his successful engagement at the Fritz Haber Institute. Professor Rudolf Brill was appointed on the Scientific Council of the MPG on October 1st, but not yet as director of the Fritz Haber Institute.

In his letter from November 4th, congratulating Lise Meitner on her eightieth birthday, Laue acknowledged her great accomplishments in radioactivity research: "What you have achieved in unabating activity and selfless devotion to science not only gave you personal gratification—which, of course, always is the main thing for a researcher—but was highly acclaimed not just by your closer colleagues, bringing you widespread fame around the world. The crowning glory was probably the Pour le mérite award, which you were the first woman scientist to have ever received (CAC, MTNR 7/3)." Then Laue mentioned the complication involved in assigning a name to the new Institute for Nuclear Research—which he had played a considerable part in establishing. Mainz already had an MPI for Chemistry bearing Otto Hahn's name. What other name could be given to the institute in West Berlin? Once again Laue had to deal with a field of research that was foreign to him in the search for suitable scientists for the leading positions at that institute and then persuade them to move to West Berlin. The negotiations with the Berlin Senate and the federal minister responsible for nuclear issues, Siegfried Balke, were successful. Thus the strong reservations that the three Western Allies had against the existence of such

---

[30] An earlier wish materialized at the same time: the publication of all of Planck's physical papers and lectures in three volumes in 1958. Max von Laue assumed the coordination of this joint undertaking of the DPG and MPG.

an institute in West Berlin could be overcome. He told Lise Meitner that the nuclear chemistry department under Karl Erik Zimen would soon be ready, but that the nuclear physics department (with a research reactor) could not be put into operation until later. No head for this department had been found yet either. He then recollected their many excursions together, ending his letter on a melancholical note: "Now I sit here, plagued by all sorts of little aches and pains, such as gouty feet, a rheumatic back, and above all fatigue, which makes it inadvisable for me to take a trip to Stockholm, for instance, and mull over the past (ibid.)." Mrs. von Laue often appended short greetings to Lise Meitner at the end of her husband's letters.[31] On a card dated November 22nd with her photo, Lise Meitner expressed her thanks: "You wrote me such an especially amicable letter with so many so dear reminiscences and those very clear 'Do you remember…?' occasions & indications (AMPG, III rep. 50 no. 1331 fols. 16/17)."

Laue's letter to Theo on 7 December 1958 mentions Nikita Khrushchev's threat to make Berlin a "free city," but without taking it too seriously (AMPG, III rep. 50 suppl. 7/18). He had spoken at the "German Academy" in East Berlin about the research at the Fritz Haber Institute but had declined another invitation to speak in March of the coming year. On that occasion Laue had also presented his own findings from 1912, however, and on how the understanding of X-ray interference had developed. His didactically excellent talk on 'X-ray wave fields in crystals' began with the experiments of 1912. Those experiments were set up with the X-ray source quite far away from the crystal. With reference to his theory, he explained, "The theory completely relied on the prediction that where there is an interference maximum the partial scattered waves of all the atoms converge in the same phase, that is, on paths differing by integer multiples of the wavelengths. Since this is a question of geometry, this theory is called the geometric theory. It was clear from the outset

---

[31] Unfortunately, the letters and greetings by Magda von Laue are mostly written in pencil, and many passages are illegible.

Ruth Lewin Sime (2001: 471) quotes another letter from Max von Laue to Lise Meitner with good wishes on her 80th birthday: "During the postwar years Laue suffered from depression. He was preoccupied with the past and his own actions. His letter on Meitner's eightieth birthday seems almost like a confession: 'We all knew that injustice was being done, but we did not want to see it, we deceived ourselves and now we shouldn't be surprised that we have to pay for it!! […] That was the year 1933, when I was following a flag that we should have ripped down immediately. I did not do it, and must now also bear the responsibility for it.'" Ms. Sime cites the Lise Meitner collection at the Churchill Archives Centre (CAC) as her source.

However, such a letter by Max von Laue could not be found in that collection. A style comparison against Laue's many other letters to Lise Meitner indicates with great certainty that the writer of the quoted letter cannot be Max von Laue. Ms. Meitner's birthday letter to Laue on 5 Oct. 1949 also speaks against this, in which she wrote: "That I was able to endure the years 33–38 in Germany, I really owe to a considerable part to you and your wife. Not just that I was warmly welcomed whenever I wanted to come and could count on your understanding for my principal attitudes. It meant almost more to me that both of you saw so clearly the unfortunate effects of Nazi ideology and did not allow yourselves to be distracted throughout the war years and tried to help wherever it was possible to help. This greatly bolstered my confidence in people and helped me to maintain the moral ground upon which we must stand if life is to retain any deeper meaning." (CAC, MTNR 5/32).

that it could only apply approximately; for it completely neglects further diffraction of these partial waves by other atoms. Nevertheless, it has done astonishingly good service in determining crystal structure; for a peculiar reason: the crystals which Nature offers have such defective space lattices that—one would almost like to say—they deserve nothing better (Laue 1959: 3)." Laue mentioned the first experimental results obtained by Gerhard Borrmann in the 1930s using very good crystals and then discussed wave fields in empty space and inside crystals. The proof of the anomalous absorption in ideal crystals by Borrmann during World War II, in 1941, and the subsequent experiments on crystals as close to the ideal as possible were illustrated by numerous impressive images and explanatory diagrams on the ray paths (ibid.).

The official and private correspondence at the beginning of 1959 was undiminished. On January 5th a letter went out to Lise Meitner. Its topic was again the naming of the Institute for Nuclear Research Institute in Berlin, as the general administration of the MPG had objected to having the chemical department named after Otto Hahn and the physical department named after Lise Meitner. At a meeting of the Berlin Senate, which Laue could not attend due to illness, his proposal was approved, "of calling the entire Berlin nuclear research institute the 'Hahn–Meitner Institute.' The scientific bond in which you have collaborated with Otto Hahn for most of your life cannot be lent better expression than by this name (AMPG, III rep. 50 no. 1332)." He hoped that she would agree to this high-handedness of his. Lise Meitner replied from Stockholm on January 27th: "Of course I agree with anything to which Hahn gives his consent as concerns the naming (AMPG, III rep. 50 no. 1332)."

On March 1st Rudolf Brill officially took over the reins from Max von Laue. The ceremonial handover from the rather reluctant departing director took place with official speeches on March 16th.[32] (Fig. 13.14) Laue was assigned a small office near the old library and a "half" secretarial position. But he initially could not set to work there, because he was scheduled from March 2nd to 3rd to attend the meeting of the PTB's board of trustees in Braunschweig, and on March 7th the couple had to be in Göttingen in order to join the celebration of Otto Hahn's eightieth birthday on March 8th. In a special jubilee session the senate of the MPG presented Hahn with the Harnack Medal in gold; the medal was normally bronze. The first of three texts on the bestowal certificate read: "This distinction is for the researcher who, by his investigations conducted with Lise Meitner and Fritz Strassmann, made a decisive contribution to the reputation of the Kaiser Wilhelm Society, primarily by the discovery of uranium fission, which paved the way to the age of atomic energy (AMPG, MPG senate minutes on 8 Mar. 1959)." The second text acknowledged the organizer, and the third cited Adolf von Harnack, who had envisioned the following goals for the society: development of the mind, promotion of prosperity, and preservation of life. In his laudatory speech Laue reminisced about shared experiences with Hahn, their first meeting at the event in Vienna as young husbands in the company of their wives, then the Alpine tours, the period after 1933, in Tailfingen, and at Farm Hall, and the Nobel prize awards. Then Lise Meitner spoke (Fig. 13.15). She

---

[32] AMPG, II rep. 1A, IB, Fritz Haber Institute files, general, vol. 1, 15 Sep. 1958 and 12 Mar. 1959.

Fig. 13.14 Lise Meitner and Max von Laue in the restaurant Wannsee-Terrassen, Berlin, celebrating the handover of the Fritz Haber Institute of the Max Planck Society to the new director Rudolf Brill on 16 March 1959. *Source* Archive of the Max Planck Society, Berlin-Dahlem

Fig. 13.15 Lise Meitner and Otto Hahn in Göttingen, on the occasion of Hahn's eightieth birthday, 8 March 1959. *Source* Archive of the Max Planck Society, Berlin-Dahlem

had come with Nelly Planck; and Theodor Heuss and Konrad Adenauer were also present, all sitting shoulder to shoulder in the social venue ("Kameradschaftshaus") of the MPG in Göttingen (ibid.).

Khrushchev's threat had consequences for West Berlin universities and for the Hahn–Meitner Institute, as no new appointments were being made. Therefore, Léo Szilárd also had to be written off, as Laue informed him in a letter on April 17th (AMPG, III rep. 50). But the construction work had made good progress, so the official "baptism" of the building could be held on March 14th. Federal Minister Siegfried Balke had spoken on the preceding day and had presented a check for DM 500,000 as a gift.

Laue was still busy with the correction proofs and the figures for the third edition of his book *Röntgenstrahlinterferenzen* and was glad to have Dr. Ernst Heinz Wagner's assistance. Laue informed Paul Peter Ewald about his further plans in another letter on April 17th (AMPG, III rep. 50, no. 565). Instead of leaving on vacation that summer, the Laues intended to go to Tenerife to observe the total solar eclipse on October 2nd together with their daughter and the three girls: "Many circumstances push us to make this sacrifice. My wife—as you know—has been unwell for some years, hotel food doesn't agree with her, and she certainly cannot go on long walks, and I have had a foot condition for a year now, gout, which was very painful for a while, and even though it is much improved now, I cannot go on any mountain tours, even short ones, because of my heart. But I cannot bring myself 'to sit around' in a hotel. There still is the car. But I don't risk driving longer stretches anymore, and I don't like to keep the driver, whom the Institute places at my disposal, away from the Institute for weeks at a time (ibid.)." The couple left town nevertheless in June to attend the MPG's general assembly in Saarbrücken. Then they continued via Trier into the Eifel mountains, driving on after a few days via Aachen to Düsseldorf and then Bonn, where the Order chapter of Pour le Mérite was convening. Magda wrote a detailed report on June 6th from Berlin for Theo and his family about the landscape and vegetation they had seen, as well as all her dietary problems (AMPG, III rep. 50 suppl. 7/19 fol. 35).

At the end of June the Laues again attended the Nobel laureate meeting in Lindau, but this time the route there led via Strasbourg, probably to refresh some memories. Arrived at the hotel in Bad Schachen in the Lindau environs, they met Otto Hahn and the Hertz couple. At the conference Max Born spoke first, recalling his book on *Optics*, which had originally been published in 1933 and then for political reasons had not been circulated, before mentioning all sorts of half-serious details about the recent English edition. He was followed by Werner Heisenberg, who reported on his progress in finding a "world formula"—a theory of everything—that incorporated elementary particles and allayed objections that Wolfgang Pauli had raised. The results of Borrmann's experiments on crystals virtually free of defects were the topic of Max von Laue's talk.

At the beginning of August the couple traveled again. They met with the Lemcke family in Tyrol, where Laue admired the magnificent mountain scenery. Lise Meitner was informed about upcoming events so that she could arrange her travel plans accordingly; she would be receiving an invitation from the Fritz Haber Institute for October 9th. The Physicists' Convention in Berlin was scheduled to end on October 4th. The Laue couple wanted to go to southern Bavaria in mid-October, as the solar eclipse excursion had to be cancelled after all. Laue had inadvertently already received word on September 9th that he would be awarded the "Great Cross of Merit with Star and Shoulder Riband" for his eightieth birthday, and Federal Chancellor Adenauer had already sent his congratulations, as we learn from Laue's letter to Theo from Berlin on September 11th (AMPG, III rep. 50, suppl. 7/19 fol. 48. Cf. the notice MPG 1959: 329). Theo wrote back to his father at the end of September that he wanted to be there for his eightieth birthday, but his father was not so pleased about that. On October 1st he wrote: "Mama and I would advise against that, though. If only because it's impossible to predict when we're going to be home on the day itself or on any of the other adjacent ones (AMPG, III rep. 50, suppl. 7/19 fol. 51)." Laue noted whistfully: "The Physicists' Convention has been underway here since the day before yesterday. It's very interesting so far. What a pity that one isn't young enough anymore to be able to act on the multifarious stimuli. Many acquaintances are taking part and some of them are staying on here until the 9th, e.g., the Meissners (ibid.)."

The "festive hour" in honor of Professor Max von Laue lasted considerably longer than one hour, as we see from the program. The American military authority made available Harnack House, the former guest house and venue of the KWG, for this birthday celebration. The first speaker, Rudolf Brill, gave a sketch of Laue's life based on his autobiography (1944/52). Otto Hahn described the many mountain hikes and ascents they had undertaken together, spiced throughout with charming anecdotes. In serious earnest he then recalled Laue's resistance from 1933 on and the help he willingly gave to persecuted individuals, but also how this stance of his was acknowledged after 1946. As a gift Hahn wanted to present to him a bust of Einstein, but it had not been completed on time. The governing mayor of West Berlin, Willy Brandt, bestowed on Laue the "Cross with Star and Shoulder Riband of the Order of Merit of the Federal Republic of Germany" in the name of Former Federal President Theodor Heuss. The next speaker was Werner Hartke, president of the East German Academy of Sciences. He called to mind that Laue had become a member of the academy 39 years ago. He combined his congratulations with an address written by Gustav Hertz about the importance of the discovery of X-ray interference. Paul Peter Ewald began his address on a very personal note, with a reminiscence of Laue's participation in the conference of crystallographers in London in 1946 and recalled Laue's words of thanks, "that just as it is the greatest joy for a father to see his children grow up well, so also it is the greatest joy for a researcher to see his intellectual children develop and flourish." Then he brought greetings from the community of crystallographers throughout the world: from Paris he brought a medal for Laue, from England the X-Ray Analysis Group, Charles W. Bunn, Dame Kathleen Lonsdale, and Sir Lawrence Bragg extended their congratulations. In closing, Ewald recalled their first meeting exactly fifty years ago in Munich: "It would be wrong to let this day pass without remembering with love and veneration your wife, who has been a faithful lifelong companion of yours, whose untiring hand has everywhere left its charitable mark. To us young students, then in Munich, she was a generously kind lady of the house, and the memory of many an amicable hour in the evening are gratefully attached to her society for me and the many guests of your home in a later period (MPG 1959: 347 f.)." The next speaker was the astronomer and then president of the "Göttingen Academy," Paul ten Bruggencate: "We remember with joy and gratitude that you were able to work in Göttingen during those five postwar years, in the lecture hall and at the physics colloquium as well as in the meetings of the Academy of Sciences. Those years, in which it was necessary to give those returning home from the war a new purpose in life and to carry forward the reconstruction, those years count twice in the life of an individual and in the life of a university. You enriched Göttingen's scientific life during that difficult time and by your attitude gave many new courage (p. 348)." He mentioned Laue's membership and his later chairmanship of the board of trustees of the Astrophysical Observatory in Potsdam, from 1923 to 1934, with all the associated political obstacles.

Walther Meissner spoke on behalf of the President of the Bavarian Academy of Sciences, weaving into his narrative how pleased he and his wife personally were that the Laues were coming to Munich more often now to visit their daughter, son-in-law, and their children, and providing plenty of occasion to refresh their bond of friendship. He then read out an address by the academy president, which also acknowledged Laue's research on superconductivity theory. The rector of the Free University of Berlin, Professor Gerhard Schenck, extended congratulations from both the Technical and Free Universities. He quoted the motto above the hall doorway: "The finest joy of the circumspect person is to have explored the explorable and to revere the unexplorable in peace." On behalf of the federal bureau of standards, the PTB, its president Richard Vieweg extended his congratulations and brought along an album that included the photo of Max von Laue by the oak tree under which he had read out that last-minute telegram preventing the demolition of the building. "That from ruins so much new and so much good could rise, we have you to thank to a great measure, to your unflagging kindness, your encouragement, your shielding hand (p. 353)." Ferdinand Trendelenburg, in his capacity as vice-chairman of the Federation of German Physical Societies, thanked Laue for his intensive, self-sacrificing labors as a member of the society. He recalled the fall meeting in Würzburg in 1933, where Laue had drawn the comparison with Galileo's recanting of his forbidden doctrine: "With bated breath the whole auditorium listened as you quietly, but perfectly understood by all, concluded, 'And yet, it moves!' (p. 355)" Trendelenburg also mentioned Laue's efforts to preserve the German Physical Society: "This was no small task, for one must grant scientists a certain individuality, and then it is often not quite easy to bring about unanimity (ibid.)." He lauded Laue's aid for the younger generation of scientists and ended by recalling a wonderful car ride in Laue's Steyr through the autumnal countryside. The keynote speaker, Professor Fritz Laves of the German Mineralogical Society, bore greetings from the German crystallographers and mineralogists from both East and West as well as their thanks, realizing "how difficult our efforts would be, and how poor our existence, viewed from the present point of view, had you not given us roentgenography in addition to our former analytic methods of mallet and magnifying glass (p. 357)." The University of Zurich and the Swiss polytechnic, ETH, congratulated Laue bringing along a twisted quartz crystal as a gift. Clemens Schaefer offered congratulations on behalf of the German Bunsen Society. He too commemorated Max von Laue's upright stance during the Third Reich. The chairman of the Society of German Chemists, Carl Wurster, had his greetings conveyed by Professor Josef Schormüller. Laue's former coworker Georg Menzer offered his congratulations as editor of *Zeitschrift für Kristallographie*. He recalled the years 1944 and 1954, when they had been able to present special jubilee issues to Laue. On the occasion of Laue's 80th birthday, the editors decided to solicit contributions for a 1959 birthday issue. Sixty papers had arrived, 35 of which already appeared in the first volume (1959/60), part one of the *Von Laue-Festschrift*. Friedrich Beck also presented him with a "birthday issue" of *Zeitschrift für Physik*, together with congratulations from the authors, the editors, and Dr. Ferdinand Springer (pp. 362 f.).

"It is not easy," Max von Laue began, "when one has been honored in such abundance, to find appropriate words of thanks; this abundance was too overwhelming." He then thanked the federal president for the high distinction, the governing mayor for the award of the "Ernst Reuter Medal," and Otto Hahn for the Einstein bust. The Helmholtz Medal had impressed him most of all, he said. Laue explained the significance of Helmholtz's writings to him personally, quoting a long passage from an address that Helmholtz had held on his own 70th birthday, which Laue had read as a student. Helmholtz had spoken of his initial aspirations to acquire a post in the profession throughout the first half of his life. Then, for those who wish to continue to pursue research, a higher conception of the relationship between science and humanity comes to the fore, the effect of instruction on students, and how the thoughts of students and contemporaries eventually begin to lead lives of their own. "An individual's own thoughts are naturally more closely attached to his own intellectual visual scope as a whole than to foreign ones; and he feels more encouragement and satisfaction when he sees his own views develop more richly than those of others. Finally, a kind of paternal love develops for such an intellectual child, which drives him to foster and protect the advancement of such offsprings as much as for his natural ones (p. 365)." These ideas had accompanied Laue throughout his studies, especially while auditing Planck's lectures. The closing word was another quotation from Helmholtz: "That these contemporaries also thanked and thank me for what I was able to give them is, besides my own intellectual gratification, the finest reward for my labors (p. 366)."

Laue's junior colleagues, Friedrich Beck and Gerhard Borrmann, also paid tribute to Laue's scientific oeuvre in longer papers. Beck described how he had experienced Laue's personality when in charge of informing Laue about the latest results in quantum mechanics. They then also discussed papers by John Bardeen, Leon N. Cooper, and John Robert Schrieffer on superconductivity (Beck 1959). Borrmann first acknowledged Laue's organizational achievements after 1946. He then described in detail Laue's research on wave fields as presented in his talk before the German Academy. He explained the difference between irradiating crystals with X-rays and with an electron beam (Borrmann 1959). There was another birthday present: a collection of essays edited by Otto Robert Frisch, Friedrich A. Paneth, Friedrich

Fig. 13.16 Max von Laue,
late 1950s. *Source*
Photograph in the author's
estate

Laves, and Paul Rosbaud under the title 'Contributions to Physics and Chemistry of the 20th Century: Lise Meitner—Otto Hahn—Max von Laue on His 80th Birthday' (1959). William Lawrence Bragg, Friedrich Laves, André Guinier, Heinz Jagodzinski, Kathleen Lonsdale, Walter Hoppe, Johannes Martin Bijvoet, Rolf Hosemann, and Gerhard Borrmann wrote scientific essays about Max von Laue, partly with personal recollections, partly about their research results and those achieved by their fellow scientists (ibid.).

It wasn't until October 29th that Laue was able to find the time to compose a long letter to Theo and his wife Hilli. He had counted 470 congratulatory notes. He was sorry that they had not been there. The Lemcke children would have been very amused that he wore two different ties during the television interview, because the recording had been made in two separate sessions at different times. The exertions exhausted Magda very much and she was also trying in vain to apply the brakes on her husband's constant activities (AMPG, III rep. 50 suppl. 7/19).

Max von Laue had not been able to accept the invitation by the University of Leipzig to speak on its 550th anniversary, because it had been scheduled on his birthday (Fig. 13.16). He had nevertheless chosen 'Epistemology and relativity' as his topic, a tribute to Einstein's work; and Friedrich Herneck read out the manuscript for him. The introduction stated: "I did not arrive at a satisfactory understanding of the theory of relativity until I succeeded in relating it to Kant's doctrine of space and time (Laue 1961b: 153)." Regarding Kant's ideas that space and time were notions impressed upon one's mind and that a person could imagine a space void of all matter, Laue argued that "without space no such event" could happen and mentioned the extension of Euclidean geometry by Gauss and Riemann. As for time, he pointed out that it was one-dimensional and ran in only one direction. The measurement of time with devices were subject to physical laws. Then he explained the concept of the inertial system and the Galilean and Lorentz transformations. "The special theory of relativity has an immense range of validity. It actually covers everything needed to interpret terrestrial experiments, as long as gravity is not considered, including electrodynamics (ibid.)." Laue continued to discuss the general theory of relativity, in which Einstein incorporated gravitation. It requires that all physical laws be brought into a form that doesn't change (is covariant) for all transformations. Laue retraced Einstein's line of thinking. He concluded: "It is by no means a mathematical invention, but a reality that underlies all physical processes. This insight is Albert Einstein's greatest achievement; it proves Schiller's words true: 'With genius nature stands in eternal alliance.'" (ibid.)

Laue's own birthday present, the third edition of 'X-ray Interference,' appeared in 1960. Laue's new preface was written in September of the foregoing year. "Since the appearance of the second edition, research into the structures of crystals, and indeed of other bodies, has advanced tumultuously; indeed, one might almost say, a new science has arisen on defects in crystalline structure and their consequences. However the fundamentals of the theory of X-ray interference have not been touched by this, and so this book can retain the old form and division (Laue 1960a)." He thanked Gerhard Borrmann, Wilhelm Hartwig, Edmund Hess, Gerhard Hildebrandt, Rolf Hosemann, Heinz Niehrs, and Herbert Wagner from the Institute.

Max von Laue had promised Friedrich Herneck at the Humboldt University in East Berlin that he would deliver a speech about Hermann von Helmholtz for the university's 150th anniversary. He had begun writing the text and corresponded with Herneck in March 1960 about Helmholtz's living dates. The introduction up to his entry among the circle of students who would later found the German Physical Society was ready (Laue 1960b).

Laue's curiosity about the latest results of physical research was undiminished. One piece of news from the USA interested him extraordinarily: A nuclear resonance line with very narrow line width had been found (i.e., nuclear resonance fluorescence, discovered by Rudolf L. Mössbauer). He immediately recognized manifold possible applications for the analysis of crystals and solids. Rudolf Brill had not taken over all of Laue's obligations, such as the negotiations about the design of the "Hahn–Meitner Institute." Another meeting was scheduled for April 8th. As usual, Laue was at the Institute and wanted the institute chauffeur Mr. Ahl to drive him to a meeting at 10 o'clock. On April 9th, Magda von Laue described the subsequent events in a letter to Theo and Hilli:

Lest you hear the erroneous radio broadcast or the many false newspaper reports, I would like to set you straight about everything that the press has irresponsibly exaggerated so much. Max wanted to attend the meeting of the Hahn–Meitner Nuclear Research Institute in Wannsee on Friday at 10 o'clock and, since Rohr had to go to the airport with Brill, Ahl was supposed to drive him. But the more I and Ahl advised him not to drive himself, the more he stiffened his resolve and then drove to Wannsee alone without Ahl. He had already been a bit excited for two days and drove along the Avus highway. It had just rained again after months; whether the road was slippery or a motorcyclist got in his way, there is no witness—in any case, the motorcycle with the dead young man is said to have been lying in the ditch and the Mercedes also fell down the embankment overturning 3 times (AMPG, III rep. 50 suppl. 7/20).

She also mentioned a large swelling on his forehead, he was still asleep (ibid.). When she visited him in hospital on April 19th, it looked as if he would recover, she assured them in another letter. He was lively and had a good appetite (ibid.). On April 21st, Magda wrote very emotionally to Mrs. Meissner:

Heartfelt thanks for your and your husband's sympathetic lines. The first week everything was advancing so well. After 10 days, due to the long bedrest, a circulatory disturbance must have set in, which is still persisting. Since then he doesn't eat anymore, doesn't speak anymore, hardly recognizes you, and is only very apathetic. It's not known yet whether the circulatory injections will help. One small reassurance for all of us is that the police determined that the motorcyclist held a driving license that was just one day old, rode a homemade bike, and had overtaken the car from the left cutting it off, unaware that on the Avus the speeds are higher than in the city at 50 kil[ometers]., and turned right too early. Unfortunately, the chief physician had unbelievably already told him that he had fatally driven over someone. Was that necessary?—but he probably hadn't grasped it for long, otherwise it would have been terrible for him to know (DMA, NL 45/006).[33]

Three days later Max von Laue died without regaining consciousness (Fig. 13.17). There were very many obituaries, detailed ones by Paul Peter Ewald and by Walther Meissner. Lise Meitner mourned:

Max von Laue died in Berlin on the 24th of April, 1960. With him the world has lost one of its most valuable and most outstanding scientists and one of its noblest persons. He himself characterized his attitude toward his scientific work by the words: 'Theory means, literally translated, God-viewing. And with due reverence I have sometimes countenanced it when it had cast unhoped-for light on a hitherto incomprehensible fact.' […]

 Max von Laue has certainly received the international scientific recognition in the form of countless distinctions that his great achievements deserved. And he also possessed the amiable quality of being able to be pleased about them.

 Nevertheless, as a result of his very sensitive soul, he lived a life that although always rich in content was not always easy. In part, this was due to the restless developments of the times. Laue, as a true friend, with his ever aware sense of justice and liberty and his active helpfulness, experienced very intensely all the problems of those close to him (Meitner 1960: 196 f.).

She also discussed Laue's empathy when Fritz Haber was struggling with the decision about whether to resign in protest and his speech in Würzburg in 1933: "Sometimes Laue, with his mental vulnerability, would have been in serious emotional trouble but for the kind balancing hand of his sensitive, warm-hearted wife, wise in the ways of man (p. 197)." Lise Meitner quoted Laue, who cited Friedrich Schiller's

---

[33] Magda von Laue's niece was with her with a small grandchild. She was driven daily to the hospital in Wannsee.

Fig. 13.17 Max von Laue,
1959. *Source* Photograph in
the author's estate

words in his autobiographical 'Physical career' (1944/52): "Speaks the soul, thus
alas speaks the soul no more." Laue had added to this passage in his copy, "I have
felt this sigh throughout my life." She corrected this remark, citing how he was able
to express his affection and reverence in an almost poignant manner in his obituary
of Einstein, as well as in 1958 in his speech on Planck. "But Laue's image would
be very incomplete without an indication of how very essentially cheerful and full
of humor he could be. Who doesn't have his happy, bright laugh in his ear? How
pleased he could be about a good literary work, about a beautiful landscape, about
tours in the high mountains and the like. I remember a glacier tour in Switzerland in
1927 from Silvaplana to the Caputschin with Laue, Ladenburg, and Mark (the latter
as guide), when Laue just could not stop joking and being jolly (p. 198)." This was
followed by reminiscences of excursions to the March of Brandenburg and his love
of nature. Lise Meitner concluded with a Goethe verse:[34]

---

[34] "Sucht Ihr das menschliche Ganze? Oh, sucht es ja nicht beim Ganzen. Nur in dem schönen
Gemüt bildet das Ganze sich ab." Goethe.

Do you seek human integrity? Oh, seek it not in integrity.

Fine temper alone embodies the whole (p. 199).

A small circle of the closest family members attended the burial at the municipal cemetery in Göttingen. Max von Laue's grave is located near those of Planck and Nernst. The gravestone bears a Greek inscription, the final words from 'Heracles' by Euripides after the translation by Wilamowitz.

So we too go, painfully, tearfully,

The one we lost was the truest of friends.

(Magda von Laue's health unfortunately soon failed her. When she couldn't care for herself anymore the apartment in the Haber villa had to be given up. She stayed in a home in Tutzing for some time before passing away on 15 July 1961 and was buried next to her husband.)

References

Beck, Friedrich. 1951. Das elektrodynamische Potential in der erweiterten phänomenologischen Theorie der Supraleitung. Z. Phys. A 3: 246–274

Beck, Friedrich. 1952a. Eine spezielle Form des elektrodynamischen Potentials von Supraleitern. Ann. Phys. 4/5: 310–313

Beck, Friedrich. 1952b. Zur Phasenumwandlung zwischen Supra- und Normalleiter im kritischen Magnetfeld. Ann. Phys. 6/7: 317–326

Beck, Friedrich. 1959. Max von Laue zum 80. Geburtstag. Phys. Bl. 15: 446–452

Beck, Friedrich. 1979. Mit Max von Laue in Berlin 1951–1953. In Boeters and Lemmerich 1979: 141–144

Becker, Christine. 1995. Zur Geschichte des Magnus-Hauses. In Hoffmann 1995a: 99–121

Blatt, John M. 1964. Theory of Superconductivity. New York, London: Academic Press

Boeters, Karl E., and Jost Lemmerich, eds. 1979. Gedächtnisausstellung zum 100. Geburtstag von Albert Einstein, Otto Hahn, Max von Laue, Lise Meitner in der Staatsbibliothek Berlin Preußischer Kulturbesitz. Exhibition catalogue. Berlin: Staatsbibliothek Preußischer Kulturbesitz

Born, Max, and Emil Wolf. 1959. Principles of Optics. Electromagnetic Theory of Propagation, Interference and Diffraction of Light. London: Pergamon Press

Born, Max. 1933. Optik. Ein Lehrbuch der elektromagnetischen Lichttheorie. (cf. Born and Wolf 1959) Berlin: J. Springer

Born, Max. 1955. Ist die klassische Mechanik tatsächlich deterministisch? Phys. Bl. 2: 49–54

[Born, Max]. 1953. Scientific Papers Presented to Max Born on His Retirement from the Tait Chair of the Natural Philosophy in the University of Edinburgh. Edinburgh: Oliver & Boyd

Borrmann, Gerhard. 1959. Max von Laue und das Fritz-Haber-Institut. Phys. Bl. 15: 453–456

Cooper, Leon N., and Dmitri Feldman, eds. 2011. BCS: 50 Years. New Jersey, London: World Scientific

Dahl, Per Fridtjof. 1992. Superconductivity. Its Historical Roots and Development from Mercury to the Ceramic Oxides. New York: American Institute of Physics

DPG. 1952: Max-Planck-Medaille an Paul Adrien Maurice Dirac. Phys. Bl. 11: 512 f

DPG. 1953: Physikertagung in Innsbruck. Phys. Bl. 9: 514 f

Einstein, Albert. 1905a. Zur Elektrodynamik bewegter Körper. Ann. Phys. 17: 891–921. English trans. by Anna Beck: Einstein, A.: On the Electrodynamics of Moving Bodies. The Collected

Papers of Albert Einstein. Vol. 2: *The Swiss Years: Writings, 1900–1909*, ed. by John Stachel et al. 1989: 140–171

Einstein, Albert. 1905b. Über einen die Erzeugung und Verwandlung des Lichtes betreffenden heuristischen Gesichtspunkt. Ann. Phys. 17: 132–148, Vol. 2: Engl. Trans by Anna Beck: Einstein, A.: On a Heuristic Point of View Concerning the Production and Transformation of Light. *The Collected Papers of Albert Einstein*. Vol. 2: *The Swiss Years: Writings, 1900–1909*, ed. by John Stachel et al. 1989: 86–103. Princeton: Univ. Press

Einstein, Albert. 1911. Die Relativitäts Theorie. *Naturforschende Gesellschaft in Zürich. Vierteljahrsschrift* 56: 1–14; English trans. by Anna Beck: Einstein, A. The Theory of Relativity. *The Collected Papers of Albert Einstein*. Vol. 3: *The Swiss Years: Writings, 1909–1911*, ed. by Martin J. Klein et al. 1993: 340–350. Princeton: Univ. Press

Festschrift. 1959/60. *Von-Laue-Festschrift*. Vol. 1. Special issue of Zeitschrift für Kristallographie, no. 112 (1959); vol. 2, no. 113 (1960). Frankfurt am Main: Akademischc Verlagsgesellschaft

Frank, Tibor. 2005. Ever ready to go: The multiple exiles of Léo Szilárd. *Physics in Perspective* 7: 204–252

Friedmann, Aleksander A. 1924. *Über die Möglichkeit einer Welt mit konstanter negativer Krümmung des Raumes. Z. Physik* 21: 326–332

Friedrich, Walter, Paul Knipping, and Max von Laue. 1952. Interferenz-Erscheinungen bei Röntgenstrahlen. *Naturw.* 39: 361–371

Frisch, Otto Robert, Friedrich A. Paneth, Friedrich Laves, and Paul Rosbaud, eds. 1959. *Beiträge zur Physik und Chemie des 20. Jahrhunderts: Lise Meitner – Otto Hahn – Max von Laue zum 80. Geburtstag*. Braunschweig: Vieweg

George, André, ed. 1953. *Louis de Broglie physicien et penseur*. (A collection of contributions for his 60th birthday on 15 Aug. 1952) Paris: Albin Michel

Hahn, Otto. 1968. *Mein Leben*. Munich: Bruckmann

Hahn, Dietrich, ed. 1975. *Otto Hahn. Erlebnisse und Erkenntnisse*. With an introduction by Prof. Dr. Karl-Erik Zimen. Düsseldorf, Vienna: Econ-Verlag

Hentschel, Klaus, and Ann M. Hentschel, eds. 1996. *Physics and National Socialism. An Anthology of Primary Sources*. Rev. reprint 2010. Basel, Boston, Berlin: Birkhäuser

Hoffmann, Dieter, and Mark Walker, eds. 2012. *The German Physical Society in the Third Reich. Physicists between Autonomy and Accommodation*. Trans. by Ann M. Hentschel (without the documentary appendix). German orig. 2007. Cambridge: Univ. Press

Hoffmann, Dieter. 1995a. Die Physikalische Gesellschaft (in) der DDR. In *150 Jahre Deutsche Physikalische Gesellschaft*. Ed. by Theo Mayer-Kuckuk. (Special issue of the journal Physikalische Blätter.) pp. 157 ff. Weinheim: VCH-Verlags-Gesellschaft

Hoffmann, Dieter. 1995b. Ein Jubiläum wird gefeiert: Die Planck-Feier(n) in Berlin 1958. New projects, preprint no. 30. Berlin: Forschungsschwerpunkt für Wissenschaftsgeschichte und Wissenschaftstheorie

Krebs, Hans. 1979. *Otto Warburg. Zellphysiologe, Biochemiker, Mediziner, 1883–1970*. Stuttgart: Wissenschaftliche Verlags-Gesellschaft

Lamla, Ernst. 1952. Vor 40 Jahren. *Naturw.* 39: 361

Laue, Max von. 1956b. Heinrich Hertz (1857–1894). *Die großen Deutschen*. Vol. 1, 103–112. Ed. by Hermann Heimpel. Berlin: Propyläen-Verlag bei Ullstein; reprint in Laue 1961a, vol. 3: 247–256

Laue, Max von. 1934. Fritz Haber gestorben. *Naturw.* 22: 97

Laue, Max von. 1944/52. Mein physikalischer Werdegang. (November 1944) Reprinted in Hans Hartmann: *Schöpfer des neuen Weltbildes. Große Physiker unserer Zeit*. 1952: 37–70. Bonn: Athenäum-Verlag

Laue, Max von. 1949. *Theorie der Supraleitung*. 2nd ed. (1st ed. 1947) Berlin, Göttingen: Springer

Laue, Max von. 1950. *Geschichte der Physik*. Geschichte der Wissenschaften, no. 2. Naturwissenschaften. 3rd rev. ed. (1st ed. 1946) Bonn: Athenäum-Verlag; English edition: Laue, M. von, 1950: *History of Physics* (trans: Ralph E. Oesper) New York: Academy Press

Laue, Max von. 1951. Sommerfelds Lebenswerk. *Naturw.* 38: 513–518

Laue, Max von. 1951/52a. Kritische Bemerkungen zur Theorie der Supraleitung. (received 2 Nov. 1951) *Ann. Phys.* 4/5: 296–301

Laue, Max von. 1951/52b. Der Atombegriff in der Physik. *Abhandlungen der Braunschweiger Wissenschaftlichen Gesellschaft* 1951/52, 3/4: 228–237

Laue, Max von. 1952/53. Die Interferenz-Absorption von Röntgenstrahlen in Kristallen. *Proceedings of the International Symposium on the Reactivity of Solids in Gothenburg 1952* (at Chalmers Univ. of Technology), 215–233. Gothenburg: Swedish Royal Academy

Laue, Max von. 1952a. *Theory of Superconductivity*. Trans. by Lothar Meyer and William Band. New York: Academic Press

Laue, Max von. 1952b. Bemerkungen zur Theorie der Supraleitung II. *Ann. Phys.* 4/5: 305–309

Laue, Max von. 1952c. Fritz London: 'Superfluids.' *Naturw.* 39: 287 f

Laue, Max von. 1952d. *Die Relativitätstheorie*. Vol. 1: *Die spezielle Relativitätstheorie*. 5th rev. ed. (1st ed. 1921, 6th ed. 1955) Braunschweig: Vieweg

Laue, Max von. 1952e. Die Energiestrom bei Röntgenstrahl-Interferenzen in Kristallen. *Acta Cryst.* 5: 619–625

Laue, Max von. 1953a. *Die Relativitätstheorie*. Vol. 2: *Die Allgemeine Relativitätstheorie und Einsteins Lehre von der Schwerkraft*. 3rd, rev. ed. (1st ed. 1921, 4th ed. 1956) Braunschweig: Vieweg

Laue, Max von. 1953b. Der Teilchenstrom bei Raumgitterinterferenzen von Materiewellen. *Acta Cryst.* 2: 217

Laue, Max von. 1953c. Am 9. Dezember 1952 zur Enthüllung der Haber-Gedenktafel von Professor R. Scheibe im Kaiser-Wilhelm-Institut für Physikalische Chemie und Elektrochemie, Berlin Dahlem: Eröffnungsrede. *Zeitschrift für Elektrochemie* 1: 1 f

Laue, Max von. 1953d. Walter Friedrich 70 Jahre. *Phys. Bl.* 12: 561

Laue, Max von. 1954a. Relativitätstheorie, Doppler- und andere spektrale Verschiebungseffekte. *Naturw.* 41: 25–29

Laue, Max von. 1954b. Le Chatelier-Braunsches Prinzip und Relativitätstheorie. *Z. Phys.* 137: 113–116

Laue, Max von. 1954c. Neues über Röntgenstrahlinterferenzen. *Verh. DPG* 3: 190

Laue, Max von. 1955a. Der Strahlengang der Röntgenstrahlen in Kristallen. *Vakuum-Technik* 2: 2

Laue, Max von. 1955b. Albert Einstein. *Frankfurter Allgemeine Zeitung*, 23 Apr. 1955

Laue, Max von. 1955c. Ist die klassische Physik tatsächlich deterministisch? *Phys. Bl.* 11: 269 f

Laue, Max von. 1956a. Einstein und die Relativitätstheorie. *Naturw.* 43: 1–7

Laue, Max von. 1956c. Gedenkworte für Friedrich Schmidt-Ott. In Laue 1961a, vol. 3: 238–242

Laue, Max von. 1957a. Von Kopernikus bis Einstein. *Naturw. Rundschau* 3: 83–89

Laue, Max von. 1957b. Erinnerung an die Entdeckung der Kernspaltung. Ansprache zur Enthüllung einer Gedenktafel. Ansprache. *Mitteilungen der Max-Planck-Gesellschaft* 1: 9 f

Laue, Max von. 1958. Zu Max Plancks 100. Geburtstag. *Naturw.* 4: 221–226

Laue, Max von. 1959. Röntgenwellenfelder in Kristallen. *Sitzungsberichte der Deutschen Akademie der Wissenschaften Berlin*, math. phys. techn. class, proceedings 1959: 3

Laue, Max von. 1960a. *Röntgenstrahl-Interferenzen*. With the collaboration of Ernst Heinz Wagner. 3rd exp. ed. (1st ed. 1941, 2nd ed. 1948) Frankfurt am Main: Akademische Verlags-Gesellschaft

Laue, Max von. 1960b. [On Helmholtz]. In *Forschen und Wirken. Festschrift zur 150-Jahr-Feier der Humboldt-Universität zu Berlin 1810–1960*, ed. by Willi Göber and Friedrich Herneck. Vol. 1: 359–366, Berlin: Deutscher Verlag der Wissenschaft

Laue, Max von. 1961a. *Gesammelte Schriften und Vorträge*. 3 vols. Braunschweig: Vieweg

Laue, Max von. 1961b. Erkenntnistheorie und Relativitätstheorie. *Phys. Bl.* 4: 153–159

Lemmerich, Jost. 1988. *Die Geschichte der Entdeckung der Kernspaltung*. Exhibitions at the Technische Universität Berlin 2 Dec. 1988–4 Feb. 1989, and Deutsches Museum 18 Feb.–10 Aug. 1989. Meisterwerke der Naturwissenschaft und Technik, Deutsches Museum and TU Berlin. Berlin: Technische Univ.

Lemmerich, Jost. 2011. *Science and Conscience. The Life of James Franck*. Trans. by Ann M. Hentschel. (German orig. 2007) Stanford: Univ. Press

London, Fritz. 1950. *Superfluids*. Part I: *Macroscopic Theory of Superconductivity*. New York: Wiley

London, Fritz. 1952a. Bemerkungen zu vorstehender Arbeit 'Theorie der Supraleitung' von Max von Laue. *Ann. Phys.* 4/5: 302 ff

London, Fritz. 1952b. Zu den weiteren 'Bemerkungen zur Theorie der Supraleitung' des Herrn von Laue. *Ann. Phys.* 4/5: 314 ff

Lorenz, Robert. 2011. *Protest der Physik: Die 'Göttinger Erklärung' von 1957*. Bielefeld: Transkript Verlag

Marshall, Bruce. 1953. *Das Wunder des Malachias*. Trans. by Jakob Hegner. Frankfurt am Main: Fischer Bücherei. English version: Marschall, B. 1949. *Father Malachy's Miracle. A Heavenly Story with an Earthly Meaning*. London: Constable

Meinecke, Friedrich. 1946. *Die deutsche Katastrophe. Betrachtungen und Erinnerungen*. Wiesbaden: Brockhaus

Meitner, Lise. 1960. Nachruf auf Max von Laue. *Mitteilungen der Max-Planck-Gesellschaft* 4: 196–199

Mercier, André, and Michel Kervaire. 1956. *Fünfzig Jahre Relativitätstheorie*. Proceedings of the Swiss Physical Society in Bern: 11–16 July 1955. Helvetica physica acta, suppl. no. 4

Meshkovski, A., and A. Shalnikov. 1947. Surface effects in super-conductors in the transitional region. *Journal of Physics* 1: 1; cf. *Journal of Experimental and Theoretical Physics* 17, no. 10: 851–861 (in Russian)

MPG. 1959. Feierstunde zu Ehren von Max von Laue an seinem 80. Geburtstag am 9. Oktober 1959 in Berlin. *Mitteilungen der Max-Planck-Gesellschaft* 6: 323–366

Murdock, Carleton C. 1934. Multiple Laue spots. *Phys. Rev.* 45: 117

Pauli, Wolfgang. 1999. *Wissenschaftlicher Briefwechsel mit Bohr, Einstein, Heisenberg u.a.* Vol. 4, Part 2: *1953–1954*. Berlin, Heidelberg: Springer

Planck, Max. 1958. *Physikalische Abhandlungen und Vorträge. Aus Anlass seines 100. Geburtstages (23. April 1958)*. Published by Verband Deutscher Physikalischer Gesellschaften and Max-Planck-Gesellschaft zur Förderung der Wissenschaften. 3 vols. Braunschweig: Vieweg

Ritter, Gerhard. 1954. *Carl Goerdeler und die deutsche Widerstandsbewegung mit einem Brief Goerdelers in Faksimile*. Stuttgart: Deutsche Verlags-Anstalt

Rosenthal-Schneider, Ilse. 1988. *Begegnungen mit Einstein, von Laue und Planck*. Braunschweig: Vieweg

Sime, Ruth Lewin. 2001. *Lise Meitner. Ein Leben für die Physik*. Frankfurt am Main, Leipzig: Insel-Verlag. English orig. Sime, R.L. 1996 *Lise Meitner. A Life in Physics*. Berkeley: Univ. California Press

Sommerfeld, Arnold. 1919. *Atombau und Spektrallinien*. Braunschweig: Vieweg

Steinhauser, Thomas, Jeremiah James, Dieter Hoffmann, and Bretislav Friedrich. 2011. *Hundert Jahre an der Schnittstelle von Chemie und Physik. Das Fritz-Haber-Institut der Max-Planck-Gesellschaft zwischen 1911 und 2011*. Berlin: de Gruyter

Sterio, Alexandre Dées de. 1963. *Nobelpreisträger in Lindau*. Solothurn: Vogt-Schild

Stuart, Herbert Arthur. 1952. *Die Physik der Hochpolymeren*. Vol. 1. Berlin: Springer

Weiss, Burghard. 1994. *Großforschung in Berlin. Geschichte des Hahn-Meitner-Instituts 1955–1980*. Frankfurt am Main, New York: Campus-Verlag

Wolff, Stefan. 2012. Marginalization and Expulsion of Physicists under National Socialism: What Was the German Physical Society's Role? In Hoffmann and Walker, eds. 2012: 50–95

Zeitz, Katharina. 2006. *Max von Laue (1879–1960). Seine Bedeutung für den Wiederaufbau der deutschen Wissenschaft nach dem Zweiten Weltkrieg*. Stuttgart: Steiner

Advertisement

Index

© Basilisken-Presse, Natur+Text GmbH 2022
J. Lemmerich, *Max von Laue*, Springer Biographies,
https://doi.org/10.1007/978-3-030-94699-9

Printed in the United States
by Baker & Taylor Publisher Services